EXS 86

Fish Ecotoxicology

Edited by T. Braunbeck
           D. E. Hinton
           B. Streit

Springer Basel AG

Editors

Dr. Thomas Braunbeck
Department of Zoology I
University of Heidelberg
Im Neuenheimer Feld 230
D-69120 Heidelberg
Germany

Prof. Dr. David E. Hinton
Department of Anatomy, Physiology and Cell
Biology
School of Veterinary Medicine
University of California
Davis, CA 95616
USA

Prof. Dr. Bruno Streit
Department of Ecology and Evolution
University of Frankfurt
Siesmayerstr. 70
D-60054 Frankfurt
Germany

Library of Congress Cataloging-in-Publication Data
Fish ecotoxicology / edited by T. Braunbeck, D. E. Hinton, B. Streit.
    p. cm. -- (EXS ; 86)
    Includes bibliographical references and index.
    ISBN 978-3-0348-9802-7   ISBN 978-3-0348-8853-0 (eBook)
    DOI 10.1007/978-3-0348-8853-0
(hardcover : alk. paper)
    1. Fishes--Effect of water pollution on.   2. Water--Pollution-
-Toxicology.   I. Braunbeck, T. (Thomas)   II. Hinton, David E.
III. Streit, Bruno.   IV. Series.
SH174.F558   1998
571.9'517--dc21

Deutsche Bibliothek Cataloging-in-Publication Data
Fish exotoxicology / ed. by T. Braunbeck ... - Basel ; Boston ; Berlin
: Birkhäuser, 1998
    (EXS ; 86)
    ISBN 978-3-0348-9802-7

86. Fish exotoxicology. – 1998
EXS. - Basel ; Boston ; Berlin : Birkhäuser
    Früher Schriftenreihe
    Fortlaufende Beil. zu: Experientia

Contents

List of Contributors

Thomas Braunbeck, Department of Zoology I, Morphology/Ecology,
University of Heidelberg, Im Neuenheimer Feld 230,
D-69120 Heidelberg, Germany.
e-mail braunbeck@urz.uni-heidelberg.de

John Couch, 4703 Soule Place, Gulf Breeze, Florida 32561, USA.

Jean-Pierre Cravedi, Institut National de la Recherche Agronomique,
Laboratoire des Xénobiotiques, 180 chemin de Tournefeuille, BP 3,
F-31931 Toulouse Cedex, France.
e-mail jcravedi@toulouse.inra.fr

Alain Devaux, Ecole Nationale des Travaux Publics de l'Etat,
Laboratoire des Sciences de l'Environnement, rue Maurice Audin,
F-69518 Vaulx-en Velin Cedex, France.
e-mail alain.devaux@entpe.fr

Karl Fent, EAWAG, Überlandstrasse 133, CH-8699 Dübendorf.
e-mail fent@eawag.ch

Anders Goksøyr, Department of Molecular Biology, University of Bergen,
HIB, N-5020 Bergen, Norway.
e-mail anders.goksoyr@mbi.uib.no

Karla Isberner, Institute of Hydrobiology, Technical University of Dresden,
Zellerscher Weg 40, D-01062 Dresden, Germany.
e-mail kisberner@rcs.urz.tu-dresden.de

David Hinton, Department of Anatomy, Physiology and Cell Biology,
School of Veterinary Medicine, University of California, Davis,
CA 95616, USA.
e-mail dehinton@ucdavis.edu

Rudolf Hofer, Department of Zoology and Limnology,
University of Innsbruck, Technikerstr. 25, A-6020 Innsbruck, Austria.
e-mail rudolf.hofer@uibk.ac.at

Astrid-Mette Husøy, Bergen College, N-5008 Bergen, Norway.
e-mail Astrid.M.Husoy@lkb.haukeland.no

Günter Köck, Department of Zoology and Limnology,
University of Innsbruck, Technikerstr. 25, A-6020 Innsbruck, Austria.
e-mail guenter.koeck@uibk.ac.at

Reinhard Lackner, Department of Zoology and Limnology,
    University of Innsbruck, Technikerstr. 25, A-6020 Innsbruck, Austria.
    e-mail Reinhard.Lackner@uibk.ac.at

Peter Matthiessen, Centre for Environment, Fisheries and Aquaculture
    Science, Remembrance Avenue, Burnham-on-Crouch,
    Essex, CM0 8HA, UK.
    e-mail p.matthiessen@cefas.co.uk

Gilles Monod, Institut National de la Recherche Agronomique,
    Unité d'Écotoxicologie Aquatique, IFR 43, Campus de Beaulieu,
    F-35042 Rennes Cedex, France.
    e-mail monod@beaulieu.rennes.inra.fr

Roland Nagel, Institute of Hydrobiology, Technical University of Dresden,
    Zellerscher Weg 40, D-01062 Dresden, Germany.
    e-mail rnagel@rcs.urz.tu-dresden.de

Helmut Segner, Centre for Environmental Research,
    Department of Chemical Ecotoxicology, Permoserstr. 15,
    D-04318 Leipzig, Germany.
    e-mail hs@uoe.ufz.de

Bruno Streit, Department of Ecology and Evolution, Biologie-Campus,
    Siesmayerstrasse, D-60054 Frankfurt, Germany.
    e-mail streit@zoology.uni-frankfurt.de

John P. Sumpter, Department of Biology and Biochemistry,
    Brunel University, Uxbridge, Middlesex UB8 3PH, UK.
    e-mail john.sumpter@brunel.ac.uk

Yves Valotaire, Equipe d' Endocrinologie Moléculaire de la Reproduction,
    UPRES-A CNRS 2026, IFR 43,
    Campus de Beaulieu, F-35042 Rennes Cedex, France.
    e-mail valotaire@univ-rennes1.fr

Preface

In the last twenty years, ecotoxicology has successfully established its place as an interdisciplinary and multidisciplinary science concerned with the effects of chemicals on populations and ecosystems, thus bridging the gap between biological and environmental sciences, ecology, chemistry and traditional toxicology. In modern ecotoxicology, fish have become the major vertebrate model, and a tremendous body of information has been accumulated. A major challenge to ecotoxicology is the integration of this vast amount of diverse knowledge from environmental chemistry, toxicology, ecology and general environmental sciences. Critical selection and compilation of the most important data has thus received increasing importance. This volume of the EXS series is dedicated to a selection of novel trends in fish ecotoxicology and makes an attempt to summarize our present knowledge in a series of fields of primary ecotoxicological interest ranging from the use of (ultra)structural modifications of selected cell systems as sources of biomarkers for environmental impact over novel approaches to monitor the impact of xenobiotics with fish *in vitro* systems such as primary and permanent fish cell cultures, the importance of early life-stage tests with fish, the bioaccumulation of xenobiotics in fish, the origin of liver neoplastic lesions in fish related to the basic architectural patterns of fish liver, immunochemical approaches to monitor effects in cytochrome P450-related biotransformation, the impact of heavy metals in soft water systems, the environmental toxicology of organotin compounds, oxidative stress in fish by environmental pollutants, and complex contamination in field populations to effects by estrogenic substances in aquatic systems. As editors, we are grateful to the colleagues, who made this volume possible by their contributions from a specialized field in fish ecotoxicology. Part of them also kindly served as reviewers for other contributions; in addition, we express our gratitude to Drs. W. Ahne, H. Babich, R. Baudo, S. Baksi, P. Bannasch, N. DiGiulio, W. Klein, P. Lemaire, J. Mosello, P. Paert, H.J. Pluta, J.D. Reis, R.B. Spies, K. Steinhäuser, R. Triebskorn, P. Weis and P.W. Wester for their valuable comments and suggestions. In the editorial phase, the tireless assistance of M. Behncke-Braunbeck has helped a lot to improve the formal standards of this volume. Finally, particular thanks are due to Birkhäuser Publishers, namely Janine Kern and Dr. Petra Gerlach for the excellent cooperation and their constant support during the entire publication process.

   In the final analysis, however, the editors will have to bear the blame for any errors and mistakes. This collection of reviews has been designed to serve as a reference not only for specialists in fish ecotoxicology, but also for scientists and students interested in related fields of environmental sciences; in case we should not have been able to meet the expectations by this broad spectrum of readers, both contributors and editors will be happy to receive any comments that might help to further improve our understanding the various functions fish play in modern ecotoxicology.

Thomas Braunbeck
Bruno Streit
David E. Hinton

Fish Ecotoxicology
ed. by T. Braunbeck, D. E. Hinton and B. Streit
© 1998 Birkhäuser Verlag Basel/Switzerland

Fish cell lines as a tool in aquatic toxicology

H. Segner

*Department of Chemical Ecotoxicology, Centre for Environmental Research,
Permoserstrasse 15, D-04318 Leipzig, Germany*

Summary. In aquatic toxicology, cytotoxicity tests using continuous fish cell lines have been suggested as a tool for (1) screening or toxicity ranking of anthropogenic chemicals, compound mixtures and environmental samples, (2) establishment of structure-activity relationships, and (3) replacement or supplementation of *in vivo* animal tests. Due to the small sample volumes necessary for cytotoxicity tests, they appear to be particularly suited for use in chemical fractionation studies. The present contribution reviews the existing literature on cytotoxicity studies with fish cells and considers the influence of cell line and cytotoxicity endpoint selection on the test results. Furthermore, *in vitro/in vivo* correlations between fish cell lines and intact fish are discussed.

During recent years, fish cell lines have been increasingly used for purposes beyond their meanwhile established role for cytotoxicity measurements. They have been successfully introduced for detection of genotoxic effects, and cell lines are now applied for investigations on toxic mechanisms and on biomarkers such as cytochrome P4501A. The development of recombinant fish cell lines may further support their role as a bioanalytical tool in environmental diagnostics.

Introduction

The application of cell cultures in toxicological research offers a number of scientific, technical, and ethical advantages. Scientifically, cells as the site of primary interaction between toxic chemicals and biological systems provide the opportunity to study toxic mechanisms and interactions underlying complex whole animal responses and to learn of the factors that control the balance between protective and pathological processes. The proximity to toxic modes of action and the absence of complicated toxicokinetics make cells in culture a suitable experimental basis for establishing structure-activity relationships as well as to classify the toxic action of individual chemicals or complex environmental samples for "effect specific" or "compound class specific" effects. Technically, the ability of *in vitro* tests to rapidly screen large numbers of samples at reasonable costs is of particular value in the assessment of the growing load of man-made chemicals or in the assessment of the toxicity of environmental samples. Finally, the *in vitro* approach has an ethical justification because the use of cell cultures can significantly reduce the number of animals sacrificed for toxicological research.

In aquatic toxicology, two different types of cell cultures have been used for *in vitro* investigations: (1) primary cells that are freshly isolated from the animal and incubated for some days to weeks without propagation;

these cells conserve many of the *in vivo* features; (2) permanent cell lines that propagate indefinitely in culture; these cells often have lost functional, structural or metabolic properties of the originating tissues or cells. The properties of piscine primary cell cultures and their use for the purposes of aquatic toxicology have been reviewed recently (Baksi and Frazier, 1990; Segner, 1998). The present work is focused on the consideration of fish cell lines only.

Apparently, the first use of established fish cell lines for toxicological research was by Rachlin and Perlmutter (1968) who measured the cytotoxic action of zinc on fathead minnow (FHM) cells. In 1979, the first report on the application of fish cell lines to genotoxicity measurements was published by Barker and Rackham. Whereas toxicological studies with fish cell lines remained rare during the first half of the 1980s, from 1985 the potential of fish cell lines for cyto- and genotoxicity screening has become more widely recognized. Much of the pathfinding work in this field has been done by Ellen Borenfreund and Harvey Babich (cf. list of references). During recent years, piscine cell lines have increasingly been utilized for applications other than cyto- and genotoxicity assays, for instance the determination of biomarker induction responses. At this stage, it appears justified to review the available literature on toxicological studies with fish cell lines, to summarize for which purposes and problems cell lines have been applied for, to ask which lessons aquatic toxicologists can learn from cell line studies and which future applications may be expected.

Culture of fish cell lines

Originally, fish cell lines were developed to study fish viruses (Wolf and Quimby, 1969), which continues to be an important use. Cell lines have been established from a rather small number of teleost species. In their classic paper on available fish cell lines, Wolf and Mann (1980) listed 61 cell lines originating from 36 fish species, representing 17 families. More recently, updated lists have been compiled by Hightower and Renfro (1988) as well as Fryer and Lannan (1994). Bols and Lee (1991) discussed various fish cell lines with respect to their tissue and organ of origin.

Nearly all known fish cell lines are anchorage-dependent, and only a few lines have been adapted to grow in suspension (Bols and Lee, 1991). The cells are grown in media originally developed for mammalian cells (Bols and Lee, 1994; Lannan, 1994; Nicholson, 1985; Wolf and Ahne, 1982). For passaging, fish cells are removed from the growth substrate by use of trypsin or trypsin/EDTA solutions, as it has been described for mammalian cell culture (Bols and Lee, 1994).

As culture substrate for growth of fish cells, conventional tissue culture plastic ware is used. To date, extracellular matrices, attachment or spreading factors have been applied almost exclusively for primary cultures, but not for permanent fish cell lines. However, it should be emphasized that the *in vitro* differentiation of fish cell lines might depend on (piscine) extra-cellular matrices (Bols and Lee, 1991; Kaneko et al., 1995).

To support *in vitro* growth of fish cell lines, the basal culture medium is supplemented with mammalian sera. A consequence of this procedure is that the cells acquire a fatty acid composition that closely reflects serum composition of mammals, but is clearly different to fatty acid composition of fish cells *in vivo* (Tocher and Dick, 1990; Tocher et al., 1988). The use of fish sera, on the other hand, suffers from problems of reliable supply and reproducible quality; in addition, in some cases piscine sera were found to exert cytotoxic activities (Collodi and Barnes, 1990; Fryer et al., 1965).

Several authors have made attempts to replace the serum by (partly) defined media supplements. Shea and Berry (1983) supplemented medium 199 with an autoclaved mixture of undefined components, including lactalbumin hydrolysate, bacto-peptone, trypticase soy broth, dextrose and yeastolate. Five fish cell lines grew in this medium with limited success. Collodi and Barnes (1990) obtained a growth-promoting effect on fish cells when supplementing the medium with a rainbow trout embryo extract. For the chinook salmon embryo cell line CHSE-214, supplementation of the medium with bovine serum albumins instead of full serum was found to support cell growth (Barlian and Bols, 1991; Barlian et al., 1993). Recently, Kohlpoth and Rusche (1997) reported the successful adaptation of the rainbow trout gonad cell line, RTG-2, to media containing artificial serum replacements (BMS from Biochrom and Ultroser G from Life Technologies).

A major difference between fish and mammalian cell culture is the propagation temperature (Bols et al., 1992). Although a few fish cell lines have been described to grow at temperatures higher than 30°C, most piscine cell lines proliferate at lower temperatures, resulting in much slower growth rates than observed with mammalian cells. Generally, temperature ranges of cultured cells and donor species are comparable, but the cells tend to be slightly more heat resistant than the whole organisms. For example, the rainbow trout gonad cell line RTG-2 grows at temperatures from 5° to 26°C (Mosser et al., 1986; Plumb and Wolf, 1971), whereas rainbow trout hardly tolerates temperature higher than 20°C. Various fish cell lines retain viability during long-term low temperature storage and grow normally after return to the temperature optimum (Araki et al., 1994; Nicholson, 1989). The RTG-2 cell line has been reported to survive and remain viable for 2 years at 4°C without any medium change (Wolf and Quimby, 1969).

Cytotoxicity studies

For the assessment of the potential hazard of chemicals, effluents, sediments, or marine and freshwaters to aquatic biota, a number of bioassays using bacteria, plants and animals have been developed (e.g., OECD, 1982; Soares and Calow, 1993). Usually a tier-structured approach is selected, with relatively simple bioassays such as acute lethality tests for initial (tier I) screening of the relative toxicity of chemicals or environmental samples, followed by a more extensive toxicity characterization on the basis of (sub)chronic and reproduction tests. One of the most commonly used tier I bioassays is the short-term fish lethality test. This test, however, has repeatedly been criticized on scientific, technical and ethical grounds (e.g., Ahne, 1985; Babich and Borenfreund, 1987a, 1991; Douglas et al., 1986; Fentem and Balls, 1993; Isomaa and Lilius, 1995; Isomaa et al., 1994; Rusche and Kohlpoth, 1993). As an alternative or supplementary bioassay for hazard ranking, *in vitro* assays that measure the cytotoxic action of chemicals or environmental samples on established fish cell lines have been suggested (Agius and Parkinson, 1991; Ahne, 1985; Babich and Borenfreund, 1991; Braunbeck, 1995; Castano and Tarazona, 1995; Rusche and Kohlpoth; 1993, Segner and Lenz, 1993; Zahn and Braunbeck, 1993).

Cytotoxicity endpoints

The toxic action of chemicals on cells can be assessed by a variety of endpoints, including measures of cell death, viability and functionality, morphology, energy metabolism, or cell growth and proliferation (for a more detailed treatment cf., e.g., Freshney, 1987; Walum et al., 1990). Rachlin and Perlmutter (1968), in their pioneering work on fish cell lines, analyzed the cytotoxicity of zinc on fathead minnow FHM epitheloid cells by monitoring the mitotic index. Mitotic index as an endpoint was also used by Bourne and Jones (1973) in their study on 7,12-dimethylbenz[a]anthracene cytotoxicity to rainbow trout RTG-2 fibroblasts and FHM cells. Later studies with fish cell lines determined cytotoxic effects by measuring changes of total cell protein, RNA and DNA (Marion and Denizeau, 1983a, b), cell attachment (Ahne, 1985; Bols et al., 1985), cell morphology (Vosdingh and Neff, 1974; Zahn and Braunbeck, 1993; Zahn et al., 1995, 1996), reduction in cell numbers of exposed populations (Kocan et al., 1979), release of lactate dehydrogenase as a result of rupture to the cell membrane (Ahne, 1985, 1993; Saito et al., 1994; Zahn and Braunbeck), changes in population growth (Babich et al., 1986a, Mitani, 1985), decrease in ATP levels (Castano and Tarazona, 1994), and alterations in dye accumulation or exclusion (Borenfreund and Puerner, 1985; Halder and Ahne, 1990; Lenz et al., 1993).

To determine lethal effects of toxicants on cells, measures of cell death and cell viability that can be easily quantified in 96 well culture plates in the plate reader have found wide spread acceptance. Cell death can be estimated, for instance, by determining the decrease of total protein per well using dye binding assays such as the kenacid blue or the crystal violet protein stain (Ahne, 1985; Clothier et al., 1988). Cell viability can be determined colorimetrically or fluorometrically by the ability of cells to take up or to metabolize vital dyes such as neutral red, tetrazolium salts, calcein, fluorescein diacetate or alamar blue (Borenfreund and Puerner, 1985; Burkey and Brenden, 1993; Denizot and Lang, 1986; Lilius et al., 1996; Mosman, 1983; Schirmer et al., 1997). For analyzing acute cytotoxicity in fish cell lines, the assays most frequently used are the neutral red test, the MTT tetrazolium test and protein stains (see example in Fig. 1).

The various assays differ in their physiological mechanisms. Neutral red (2-methyl-3-amino-7-dimethylaminophenanzine) is a weakly cationic dye that accumulates in lysosomes of living cells. Cellular uptake of neutral red is believed to occur by passive diffusion across the cell membrane (Lullman-Rauch, 1979). Since only living cells are able to accumulate and retain the dye, the neutral red staining intensity is directly related to the number of viable cells (Babich and Borenfreund, 1992, 1993). The MTT assay is based on the reduction of the yellow, soluble tetrazolium salt 3-(4,5-dimethyldiazol-2-yl)-2,5-diphenyltetrazolium bromide (MTT) into a blue, insoluble formazan product by the mitochondrial succinate dehydrogenase (Berridge and Tan, 1993; Denizot and Lang, 1986; Mosman, 1983). Again, only living, but not injured or dead cells can catalyze the MTT conversion. Therefore, the amount of MTT formed is linearly related to the number of viable cells. Protein methods rely on a different principle:

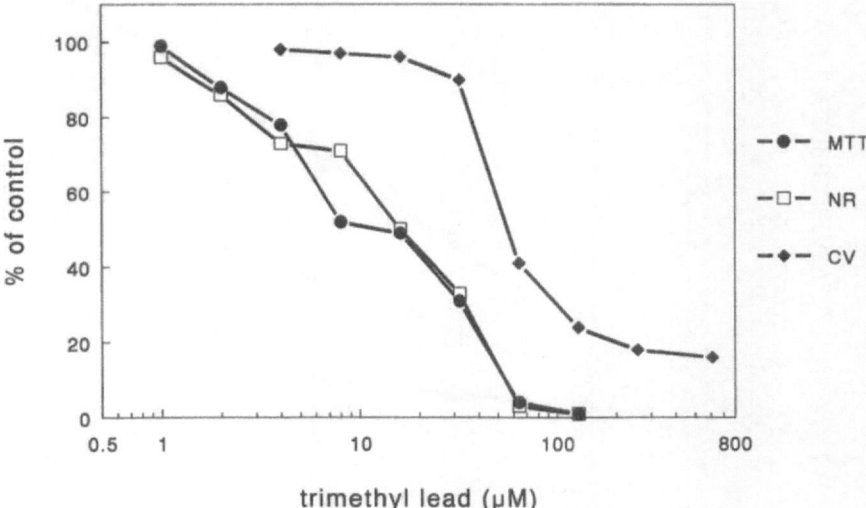

Figure 1. Concentration-dependent cytotoxicity of trimethyl lead to rainbow trout R1 cell line in the MTT tetrazolium reduction, the neutral red (NR), and the crystal violet (CV) protein assay (adapted from Lenz et al., 1993).

During toxic exposure, dead cells usually detach from the culture substratum, whereas viable cells remain attached. The number of surviving
cells is quantified – after chemical fixation of the cells – as the amount of
protein per well, which is a linear function of the number of attached cells.
Due to the different staining mechanisms, the various assays differ with
respect to their linear range for optical density readings *versus* cell number,
in their sensitivity to changes in cell number (i.e., change of optical density
with change of cell number), and in their minimum detectable cell number
(Martin and Clynes, 1993). With fish cell lines, the MTT test seems to yield
low optical density readings and low sensitivity, i.e., small changes of
optical density with changing cell numbers, when compared to the neutral
red test or to protein dyes such as crystal violet (Fig. 2; Lenz et al., 1993;
Saito et al., 1994). The low staining response of fish cell lines to MTT may
be explained by a low mitochondrial succinate dehydrogenase activity.

Considering the differences in mechanisms and staining behavior among
cytotoxicity assays, the question arises as to what extent the cytotoxic
response depends on the specific endpoint of the method selected. In relation to the mode of toxic action, chemicals may lead to different effective
concentrations in different cytotoxicity assays. For instance, the MTT assay
might be particularly sensitive to uncouplers of oxidative phosphorylation,
because the method is based on intact mitochondrial metabolic function.
Saito et al. (1994) evaluated the cytotoxicity of a series of chlorophenols,

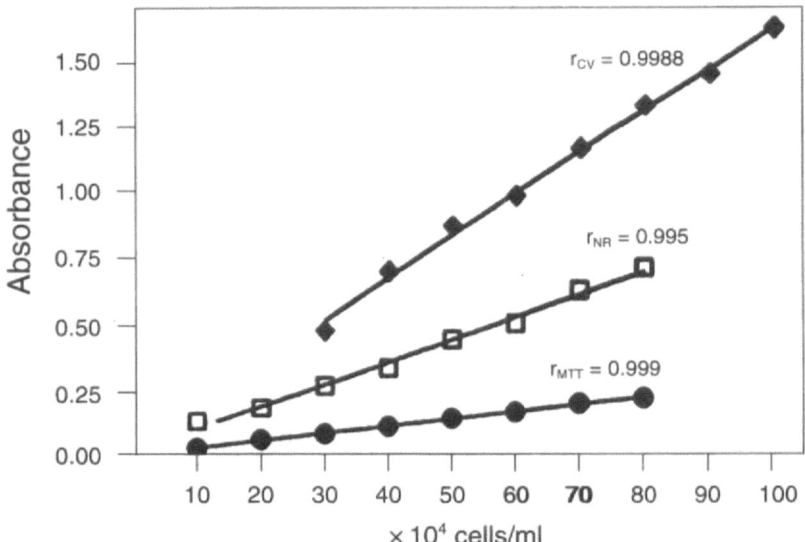

Figure 2. Optical density measurements with crystal violet (CV), neutral red (NR) and MTT
formazan (MTT) stain as a function of cell number. Absorption was read in a microtiterplate
reader at 590 nm for CV, and at 540 nm for NR and MTT (adapted from Lenz et al., 1993).

including polar narcotics and uncouplers, to goldfish scale (GFS) cells by means of three different endpoints: MTT, NR and lactate dehydrogenase (LDH) release. The midpoint cytotoxicity concentrations (IC_{50}) of the test chemicals from all three assays showed a high degree of correlation ($r > 0.97$), despite the different mechanistic basis of the assays. In other words: the various assays did not differentiate between test agents exerting different modes of toxic action. Significant correlations of cytotoxicity values as established from different endpoint methods for chemicals with different toxic mechanisms were also observed by other authors: Lenz et al. (1993), for the rainbow trout R1 cell line; Schulz et al. (1995), for rainbow trout RTG-2 cells; and Fent and Hunn (1996), for the topminnow PLHC-1 cell line. Obviously, despite different physiological mechanisms, various endpoint methods for acute cytotoxicity show good correlation in relative ranking of lethal effects for chemicals on fish cell lines; even absolute sensitivities appear to be rather similar.

Few studies with fish cell lines have used cell growth or cell proliferation as endpoint to measure cytotoxicity. This may be due to the comparatively slow growth of fish cells, with doubling times of 50 h or more (Kohlpoth and Rusche, 1997; Plumb and Wolf, 1971). Cell growth can be estimated from a variety of parameters, including the time-dependent increase in cell number (measured by using a hemocytometer or an electronic cell counter), increase in cellular macromolecules (e.g., protein, DNA – Schirmer et al., 1994), increase of metabolic activities (e.g., MTT tetrazolium reduction), or the incorporation of precursors (^{3}H-thymidine, bromodeoxyuridine) into cellular DNA (e.g., Baserga, 1989; Walum et al., 1990). Babich et al. (1986b) demonstrated an increase in cadmium cytotoxicity to BF-2 cells when shifting from short-term endpoints measuring cell death and cell viability to longer-term endpoints measuring cell growth. Whereas the 50% effect concentration in a cell detachment assay (assessed by protein staining) after 4 h of exposure was 0.21 mM, it was 0.08 mM in a 24 h neutral red cell viability assay, and 0.04 mM in a 72 h cell replication assay (assessed by cell counting in a hemocytometer). In a recent study with R1 cells from rainbow trout, Segner and Schüürmann (1997) compared the effects of 10 reference chemicals of the Multicentre Evaluation of *In Vitro* Cytotoxicity (MEIC) program on cell survival (crystal violet protein stain after 24 h exposure) and cell growth (crystal violet protein stain after 144 h exposure). By using growth instead of cell survival as an endpoint, sensitivity of the cytotoxic response of R1 cells remained unchanged, e.g., ethylene glycol, or was increased at maximum by a factor 7 for digoxin. For the same 10 substances in mammalian cell lines, the change from short-term to growth endpoints was accompanied by an enhancement of cytotoxicity by factors between 4 (isopropyl alcohol) and 44 (paracetamol; Clemedson et al., 1996).

Direct growth-inhibiting effects of a toxicant can be analyzed by monitoring cell growth in the presence of the test chemical. The colony or

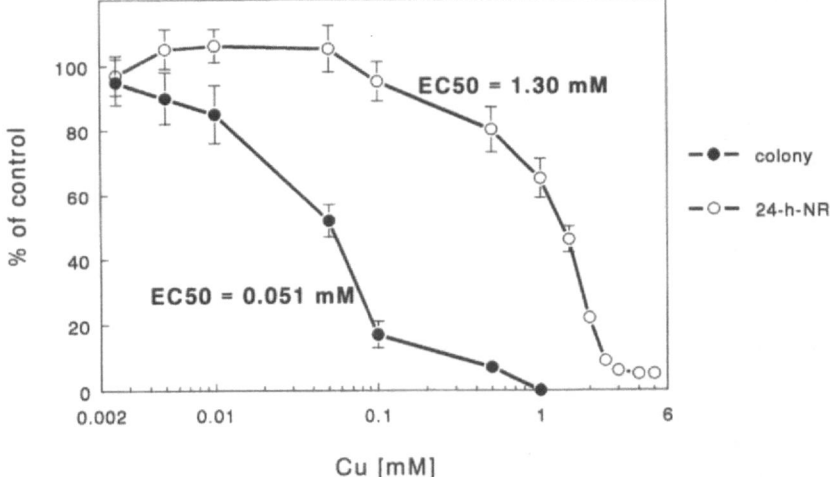

Figure 3. Cytotoxicity of copper (CuCl₂) in the short-term (24 h exposure) neutral red (NR) assay, and in the colony formation assay. For the latter test, cells were exposed to various concentrations of copper for 24 h; afterwards, the cells were trypsinized and seeded in control medium at a density of 1000 cells/7 mL into a 25 cm² culture flask. After 14 days, the number of colonies (with a minimum cell number of 50 cells/colony) was estimated under the microscope (H. Segner, unpublished data).

clonogenic assay, on the other hand, measures how many cells survive and proliferate after toxic exposure (Freshney, 1987; Puck and Marcus, 1955). In brief, a cell population is treated with a test compound for a certain (usually short-term) period, after which the compound is removed, the cells are dissociated, seeded at very low densities and are further cultivated (for several weeks in the case of fish cell lines) in the absence of the toxicant in order to determine their ability to form colonies. Whereas the highly sensitive colony formation assay is frequently used with mammalian cells, it has only occasionally been applied with fish cell lines (Babich et al., 1986b; Lyons-Alcantara et al., 1996; Mitani, 1985). Several fish cell lines such as RTG-2 and FHM are not able to form colonies, but other lines including bluegill sunfish BF-2 cells and carp EPC cells (*Epithelioma papillosum cyprini*) are clonogenic. Using the colony assay, Babich et al. (1986b) observed for Cd-exposed BF-2 cells a clear-cut increase of sensitivity in comparison to the 24 h neutral red test. Similar results have been obtained in our own studies with the EPC cell line (Fig. 3).

Test considerations

Since fish cell lines originate from eurythermic organisms, *temperature* is a crucial factor for the growth, physiology and cytotoxic response of these cells (Babich and Borenfreund, 1991; Bols et al., 1992). For example, the bluegill sunfish BF-2 and the fathead minnow FHM cells grow at 26° and 34°C, with the rate of replication being greater at the higher temperature. Diethyl tin chloride and 4,4'-DDE were more cytotoxic to both cell lines when the exposures occurred at 34°C, rather than at 26°C (Babich and Borenfreund, 1987d). A most pronounced influence of the incubation temperature exists with respect to the cytotoxicity of metabolism-mediated toxicants such as polycyclic aromatic hydrocarbons (Babich and Borenfreund, 1987e; Smolarek et al., 1988).

A further factor that may affect the cytotoxic response of cell lines is the medium composition. Bioavailability and, hence, cytotoxicity of chemicals can be greatly influenced by the presence of *serum* in the medium (Schirmer et al., 1997). For the BALB/c mouse 3T3 fibroblasts, Borenfreund and Puerner (1986) showed that reduction of serum levels enhanced cytotoxicity of cationic metals by a factor of 2–4 (Tab. 1). With the exception of Zn and Ni, congruent findings were obtained in own experiments comparing metal cytotoxicity towards rainbow trout RTG-2 cells exposed in medium containing 10 or 0% serum (Tab. 1). Lyons-Alcantara et al. (1996) demonstrated that the concentration of free cadmium ions differs greatly in serum-free and serum-containing medium and that this difference in ion activity is associated with a corresponding difference in cadmium cytotoxicity. Serum effects on cytotoxic activity are not restricted to metals but can also be observed for organic test agents. Kocan et al. (1983), for example, showed that uptake of benzo[a]pyrene by RTG-2 cells was reduced in serum-containing media. A systematic understanding of the physicochemical or molecular parameters that may determine whether the cytotoxicity of organic compounds is altered by the presence of serum has not been developed to date.

Choice of the test *cell line* is a further factor that can modify the outcome of cytotoxicity studies. Rational selection of a cell line according to the specific purpose of an investigation is difficult because the available

Table 1. Influence of serum on metal cytotoxicity

	BALB/c 3T3 NR_{90} (μM) 10% serum	BALB/c 3T3 NR_{90} (μM) 1% serum	RTG-2 NR_{90} (μM) 10% serum	RTG-2 NR_{90} (μM) 0% serum
Hg	8	2,5	15	3
Cd	6	1,5	30	9
Zn	110	40	260	301
Cu	170	80	470	159
Ni	300	70	9610	9049

BALB/c mouse 3T3 fibroblasts: data taken from Borenfreund and Puerner (1986).
RTG-2 fibroblasts: H. Segner, unpublished data.
NR_{90}: concentration of the metal giving 90% of neutral red optical absorption as compared to the control group (in other words: 10% loss of cell viability). Exposure period: 24 h.

knowledge on sensitivities and metabolic capacities of fish cell lines is limited. Rachlin and Perlmutter (1968, 1969) reported a pronounced difference in the sensitivities of fathead minnow (FHM) and rainbow trout (RTG-2) cell lines towards zinc: whereas in FHM cells a 50% decrease of the mitotic index was evoked by zinc concentrations between 1.8 and 7.5 mg/L, it took 18 mg zinc/L to obtain the same effect in RTG-2 cells. The cell lines RTG-2 and BF-2 showed identical relative rank order for metal cytotoxicity, but differed in absolute sensitivities: the RTG-2 cell line was clearly more tolerant than the BF-2 cell line, particularly for Ni (Agius and Parkinson, 1991; Babich et al., 1986b). For metal cytotoxicity, a high correlation was observed for the cell lines BF-2 and R1, both in terms of absolute sensitivity and of relative ranking (Segner et al., 1994). Short-term cytotoxicity data from rainbow trout R1 cells even correlate well (r^2 values from 0.77 to 0.87) with cytotoxicity data from mammalian cell lines, as shown for the heterogenous set of the 50 MEIC chemicals by Clemedson et al. (1998).

Babich and Borenfreund (1987d) compared the cytotoxicity (neutral red assay) of several classes of organic compounds (organometals, chlorinated pesticides, polychlorinated biphenyls, polynuclear aromatic hydrocarbons, phenolics) to two fish cell lines, FHM and BF-2. Toxicant exposures were done at the optimal incubation temperatures of the cell lines, i.e., 34°C for FHM and 26°C for BF-2. Ranking of cytotoxic potencies of the test agents was found to be almost identical for both cell lines, as became evident from the correlation coefficient of 0.988. On the other hand, for each class of test chemicals, the FHM cells were more sensitive than the BF-2 cells. In this specific case, the difference of sensitivity may be a reflection of the different incubation temperatures. In fact, when exposures were performed at equivalent temperatures (34°C), the difference was lessened, although the FHM cells remained slightly more responsive than did the BF-2 cells (Babich and Borenfreund, 1987d).

The question of selecting the appropriate test cell line may further be illustrated by an example with chlorophenol compounds. For this class of chemicals, acute cytotoxicity results (neutral red or MTT assay) are available from six fish cell lines: BF-2 (Babich and Borenfreund, 1987c), goldfish GFS cells (Saito et al., 1991, 1994), R1 cells (Schüürmann and Segner, 1994a, Schüürmann et al., 1997), CCB cells from carp brain (Saito and Shigeoka, 1994), OLF-136 cells from medaka fin (Saito and Shigeoka, 1994), and PLHC-1 cells from *Poeciliopsis lucida* (Fent and Hunn, 1996). Principal component analysis of the data reveals a generally good agreement in chlorophenol cytotoxicity ranking between the cell lines, except for the PLHC-1 cell line (Fig. 4). The similarity in responsiveness of different cell lines from a variety of species may be explained on the basis of the basal cytotoxicity concept (Clemedson et al., 1996; Ekwall, 1980). This concept consists of a classification of chemical toxicity into three categories: basal (general) cytotoxicity, organ specific (selective)

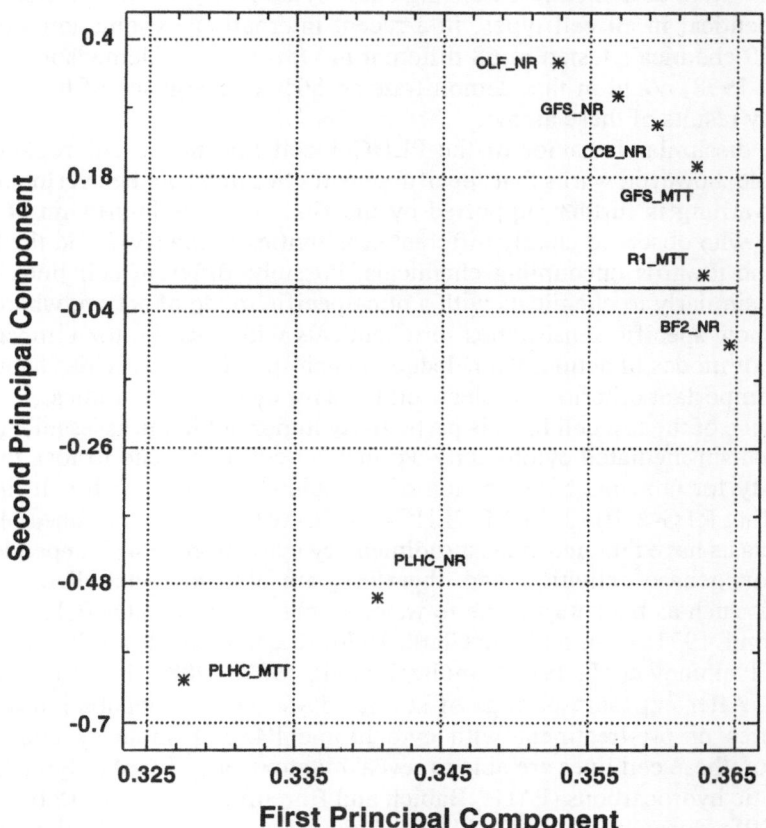

Figure 4. Principal component analysis on the cytotoxicity of six chlorophenols (phenol; 2-chlorophenol; 2,4-dichlorophenol; 2,4,6-trichlorophenol; 2,3,4,6-tetrachlorophenol; pentachlorophenol) in six fish cell lines, BF-2: Babich and Borenfreund (1987c); GFS – Saito et al. (1991, 1994); R1 – Schüürmann and Segner (1994a, Schüürmann et al., 1997); PLHC-1: Fent and Hunn (1996); CCB – Saito and Shigeoka (1994); OLF – Saito and Shigeoka (1994), using two different endpoint methods: NR uptake assay and MTT reduction assay. Except for the cell line PLHC-1, which was assessed by means of the MTT and the NR assays, all cell lines/endpoints reveal a high degree of similarity. Analysis and graph were kindly contributed by Prof. G. Schüürmann (Centre for Environmental Research, Leipzig).

toxicity, and organizatory toxicity. It is assumed that the majority of chemicals cause acute lethality to cells by basal cytotoxicity mechanisms. Basal cytotoxicity represents the interference of chemicals with fundamental functions common to all cells, irrespective of their origin (i.e., effects on plasma membrane integrity, energy metabolism, cytoskeleton, ion regulation, etc.). Due to this generality, it should be possible to use any cell type to determine the cytotoxic concentration of a chemical which causes its effects by basal cytotoxicity, or, with other words: The cytotoxic concentration of a chemical exerting basal cytotoxicity should be more or less identical in all cell types. In a recent international evaluation study with 30 chemicals tested in 68 different *in vitro* assays, Clemedson et al. (1996, 1998) could in fact demonstrate an 80% convergence of the cytotoxicity results of these assays.

The dissimilar behavior of the PLHC-1 cell line in the chlorophenol example, however, warns that specific sensitivities must not be overlooked. This warning is further supported by the findings of Schüürmann et al. (1997) who observed clearly different sensitivities of the BF-2 and the R1 cell line towards uncoupling chemicals. Possibly, different cell lines respond similarly to chemicals with a non-specific mode of action, whereas they show specific sensitivities for chemicals with specific toxic mechanisms or modes of action. Knowledge of such specific sensitivities should be an important criterion to select cell lines for cytotoxicity studies.

Choice of the test cell lines is particularly important in the assessment of metabolism-mediated cytotoxicity. Permanent cell lines tend to lose their capacity for enzymatic conversion of xenobiotics. Several fish cell lines including RTG-2, BF-2, FHM, PLHC-1, RTL-W1 and brown bullhead BB fibroblasts have retained at least rudimentary cytochrome P450-dependent monooxygenase activities, and, therefore, are able to metabolize compounds such as benzo(a)pyrene to water-soluble intermediates (Clark and Diamond, 1971; Diamond and Clark, 1970; Kocan et al., 1983; Lee et al., 1993; Plakunov et al., 1987; Smolarek et al., 1987, 1988; Thornton et al., 1982). After exposure periods of several days, elevated incubation temperatures or pre-treatment with cytochrome P4501A inducers such as Aroclor, these cell lines are able to reveal cytotoxic activity of polynuclear aromatic hydrocarbons (PAHs; Babich and Borenfreund, 1987d; Babich et al., 1991; Kocan et al., 1979; Martin-Alguacil et al., 1991). On the other hand, fish cell lines such as the rainbow trout hepatoma RTH-149 cells or R1 cells that have apparently lost intrinsic levels of P450 activity, do not show any cytotoxic response even after prolonged PAH exposure (Babich et al., 1989a; Braunbeck et al., 1995).

Alternatively to relying on the endogenous metabolic capacities of cell lines, exogenous metabolic activation systems such as hepatic S9 microsomal fractions can be employed. For example, incorporation of rat liver S9 preparations into the neutral red cytotoxicity assay with BF-2 cells potentiated the cytotoxic effects of PAHs (Babich and Borenfreund, 1987e). Moreover, compounds such as cyclophosphamide that do not evoke cytotoxicity in both metabolism-competent (BF-2) and non-competent (R1) fish cells show cytotoxicity when incubated in the presence of S9 (Fig. 5; Babich and Borenfreund, 1987e; Braunbeck et al., 1995).

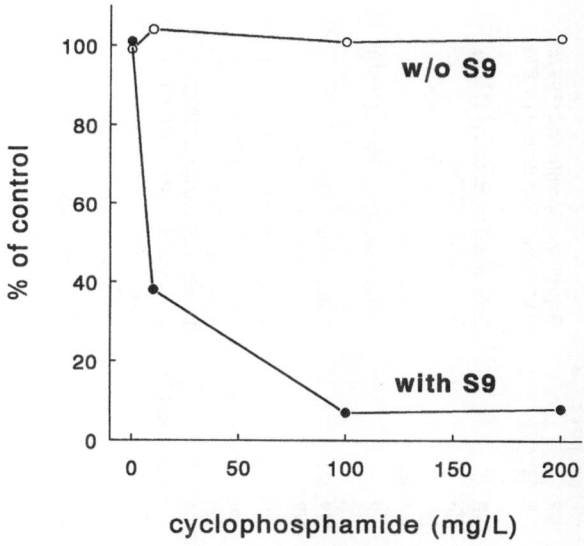

Figure 5. Cytotoxicity of cyclophosphamide to the R1 cell line, as estimated from the 24 h neutral red assay, with or without addition of rat liver S9 preparations (adapted from Braunbeck et al., 1995).

Applications of cytotoxicity tests with fish cell lines in aquatic toxicology

In Table 2, examples for cytotoxicity studies using established fish cell lines are given. *In vitro* tests are mainly utilized for the following purposes (cf. Fig. 6):

Screening and relative toxicity ranking of chemicals or environmental samples

Cytotoxicity assays are particularly suitable for screening (tier I) purposes, because they can be performed on a small scale with many replicates in 96-well cell culture plates, thus allowing for simple and rapid measurement of large sample numbers, be it single chemicals, mixtures of toxic compounds or complex environmental samples and extracts.

Each year, 1000 to 1500 new chemicals are added to about 100 000 already existing substances (Isomaa and Lilius, 1995; Johnston et al., 1996; Nagel, 1993). Even given that only a minority of these chemicals will reach the aquatic environment in significant amounts, we still lack sufficient (eco)toxicological data for most of them. In this situation, *in vitro* cytotoxicity tests could be valuable as screening tools for initial hazard assessment. Although such a reductionistic approach is always subject to criticism, it has to be discussed which loss of information would result from the introduction of *in vitro* tests when compared to the present practice of tier

Table 2. Examples of cytotoxicity studies with continuous fish cell lines

Cell line	Origin	Toxicants tested	Endpoint	Reference
BF-2	bluegill sunfish	Ag, Hg, Cd, Zi, Cu, Co, Ni, Mn, Pb, Cr, Sn (chlorides), Na-arsenic compounds, K$_2$Cr$_2$O$_7$, K$_2$Cr$_2$O$_7$, KMnO$_4$ Cd, Ni, Zn, Cu (chlorides)	NR NR, CP, replication, uridine uptake, CF	Babich et al. (1986b) Babich et al. (1986a)
		(CH$_3$)MgCl, m-cresol, (substituted) phenol(s) and toluene(s) organometal-Cl, chlordane, heptachlor, aldrin, DDD, DDT, Aroclors, acenaphthalene, 3-MC, B[a]P, phenolics	NR NR, CP, CD	Babich and Borenfreund (1987a) Babich and Borenfreund (1987d)
		chlorinated benzenes and anilines, diorganotin compounds EMS, MNNG, B[a]P, 3-hydroxy-B[a]P, 3-MC, 9-AA, 2-AF Cd, Cu, Mn, Ni, Zn, Pb, Co Puget Sound sediments	NR CN NR cell death	Babich and Borenfreund (1988) Kocan et al. (1979) Magwood and George (1996) Kocan et al. (1985)
BG/F	bluegill sunfish	inorganic lead and di- and trialkyl lead compounds	NR	Babich and Borenfreund (1990)
		phenyl-, methyl-, ethylHgCl, inorganic HgCl$_2$ arsenic and selenium interactions	NR, MF NR	Babich et al. (1990) Babich et al. (1989b)
R1	rainbow trout	sewage waters, CuSO$_4$, NaNO$_2$, ZnCl$_2$, HgCl$_2$, CdNO$_3$, Pb-acetate, BaCl$_2$ AgNO$_3$, EDTA, TCA, citric acid, trition, tensides EDTA, copper, sodium nitrite, urea 362 sewage water samples 4-CA	CM, CG, CA, enzymes, TP, LDH EM CA EM, CA, NR, LDH	Ahne (1985), Ahne and Halder (1991), Halder and Ahne (1990) Mayer et al. (1988) Rusche and Kohlpoth (1993) Zahn and Braunbeck (1993)
		MgCl$_2$, PbNO$_3$, CdCl$_2$, ZnCl$_2$, CuSO$_4$, NiCl$_2$, isopropanol, TCA, brenzcatechine, phenol, 2,4-di-CO, penta-CO, triethyl lead, diethyl lead, trimethyl lead	CA,NR, MTT	Lenz et al. (1993), Segner and Lenz (1993)
		NaF, HgCl$_2$, CdCl$_2$, ZnCl$_2$, CuSO$_4$, NiCl$_2$, PbNO$_3$, Pb-acetate, penta-CO dodecyllhydrogensulfate, phenol, 2-CO, 3-CO, 2,4-di-CO, 2,3-di-CO, 4-CO, EDTA, 4-nitroaniline, endosulfan, 4-CA, TCA, citric acid, oxalic acid, chlorobenzene, acetone, chloramphenicol, formaldehyde, ethanol, urea, isopropanol chloro- and nitrophenols	CA	Segner and Lenz (1993)
		Cd, Ag, Hg, Zi, Pb, Ni organometal chlorides, chlordane, heptachlor, aldrin, DDD, DDT, DDE, PCBs, acenaphthalene, 3-MC, B[a]P, phenols	MTT NR, CA NR, CP, CD	Schürmann et al. (1997) Segner et al. (1994) Babich and Borenfreund (1987a)

Cell line	Origin	Toxicants tested	Endpoint	Reference
FHM	fathead minnow	organometals, chlordane, heptachlor, aldrin, DDD, DDT, DDE, PCBs, acenaphthalene, 3-MC, B[a]P, phenolics	NR, CP	Babich and Borenfreund (1987d)
		Hg, Zn, Ni, n-octanol, ethanolamine, n-heptanol, phenol, chloral hydrate, citric acid, oxalic acid, diethanolamine, TCA, benzoid acid	NR	Dierickx and van den Vyver (1991), Brandao et al. (1992), Dierickx and Bredael-Rozer (1996)
RTG-2	rainbow trout	penta-CO, p-methylaminophenol, 2,4-di-CO, p-CO, p-cyanophenol, p-nitrophenol, benzene, p-methylphenol, aniline, phenol, p-methoxyphenol, 1,2,4-trichlorobenzene	CA	Bols et al. (1985)
		EMS, MNNG, B[a]P, 3-OH-B[a]P, 3-MC, 9-AA, 2-AF	CN	Kocan et al. (1979)
		Puget Sound sediments	cell death	Kocan et al. (1985)
		Cu, Cd, Zn, Pb, aniline, nitrobenzene, styrene, p-nitrophenol, 2,4-DCA, phenol, 4-CO, 4-methylphenol, penta-CO, malathion	NR	Castano et al. (1996)
		Puget Sound sediment extracts	CN	Landolt and Kocan (1984)
		industrial effluents	NR	Castano et al. (1994)
		Cd, Pb	DNA synth., CP	Marion and Denizeau (1983a, b)
		aniline, Cd, di-CO, $K_2Cr_2O_7$, Cu, phenol, triton X-100	NR, CP	Schulz et al. (1995)
		PAHs, DDD, DDT, DDE, aldrin	NR	Martin-Alguacil et al. (1991)
GSF	goldfish	COs	NR, MTT	Saito et al. (1991, 1994), Saito and Shigeoka (1994)
		alcohols, benzenes, chloro- and nitrophenols, pesticides	NR	Saito et al. (1993)
PLHC-1	Poecil. lucida	alkylbenzenes, malathion, diazinon, methoxychlor, alachlor, phtalate diesters, B[a]P	NR	Babich et al. (1991)
		organotins	NR	Brüschweiler et al. (1995)
		COs, nitrophenols, dinoseb, phenol, sulfonic acids, OP, NP	NR	Fent and Hunn (1996)
EPC	carp	DDT	CP	Parkinson and Agius (1986, 1988)
		cadmium	CF	Lyons-Alcantara et al. (1996)

Abbreviations: 9-AA – 9-aminofluorene; 2-AF – 2-aminofluorene; B[a]P – Benzo[a]pyrene; CO-chlorophenol; CA – chloroaniline; EMS – ethylmethanesulfonic acid; 3-MC – 3-methylcholanthrene; MNNG – N-methyl-N'-nitro-N-nitrosoguanidine, NP – nonylphenol; OP – octylphenol; PAHs – polycyclic aromatic hydrocarbons; TCA – trichloroacetic acid.
CA – cell adhesion; CD – cell detachment; CF – colony formation; CG – cell growth; CM – cell morphology; CN – cell number; CP – cell protein; EM – electron microscopy; LDH – lactate dehydrogenase leakage; NR – neutral red retention; MF – micronucleus formation; MTT – 3-(4,5-dimethyldiazol-2-yl)-2,5-diphenyltetrazolium bromide [tetrazolium salt]; TP – trypane blue exclusion.

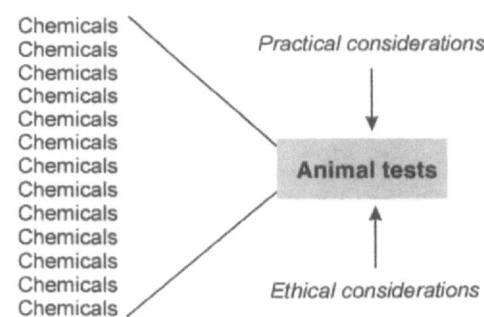

Figure 6. Applications for cytotoxicity tests with fish cell lines in aquatic toxicology. The use of animal tests represents, due to practical and ethical considerations, a bottleneck for sufficient ecotoxicity testing. This problem could be overcome by application of cytotoxicity assays.

I testing, i.e., the use of short-term fish lethality tests that are of doubtful ecotoxicological relevance as well (cf. Isomaa and Lilius, 1995).

The simplicity, rapidity as well as the requirement for small sample volumes in fish cell line assays allow the addressing of problems beyond the mere estimation of midpoint toxicity of individual chemicals or environmental samples:

— Understanding and evaluating mixture toxicity: In the environment, animals are not exposed to individual chemicals but to complex chemical mixtures. Presently, accurate prediction of the joint effects of complex mixtures of substances is not possible (Altenburger et al., 1993; Johnston et al., 1996). The necessary systematic investigation of combinations of chemical compounds is, due to the large number of tests required, difficult to realize by means of *in vivo* tests with fish. Again, *in vitro* cytotoxicity assays are a pragmatic and sufficiently robust tool for evaluating chemical mixtures at reasonable costs and workloads. Surprisingly, this potential application of fish cell lines has been hardly explored to date.
— More differentiated toxicity evaluation: *in vitro* tests with fish cell lines are not necessarily restricted to the analysis of cytotoxicity, but can easily include the assessment of further parameters, e.g., genotoxicity or induction of biomarkers (see below). Thus, *in vitro* approaches offer a relatively easy access to a more differentiated effect assessment in aquatic toxicology.
— Effect analysis in the frame of bioassay-directed toxicity identification and evaluation (TIE) techniques: TIE procedures which combine chemical fractionation and chemical analytics of environmental samples with toxicity testing (e.g., Ankley et al., 1992; Ho et al., 1996) often suffer

constraints from the small amounts of fractioned material available. The much smaller sample volume required for an *in vitro* test than for a conventional *in vivo* fish assay is a clear advantage of the use of cytotoxicity assays in TIE applications (Castano et al., 1994).

Establishment of structure-activity relationships (SARs, QSARs)
The ease and rapidity of performance of cytotoxicity assays as well as the absence of complicated, whole organism toxicokinetics make cultured cells suitable to determine the (quantitative) relationship between molecular or physicochemical properties of chemical agents and their biological activity. For lipophilic chemicals with a non-specific, narcotic mode of action (alcohols, benzenes, phenolics, anilines, etc.), acute cytotoxicity to fish cell lines can be modeled – in accordance with *in vivo* findings – by the logarithm of the octanol/water partition coefficient, log K_{OW} (Fig. 7), reflecting that the uptake of these compounds is governed by partitioning into the cell membrane lipid (Babich and Borenfreund, 1987b, 1988a; Bols et al., 1985; Fent and Hunn, 1996; Saito et al., 1991, 1993; Schüürmann and Segner, 1994a; Segner and Lenz, 1993). In the case of organolead or organotin compounds, although lipophilicity contributes to toxicity (Babich and Borenfreund, 1988a; Brüschweiler et al., 1995), additional factors are involved. This is evident from the existence of a parabolic relationship between cytotoxicity and lipophilicity within a homologous series of triorganotins (trimethyl- to triphentyltin) (Schüürmann and Segner, 1994b).

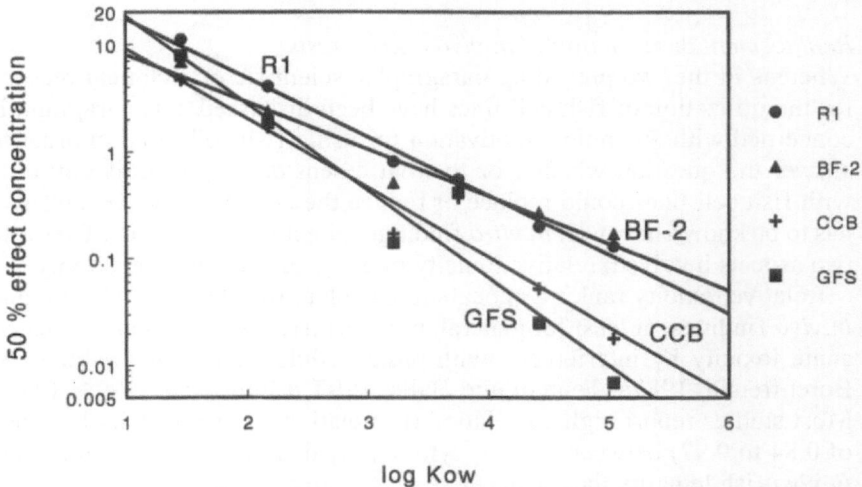

Figure 7. Structure-activity relationship between cytotoxicity of chlorophenols and their lipophilicity, as expressed by the logarithm of the octanol/water partition coefficient, log K_{ow}. Cytotoxicity data are from R1 cells (MTT test, Schüürmann et al., 1997), BF-2 cells (Babich and Borenfreund, 1987c), CCB cells (Saito and Shigeoka, 1994), and GFS cells (Saito et al., 1991).

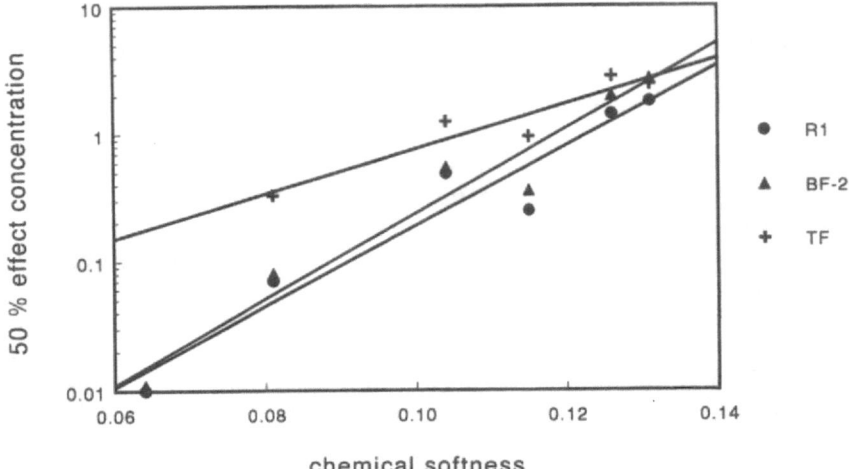

chemical softness

Figure 8. Structure activity-relationship between cytotoxicity of metal cations (Hg, Cd, Cu, Zn, Ni, Pb) and their chemical softness parameter (cf. Segner et al., 1994). Cytotoxicity data are from R1 cells (NR test, Segner et al., 1994), BF-2 cells (NR test; Babich et al., 1986a), and turbot TF cells (NR test, George, 1986).

Structure-activity relationships could also be established for the cytotoxic action of bivalent metal cations on several fish cell lines (Fig. 8). Metal cytotoxicity was found to be related to the Irving-Williams-series, i.e., chemical hardness (Babich et al., 1986a, b; George, 1996; Segner et al., 1994).

Replacement or reduction of in vivo animal tests
Whereas in the two preceding paragraphs, scientific or technical reasons for the utilization of fish cell lines have been discussed, this paragraph is concerned with the ethical motivation for using fish cell tests. In order to answer the question whether or to what extent *in vitro* cytotoxicity tests with fish cell lines could replace or reduce the use of *in vivo* fish tests, it has to be known how well *in vitro* findings reflect *in vivo* toxicity. There are two aspects involved: relative toxicity ranking and absolute sensitivity.

Relative toxicity ranking appears to correlate well between *in vitro* and *in vivo* findings, at least for general, non-specific toxicants that may cause acute toxicity by interference with basal cellular functions (Babich and Borenfreund, 1987a; Fentem and Balls, 1993; Lilius et al., 1994, 1995). Most studies report high correlation (correlation coefficients in the range of 0.84 to 0.97) between *in vitro* cytotoxicity data from fish cell lines and *in vivo* fish lethality data. Bols et al. (1985) studied the cytotoxicity of 12 aromatic hydrocarbons to RTG-2 cells and found the cytotoxicity of the compounds to be significantly correlated to their acute toxicity in rainbow trout. In a study using the GFS cell line, Saito et al. (1991) demonstrated a good correlation between *in vitro* NR_{50} values of chlorophenols and their *in vivo* LC$_{50}$ values of several fish species. High correlation rates were also observed if the comparison was made for a more heterogenous set of test chemicals instead of one specific chemical class (Babich and Borenfreund, 1987c; Brandao et al., 1992; Castano et al., 1996; Dierickx and van den Vyver, 1991; Segner and Lenz, 1993). Likewise, toxicity ranking of complex effluent samples agreed well between *in vitro* and *in vivo* tests (Rusche and Kohlpoth, 1993).

In vitro/in vivo correlations, however, become less satisfying, if the comparison includes chemicals exerting their toxicity by interference with tissue-specific biochemical or physiological functions. Metals provide an example; for this class of chemicals, the *in vitro/in vivo* correlation coefficients are lower than 0.70 (Babich et al., 1986b; Segner et al., 1994). For anionic metals such as arsenite, selenite or chromate, the great difference between *in vitro* and *in vivo* data could be related to an unusually high tolerance of intact fish to these compounds. For cationic metals, copper is ranked much less toxic *in vitro* than *in vivo* (Babich et al., 1986b; Segner et al., 1994). Whereas in golden ide, *Leuciscus idus melanotus*, rainbow trout, and bluegill sunfish, the sequence of toxicity is copper > cadmium > zinc, in BF-2, R1, and the turbot cell line TF, the rank order is cadmium > zinc > copper (Babich et al., 1986b; Magwood and George, 1996; Segner et al., 1994). The difference might be explained by the fact that acute metal toxicity in intact fish is mainly executed by rapid metal accumulation in the gill tissue and associated damage to branchial ion- and osmoregulatory function (McDonald and Wood, 1993; Segner, 1987; Wilson and Taylor, 1993) – an effect that finds no correspondence in an *in vitro* cytotoxicity assay with fibroblastic cell lines such as R1 and BF-2.

By means of multivariate analysis, Schüürmann et al. (1997) characterized the acute toxicity of eight phenols containing polar narcotics and uncouplers, towards a test battery including three fish species (rainbow trout, fathead minnow, bluegill sunfish) and two fish cell lines (BF-2 from bluegill sunfish and R1 from rainbow trout). The results revealed pronounced differences in the response pattern both between the two fish cell lines as well as between fish cells and fish. Whereas the R1 cell line was sensitive to uncoupling activity of the test chemicals, the BF-2 cells showed no corresponding sensitivity. Interestingly, the behavior of the rainbow trout R1 cell line was much closer to that of the bluegill sunfish than was the behavior of the BF-2 cells. Overall, multivariate analysis demonstrated that each cell line had their value within the test battery by displaying a specific, characteristic response to the eight phenols, but cytotoxicity data did not substitute for fish toxicity data.

The absolute sensitivity of fish cell lines seems to be lower than that of the intact fish. Bols et al. (1985) noted good correlation in toxicity ranking between the EC$_{50}$ values from the *in vitro* cell attachment assay and LC$_{50}$ data measured *in vivo*, but at the same time they recognized a two to three orders of magnitude difference in the absolute values. Similarly, for six out of 21 chemicals studied, Segner and Lenz (1993) found that the *in vivo* LC_{50} value was at least 10 times lower than the *in vitro* midpoint toxicity concentration. Of course, due to different toxicokinetics comparison of cytotoxicity and whole animal data are difficult. However, if cell line tests should be applied for, for example, monitoring of effluent toxicity or determination of environmentally safe toxicant concentrations, absolute sensitivity becomes an important issue.

As pointed out by Fentem and Balls (1993), on the basis of the presently available data base, it is premature to arrive at definitive conclusions about the replacement of *in vivo* fish tests by *in vitro* cytotoxicity assays. Future *in vitro/in vivo* comparisons should focus on chemicals with different modes of action or with specific toxic mechanisms in order to understand more clearly under what conditions *in vitro* findings can be a good predictor of an *in vivo* effect. Moreover, attention should be given to the question of whether a cytotoxicity assay could contribute to ecotoxicological test batteries, not necessarily as an alternative to fish tests, but as a test system on its own rights and with its own response profile.

Genotoxicity studies

Genetic toxicology investigates the interaction of chemical or physical agents with genetic material, and the relation to subsequent adverse effects such as cancer (in the case of alterations in somatic cells) or genetic disease in future generations (in the case of alterations in germ cells). Substances that produce alterations in genetic material at non-lethal, non-cytotoxic concentrations are classified as genotoxins (Shugart, 1995). A considerable number, but not all chemical genotoxins are inactive as such and require conversion (metabolic activation) to metabolites which then are the ultimate genotoxic form.

Within the discipline of genetic toxicology, research has taken place mainly at three levels: (1) identifying the kind and determining the frequency of genetic diseases, (2) studying the mechanisms of how chemical and physical agents cause genetic disorders, and (3) evaluating agents for their potential to cause genetic disease (Maccubin, 1994). It is the latter field where *in vitro* genotoxicity tests find their application.

A wide variety of assays for genotoxic effects has been developed, including bacterial assays (e.g., Ames test), plant assays (e.g., *Tradescantia* test), invertebrate assays (e.g., sex-linked recessive lethal test with *Drosophila*), and a large number of assays using primary cells and cell lines from mammals (Hoffmann, 1996). Fish cells respond to genotoxins much as mammalian cells do. The use of fish cell lines in aquatic genotoxicity assessment is based, in part, on the observation that they retain some fish-specific traits, for instance the poikilothermic nature, peculiarities of biotransformation processes, or low rates of DNA repair (Kocan et al., 1985;

Maccubbin, 1995; Shugart, 1995). Questions and problems that have been discussed in connection with *in vitro* cytotoxicity testing also apply for *in vitro* genotoxicity testing, i.e., selection of suitable endpoints, selection of appropriate cell lines (particularly relevant is the question of their metabolic capacity to convert pro-mutagens into the genotoxic agent), or extrapolation between *in vitro* and *in vivo* findings.

In Table 3, examples of the use of fish cell lines for *in vitro* genotoxicity assays are given. Most studies with fish cell lines have utilized metabolism-competent cell lines such as BF-2; however, experiments with mammalian S9 supplementation have also been done (e.g, Walton et al., 1984, 1987).

Chemical agents can induce the three principal categories of genetic alterations in fish cells: gene mutations, chromosome aberrations, or genomic changes (Hoffmann, 1996). *In vitro* genotoxicity assays with fish cells can basically rely on the same endpoints as used in mammalian *in vitro* systems, although some specificities do exist. Chromosomes in fish cells are usually of small size and occur in large numbers (e.g., RTG-2 cells have more than 80 chromosomes: personal communication by H.G. Miltenburger, University of Darmstadt, FRG), which makes the cytogenetic analysis of chromosomal aberrations difficult (Babich and Borenfreund, 1991; Landolt and Kocan, 1983). Also the UDS (unscheduled DNA synthesis) assay may be difficult to perform with fish cell lines because their rates of DNA repair are rather low (Fong et al., 1988; Sikka et al., 1990; Walton et al., 1983, 1984).

Borenfreund et al. (1989) attempted to apply the fish cell line BG/F to a cell transformation technique that is frequently used for the elucidation of stepwise *in vitro* carcinogenic transformation of mammalian cells. Upon exposure to carcinogens, the BG/F cells exhibited early indications of cell transformation including induction of polyploidy, increased colony-forming efficiency, loss of contact inhibition, and formation of transformed foci. However, a later criterion of transformation, anchorage-independent growth, was not observed. Possibly, the criteria for assessment of transformation in mammalian cells are not applicable for fish cells.

Recently, the single cell gel electrophoresis or "comet" assay has been developed for use with mammalian cells (McKelvey-Martin et al., 1993). The genotoxic endpoint of this method is the measurement of toxicant-induced single and/or double strand breaks in DNA. If exposed cells are subjected to electrophoresis, the DNA fragments migrate out of the nucleus and form a tail behind the nucleus. This comet-like tail can be visualized by fluorescent staining of DNA (ethidium bromide), with the length of the comet indicating the degree of genotoxic activity. The test is considered to be very sensitive and allows the use of any cell type, no matter whether proliferating or not. Preliminary studies have shown that the comet assay can also be applied to both primary cells and cell lines from fish, and a number of laboratories are on the way to explore the potential applications of this method for problems in aquatic toxicology (Braunbeck and Neumüller,

Table 3. Examples of genotoxicity studies with permanent fish cell lines

Endpoint	Agent(s)	Cell line	Reference
Induction of ouabain-resistant mutants	B(a)P, MNNG	BF-2 (bluegill sunfish)	Kocan et al. (1981)
	MNNG	GEM 199	Mitani (1983)
DNA adduct formation	B[a]P, DMBA	BF-2, RTG-2, BB	Smolarek et al. (1987, 1988)
Chromosomal damage: anaphase aberration	B[a]P, MNNG, MMC, AA, 3-MC	RTG-2	Kocan et al. (1982, 1985), Koccan and Powell (1985)
	Marine sediments	RTG-2	Kocan et al. (1985), Landolt and Kocan (1984)
	Atrazine, metuxoron, 4-chloroaniline, lindane, pentachlorophenol, alachlor, carbofuran	R1	Ahne and Schweitzer (1993)
Comet assay	MNNG, H_2O_2	R1, RTG-2	Braunbeck and Neumüller (1996)
	B[a]P, 4NQO	RTG-2, RTL-W1	Nehls and Segner (1998)
Micronucleus induction	organomercurials (phenyl, ethyl, methyl)	BG/F	Babich et al. (1990)
	MNNG, 4NQO	U1-H	Walton et al. (1984)
Sister chromatid exchange (SCE)	MMC, MNNG, methyl methane sulfonate	Ameca splendens cell line	Barker and Rackham (1979)
Unscheduled DNA repair synthesis (UDS)	MNNG	ULF-23	Suyamah and Etoh (1988)
	N-nitroso-N-methylurea	ULF-23	Park et al. (1989)
	MNNG, 4NQO, aflatoxin B1	RTG-2	Walton et al. (1983)
	MNNG, 4NQO	U1-H	Walton et al. (1984)
	B[a]P, aflatoxin B1	RTG-2 + S9	Walton et al. (1987)
	sediments	BB	Ali et al. (1993)

Abbreviations: AA – aminoacridine; B[a]P – benzo[a]pyrene; MMC – mitomycin C; DMBA – 7,12-dimethylbenz(a)anthracene; MNNG – N-methyl-N′-nitro-N-nitrosoguanidine; 4NQO – nitroquinoline-1-oxide.
Origin of cell lines: BB – brown bullhead (*Ameriurus nebulosus*); BF-2 – bluegill sunfish (*Lepomis macrochirus*); BG/F – bluegill sunfish (*Lepomis macrochirus*); GEM 199 – goldfish (*Carassius auratus*); R1 – rainbow trout (*Oncorhynchus mykiss*); RTG-2 – rainbow trout (*Oncorhynchus mykiss*); RTL-W1 – rainbow trout (*Oncorhynchus mykiss*); U1-H – Central mudminnow (*Umbra limi*); ULF-23 – Central mudminnow (*Umbra limi*).

1996; Deventer, 1995; Monod et al., 1998; Nehls and Segner, 1998; Segner and Braunbeck, 1998).

Short-term genotoxicity assays with fish cell lines can screen individual chemicals, chemical mixtures and environmental samples or their fractions for their genotoxic potentials. These bioassays may have their value – comparable to the cytotoxicity tests – in a tier-structured testing approach, and may prioritize substances for more elaborated animal testing (Shugart, 1995).

Mechanistic studies

The initial interaction between chemicals and biological systems occurs at the cellular level. A variety of cellular targets can be affected by chemicals, including the life span of cells, information exchange with the cellular environment *via* receptors or membrane channels, membrane structures and functions, nuclear structures and functions, cellular metabolic processes, or general cell functions such as energy metabolism and volume regulation (DiGiulio et al., 1995; Segner and Braunbeck, 1998). Toxicological studies using cellular experimental systems are suited for the early detection of chemical exposure as well as for the investigation of toxic mechanisms, i.e., the adaptive or pathological changes occurring in response to the presence of the toxicant. Moreover, cultured cells allow to elucidate under controlled conditions concentration-response relationships for a toxicant-induced biological response, the modifying effects of mixtures, inhibitors, etc., and the effects of physiological regulators such as hormones.

The application of fish cell lines to investigations on the mode of action of toxicants and factors controlling expression of metabolism and detoxifying processes is relatively recent. Of course, the actual use of *in vitro* techniques depends greatly on the availability of cell lines exhibiting the required metabolic properties and of appropriate methods. Studies on the regulation of the enzyme cytochrome P4501A (CYP1A) provide an example. CYP1A plays an important role in the biotransformation of polynuclear aromatic hydrocarbons (PAHs) and halogenated aromatic hydrocarbons (HAHs) such as polychlorinated dibenzo-p-dioxins, dibenzofurans or polychlorinated biphenyls (Stegeman and Hahn, 1994). Exposure to these compounds leads to the induction of the CYP1A gene and related gene products via a process that is mediated by an intracellular receptor, the Ah receptor (Hankinson, 1995). With the establishment of fish cell lines that express comparatively high levels of constitutive CYP1A activity (PLHC-1, derived from hepatocellular carcinoma of *Poeciliopsis lucida*: Hahn et al., 1993; Hightower and Renfro, 1988; RTL-W1, derived from rainbow trout liver: Bechtel and Lee, 1994; Lee et al., 1993), together with the development of simplified techniques for assessing the CYP1A-associated catalytic activity, 7-ethoxyresorufin-O-deethylase (EROD), in cultured cells (Behrens et al., 1998; Hahn et al., 1996; Kennedy et al., 1995), applicability of fish cell lines as experimental tool for the analysis of CYP1A regulation has clearly increased. Fish cell line studies were among the first reports to demonstrate the existence of an Ah receptor protein in the class of teleosts (Hahn and Stegeman, 1992; Lorenzen and Okey, 1990; Swanson and Perdew, 1991). A detailed analysis of CYP1A was done for the PLHC-1 cell line (Hahn et al., 1993). Specific binding of the photoaffinity ligand 2-azido-3-[^{125}I]iodo-7,8-dibromodibenzo-p-dioxin to proteins in PLHC-1 cytosol indicated the presence of the Ah receptor. 3,3′,4,4′-tetrachlorobiphenyl (TCB) induced CYP1A in a dose-dependent manner, as detected immunochemically and by enzyme activity measurements. At high concentrations of TCB, CYP1A catalytic activity was inhibited or inactivated, similar to effects observed *in vivo*. The results of the study demonstrated that PLHC-1 retains all the characteristic features of the CYP1A system known from intact fish. Subsequent studies with this cell line have evaluated modulation of the CYP1A induction response by endocrine factors or by exposure to chemical mixtures. Combined exposure of PLHC-1 to 3-methylcholanthrene and triorganotins resulted in an inhibition of CYP1A enzyme activity, whereas CYP1A protein synthesis appeared not to be affected (Brüschweiler et al., 1996). Endocrine regulation of CYP1A in PLHC-1 was considered by Celander et al. (1996, 1997). The authors showed that glucocorticoids, although not affecting basal CYP1A levels, have a potentiating effect on xenobiotic induction of CYP1A in PLHC-1 cells. The hormonal influence was found to be mediated through a classical glucocorticoid receptor pathway.

Fish cell lines are used to address the question of species- or taxon-specificity of CYP1A induction. In studies with rat and fish cell lines, Clemons and co-workers (1994; 1996; 1997) compared CYP1A induction potencies of xenobiotics. For several planar polychlorinated hydrocarbons, the induction response differed strongly between mammalian and piscine systems; for instance, 2,3,7,8-tetrachlorodibenzofuran was approximately seven-fold more potent in RTL-W1 cells than in the rat hepatoma cell line, H4IIE. In this case, differences in the metabolism of the inducing compounds by the two cell lines might account for the difference in potency (Clemons et al., 1997). Another example pointing to possible species differences in induction responses is provided from investigations with a cell line established from the liver of zebrafish, *Brachydanio rerio*, that showed an intriguing specificity with respect to CYP1A inducers. Whereas TCDD evoked the induction of a CYP1A-related protein, no response was obtained after exposure of the cells to β-naphthoflavone (Collodi et al., 1992; Miranda et al., 1993). The authors speculated that the Ah receptor in the zebrafish cells may have strict requirements for recognition of structures of chemicals serving as receptor ligands; this would make the zebrafish cell line a unique model to reveal structure-activity relationships in the induction of CYP1A.

A number of studies with fish cell lines investigated processes and mechanisms of metal homeostasis and toxicity. The intracellular fate and effect of metals is the result of the interaction between free metal ions, their binding to a diverse group of cytoplasmic, soluble ligands and their sequestration into vesicle-bound granules and lysosomes (Roesijadi and Robinson, 1994). These cytoplasmic binding sites enable the cells to actively regulate the availability of both essential and non-essential metals, and, as a consequence, cells are able to handle the ambivalency of metals in biological systems – their essentiality for many biochemical reactions or metalloenzymes, and their potential toxicity by non-specific binding to oxygen, nitrogen and sulphur.

One group of cytoplasmic ligands involved in metal regulation are the metallothioneins (MT), a family of low-molecular mass proteins with numerous cysteinyl thiol groups (Roesijadi, 1992). Cell line studies have greatly facilitated the elucidation of the mechanisms of MT induction (George, 1996; Olson, 1996). Administration of the metals cadmium, copper, mercury and particularly zinc induce MT synthesis in cultured fish cells (George et al., 1989, 1992; Olson et al., 1990; Zafarullah et al., 1990). For instance, induction of MT mRNA in RTH-149 cells at 20°C results in detectable transcription and translation within 3–6 h after metal treatment (Price Haughley et al., 1986; Zafarullah et al., 1989). The MT response is temperature-dependent (Hyllner et al., 1989), and is modified by endocrine factors such as glucocorticoids, progesterone and noradrenaline (Burgess et al., 1993; George et al., 1992; Hyllner et al., 1989; Olson et al., 1990). Estrogen has been suspected, on the basis of *in vivo* findings, to act as a MT inducer, however, as shown by *in vitro* experiments, it appears to affect MT only indirectly, by mobilization of Zn from peripheral tissues to the liver (George, 1996).

The tripeptide glutathione (GSH) functions as another cytoplasmic ligand for metal cations (DiGiulio et al., 1995; Segner and Braunbeck, 1998). The function of GSH for metal homeostasis in fish is poorly characterized. Results from a recent study with the RTG-2 line support a protective function of GSH against metal damage in fish (Maracine and Segner, 1998). RTG-2 cells were exposed to non-cytotoxic concentrations of buthionine sulfoximine (BSO) which causes a sustained decrease in cellular GSH levels; afterwards the cells were exposed to metal ions and cytotoxicity was determined by means of the neutral red test. The BSO-caused reduction of cellular GSH levels lead to enhanced sensitivity of the cells towards heavy metals (Fig. 9). Particularly the "soft" metals Hg, Cd and Cu were sensitive to GSH depletion, whereas the effect was less pronounced for the "hard" metals Pb, Zn and Ni. Further studies utilizing the fish cell line CHSE which is not able to express MT because of hyper-methylation of the MT gene (Price-Haughley et al., 1987) may provide insight into the interplay of GSH and MT in cellular metal acclimation, as well as into possible toxic consequences, for instance, the onset of lipid

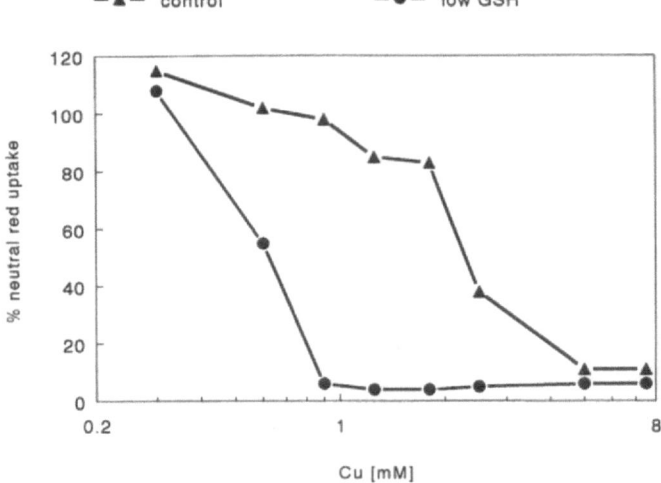

Figure 9. Cytotoxicity of copper (CuCl₂) to control RTG-2 cells and to glutathione-depleted RTG-2 cells. Glutathione depletion was attained by 24 h pre-treatment of the cells with 1 mM buthionine sulfoxide. Cytotoxicity was assessed using the neutral red assay.

peroxidation. The suitability of fish cell lines to analyzing oxidative stress has been shown by Babich et al. (1993, 1994).

Biomarker studies

The findings discussed in the previous section suggest that fish cell lines, in addition to measuring general toxicity by means of cytotoxicity assays, may also be suitable to screen for the induction of biomarkers. Biomarkers have been defined as a "biological response that can be related to an exposure to, or toxic effect of, an environmental chemical or chemicals" (Peakall, 1994). In general, application of biomarkers has focused on lower levels of biological organization, i.e., molecular, biochemical and cellular responses. Examples are inhibition of acetylcholine esterase, induction of CYP1A, formation of DNA adducts, or induction of stress proteins. Biomarkers have been used for field monitoring studies in order to recognize polluted sites. However, biomarkers can also be applied in cell culture assays in the laboratory to evaluate the potential environmental impact of individual chemicals, or for identifying the presence of pollutants in samples taken from the environment (Bols et al., 1997; Ryan and Hightower, 1996). One of the first studies using the latter approach was the work of Tillit and colleagues (1991), who introduced the H4IIE rat hepatoma cell line as a tool for assessing the dioxin-like toxicity of halogenated aromatic hydrocarbons (HAHs) present in extracts from bird eggs sampled at polluted sites. Induction of CYP1A-associated catalytic activity in H4IIE cells was used as integrative measure to detect the presence of HAHs in the extracts. The observed induction potency of the extracts were compared relative to a standard, 2,3,7,8-tetrachlorodibenzo-*p*-dioxin (TCDD), and expressed as TCDD equivalents or toxic equivalency factors (TEF; Safe, 1990). Tillit et al. (1991) found the highest levels of TCDD equivalents in bird eggs from the most polluted sites; and these levels could be linked with the reproductive success rates of the birds from these areas.

The cell-based approach for detection of the HAH- or PAH-contamination of environmental samples is increasingly used in environmental toxicology (Kennedy et al., 1996). The use of fish cell lines is preferred to the use of mammalian cells (Bols et al., 1997; Clemons et al., 1994, 1996; Zabel et al., 1996), because there appear to exist taxon-specific differences in the CYP1A induction potencies of HAHs (Clemons et al., 1996; Kennedy et al., 1996).

Cell cultures can be used to screen for new biomarker candidates. Ryan and Hightower (1994) exposed two fish cell lines to heavy metals and analyzed the toxicant-related, dose-dependent changes in cellular stress protein levels. In parallel, the neutral red cytotoxicity assay was performed to measure the stressor's effect on cellular physiology. There was a direct concentration-dependent relationship between sublethal cytotoxic effects and the increase of stress protein levels. The evolutionarily conserved heat-shock protein hsp70 was detected in all experiments, in addition, a 27-kDa stress-inducible protein that has not been characterized previously could be observed. The elevations in protein levels were reversible after removal of the toxic agent.

A new development in the use of cell lines for biomarker studies is the genetic engineering of cell lines to make them an easy-to-use diagnostic tool in environmental monitoring and surveillance. For instance, Aarts et al. (1993) developed a recombinant H4IIE cell line (H4IIE-luc) that exhibits Ah receptor-mediated luciferase expression. This cell line has been permanently transfected with a luciferase transporter gene under transcriptional control of several dioxin-responsive elements from mouse. The recombinant cell line showed comparable sensitivity and structure-activity relationship to HAHs as the wild-type cell line (Sanderson et al., 1996). Similar cell-line based reporter gene systems for dioxin-like chemicals have been developed by Anderson et al. (1995), El-Fouly et al. (1994) and Murk et al. (1996). Fischbach and co-workers (1993) developed an *in vitro* test system by permanently transfecting a mouse cell line with a plasmid construct carrying the human hsp70 promoter coupled to the human growth hormone gene. The recombinant cell line expressed the reporter gene only when subjected to stress; the toxicant-induced synthesis of growth hormone could be easily measured by a sensitive, commercially available enzyme immunoassay. The importance of genetically engineered cells as a bioanalytical tool will certainly increase in the future, thereby supporting the development of environmental diagnostics as its own discipline between analytical chemistry and aquatic toxicology.

Conclusions

In contrast to mammalian toxicology, where *in vitro* systems have found wide-spread acceptance and application, the use of cellular methods in aquatic toxicology is more recent and still in its infancy. This appears to be related in part to the different objectives of human toxicology and ecotoxicology – protection of man *versus* protection of ecosystems. Certainly, the prediction of ecosystem changes from cellular effects is not realistic. Cell toxicological approaches, however, have their value within the interdisciplinary and tier-structured approach as it is characteristic for the analysis of ecotoxicological problems. Simplification – which is a major objective behind the use of *in vitro* systems – can be helpful for addressing complex ecotoxicological problems. Cytotoxicity and genotoxicity assays are an ethically defendable tool that provides initial information on hazardous potencies of individual chemicals, chemical mixtures or environmental samples. This screening approach can be strengthened by using the cell lines for detection of biomarker responses. Moreover, fish cell lines can be instrumentalized for investigations into toxic mechanisms. In fact, a continuous development from a rather restricted application of fish cell lines – mainly cytotoxicity screening – to an increasing use of more mechanism-oriented studies or sublethal effects assessment can be observed in aquatic toxicology.

A serious obstacle for more intensive application of fish cell lines is the insufficient knowledge of their metabolic capabilities and toxicological response profiles. If results from cytotoxicity screening shall be valid, we need information on specific sensitivities of the cells in order to avoid under- or overestimation of toxicities. If we want to employ cell lines in mechanism-oriented or biomarker studies, we need detailed knowledge of their metabolic properties. To date, the selection of a cell line for toxicological studies largely relies on practical considerations, for instance, availability of the cell line, but not on scientific considerations, i.e., what cell line is likely to fit best because of its specific properties for the intended study purpose. It is mainly this aspect in which cell culture applications of aquatic toxicology lag behind the state of the art in human toxicology.

Today, aquatic toxicology is being challenged to develop methods for detecting low-dose, chronic contamination with complex chemical mixtures. In this situation, establishing causative relationships between biological effects and chemical exposure, as well as the ultimate identification of the causative toxicants, is particularly difficult and will be impossible without utilizing the potential of powerful toxicological tools such as *in vitro* cellular models.

Acknowledgement
Supported by the European Commission, contract No ENV4-CT96-0223, and by the German Ministry of Research BMBF, project No 0310728.

References

Aarts, J.M., Denison, M.S., DeHaan, L.H.J., Schalk, J.A.C., Cox, M.A. and Brouwer, A. (1993) Ah receptor-mediated luciferase expression: a tool for monitoring dioxin-like toxicity. *Dioxin* 93:361–364.

Agius, C. and Parkinson, C. (1991) A place for *in vitro* methods in fish toxicology. *In*: Abel, P.D. and Axiak, V. (Eds) *Ecotoxicology and the marine environment*. Ellis Horwood Ltd, Chichester, pp 147–156.

Ahne, W. (1985) Untersuchungen über die Verwendung von Fischzellkulturen für Toxizitätsbestimmungen zur Einschränkung und Ersatz des Fischtests. *Zentralbl. Bakt. Hyg. I Abteilung B* 180:480–504.

Ahne, W. and Halder, M. (1991) Die R1-Fischzellkultur zur Ermittlung der Abwassertoxizität. *Münchener Beiträge zur Abwasser-, Fischerei- und Flussbiologie* 45:145–158.

Ahne, W. and Schweitzer, U. (1993) Durch Pestizide induzierte chromosomale Strukturveränderungen (Anaphase-Aberrationen) bei Fischzellkulturen (R1-Forellenzellen). *Wasser Abwasser* 134:745–748.

Ali, F., Lazar, R., Haffner, D. and Adeli, K. (1993) Development of a rapid and simple genotoxicity assay using brown bullhead fish cell line: application to toxicological surveys of sediments in the Huron-Erie corridor. *J. Great Lakes Res.* 19:342–351.

Altenburger, R., Boedeker, W., Faust, M. and Grimme, L.H. (1993) Aquatic toxicology, analysis of combination effects. *In*: Corn, M. (Ed.) *Handbook of hazardous materials*. Academic Press, San Diego, pp 15–27.

Anderson, J.W., Rossi, S.S., Tukey, R.H., Vu, T. and Quattrochi, L.C. (1995) A biomarker, P450 RGS, for assessing the induction potential of environmental samples. *Environ. Toxicol. Chem.* 14:1159–1169.

Ankley, G.T. Schubauer-Berigan, M.K. and Hoke, R.A. (1992) Use of toxicity identification evaluation techniques to identify dredged material disposal options: proposed approach. *Environ. Man.* 16:1–6.

Araki, N., Yamaguchi, M., Nakano, H., Wada, Y. and Hasobe, M. (1994) A fish cell line CHSE-sp exposed to long-term cold temperature retains viability and ability to support viral replication. *In Vitro Cell. Dev. Biol.* 30A:148–150.

Babich, H. and Borenfreund, E. (1987a) Cultured fish cells for the ecotoxicity testing of aquatic pollutants. *Tox. Assess.* 2:119–133.

Babich, H. and Borenfreund, E. (1987b) Structure-activity relationships (SAR) models established *in vitro* with the neutral red cytotoxicity assay. *Toxicol. in Vitro* 1:3–9.

Babich, H. and Borenfreund, E. (1987c) *In vitro* cytotoxicity of organic pollutants to bluegill sunfish (BF-2) cells. *Environ. Res.* 42:229–237.

Babich, H. and Borenfreund, E. (1987d) Fathead minnow FHM cells for use in *in vitro* cytotoxicity assays of aquatic pollutants. *Ecotox. Environ. Safety* 14:78–87.

Babich, H. and Borenfreund, E. (1987e) Polycyclic aromatic hydrocarbon *in vitro* cytotoxicity to bluegill BF-2 cells: mediation by S-9 microsomal fraction and temperature. *Tox. Lett.* 36:107–116.

Babich, H. and Borenfreund, E. (1988a) Structure-activity relationships for diorganotins, chlorinated benzenes, and chlorinated anilines established with bluegill sunfish BF-2 cells. *Fund. Appl. Toxicol.* 10:295–301.

Babich, H. and Borenfreund, E. (1988b) *In vitro* cytotoxicity of polychlorinated biphenyls and toluenes to cultured bluegill sunfish BF-2 cells. *In*: Adams, W.J., Chapman, G.A. and Landis, W.G. (Eds) *Aquatic toxicology and hazard assessment*. Vol. 10. American Society for Testing and Materials, Philadelphia, pp 454–462.

Babich, H. and Borenfreund, E. (1990) *In vitro* cytotoxicities of inorganic lead and di- and trialkyl lead compounds to fish cells. *Bull. Environ. Contam. Toxicol.* 44:456–460.

Babich, H. and Borenfreund, E. (1991) Cytotoxicity and genotoxicity assays with cultured fish cells: a review. *Toxicol. in Vitro* 5:91–100.

Babich, H. and Borenfreund, E. (1992) Neutral red assay for toxicology *in vitro*. *In*: Watson, R.R. (Ed.) *In vitro methods in toxicology*. CRC Press, Boca Raton, pp 237–251.

Babich, H. and Borenfreund, E. (1993) Applications of the neutral red cytotoxicity assay to risk assessment of aquatic contaminants: an overview. *In*: Landis, W.G., Hughes, J.S. and Lewis, M.A. (Eds) Environmental toxicology and risk assessment. American Society for Testing and Materials, Philadelphia, pp 215–229.

Babich, H., Puerner, J.A. and Borenfreund, E. (1986a) *In vitro* cytotoxicity of metals to bluegill (BF-2) cells. *Arch. Environ. Contam. Toxicol.* 15:31–37.

Babich, H., Shopsis, C. and Borenfreund, E. (1986b) *In vitro* cytotoxicity testing of aquatic pollutants (cadmium, copper, zinc, nickel) using established fish cell lines. *Ecotox. Environ. Safety* 11:91–99.

Babich, H., Martin-Alguacil, N. and Borenfreund, E. (1989a) Use of the rainbow trout hepatoma cell line, RTH-149, in a cytotoxicity assay. *ATLA* 17:67–71.

Babich, H., Martin-Alguacil, N. and Borenfreund, E. (1989b) Arsenic-selenium interactions determined with cultured fish cells. *Toxicol. Lett.* 45:157–164.

Babich, H., Goldstein, S.H. and Borenfreund, E. (1990) *In vitro* cyto- and genotoxicity of organomercurials to cells in culture. *Toxicol. Lett.* 50:143–149.

Babich, H., Rosenberg, D.W. and Borenfreund, E. (1991) *In vitro* cytotoxicity studies with the fish hepatoma cell line, PLHC-1 (*Poeciliopsis lucida*). *Ecotox. Environ. Safety* 21:327–336.

Babich, H., Palace, M.R. and Stern, A. (1993) Oxidative stress in fish cells: *in vitro* studies. *Arch. Environ. Contam. Toxicol.* 24:173–178.

Babich, H., Palace, M.R., Borenfreund, E. and Stern, A. (1994) Naphthoquinone cytotoxicity to bluegill sunfish BF-2 cells. *Arch. Environ. Contam. Toxicol.* 27:8–13.

Baksi, S.M. and Frazier, J.M. (1990) Isolated fish hepatocytes – model systems for toxicology research. *Aquat. Toxicol.* 16:229–256.

Baserga, R. (Ed.) (1989) *Cell growth and division – a practical approach*. IRL Press, Oxford.

Barker, C.J. and Rackham, B.D. (1979) The induction of sister-chromatid exchanges in cultured fish cells (*Ameca splendens*) by carcinogenic mutagens. *Mutat. Res.* 68:381–387.

Barlian, A. and Bols, N.C. (1991) Identification of bovine serum albumins that support salmonid cell proliferation in the absence of serum. *In Vitro Cell. Dev. Biol.* 27A:439–441.

Barlian, A., Ganassin, R.C., Tom, D. and Bols, N.C. (1993) A comparison of bovine serum albumin and chicken ovalbumin as supplements for the serum-free growth of chinook salmon embryo cells, CHSE-214. *Cell Biol. Int.* 17:677–684.

Bechtel, D.G. and Lee, L.E.J. (1994) Effects of aflatoxin B1 in a liver cell line from rainbow trout (*Oncorhynchus mykiss*). *Toxicol. in Vitro* 8:317–328.

Behrens, A., Schirmer, K., Bols, N.C. and Segner, H. (1997) Microassay for rapid measurement of 7-ethoxyresorufin-O-deethylase activity in intact fish hepatocytes. *Mar. Environ. Res.*: in press.

Berridge, M.V. and Tan, A.S. (1993) Characterization of the cellular reduction of 3-(4,5-dimethylthiazol-2-yl)-2,5-diphenyltetrazolium bromide (MTT): subcellular localization, substrate dependence, and involvement of mitochondrial electron transport in MTT reduction. *Arch. Biochem. Biophys.* 303:474–482.

Bols, N.C. and Lee, L.E.J. (1992) Technology and uses of cell cultures from the tissues and organs of bony fish. *Cytotechnol.* 6:163–187.

Bols, N.C. and Lee, L.E.J. (1994) Cell lines: availability, propagation and isolation. *In*: Hochachka, P.W. and Mommsen, T.P. (Eds) *Biochemistry and molecular biology of fishes*. Vol. III. *Analytical techniques*. Elsevier, Amsterdam, pp 145–159.

Bols, N.C., Boliska, S.A., Dixon, D.G., Hodson, P.V. and Kaiser, K.L.E. (1991) The use of fish cell cultures as an indication of contaminant toxicity to fish. *Aquat. Toxicol.* 6:(1985) 147–155.

Bols, N.C., Mosser, D.D. and Steels, G.B. (1992) Temperature studies and recent advances with fish cells *in vitro*. *Comp. Biochem. Physiol.* 103A:1–14.

Bols, N.C., Whyte, J., Clemons, J.H., Tom, D.J., van den Heuvel, M.R. and Dixon, D.G. (1997) Use of liver cell lines to develop TCDD equivalency factors and to derive TCDD equivalent concentrations in environmental samples. *In*: Zelikoff, J.T., Shepers, J., Lynch, J. (Eds) *Ecotoxicology – responses, biomarkers, and risk assessment*. Chapter 23. SOS Publications, Fair Haven, NJ, USA, pp 329–350.

Borenfreund, E. and Puerner, J.A. (1985) Toxicity determined *in vitro* by morphological alterations and neutral red absorption. *Toxicol. Lett.* 24:119–124.

Borenfreund, E. and Puerner, J.A. (1986) Cytotoxicity of metals, metal-metal and metal-chelator combinations assayed *in vitro*. *Toxicology* 39:121–134.

Borenfreund, E., Babich, H. and Martin-Alguacil, N. (1988) Comparisons of two *in vitro* cytotoxicity assays – the neutral red (NR) and tetrazolium MTT tests. *Toxicol. in Vitro* 2:1–6.

Borenfreund, E., Babich, H. and Martin-Alguacil, N. (1989) Effect of methylazoxymethanol acetate on bluegill sunfish cell cultures *in vitro*. *Ecotox. Environ. Safety* 17:297–307.

Bourne, E.W. and Jones, R.W. (1973) Effects of 7,12-dimethylbenz[a]anthracene (DMBA) in fish cells *in vitro*. *Trans. Am. Microsc. Soc.* 92:140–142.

Brandao, J.C., Bohets, H.H.L., van de Vyver, I.E. and Dierickx, P.J. (1992) Correlation between the *in vitro* cytotoxicity to cultured fathead minnow fish cells and fish lethality data for 50 chemicals. *Chemosphere* 25:553–562.

Braunbeck, T. (1995) *Zelltests in der Ökotoxikologie*. Veröffentlichungen Projekt Angewandte Ökologie, Vol. 11, Karlsruhe, Germany.

Braunbeck, T. and Neumüller, D. (1996) The comet assay in permanent and primary fish cell cultures – a novel system to detect genotoxicity. *In Vitro Cell. Dev. Biol.* 32:61.

Braunbeck, T., Hauck, C., Scholz, S. and Segner, H. (1995) Mixed function oxygenases in cultured fish cells: contributions of *in vitro* studies to the understanding of MFO induction. *Z. Angew. Zool.* 81:55–72.

Brüschweiler, B.J., Würgler, F.E. and Fent, K. (1995) Cytotoxicity *in vitro* of organotin compounds to fish hepatoma cells PLHC-1 (*Poeciliopsis lucida*). *Aquat. Toxicol.* 32:143–155.

Brüschweiler, B.J., Würgler, F.E. and Fent, K. (1996) Inhibition of cytochrome P4501A by organotins in fish hepatoma cells PLHC-1. *Environ. Toxicol. Chem.* 15:728–735.

Burgess, D., Frerichs, N. and George, S.G. (1993) Control of metallothionein expression by hormones and stressors in cultured fish cells. *Mar. Environ. Res.* 35:25–28.

Burkey, J.L. and Brenden, R.A. (1993) A semi-quantitative direct contact assay using fluorescein diacetate and L929 mouse fibroblasts in monolayer culture. *J. Tissue Cult. Meth.* 15:165–170.

Castano, A.M. and Tarazona, J.V. (1994) ATP assay on cell monolayers as an index of cytotoxicity. *Bull. Environ. Contam. Toxicol.* 53:309–316.

Castano, A. and Tarazona, J.V. (1995) The use of cultured fish cells in environmental toxicology: *in vitro* toxicity tests. *In*: Carajaville, M.P. (Ed.) *Cell biology in environmental toxicology*. University of the Basque Country Press Service, Bilbao, pp 279–288.

Castano, A., Vega, M., Blazquez, T. and Tarazona, J.V. (1994) Biological alternatives to chemical identification for the ecotoxicological assessment of industrial effluents: the RTG-2 *in vitro* cytotoxicity test. *Environ. Toxicol. Chem.* 13:1607–1611.

Castano, A., Cantarino, M.J., Castillo, P. and Tarazona, J.V. (1996) Correlations between the RTG-2 cytotoxicity test and *in vivo* LC$_{50}$ rainbow trout bioassay. *Chemosphere* 32:2141–2157.

Celander, M., Hahn, M.E. and Stegeman, J.J. (1996) Cytochromes P450 (CYP) in the *Poeciliopsis lucida* hepatocellular carcinoma cell line (PLHC-1): dose- and time-dependent glucocorticoid potentiation of CYP1A induction without induction of CYP3A. *Arch. Biochem. Biophys.* 329:113–122.

Celander, M., Bremer, J., Hahn, M.E. and Stegeman, J.J. (1997) Glucocorticoid-xenobiotic interactions: dexamethasone-mediated potentiation of cytochrome P4501A induction by β-naphthoflavone in a fish hepatoma cell line (PLHC-1). *Environ. Toxicol. Chem.* 16:900–907.

Clark, H.F. and Diamond, L. (1971) Comparative studies on the interaction of benzpyrene with cells derived from poikilothermic and homoiothermic vertebrates. II. Effect of temperature on benzo[a]pyrene metabolism and cell multiplication. *J. Cell. Physiol.* 77:385–392.

Clemedson, C., McFarlane-Abdullah, E., Andersson, M., Barile, F.A., Calleja, M.C. et al. (1996) MEIC evaluation of acute systemic toxicity. Part II. *In vitro* results from 68 toxicity assays used to test the first 30 reference chemicals and a comparative cytotoxicity analysis. *ATLA* 24:273–311.

Clemedson, C., Barile, F.A., Ekwall, B., Gomez-Lechon, M.J., Hall, T. et al. (1998) MEIC evaluation of acute systemic toxicity. Part III. *In vitro* results from 16 additional methods used to test the first 30 reference chemicals and a comparative cytotoxicity analysis. *ATLA*: 26:93–129.

Clemons, J.H., van den Heuvel, M.R., Stegeman, J.J., Dixon, D.G. and Bols, N.C. (1994) Comparison of toxic equivalent factors for selected dioxin and furan congeners derived using fish and mammalian liver cell lines. *Can. J. Fish. Aquat. Sci.* 51:1577–1584.

Clemons, J.H., Lee, L.E.J., Myers, C.R., Dixon, D.G. and Bols, N.C. (1996) Cytochrome P4501A1 induction by polychlorinated biphenyls (PCBs) in liver cell lines from rat and trout and the derivation of toxic equivalency factors. *Can. J. Fish. Aquat. Sci.* 53:1177–1183.

Clemons, J.H., Dixon, D.G. and Bols, N.C. (1997) Derivation of 2,3,7,8-TCDD toxic equivalent factors (TEFs) for selected dioxins, furans and PCBs with rainbow trout and rat liver cell lines and the influence of exposure time. *Chemosphere* 34:1105–1119.

Clothier, R.H., Hulme, L., Ahmed, A.B., Reeves, H.L., Smith, M. and Blass, M. (1988) *In vitro* cytotoxicity of 150 chemicals to 3T3-L1 cells, assessed by the FRAME kenacid blue method. *ATLA* 16:84–95.

Collodi, P. and Barnes, D.W. (1990) Mitogenic activity from trout embryos. *PNAS* 87: 3498–3502.

Collodi, P., Kamei, Y., Ernst, T., Miranda, C., Buhler, D.R. and Barnes, D.W. (1992) Culture of cells from zebrafish (*Brachydanio rerio*) embryo and adult tissues. *Cell. Biol. Toxicol.* 8:43–61.

Denizot, F. and Lang, R. (1986) Rapid colorimetric assay for cell growth and survival. *J. Immunol. Meth.* 89:271–277.

Deventer, K. (1996) Detection of genotoxic effects on cells of liver and gills of *Brachydanio rerio* by means of single cell gel electrophoresis. *Bull. Environ. Contam. Toxicol.* 56:911–918.

Diamond, L. and Clark, H.F. (1970) Comparative studies on the interaction of benzo(a)pyrene with cells derived from poikilothermic and homoiothermic vertebrates. I. Metabolism of benzo(a)pyrene. *J. Natl. Cancer Inst.* 45:1005–1012.

Dierickx, P.J. and Bredael-Rozen, E. (1996) Correlation between the *in vitro* cytotoxicity of inorganic metal compounds to cultured fathead minnow fish cells and the toxicity to *Daphnia magna. Bull. Environ. Contam. Toxicol.* 57:107–110.

Dierickx, P.J. and van den Vyver, I.E. (1991) Correlation of the neutral red uptake inhibition assay of cultured fathead minnow fish cells with fish lethality data. *Bull. Environ. Contam. Toxicol.* 46:649–653.

DiGiulio, R.T., Benson, W.H., Sanders, B.M. and Van Veld, P.A. (1995) Biochemical mechanisms: metabolism, adaptation and toxicity. *In*: Rand, G.M. (Ed.) *Aquatic toxicology.* 2nd ed. Taylor & Francis, Washington, pp 523–561.

Douglas, M.T., Chanter, D.O., Pell, I.B. and Burney, G.M. (1986) A proposal for the reduction of animal numbers required for the acute toxicity testing to fish (LC_{50} determination). *Aquat. Toxicol.* 8:243–250.

Ekwall, B. (1980) Preliminary studies on the validity of *in vitro* measurement of drug toxicity using HeLa cells: III. Lethal action to man of 43 drugs related to the HeLa cell toxicity of the lethal drug concentration. *Toxicol. Lett.* 5:319–331.

El-Fouly, M.H., Richter, C., Giesy, J.P. and Denison, M.S. (1994) Production of a novel recombinant cell line for use as a bioassay system for detection of 2,3,7,8-tetrachlorodibenzo-*p*-dioxin-like chemicals. *Environ. Toxicol. Chem.* 13:1581–1588.

Fent, K. and Hunn, J. (1996) Cytotoxicity of organic environmental chemicals to fish liver cells (PLHC-1). *Mar. Environ. Res.* 42:377–382.

Fentem, J. and Balls M. (1993) Replacement of fish in ecotoxicology testing: use of bacteria, other low organisms and fish cells *in vitro*. *In*: Richardson, M. (Ed.) *Ecotoxicology monitoring.* VCH, Weinheim, pp 71–81.

Fischbach, M., Sabbioni, E. and Bromley, P. (1993) Induction of the human growth hormone gene placed under human hsp70 promoter control in mouse cells: a quantitative indicator of metal toxicity. *Cell. Biol. Toxicol.* 9:177–188.

Fong, At., Hendricks, J.T., Dashwood, R.H., van Winkle, S. and Bailey, G.S. (1988) Formation and persistence of ethylguanine in liver DNA of rainbow trout (*Salmo gairdneri*) treated with – diethylnitrosamine by water exposure. *Food Chem. Toxicol.* 26:699–706.

Freshney, R.I. (1987) *Culture of animal cells.* 2nd edition. Alan R. Liss Inc., New York, 397 pp.

Fryer, J.L. and Lannan, C.N. (1994) Three decades of fish cell culture: a current listing of cell lines derived from fishes. *J. Tissue Cult. Meth.* 16:87–94.

Fryer, J.L., Yusha, A. and Pilcher, K.S. (1965) The *in vitro* cultivation of tissues and cells of Pacific salmon and steelhead trout. *New York Acad. Sci.* 126:566–586.

George, S.G. (1996) *In vitro* toxicology of aquatic pollutants: use of cultured fish cells. *In*: Taylor E.W. (Ed.) *Toxicology of aquatic pollution.* Cambridge University Press, Cambridge, pp 253–265.

George, S.G. Leaver, M., Frerichs, N. and Burgess, D. (1989) Fish metallothioneins: molecular cloning studies and induction in cultured cells. *Mar. Environ. Res.* 28:173–177.

George, S.G., Burgess, D., Leaver, M. and Frerichs, N. (1992) Metallothionein induction in cultured fibroblasts and liver of marine flatfish, *Scophthalmus maximus. Fish Physiol. Biochem.* 10:43–54.

Hahn, M.E. and Stegeman, J.J. (1992) Phylogenetic distribution of the Ah receptor in non-mammalian species: implications for dioxin toxicity and the Ah receptor evolution. *Chemosphere* 25:931–937.

Hahn, M.E., Lamb, T.M., Schultz, M.E., Smolowitz, R.M. and Stegeman, J.J. (1993) Cytochrome P4501A induction and inhibition by 3,3′,4,4′-tetrachlorobiphenyl in an Ah receptor-containing fish hepatoma cell line, PLHC-1. *Aquat. Toxicol.* 26:185–208.

Hahn, M.E., Woodward, B.L., Stegeman, J.J. and Kennedy, J.W. (1996) Rapid assessment of induced cytochrome P4501A protein and catalytic activity in fish hepatoma cells grown in multiwell plates: response to TCDD, TCDF, and two planar PCBs. *Environ. Toxicol. Chem.* 15:582–591.

Halder, M. and Ahne, W. (1990) Evaluation of waste water toxicity with three cytotoxicity tests. *Z. Wasser- Abw.-forsch.* 23:233–236.

Hankinson, O. (1995) The aryl hydrocarbon receptor complex. *Ann. Rev. Pharm. Toxicol.* 35:307–340.

Hightower, L.E. and Renfro, J.L. (1988) Recent applications of fish cell culture to biomedical research. *J. Exp. Zool.* 248:290–302.

Ho, K.T., McKinney, R.A., Kuhn, A., Pelletier, M.C. and Burgess, R.M. (1997) Identification of acute toxicants in New Bedford harbor sediments. *Environ. Toxicol. Chem.* 16:551–558.

Hoffmann, G.R. (1996) Genetic toxicology. *In:* Klaassen, C.D. (Ed.) Casarett and Doull's toxicology. 5th ed. McGraw-Hill, New York, pp 269–300.

Hyllner, S.J., Andersson, T., Hauux, C. and Olson, P.E. (1989) Cortisol induction of metallothionein in primary culture of rainbow trout hepatocytes. *J. Cell. Physiol.* 139:24–28.

Isomaa, B. and Lilius, H. (1995) The urgent need for *in vitro* tests in ecotoxicology. *Toxicol. in Vitro* 9:821–825.

Isomaa, B., Lilius, H. and Rabergh, C. (1994) Aquatic toxicology *in vitro*: a brief review. *ATLA* 22:243–253.

Johnston, P.A., Stringer, R.L. and Santillo, D. (1996) Effluent complexity and ecotoxicology: regulating the variable within varied systems. *TEN* 3:115–120.

Kaneko, Y., Igarashi, M., Iwashita, M., Suzuki, K., Kojima, H., Kimura, S. and Hasobe, M. (1995) Effects of fish and calf type I collagens as culture substrate in the adhesion and spreading of established fish cells. *In Vitro Cell. Dev. Biol.* 31:178–182.

Kennedy, S.W., Jones, S.P. and Bastien, L.P. (1995) Efficient analysis of cytochrome P4501A catalytic activity, porphyrins, and total protein in chicken embryo hepatocyte cultures with a fluorescence plate reader. *Anal. Biochem.* 226:362–370.

Kennedy, S.W., Lorenzen, A., Norstrom, R.J. (1996) Chicken embryo hepatocyte bioassay for measuring cytochrome P4501A-based 2,3,7,8-tetrachlorodibenzo-*p*-dioxin equivalent concentrations in environmental samples. *Environ. Sci. Technol.* 30:706–715.

Kocan, R.M. and Powell, D.B. (1985) Anaphase aberrations: an *in vitro* test for assessing the genotoxicity of individual chemicals and complex mixtures. *In:* Waters, M.D., Sandhu, S.D., Lewtas, J., Claxton, L., Straus, G. and Nesnow, S. (Eds) *Short-term bioassays in the analysis of complex environmental mixtures.* Vol. 14. Plenum Press, New York, pp 75–86.

Kocan, R.M., Landolt, M.L and Sabo, K.M. (1979) *In vitro* toxicity of eight mutagens/carcinogens for three fish cell lines. *Bull. Environ. Contam. Toxicol.* 23:269–274.

Kocan, R.M., Landolt, M.L., Bond, J. and Beneditt, E.P. (1981) *In vitro* effects of some mutagens/carcinogens on cultured fish cells. *Arch. Environ. Contam. Toxicol.* 10:663–671.

Kocan, R.M., Landolt, M.L. and Sabo, K.M. (1982) Anaphase aberrations: a measure of genotoxicity in mutagen-treated fish cells. *Environ. Mut.* 4:181–189.

Kocan, R.M., Chi, E.Y., Eriksen, N., Benditt, E.P. and Landolt, M.L. (1983) Sequestration and release of polycyclic aromatic hydrocarbons by vertebrate cells *in vitro*. *Environ. Mut.* 5:643–656.

Kocan, R.M., Sabo, K.M. and Landolt, M.L. (1985) Cytotoxicity/genotoxicity: the application of cell culture techniques to the measurement of marine sediment pollution. *Aquat. Toxicol.* 6:165–177.

Kohlpoth, M. and Rusche, B. (1997) Kultivierung einer permanenten Fischzellinie in serumfreien Medium: spezielle Erfahrungen mit einem Zytotoxizitätstest für Abwasserproben. *ALTEX* 14:16–20.

Landolt, M.L. and Kocan, R.M. (1983) Fish cell cytogenetics: a measure of the genotoxic effects of environmental pollutants. *In:* Nriagu, J.O. (Ed.) *Aquatic toxicology.* Wiley, New York, pp 336–353.

Landolt, M.L. and Kocan, R.M. (1984) Lethal and sublethal effects of marine sediment extracts on fish cells and chromosomes. *Helgol. Wiss. Meeresunters.*, 37:479–491.

Lannan, C.N. (1994) Fish cell culture: a protocol for quality control. *J. Tissue Cult. Meth.* 16:95–98.

Lee, L.E.J., Clemons, J.H., Bechtel, D.G., Caldwell, S.J., Han, K.B., Pasitschniak-Arts, M.,Mosser, D.D. and Bols, N.C. (1993) Development and characterization of a rainbow trout liver cell line expressing cytochrome P450-dependent monooxygenase activity. *Cell. Biol. Toxicol.* 9:279–294.

Lenz, D., Segner, H. and Hanke, W. (1993) Comparison of different endpoint methods for acute cytotoxicity tests with the R1 cell line. *In*: Braunbeck, T., Hanke, W. and Segner, H. (Eds) Fish ecotoxicology and ecophysiology. VCH, Weinheim, pp 93–102.

Lilius, H., Isomaa, B. and Holmström, T. (1994) A comparison of the toxicity of 50 reference chemicals to freshly isolated rainbow trout hepatocytes and *Daphnia magna*. *Aquat. Toxicol.* 30:47–60.

Lilius, H., Sandbacka, M. and Isomaa, B. (1995) The use of freshly isolated gill epithelial cells in toxicity testing. *Toxicol. in Vitro* 9:299–305.

Lilius, H., Hästbacka, T. and Isomaa, B. (1996) A combination of fluorescent probes for evaluation of cytotoxicity and toxic mechanisms in isolated rainbow trout hepatocytes. *Toxicol. in Vitro* 10:341–348.

Lorenzen, A. and Okey, A.B. (1990) Detection and characterization of [^{3}H]2,3,7,8-tetrachlorodibenzo-*p*-dioxin binding to Ah receptor in a rainbow trout hepatoma cell line. *Toxicol. Appl. Pharmacol.* 106:53–62.

Lullmann-Rauch, R. (1979) Drug-induced lysosomal storage disorders. *In*: Dingle, J.J., Jaques, P.J. and Shaw, I.H. (Eds) *Lysosomes in applied biology and therapeutics*. North Holland Publishers, Amsterdam, pp 49–129.

Lyons-Alcantara, M., Tarazona, J.V. and Mothersill, C. (1996) The differential effects of cadmium exposure on the growth and survival of primary and established cells from fish and mammals. *Cell. Biol. Toxicol.* 12:29–38.

Maccubbin, A.E. (1994) DNA adduct analysis in fish: laboratory and field studies. *In*: Malins, D.C. and Ostrander, G.K. (Ed.) *Aquatic toxicology*. Lewis Publ., CRC Press, Boca Raton, pp 267–294.

MacDonald, D.G. and Wood, C.M. (1993) Branchial mechanisms of acclimation to metals in freshwater fish. *In*: Rankin, J.C. and Jensen B. (Eds) *Fish ecophysiology*. Chapman and Hall, London, pp 297–321.

Magwood, S. and George, S. (1996) *In vitro* alternatives to whole animal testing. Comparative cytotoxicity studies of divalent metals in established cell lines derived from tropical and temperate water fish species in a neutral red assay. *Mar. Environ. Res.* 42:1–4.

Maracine, M. and Segner, H. (1998) Cytotoxicity of metals in isolated fish cells: the role of glutathione. *Comp. Biochem. Physiol.*: in press.

Marion, M. and Denizeau, (1983a) F. Rainbow trout and human cells in culture for the evaluation of the toxicity of aquatic pollutants: a study with lead. *Aquat. Toxicol.* 3:47–60.

Marion, M. and Denizeau, F. (1983b) Rainbow trout and human cells in culture for the evaluation of the toxicity of aquatic pollutants: a study with cadmium. *Aquat. Toxicol.* 3:329–343.

Martin, A. and Clynes, M. (1993) Comparison of five microplate colorimetric assays for *in vitro* cytotoxicity testing and cell proliferation assays. *Cytotechnol.* 11:49–58.

Martin-Alguacil, N., Babich, H., Rosenberg, D.W. and Borenfreund, E. (1991) *In vitro* response of the brown bullhead catfish cell line, BB, to aquatic pollutants. *Arch. Environ. Contam. Toxicol.* 20:113–117.

Mayer, D., Ahne, W. and Storch, V. (1988) Cytotoxicity of chemicals to fibroblastic fish cell cultures (R1 cells) investigated by electron microscopy. *Z. angew. Zool.* 75:147–157.

McKelvey-Martin, V.J., Green, M.H.L., Schmezer, P., Pool-Zobel, B.L., DeMeo, P. and Collins, A. (1993) The single cell gel electrophoresis (comet assay): a European view. *Mutat. Res* 288:47–63.

Miranda, C.L., Collodi, P., Zhao, X., Barnes, D.W. and Buhler, D.R. (1993) Regulation of cytochrome P450 expression in a novel liver cell line from zebrafish (*Brachydanio rerio*). *Arch. Biochem. Biophys.* 305:320–327.

Mitani, H. (1983) Lethal and mutagenic effects of radiation and chemicals on cultured fish cells derived from erythrophoroma of goldfish (*Carassius auratus*). *Mutat. Res* 107:279–283.

Mitani, H. (1985) Difference in the lethal effects on carcinogens/mutagens among three cultured goldfish (*Carassius auratus*) cell lines. *J. Radiat. Res.* 26:459–463.

Mosman, T. (1983) Rapid colorimetric assay for cellular growth and survival: application to proliferation and cytotoxicity assays. *J. Immunol. Meth.* 65:55–63.

Monod, G., Devaux, A., Valotaire, Y. and Cravedi, J.P. (1998) Primary cell cultures from fish in ecotoxicology. *This volume.*

Mosser, D.D., Heikkila, J.J., and Bols, N.C. (1986) Temperature ranges over which rainbow trout fibroblasts survive and synthesize heat-shock proteins. *J. Cell. Physiol.* 128:432–440.

Murk, A.J., Legler, J., Denison, M.S., Giesy, J.P., Vandeguchte, C. and Brouwer, A. (1996) Chemical-activated luciferase gene expression (CALUX): a novel *in vitro* bioassay or Ah receptor active compounds in sediments and pore water. *Fund. Appl. Toxicol.* 33:149–160.

Nagel, R. (1993) Fish and environmental chemicals – a critical evaluation of tests. *In*: Braunbeck, T., Hanke, W. and Segner, H. (Eds) Fish ecotoxicology and ecophysiology. VCH, Weinheim, pp 146–156.

Nehls, S. and Segner, H. (1997) Application of the comet assay to detect genotoxic effects in cultured fish cells: in preparation.

Nicholson, B.L. (1985) Techniques in fish cell culture. *In*: Kurstak, E. (Ed.) *Techniques in the life sciences.* Vol. C1. Elsevier, Amsterdam, pp C015/1–16.

Nicholson, B.L. (1989) Fish cell culture: an update. *Adv. Cell Cult.* 7:1–18.

OECD (1982) *Guidelines for ecotoxicological testing of chemicals.* Organization for Economic Cooperation and Development, OECD, Paris.

Olson, P.E. (1996) Metallothioneins in fish: induction and use in environmental monitoring. *In*: Taylor, E.W. (Ed.) *Toxicology of aquatic pollution.* Cambridge University Press, Cambridge, pp 187–203.

Olson, P.E., Hyllner, S.J., Zafarullah, M., Andersson, T. and Gedamu, L. (1990) Differences in metallothionein gene expression in primary cultures of rainbow trout hepatocytes and the RTH-149 cell line. *Biochim. Biophys. Acta* 1049:78–82.

Park, E.H., Lee, J.S., Yi, A.E. and Etoh, H. (1989) Fish cell line (ULF-23Hu) derived from the fin of the central mudminnow (*Umbra limi*): suitable characteristics for the clastogenicity assay. *In Vitro Cell. Dev. Biol.* 11:987–994.

Parkinson, C. and Agius, C. (1986) Acute toxicity of DDT to fish cells *in vitro*. *Food Chem. Toxicol.* 24:591.

Parkinson, C. and Agius, C. (1988) Acute toxicity of DDT to tilapia (*Oreochromis spilurus* Gunther) *in vivo* and *in vitro*. *ATLA* 15:298–302.

Peakall, D.B. (1994) Biomarkers. *TEN* 1:55–60.

Price-Haughley, J., Bonham, K. and Gedamu, L. (1986) Heavy metal induced gene expression in fish and fish cell lines. *Environ. Health Persp.* 65:141–147.

Price-Haughley, J., Bonham, K. and Geamu, L. (1987) Metallothionein gene expression in fish cell lines: its activation in embryonic cells by 5-azacytidine. *Biochim. Biophys. Acta* 908:158–168.

Plakunov, I., Smolarek, T.A., Fischer, D.L., Wiley, J.C. and Baird, W.M. (1987) Separation by ion-pair high performance liquid chromatography of the glucuronide, sulfate and glutathione conjugates formed from benzo(a)pyrene in cell cultures from rodents, fish and humans. *Carcinogenesis* 8:59–66.

Puck, T.T. and Marcus, P.I. (1955) A rapid method for viable cell titration and clone production with HeLa cells in tissue culture: the use of X-irradiated cells to supply conditioning factors. *PNAS* 41:432–437.

Plumb, J.A. and Wolf, K. (1971) Fish cell growth rates. Quantitative comparison of RTG-2 cell growth at 5–25°C. *In Vitro* 7:42–44.

Rachlin, J.W. and Perlmutter, A. (1968) Fish cells in culture for study of aquatic toxicants. *Water Res.* 2:409–414.

Rachlin, J.W. and Perlmutter, A. (1969) Response of rainbow trout cells in culture to selected concentrations of zinc sulphate. *Progr. Fish Cult.* 18:94–98.

Roesijadi, G. (1992) Metallothionein in metal regulation and toxicity in aquatic animals. *Aquat. Toxicol.* 22:81–114.

Roesijadi, G. and Robsinson, W.E. (1994) Metal regulation in aquatic animals: mechanisms of uptake, accumulation and release. *In*: Malins, D.C. and Ostrander, G.K. (Ed.) *Aquatic toxicology.* Lewis Publ., CRC Press, Boca Raton, pp 387–420.

Ryan, J.A. and Hightower, L.E. (1994) Evaluation of heavy metal ion toxicity in fish cells using a combined stress protein and cytotoxicity assay. *Environ. Chem. Toxicol.* 13:1231–1240.

Ryan, J.A. and Hightower, L.E. (1996) Stress proteins as molecular biomarkers for environmental toxicology. *In*: Feige, U., Morimoto, R.I., Yahara, I. and Polla, B.S. (Eds) *Stress-inducible cellular responses*. Birkhäuser, Basel, pp 411–424.

Rusche, B. and Kohlpoth, M. (1993) The R1-cytotoxicity test as a replacement for the fish test stipulated in the German Waste Water Act. *In*: Braunbeck, T., Hanke, W. and Segner, H. (Eds) *Fish ecotoxicology and ecophysiology*. VCH, Weinheim, pp 81–92.

Safe, S. (1990) Polychlorinated biphenyls (PCBs), dibenzo-*p*-dioxins (PCDDs), dibenzofurans (PCDFs) and related compounds: environmental and mechanistic considerations which support the development of toxic equivalency factors (TEFs). *Crit. Rev. Toxicol.* 21:51–88.

Saito, H. and Shigeoka, T. (1994) Comparative cytotoxicity of chlorophenols to cultured fish cells. *Environ. Toxicol. Chem.* 13:1649–1650.

Saito, H., Sudo, M., Shigeoka, T. and Yamauchi, F. (1991) *In vitro* cytotoxicity of chlorophenols to goldfish GF-scale (GFS) cells and quantitative structure-activity relationships. *Environ. Toxicol. Chem.* 10:235–241.

Saito, H., Koyasu, J., Yoshida, K. Shigeoka, T. and Koike, S. (1993) Cytotoxicity of 109 chemicals to goldfish GFS cells and relationship with 1-octanol/water partition coefficients. *Chemosphere* 26:1015–1028.

Saito, H., Koyasu, J., Shigeoka, T. and Tomita, I. (1994) Cytotoxicity of chlorophenols to goldfish GFS cells with the MTT and LDH assays. *Toxicol. in Vitro* 8:1107–112.

Sanderson, J.T., Aarts, J.M., Brouwer, A., Froese, K.L., Denison, M.S. and Giesy, J.P. (1996) Comparison of Ah receptor-mediated luciferase and ethoxyresorufin-O-deethylase induction in H4IIE cells: implications for their use as bioanalytical tools for the detection of polyhalogenated aromatic hydrocarbons. *Toxicol. Appl. Pharmacol.* 137:316–325.

Schirmer, K., Ganassin, R.C., Brubacher, J.-L. and Bols, N.C. (1994) A DNA fluorometric assay for measuring fish cell proliferation in microplates with different well sizes. *J. Tissue Cult. Meth.* 16:133–142.

Schirmer, K., Chan, A.G.J., Greenberg, B.M., Dixon, D.G. and Bols, N.C. (1997) Methodology for demonstrating and measuring the photocytotoxicity of fluoranthene to fish cells in culture. *Toxicol. in Vitro* 11:107–119.

Schulz, M., Lewland, B., Kohlpoth, M., Rusche, B., Lorenz, K.J.H., Unruh, E., Hansen, P.D. and Miltenburger, H.G. (1995) Fischzellinien in der toxikologischen Bewertung von Abwasserproben. *ALTEX* 12:188–195.

Schüürmann, G. and Segner, H. (1994a) Wirkungsforschung in der chemischen Ökotoxikologie. *UWSF-Z. Umweltch. Ökotox.* 6:351–358.

Schüürmann, G. and Segner, H. (1994b) Struktur-Wirkungs-Analyse von Trialkylzinnverbindungen. *Ecoinforma* 7:439–453.

Schüürmann, G., Segner, H. and Jung, K. (1997) Multivariate mode-of-action analysis of acute toxicity of phenols. *Aquat. Toxicol.* 38:277–296.

Segner, H. (1987) Response of fed and starved roach, *Rutilus rutilus*, to sublethal copper contamination. *J. Fish Biol.* 30:423–437.

Segner, H. (1998) Isolation and primary culture of teleost hepatocytes. *Comp. Biochem. Physiol.*: in press.

Segner, H. and Braunbeck, T. (1998) Cellular response profile to chemical stress. *In*: Schüürmann, G. and Markert, B. (Eds) *Ecotoxicology*. Wiley, New York, pp 521–569.

Segner, H. and Lenz, D. (1993) Cytotoxicity assays with the rainbow trout R1 cell line. *Toxicol. in Vitro* 7:537–540.

Segner, H. and Schüürmann, G. (1997) Cytotoxicity of MEIC chemicals to rainbow trout R1 cell line and multivariate comparison with ecotoxicity tests. *ATLA* 25:331–338.

Segner, H., Lenz, D., Hanke, W. and Schüürmann, G. (1994) Cytotoxicity of metals toward rainbow trout R1 cell line. *Environ. Toxicol. Water Qual.* 9:273–279.

Shea, T.B. and Berry E.S. (1983) A serum-free medium that supports the growth of piscine cell culture. *In Vitro* 19:818–824.

Shugart, R.L. (1995) Environmental genotoxicology. *In*: Rand, G.M. (Ed.) *Aquatic toxicology*. 2nd ed. Taylor & Francis, Washington, pp 405–419.

Sikka, H.C., Rutkowski, J.P., Kandaswami, C., Kumar, S., Earley, K. and Gupta, R.C. (1990) Formation and persistence of DNA adducts in the liver of brown bullheads exposed to benzo(a)pyrene. *Cancer Lett.* 49:81–86.

Smolarek, T.A., Morgan, S.L., Moyniham, C.G., Lee, H., Harvey, R.G. and Baird, W.M. (1987) Metabolism and DNA adduct formation of benzo[a]pyrene and 7,12-dimethylbenz[a]anthracene in fish cell lines in culture. *Carcinogenesis* 8:1501–1509.

Smolarek, T.A., Morgan, S. and Baird, W.M. (1988) Temperature-induced alterations in the metabolic activation of benzo(a)pyrene to DNA-binding metabolites in the bluegill fish cell line, BF-2. *Aquat. Toxicol.* 13:89–98.

Soares, A. and Calow, P. (Eds) (1993) *Progress in standardization of aquatic toxicity tests.* Lewis Publishers, Boca Raton.

Stegeman, J.J. and Hahn, M.E. (1994) Biochemistry and molecular biology of monooxygenases: current perspectives in form, function and regulation of cytochrome P450 in aquatic species. *In*: Malins, D.C. and Ostrander, G.K. (Eds) *Aquatic toxicology.* Lewis Publ., CRC Press, Boca Raton, pp 87–206.

Suyama, I. and Etoh, H. (1988) Establishment of a cell line from *Umbra limi* (Umbridae, Pisces). *In*: Kuroda, Y., Kurstak, E. and Maramorosch, K. (Eds) *Invertebrate and fish tissue culture.* Springer, New York, pp 270–273.

Swanson, H.I. and G.H. Perdew (1991) Detection of the Ah receptor in rainbow trout: use of 2-azido-3-[^{125}I]-iodo-7,8-dibromodibenzo-*p*-dioxin in cell culture. *Toxicol. Lett.* 58:85–95.

Thornton, S.C., Diamond, L. and Baird, W.M. (1982) Metabolism of benzo[a]pyrene by fish cells in culture. *J. Toxicol. Environm. Health* 10:157–167.

Tillit, D.E., Giesy, J.P. and Ankley, G.T. (1991) Characterization of the H4IIE rat hepatoma cell bioassay as a tool for assessing toxic potency of planar halogenated hydrocarbons in environmental samples. *Environ. Sci. Technol.* 25:87–92.

Tocher, D.R. and Dick, J.R. (1990) Polyunsaturated fatty acid metabolism in cultured fish cells: incorporation and metabolism of (n-3) and (n-6) series acids by Atlantic salmon (*Salmo salar*) cells. *Fish Physiol. Biochem.* 8:311–319.

Tocher, D.R., Sargent, J.R. and Frerichs, G.N. (1988) The fatty acid composition of established fish cell lines after long-term culture in mammalian sera. *Fish Physiol. Biochem.* 5:219–227.

Vosdingh, R.A. and Neff, M.J.C. (1974) Bioassay of aflatoxins by catfish cell cultures. *Toxicology* 2:107–112.

Walton, D.G., Acton, A.B. and Stich, H.F. (1983) DNA repair synthesis in cultured mammalian and fish cells following exposure to chemical mutagens. *Mutat. Res* 124:153–161.

Walton, D.G., Acton, A.B. and Stich, H.F. (1984) Comparison of DNA repair synthesis, chromosome aberrations, and induction of micronuclei in cultured human fibroblast, Chinese hamster ovary and central mudminnow (*Umbra limi*) cells exposed to chemical mutagens. *Mutat. Res* 129:129–136.

Walton, D.G., Acton, A.B. and Stich, H.F. (1987) DNA repair synthesis in cultured fish and human cells exposed to fish S9-activated aromatic hydrocarbons. *Comp. Biochem. Physiol.* 96C:399–404.

Walum, E., Stenberg, K. and Jenssen, D. (1990) *Understanding cell toxicology.* Ellis Horwood, New York, 206 pp.

Wilson, R.W. and Taylor, E.W. (1993) The physiological responses of freshwater rainbow trout, *Oncorhynchus mykiss*, during acutely lethal copper exposure. *J. Comp. Physiol.* 163B:38–67.

Wolf, K. and Ahne, W. (1982) Fish cell culture. *Adv. Cell Cult.* 2:305–328.

Wolf, K. and Mann, J.A. (1980) Poikilotherm vertebrate cell lines and viruses: a current listing of fishes. *In Vitro* 16:168–179.

Wolf, K. and Quimby, M.C. (1969) Fish cell and tissue culture. *In*: Hoar, W.S. and Randall, D.J. (Eds) *Fish physiology.* Vol. III. Academic Press, New York, pp 253–305.

Zabel, E.W., Pollenz, R. and Peterson, R.E. (1996) Relative potencies of individual polychlorinated dibenzo-*p*-dioxin, dibenzofuran, and biphenyl congeners and congener mixtures based on induction of cytochrome P4501A mRNA in a rainbow trout gonadal cell line (RTG-2). *Environ. Toxicol. Chem.* 15:2310–2318.

Zafarullah, M., Olson, P.E. and Gedamu, L. (1989) Endogenous and heavy metal ion metallothionein gene expression in salmonid tissues and cell lines. *Gene* 83:85–93.

Zafarullah, M., Olson, P.E. and Gedamu, L. (1990) Differential regulation of metallothionein gene in rainbow trout fibroblasts RTG-2. *Biochim. Biophys. Acta* 1049:318–323.

Zahn, T. and Braunbeck, T. (1993) Cytological alterations in the fish fibrocytic R1 cells as an alternative test system for the detection of sublethal effects of environmental pollutants: a case-study with 4-chloroaniline. *In*: Braunbeck, T., Hanke, W. and Segner, H. (Eds) *Fish ecotoxicology and ecophysiology.* VCH, Weinheim, pp 103–126.

Zahn, T., Hauck, C., Holzschuh, J. and Braunbeck, T. (1995) Acute and sublethal toxicity of see-
    page waters from garbage dumps to permanent cell lines and primary cultures of hepatocytes
    from rainbow trout (*Oncorhynchus mykiss*): a novel approach to environmental risk asses-
    sment for chemicals and chemical mixtures. *Zbl. Hyg. Umweltmed.* 196:455–479.
Zahn, T., Arnold, H. and Braunbeck, T. (1996) Cytological and biochemical response of R1 cells
    and isolated hepatocytes from rainbow trout (*Oncorhynchus mykiss*) to subacute *in vitro*
    exposure to disulfoton. *Exp. Toxicol. Pathol.* 48:47–64.

Fish Ecotoxicology
ed. by T. Braunbeck, D.E. Hinton and B. Streit
© 1998 Birkhäuser Verlag Basel/Switzerland

Primary cell cultures from fish in ecotoxicology

Gilles Monod[1], Alain Devaux[2], Yves Valotaire[3] and Jean-Pierre Cravedi[4]

[1] Institut National de la Recherche Agronomique, Unité d'Écotoxicologie Aquatique, IFR 43,
Campus de Beaulieu, F-35042 Rennes Cedex, France
[2] École Nationale des Travaux Publics de l'État, Laboratoire des Sciences de l'Environnement,
Rue Maurice Audin, F-69518 Vaulx-en Velin Cedex, France
[3] Équipe d'Endocrinologie Moléculaire de la Reproduction, UPRES-A CNRS 6026, IFR 43,
Campus de Beaulieu, F-35042 Rennes Cedex, France
[4] Institut National de la Recherche Agronomique, Laboratoire des Xénobiotiques, 180 chemin
de Tournefeuille, BP 3, 31931 Toulouse Cedex, France

Summary. Fish cell cultures from fish are used in studies on the fate and effects of xenobiotics at the cellular level. So far, special attention has been paid to primary cultures of hepatocytes. This cell type has been particularly useful for studies on biotransformation of xenobiotics. Among biotransformation enzymes, a number of studies have been devoted to cytochrome P4501A due to its inducibility by ecotoxicologically relevant pollutants. Biomarkers of DNA damage, i.e., DNA adducts and single strand breaks, have also been investigated. Furthermore, primary cultures of fish hepatocytes appear as promising models for studies on the effects of xenobiotics on critical steps of fish reproductive physiology. All the data are discussed in an ecotoxicological perspective.

Introduction

Ecotoxicological effects of chemical pollutants are primarily due to their interactions at the molecular level in contaminated organisms. In the area of fish physiology, growing interest has been paid to studies at the cellular level in order to improve the knowledge on the molecular basis of different functions. Thus, an attempt has been made to develop cultures of non-transformed cells. Isolated non-transformed fish cells have been used to elucidate the mechanisms of action of pollutants and to characterize bio-chemical parameters which could be used as early markers of biological alterations (biomarkers). Beside scientific interests, ecotoxicological studies at the cellular level also meet ethical concerns about the use of biological systems alternative to *in vivo* studies. Advantages of cellular systems in-clude the control of the cellular environment, elimination of interactive systemic effects, reduced variability between experiments, simplicity to study chemical interactions, and numerous practical aspects (for review, see Baksi and Frazier, 1990). In fish, isolated non-transformed cells have been obtained from several tissues, but special attention has been paid to hepatocytes.

These cells, which represent more than 90% of the liver cell volume, are involved in several critical functions including xenobiotic metabolism and reproduction (Fig. 1).

The two-step collagenase process originally described for mammals (Seglen, 1975) has been successfully applied to isolate fish hepatocytes. However, liver perfusion is difficult in species where the liver is anatomically not well defined (e.g., in cyprinids). Consequently, the majority of toxicological studies with hepatocytes has been performed in rainbow trout (*Oncorhynchus mykiss*), in which the liver can be easily perfused. In this species, isolated hepatocytes may be maintained alive for different periods of time depending on the purpose of the study. While freshly isolated hepatocytes in suspension (suspension culture) are very useful for short-term experiments (up to 1 day), hepatocytes plated in culture dishes (monolayer culture) allow the study of effects for about 1 week. Moreover, plated hepatocytes establish cell-to-cell contacts, which are generally essential for expression of cell-typical functions such as protein synthesis (Baldwin and Calabrese, 1994). Recently, shaking of isolated rainbow trout hepatocytes was shown to cause cell aggregation and allow the culture to be maintained for 1 month (Flouriot et al., 1993).

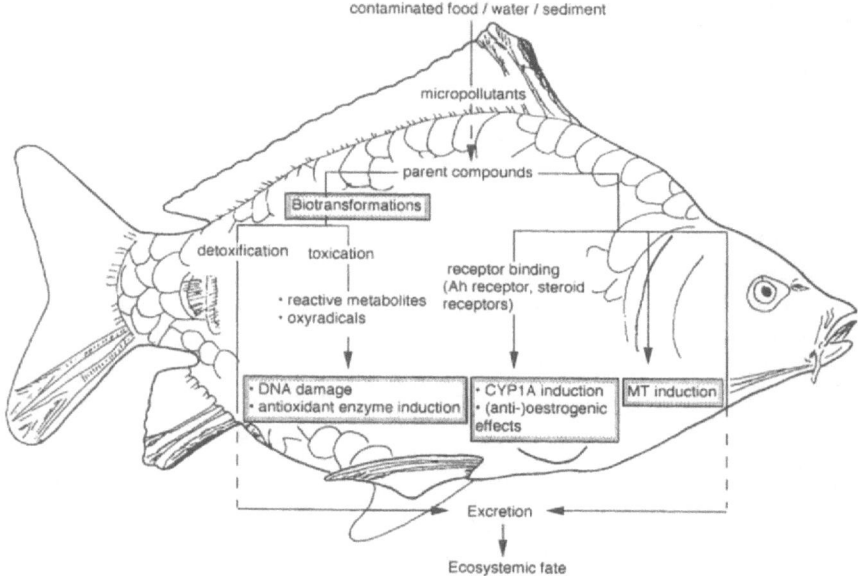

Figure 1. Schematic representation of the fate and effects of micropollutants in fish hepatocytes. Hepatocytes play a major role in the biotransformation of xenobiotics into metabolites which can be excreted from the body more readily than the parent compounds. Nevertheless, biotransformation reactions may also lead to the formation of (more) reactive species, which can interact with macromolecules, especially DNA. Several widespread micropollutants can modulate hepatic gene expression: heavy metals are inducers of metallothioneins; the affinity of certain organic micropollutants for the Ah- or steroid receptors leads to the induction of cytochrome P4501A or to (anti-)estrogenic effects, respectively. Text within frames is discussed in more detail in this review.

Since the liver is a major site of xenobiotic biotransformation, it is of considerable interest to determine whether hepatocytes retain most of the metabolic capabilities of the intact liver. Metabolic pathways of xenobiotics and corresponding kinetic data are of major importance for the evaluation of the safety of chemicals, to which organisms may be exposed. Hepatocyte cultures provide an excellent opportunity to study both the pathways and extent of metabolism of pollutants. The main specific advantages of such cultured hepatocytes are the small amounts of radioactively labeled compounds required for metabolism studies, the ease to use the culture medium for analysis of the parent compounds and metabolites, and the possibility to follow the metabolism of different chemical in the same cell type from a unique animal.

In fish, biotransformation of xenobiotics involves several enzyme systems, among which the cytochrome P450-dependent monooxygenases take a particularly important position, since they have been shown to be inducible by several organic micropollutants. This biochemical response has repeatedly been used as a biomarker for monitoring pollution in the aquatic environment. Particular attention has been paid to the induction of cytochrome P4501A1 (CYP1A), a biomarker of contamination by hazardous micropollutants such as polycyclic aromatic hydrocarbons (PAHs), polychlorinated biphenyls, polychlorinated dibenzo-p-dioxins (PCDDs), and polychlorinated dibenzofurans (PCDFs; Goksøyr and Förlin, 1992; Payne, 1976; Stegeman and Hahn, 1994). This response has recently been characterized using primary cultures of hepatocytes which were used for basic mechanistic research as well as for the development of bioassays. Induction of antioxidant enzymes and of metallothioneins are further proposed biomarkers of exposure to micropollutants which can be measured in primary cultures of fish liver cells.

Biotransformation enzymes, especially CYP1A, may be involved in activation of several chemicals into reactive metabolites, which can interact with biological macromolecules leading to cellular alterations, such as DNA damage. Formation of addition products (adducts) following interaction between electrophilic metabolites and nucleophilic sites of DNA has been studied in detail. DNA damage may also be caused by biotransformation-dependent production of oxyradicals. Due to the critical role of DNA integrity for living organisms, DNA damage can be considered as a potent biomarker of chemical contamination (Shugart, 1990). Thus, primary cultured hepatocytes from fish that express and retain biotransformation capabilities after isolation and during the culture are of great value for genotoxicological investigations.

The involvement of hepatocytes in fish reproduction makes them a critical target for reproductive toxicants. Several pollutants have been shown to alter the hepatic metabolism of steroid hormones involved in reproduction. More recently, pollutants acting as steroid agonists or antagonists have been suggested as modulators of steroid-dependent gene expression. Vitellogenin genes coding for the synthesis of the egg yolk protein precursor are specifically expressed under the control of 17-β-estradiol in the liver (Wahli et al., 1981). Therefore, primary cultured hepatocytes provide the possibility to study the modulation of estrogen-dependent gene expression by xenobiotics.

The arguments listed above explain why this chapter is mainly dedicated to studies on the biotransformation and effects of xenobiotics in primary cultures of fish hepatocytes. Some perspectives about the use of other cell types, development of cell-based bioassays, and ecotoxicological validation of studies conducted at the cellular level are also presented. In this paper, a primary culture is considered as a culture of cells either in suspension or plated for different periods of time, without cell divisions occurring. Methodological aspects are not specifically mentioned except when necessary. Cytotoxicological aspects are discussed in another chapter of this volume.

Biotransformation of xenobiotic compounds

Several studies have investigated the capacities of fish hepatocyte suspensions to metabolize xenobiotics *via* the cytochrome P450 and/or the conjugation enzyme systems (for review, see Baksi and Frazier, 1990). However, little is known about the differences between the metabolic profiles of xenobiotic compounds in isolated hepatocytes and *in vivo*. Such a comparison needs to take into account all possible routes of elimination of hepatic metabolites, i.e., bile, urine and gill excretion. However, in most *in vivo* experiments reported so far, only the biliary metabolites have been investigated. Thus, a recent study on the biotransformation of 17-methyltestosterone in rainbow trout has shown that some diol metabolites produced by hepatocyte suspensions were specifically eliminated *in vivo via* branchial excretion (Cravedi et al., 1993 and unpublished data).

Table 1 summarizes some studies conducted concurrently in the whole animal and in hepatocyte suspensions. Most of them have demonstrated that metabolic pathways are qualitatively similar in either system. Incubation of rainbow trout hepatocytes with pristane, a branched alkane, led to pristanols, pristanic acid and pristanediol glucuronide, which were also found in liver and bile of rainbow trout exposed to this saturated hydrocarbon *via* the diet (Cravedi et al., 1989).

Similarly, the metabolic pattern of pentachlorophenol, a pesticide known to produce mainly conjugated metabolites, in suspensions of rainbow trout hepatocytes closely resembled that obtained *in vivo*, with both urinary and biliary metabolites. The major route of *in vivo* biotransformation of benzo[a]pyrene in fish leads to the formation of 3-hydroxy-benzo[a]-pyrene, benzo[a]pyrene-7,8-diol and benzo[a]pyrene-9,10-diol, which may

Table 1. Xenobiotic metabolism by fish hepatocytes as compared with *in vivo* data

Compounds	Species	Metabolites		References
		in vivo	*in vitro*	
Pristane	Rainbow trout (*Oncorhynchus mykiss*)	pristan-1-ol pristan-2-ol pristanediol glucuronide pristanic acid	pristan-1-ol pristan-2-ol pristanediol glucuronide pristanic acid	Cravedi et al. (1989)
Benzo[a]pyrene	Toadfish (*Opsanus tau*)	13 phase I metabolites (tetrols, triols, diols, quinones and phenols) + glucurono-conjugates + sulfo-conjugates + putative glutathione conjugate	3 phase I metabolites (tetrols and diols) + glucurono-conjugates + sulfo-conjugates + putative glutathione conjugate	Kennedy et al. (1989), Gill and Walsh (1990)
Benzo[a]pyrene	Brown bullhead (*Ictalurus nebulosus*)	8 phase I metabolites (diols, quinones and phenols) + glucurono-conjugates + sulfo-conjugates + putative glutathione conjugate	5 phase I metabolites (triol, diols and phenols) + glucurono-conjugates + sulfo-conjugates + putative glutathione conjugate	Steward et al. (1990a, b)
Benzo[a]pyrene	English sole (*Parophrys vetulus*)	5 phase I metabolites (diols and phenols) + glucurono-conjugates + sulfo-conjugates	5 phase I metabolites (diols and phenols) + glucurono-conjugates + sulfo-conjugates	Nishimoto et al. (1992)
Pentachlorophenol	Rainbow trout (*O. mykiss*)	2 metabolites (glucuronide and sulfate conjugates)	2 metabolites (glucuronide and sulfate conjugates)	Cravedi et al. (unpublished results)
Aflatoxin B$_1$	Rainbow trout (*O. mykiss*)	2 metabolites (glucuronide and glutathione conjugates)	3 phase I metabolites + series of conjugates	Loveland et al. (1987)
Chloramphenicol	Rainbow trout (*O. mykiss*)	3 phase I metabolites + glucurono-conjugate	3 phase I metabolites + glucurono-conjugate + ethanolamine conjugate	Cravedi et al. (1985, 1991), J.-P. Cravedi et al. (unpublished results)

be converted to the corresponding glucuronide and sulphate conjugates. These metabolites were detected in the medium after incubation of benzo[a]pyrene with brown bullhead (*Ameiurus nebulosus*; Steward et al., 1990a) and English sole (*Parophrys vetulus*) hepatocytes in suspension (Nishimoto et al., 1992). Gill and Walsh (1990) found the diol isomers, but no trace of phenols in toadfish (*Opsanus beta*) hepatocytes. In the latter case, several other minor biliary metabolites including tetrols, triols, diols, diones, and phenols were not found *in vitro*. Two hypotheses were proposed to explain this discrepancy between *in vivo* and *in vitro* studies: (1) the possibility that other tissues may have produced different benzo[a]pyrene metabolites, and (2) the shorter exposure time *in vitro* (1.5 h *versus* 24 h *in vivo*), which may have resulted in lower concentrations of some, if any, metabolites.

Aflatoxin B_1 is a model substance frequently used in metabolic studies, since it is enzymatically converted to a highly electrophilic epoxide, which can subsequently react with DNA, thus inducing mutagenic and/or carcinogenic effects. Bailey and co-workers have thoroughly investigated this metabolic pathway *in vivo* in rainbow trout. Moreover, they also demonstrated the capability of hepatocytes in suspension to express both activation and detoxification processes (Bailey et al., 1984a, b; Coulombe et al., 1984; Loveland et al., 1987).

Hepatocyte suspensions have been shown to be particularly useful for studying interspecific differences in chemical metabolism (Coulombe et al., 1984; Sikka et al., 1993). Steward et al. (1989) investigated the metabolism of benzo[a]pyrene in isolated hepatocytes from mirror carp (*Cyprinus carpio*) and brown bullhead in order to determine if the greater susceptibility of brown bullhead to hepatic tumorigenesis induced by polycyclic aromatic hydrocarbon exposure resulted from metabolic factors. They found that carp hepatocytes produced larger amounts of the ultimate carcinogenic metabolite benzo[a]pyrene-7,8-dihydrodiol-9,10-epoxide and of DNA adducts than bullhead hepatocytes, suggesting that the differences in hepatic biotransformation pathways did not account for the greater susceptibility of brown bullheads to hydrocarbons.

The biotransformation of chloramphenicol *in vivo* and in suspended hepatocytes was investigated in various vertebrate species (Bories and Cravedi, 1994). In rainbow trout, chloramphenicol led to chloramphenicol glucuronide in bile and to chloramphenicol glucuronide, chloramphenicol alcohol, chloramphenicol oxamic acid, as well as chloramphenicol base in urine. All of these metabolites were also found in the medium, when freshly isolated hepatocytes were incubated for 2 h with chloramphenicol. However, when the incubation time was raised to 20 h, a new metabolite not noticed *in vivo* and amounting to approx. 2% of the total metabolites was observed and identified as chloramphenicol oxamylethanolamine (J.-P. Cravedi, unpublished results). This compound was recently detected in rat, humans and birds, and its metabolic pathway was elucidated (Cravedi et al., 1995).

Chloramphenicol oxamylethanolamine results from an initial activation of chloramphenicol to chloramphenicol acylchloride by cytochrome P450-dependent monooxygenases, followed by a covalent binding of this reactive intermediate metabolite with phosphatidyl ethanolamine in the reticulum membrane, and then by the breakdown of the phospholipid adduct and the elimination of chloramphenicol oxamylethanolamine.

Hepatocytes isolated from various mammalian species have been used to demonstrate that the relative rates of xenobiotic metabolism correlated well with published *in vivo* data (Gee et al., 1984). Nevertheless, the use of fish hepatocyte cultures for toxicokinetic studies has been poorly explored, and there is little data documenting their usability for assessing *in vivo* metabolic pathways of xenobiotic compounds. A comparative study of the kinetics of the glucuronidation of chloramphenicol in rainbow trout, rat and chicken hepatocytes in suspension (J.-P. Cravedi, unpublished results) has demonstrated a good correlation between *in vitro* and *in vivo* systems. Kinetic parameters have been used for interspecies comparison, or in order to elucidate the mechanisms of transport. Metabolism of benzo[a]pyrene by English sole hepatocytes (Nishimoto et al., 1992) showed a K_m and an apparent V_{max} similar to those obtained with rat and carp hepatocytes (Shen et al., 1980; Zaleski et al., 1991). The K_m value for benzo[a]pyrene metabolism by English sole hepatocytes was significantly higher than the K_m previously reported for English sole hepatic microsomes (Nishimoto, 1986), indicating that the transport of the hydrocarbon from the extracellular medium to the endoplasmic reticulum may have been affected.

Limited data are available regarding the influence of age, sex, strain or nutritional status of the animal donor on the metabolism of foreign compounds by isolated hepatocytes. Steward et al. (1990b) observed no consistent difference in the activities of hepatocyte suspension from male and female brown bullhead. Nevertheless, these authors suggested that the physiological changes associated with gonad maturation may explain some seasonal variations in the rate of benzo[a]pyrene biotransformation in fish hepatocytes. In contrast to the rat, starvation did not affect total cytochrome P450 contents in rainbow trout hepatocytes, and a significant decrease in 7-ethoxycoumarin deethylation was observed after only 12 weeks of starvation (Andersson, 1986). This difference between rat and rainbow trout may be explained by the capacity of the latter to retain high NADPH/NADP ratio in liver cells during starvation. Walsh and co-workers have investigated the biotransformation of model substrates including benzo[a]pyrene in hepatocyte suspensions prepared from the gulf toadfish (*Opsanus beta*) acclimated to 18° or 28°C (Gill and Walsh, 1990; Kennedy et al., 1991). Acclimation to colder temperature increased benzo[a]pyrene hydroxylation, styrene oxide hydrolysis and cytochrome P450 contents (a phenomenon not observed when metabolic rates were measured in whole liver), while decreasing *p*-nitrophenol glucuronidation. The activity of glutathione S-transferase appeared to be relatively insensitive to temperature.

Since metabolic studies in hepatocyte suspension have generally been limited to 1 day, some attempts have been made using plated fish hepatocytes. However, as recently reviewed by Skett (1994), a major disadvantage of this system may be the rapid loss of xenobiotic metabolizing capacity in a reaction-dependent manner when the cells are kept in culture for prolonged periods. However, the extent of this loss seems to be dependent on both the enzymatic activity measured and the culture conditions. Recent techniques such as aggregate cultures of rainbow trout hepatocytes provide the possibility to maintain metabolic capabilities for 1 month (Flouriot et al., 1995) at a level similar to that obtained in freshly isolated hepatocytes. Further studies on this subject need to be undertaken, in particular concerning the *in vitro-in vivo* extrapolation of biotransformation patterns.

Induction of CYP1A

Recent studies have shown the presence of an Ah receptor in fish (Hahn et al., 1994; Lorenzen and Okey, 1990), and suggested an Ah-receptor-mediated induction of CYP1A similar to that known in mammals (Stegeman and Hahn, 1994). In fish, experimental induction of CYP1A has primarily been obtained after *in vivo* exposure to chemical inducers (Goksøyr and Förlin, 1992; Stegeman and Hahn, 1994). Nevertheless, since the mid 1980s, an increasing amount of literature has become available about the detection and induction of CYP1A mediated activities in hepatocyte monolayer, mainly from rainbow trout.

Conventionally, model substrates for CYP1A such as 7-ethoxyresorufin, 7-ethoxycoumarin were added directly into culture medium. Unfortunately, in hepatocytes hydroxylated metabolites are mainly produced as conjugates which are undetectable by means of spectrofluorimetric analysis. Deconjugation, using deconjugating enzyme mixtures (glucuronidase/ sulfatase), are, thus, required after the end of incubation (Andersson and Förlin, 1985; Vaillant et al., 1989). In more recent studies, the procedure has been greatly simplified by direct measurement of hydroxylated metabolites after incubation of model substrates with whole homogenates from sonicated or potter-homogenized hepatocytes (Anderson et al., 1996a; Clemons et al., 1994; Devaux et al., 1991; Pesonen and Andersson, 1991).

Measurement of induction of CYP1A-dependent activities in monolayer culture of rainbow trout hepatocytes is facilitated, because enzyme activity is relatively constant from day 1 to day 5 in control cells. In contrast, a tremendous decrease in monooxygenase activity is usually observed in primary culture of mammalian hepatocytes (Pesonen and Andersson, 1991). In rainbow trout, only minor cytological alterations were observed for culture periods of up to 5 days (Braunbeck and Storch, 1992).

Induction of cytochrome P450-dependent enzyme activity in monolayer cultures of rainbow trout hepatocytes exposed to the model inducer β-naphthoflavone in the culture medium was first demonstrated by our group (Vaillant et al., 1989). CYP1A has been mostly assessed through measurement of associated catalytic activities: 7-ethoxyresorufin O-deethylase (EROD), aryl hydrocarbon hydroxylase (AHH), and 7-ethoxycoumarin O-deethylase (ECOD). Induction was recently observed for CYP1A mRNA and protein in monolayer cultures of rainbow trout hepatocytes (Anderson et al., 1996a; Pesonen et al., 1992). The time-course of induction was shown to be similar to that observed *in vivo* (Devaux, 1990; Pesonen and Andersson, 1991; Pesonen et al., 1992; Fig. 2). Nevertheless, in isolated hepatocytes the level of induction was generally lower than *in vivo*, suggesting the involvement of endogenous factors that were not present in the culture medium. Glucocorticoids were found to potentiate EROD induction in rainbow trout hepatocytes treated with β-naphthoflavone (Celander et al. 1995; Devaux et al., 1992).

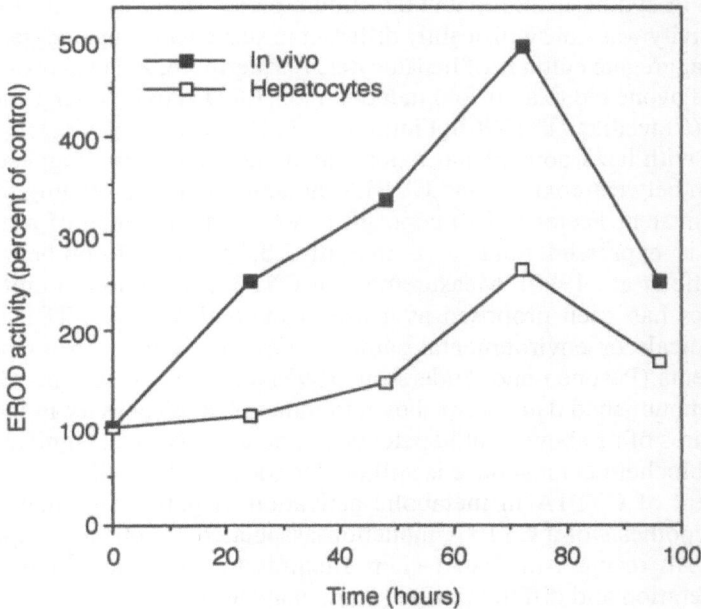

Figure 2. Kinetics of 7-ethoxyresorufin O-deethylase (EROD) induction by β-naphthoflavone in primary cultures of rainbow trout hepatocytes compared to the *in vivo* response. The individual fish used in this study were from the same batch. *In vivo* exposure was performed by i.p. injection of 50 mg β-naphthoflavone per kg body weight, and EROD activity was measured in the hepatic microsomal fraction. Isolated hepatocytes were exposed to 0.1 μg β-naphthoflavone per mL of culture medium 1 day after cell isolation, and EROD activity was determined in total cell homogenates. Values represent the mean from four individual measurements. For details on the procedures, see references in the text.

Metabolism of inducers in hepatocytes may be critical towards the extent and kinetic of CYP1A induction. For instance, a continuous increase in EROD activity was observed up to 120 h in rainbow trout hepatocytes exposed to 2,3,7,8-tetrachlorodibenzo-*p*-dioxin (TCDD) for 48 h whereas, in hepatocytes exposed to β-naphthoflavone for the same period, the induction was lower and EROD activity returned to control levels after 72 h in culture (Pesonen et al., 1992). The low metabolism of 2,3,7,8-tetrachlorodibenzo-*p*-dioxin in contrast to the rapid metabolism of β-naphthoflavone was proposed to explain these results. A recent study demonstrated CYP1A induction in monolayer cultures of rainbow trout hepatocytes exposed to phenobarbital (Sadar et al., 1996). The authors suggested that, though phenobarbital has been shown to be a poor ligand of the Ah receptor and is relatively quickly metabolized, this compound could act on receptor Ah-dependent expression of CYP1A gene through modulation of signal transduction pathways.

In an attempt to perform long-term studies, primary cultures of rainbow trout hepatocytes were cultured as spheroid aggregates which have extended longevity and exhibit differentiated functions such as albumin and vitellogenin mRNA synthesis throughout the culture period (Flouriot et al., 1993). EROD activity was not significantly different in suspension hepatocytes and in 30-day aggregate cultures of hepatocytes. Furthermore, 2-day exposure to β-naphthoflavone led to a 10-fold induction in EROD activity after 1 month in culture (Cravedi et al., 1996b; Flouriot et al., 1995a).

Studies with hazardous planar halogenated chemicals have suggested a correlation between toxicity and CYP1A induction potency, leading to the Toxic Equivalent Factor (TEF) concept, in which the potency of a given chemical is expressed relative to that of 2,3,7,8-tetrachlorodibenzo-*p*-dioxin (Safe et al., 1990). Measurement of CYP1A induction in cultured hepatocytes has been proposed as a useful method to assess TEF from pure chemicals or environmental samples. Organic extracts from paper mill effluents (Pesonen and Andersson, 1992) and aquatic sediments (A. Devaux, unpublished data) were shown to induce EROD activity in monolayer cultures of rainbow trout hepatocytes. The toxicological significance of such a biochemical response is still unclear despite the well-established involvement of CYP1A in metabolic activation of genotoxic chemicals, and the hypothesis of a CYP1A induction-associated pleiotropic response involving Ah receptor-mediated altered expression of genes controlling cell proliferation and differentiation (Stegeman and Hahn, 1994).

Induction of metallothioneins and oxidative stress enzymes

Although not exclusively expressed in fish liver, some biochemical responses to pollutants such as induction of metallothioneins or induction of antioxidant enzymes occur in this tissue. Most of the data published so far deal with the detection of these biomarkers in fish exposed *in vivo* in laboratory or field conditions. Nevertheless, some studies have also been performed using cultures of fish hepatocytes. Metallothioneins are a family of metal-binding proteins, which are widely distributed in living organisms and which are characterized by high cystein contents. They play a key role in the regulation of the essential metals zinc and copper, as well as in the detoxification of non-essential metals such as cadmium and mercury. Metallothioneins are inducible by metals, and this biochemical response has been proposed as a biomarker of exposure to these pollutants. However, the (eco)toxicological implications of metallothionein induction remain unclear. In fish, the organs in which metallothioneins have been found in highest concentrations are the liver, kidney, gill and intestine (for review, see Roesijadi, 1992). Induction of metallothioneins has been reported in monolayer cultures of rainbow trout hepatocytes exposed to cadmium (Gagné et al., 1990) and zinc (Hyllner et al., 1989). In this latter study, metallothioneins were shown to be inducible by cortisol, thus indicating a possible hormonal regulation of metallothionein level.

Antioxidant enzymes (superoxide dismutase, catalase, glutathione peroxidase) are involved in the detoxification of reactive oxygen species (oxyradicals) generated through various processes and responsible for oxidative damages such as lipid peroxidation and DNA oxidation. They are generally present in high levels in fish liver and have been shown to be inducible under oxidative stress (for review, see Winston and Di Giulio, 1991). Due to the high level of cytochrome P450-dependent enzymes in hepatocytes and to the involvement of these enzymes in oxyradical production (Lemaire and Livingstone, 1993), primary cultures of hepatocytes would be particularly suitable for the investigation of the occurrence of oxidative damage and induction of antioxidant enzymes.

Detection of DNA damage

Genotoxicity of xenobiotics has been studied in fish using various endpoints including DNA adducts, single DNA strand breaks, chromosome aberrations (anaphase aberrations and micronuclei), induction of sister chromatid exchange (SCE), and unscheduled DNA synthesis (UDS). In cultured cells, these parameters have been monitored mainly in permanent cell lines (Babich and Borenfreund, 1991; Brambilla and Martelli, 1990; Hightower and Renfro, 1988; Leadon, 1990; McQueen et al., 1988). Only recently, investigations have addressed the detection of DNA damage in primary cultures of hepatocytes.

DNA adducts are important in initiation, first stage promotion and possibly later stages of chemical carcinogenesis. As discussed above, isolated fish hepatocytes may display metabolite patterns similar to those obtained *in vivo*. They subsequently give an opportunity to study the effects of various factors on DNA adduct formation after exposure to procarcino-gens. Bailey and co-workers (1982) clearly demonstrated that the metabo-lism of aflatoxin B1 in suspensions of rainbow trout hepatocytes resulted in the production of DNA adducts linearly related to aflatoxin B1 dose and qualitatively similar to the adducts formed *in vivo*. Comparison of aflatoxin B1-DNA binding in rainbow trout and salmon (*Oncorhynchus kisutch*) hepatocyte suspensions confirmed that *in vivo* DNA adduct formation was higher in rainbow trout than in salmon (Bailey et al., 1988). This result could further be correlated to the observation that the activation efficiency was approximately three times higher in rainbow trout than in salmon, sug-gesting that the greater sensitivity of rainbow trout to aflatoxin B1 may result from differences in the metabolic pathways. Recently, we have shown that the pattern of DNA adducts in plated rainbow trout hepatocytes ex-posed to benzo[a]pyrene for 24 h was identical to that obtained after *in vivo* benzo[a]pyrene treatment of rainbow trout, whereas adducts obtained after *in vitro* incubation of DNA with benzo[a]pyrene and hepatic microsomes were qualitatively different (Masfaraud et al., 1992). Furthermore, this study suggested that DNA repair mechanisms are active in primary cultures of rainbow trout hepatocytes. DNA repair measured in suspension of rain-bow trout hepatocytes is likely to occur through a mechanism similar to that described in mammalian cells (Miller et al., 1989).

Many xenobiotics may enhance oxyradical production as a result of their conversion during biotransformation. The assessment of oxidative DNA damage in metabolically competent cells like fish hepatocytes appears as a promising approach in aquatic ecotoxicology (Lemaire and Livingstone, 1993). DNA single strand breaks may provide a meaningful indication of the degree of oxidative damage to the DNA (Imlay et al., 1988). Recently, genotoxicity of marine sediment extracts was assayed in primary cultures of rainbow trout hepatocytes, through the measure of DNA strand breaks using the nick translation and the alkaline precipitation assays (Gagné and Blaise, 1995; Gagné et al., 1995). Another approach referred to as the single cell gel electrophoresis assay (Singh et al., 1988) or comet assay (Olive et al., 1990) has recently been developed for the detection of DNA strand breaks in cells exposed to genotoxins. Using the comet assay, mono-layer cultures of rainbow trout hepatocytes exposed to hydrogen peroxide or benzo[a]pyrene as genotoxic model molecules exhibited a significant amount of DNA single strand breaks quantitatively related to the genotoxin concentration (Fig. 3; Devaux et al., 1997).

Although fish cells usually exhibit a low DNA repair activity compared to that of mammalian cells, unscheduled DNA synthesis tests have been performed in hepatocyte suspensions or monolayers of rainbow trout (Klaunig, 1984) and oyster toadfish (*Opsanus tau*; Kelly and Maddock, 1985), as well as in primary cultured cells from rainbow trout stomach and intestine (Walton et al., 1984).

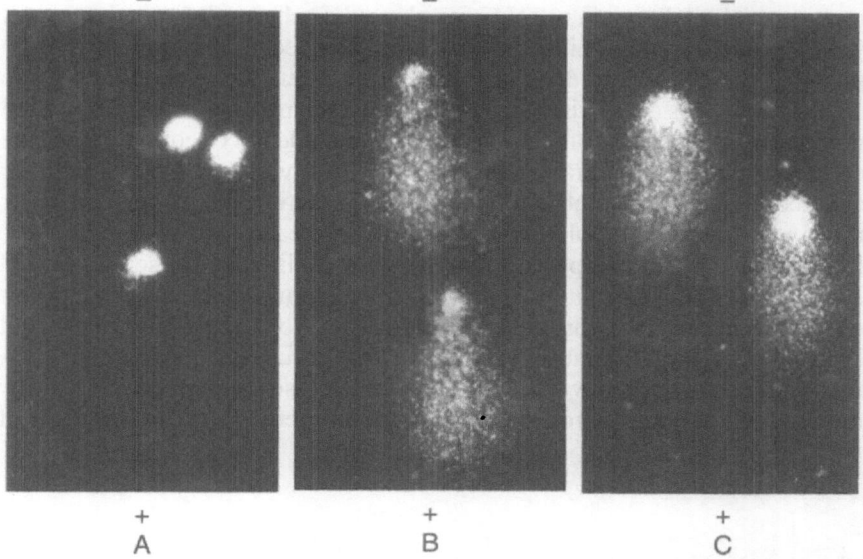

Figure 3. Fluorescence photomicrographs of rainbow trout hepatocytes processed according to the comet assay procedure. The plus and minus signs indicate the positions of the anode and cathode, respectively, during electrophoresis; negatively charged DNA migrates towards the anode. For detailed procedure, see references in the text. (A): control cells; (B): comet figures of hepatocytes treated for 2 h with 53 µM H_2O_2; (C): comet figures of hepatocytes treated for 24 h with 1 µM benzo[a]pyrene (magnification 400 ×).

Reproductive parameters

Estradiol (E2) regulates vitellogenin gene expression through the binding to a specific estrogen receptor (ER). The ER/E2 complex is able to interact with the estrogen-responsive-element (ERE) of target promoter genes to modulate their transcriptional activity (Beato, 1989). In rainbow trout, the hepatic ER level is also under estradiol control (Pakdel et al., 1991).

A significant estrogenic contamination of the aquatic environment has been suggested in a recent field survey where caged rainbow trout showed pronounced increases in plasma vitellogenin concentration after exposure to the effluents of sewage treatment plants (Purdom et al., 1994). In addition, a random screening of 20 organic chemicals extracted from the effluents revealed that 10 of them were able to interact with the estrogen receptor (Jobling et al., 1995).

The hepatocyte aggregate cultures of rainbow trout recently developed by Valotaire and co-workers maintain a stable estrogen receptors and vitellogenin gene expression over a 1-month period (Flouriot et al., 1993). In this system, estrogen receptor and vitellogenin genes were expressed, as *in vivo*, under 17-β-estradiol control. A dose-dependent induction of estrogen receptor and vitellogenin mRNA was observed, when cultures were treated with a polychlorinated biphenyl mixture, chlordecone, lindane, or nonylphenol (Flouriot et al., 1995b; Fig. 4). However, these chemicals were significantly less active than 17-β-estradiol, the "natural effector", and did not modify the activity of 17-β-estradiol when added concomitantly to the culture medium. In contrast, pentachlorophenol, which had no effect after isolated exposure, was able to reduce the induction of estrogen receptor mRNA by 17-β-estradiol and to markedly decrease vitellogenin mRNA. Interestingly, when compared to 17-β-estradiol, lindane caused a delayed estrogen receptor and vitellogenin mRNA accumulation, suggesting that the estrogenic effect of this pesticide may result from metabolites produced by hepatic biotransformation enzymes. These results demonstrate that primary cultures of rainbow trout hepatocytes represent a convenient cellular model for studying estrogenic/antiestrogenic activity of chemicals and

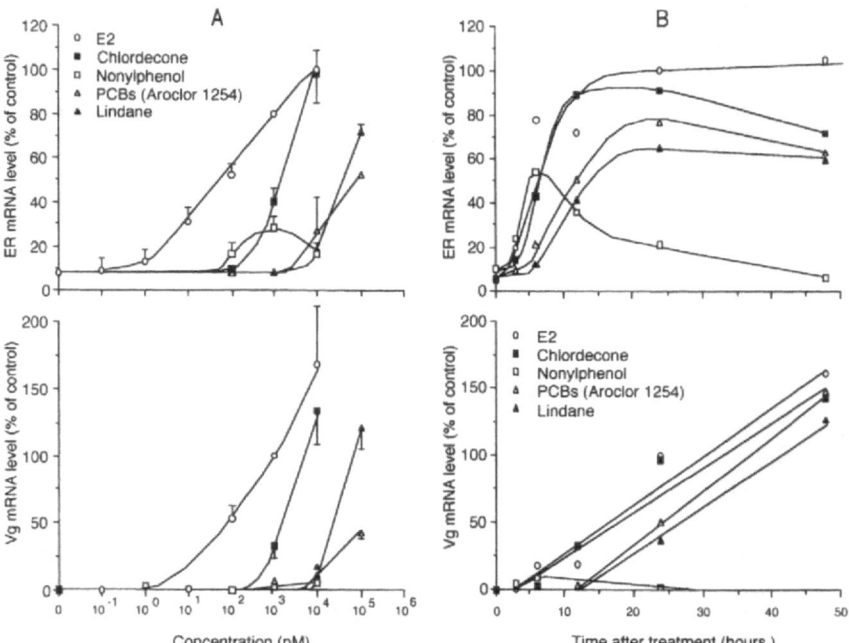

Figure 4. Effects of several xenobiotics on estradiol receptor (ER) and vitellogenin mRNA synthesis in rainbow trout hepatocytes in aggregates. Hepatocyte aggregates were incubated either during 24 h (A) with increasing concentrations of 17-β-estradiol (E2), chlordecone, nonylphenol, polychlorinated biphenyls (Aroclor 1254), lindane, or for different times (B) with 1 μM 17-β-estradiol, 10 μM chlordecone, 10 μM nonylphenol and 100 μM polychlorinated biphenyls (Aroclor 1254). Levels of estradiol receptor and vitellogenin (Vg) mRNA were determined by slot-blot hybridization. Values are expressed as percentage of estradiol receptor or vitellogenin mRNA levels following exposure 1 μM 17-β-estradiol for 24 h. Data represent the mean ± SE from three individual determinations in (A), and the mean of two determinations in (B) (modified from Flouriot et al., 1995b).

their metabolites as well as interaction of xenobiotics with 17-β-estradiol towards modulation of estrogen receptor and vitellogenin genes. Furthermore, this cellular system appears to be particularly suitable to investigate the recently suggested Ah receptor-mediated modulation of estrogenic response pathways (Anderson et al., 1996a; Safe, 1995). Thus, CYP1A inducers including β-naphthoflavone and congeners of polychlorinated dibenzo-p-dioxins, polychlorinated dibenzofurans, and polychlorinated biphenyls were shown to inhibit vitellogenin synthesis in 17-β-estradiol-treated rainbow trout hepatocytes cultured in monolayers (Anderson et al., 1996b). The authors pointed out that suppression of vitellogenin synthesis was correlated with the CYP1A1 protein inducing-potency of polychlorinated biphenyls and polychlorinated dibenzofuran congeners. Surprisingly, weak CYP1A inducers, e.g., the polychlorinated biphenyl congeners 114 and 156, appeared to potentiate vitellogenin synthesis.

In addition to vitellogenin synthesis, fish hepatocytes are also involved in other aspects of reproduction, which appear worthy to be assessed as targets for aquatic pollutants. Steroid binding proteins, which are synthesized in hepatocytes under the control of steroid hormones, play a major role in the control of plasmatic concentration and bioavailability of steroid hormones. Xenohormones, especially phytoestrogens, have been shown to stimulate steroid binding proteins synthesis in hepatocarcinoma human liver cells (Mousavi and Adlercreutz, 1993) as well as in monolayer cultures of rainbow trout hepatocytes (F. Le Gac, personal communication). Proteins of the egg envelope have also been shown to be synthesized in hepatocytes of several fish species, particularly in salmonids (Oppen-Berntsen et al., 1992; Yamagami et al., 1994). Synthesis would be under the control of 17-β-estradiol and was shown to parallel vitellogenin synthesis. Therefore, investigations on the effects of pollutants known to interact with vitellogenin synthesis should be carried on with respect to steroid binding proteins and egg envelope protein synthesis.

Perspectives

With regard to the cellular effects of pollutants in fish, special attention has been paid to hepatocytes in primary culture. Information related to xenobiotic biotransformation, biomarkers, as well as reproductive effects of pollutants can be obtained using these cells. Hepatocytes are particularly suitable to investigate the relationship between xenobiotic metabolism and toxicity. In addition, the good correspondence between data obtained in cultured cells and in *in vivo* experiments support the call for further validation of cultured hepatocytes as an alternative model for the study of mechanisms of pollutant toxicity in fish.

Other cell types are potential targets for pollutants and need to be taken into account to address more in detail the hazard of exposure to pollutants in fish. Due to the significant role of extrahepatic tissues in the biotransformation of chemicals, the use of the corresponding cells would be of particular interest for assessing the metabolic fate of xenobiotics on a whole body basis. However, isolation of these cells and their maintenance in culture is a prerequisite not always easy to achieve. In this context, germ cells appear to be of particular importance, but, in contrast to the culture of cells of mammalian origin, very few data are available about their use in fish ecotoxicology. Inhibition of 17-β-estradiol secretion was observed after exposure of rainbow trout follicular cells to imidazole fungicides in the culture medium (Monod et al., 1993). Those fungicides were also shown to alter steroid metabolism in rainbow trout testicular cells maintained in culture (M. Loir, personal communication).

Fish gill tissue is primordial for homeostasis processes including respiration, ion regulation, acid-base regulation and excretion. It also represents an important route of uptake of waterborne pollutants and has been shown to display biotransformation activities. Finally, it is known to be a critical site for the expression of metal toxicity. Conditions for primary cultures of gill epithelial cells from sea bass (*Dicentrarchus labrax*), gulf toadfish (*Opsanus beta*) and rainbow trout have recently been described (Avella et al., 1994; Kennedy and Walsh, 1994; Pärt and Bergström, 1995; Pärt et al., 1993). Preliminary data suggest that this cell type might be a relevant model for screening chemicals acting on the properties of cell membrane transport systems (Witters et al., 1996) and for studying the specific role of gills in the biotransformation of xenobiotics in fish (Cravedi et al., 1996a; Pärt, 1996).

Immunocompetent cells are another critical cellular target for pollutants (Wester et al., 1994). Immunotoxic effects of aquatic pollutants have been assessed in lymphocytes (Faisal and Hugget, 1993) and macrophages (Anderson and Brubacher, 1993) exposed to pollutants in the culture medium. Fish red blood cells appear to be further useful tools in ecotoxicity tests and in aquatic pollution monitoring programs (Nikinmaa, 1992). Easily sampled by non-destructive methods, they can be maintained in culture for several days to study the effects of pollutants. Thus, the effects of various metals on antioxidant enzyme activities have recently been studied in cultured red blood cells of sea bass (Roche and Bogé, 1993).

Future developments of basic research on fish physiology should try to further improve isolation and maintenance of different cell types and, thus, facilitate the study of the effects of aquatic pollutants on various physiological targets.

Development of standardized bioassay procedures is an important issue in ecotoxicology. Primary cultures exhibit some major pitfalls, particularly due to the difficulty to ensure similar responsiveness between cells from different individuals. Technical expertise, especially with respect to cell isolation, is another limiting factor to ensure reliability for such approaches. Cryopreservation of isolated cells urgently needs to be addressed prior to routine testing development. In this way, established cell lines, which have retained functions of non-transformed cells, turn out to be very useful to set up standardized, straightforward bioassays. The expression of xenobiotic metabolizing enzymes has been explored in fish cell lines including BF-2 (bluegill sunfish fibroblast; Babich and Borenfreund, 1987), PLHC-1 (*Poeciliopsis lucida* hepatoma cell; Hahn et al., 1993) and RTL-W1 (rainbow trout liver cell; Lee et al., 1993). Although some authors have analyzed the metabolites of polycyclic aromatic hydrocarbons in fish cell lines (Baird et al., 1988; Smolarek et al., 1987; Thornton et al., 1982), their usefulness in establishing metabolic profiles remains to be investigated. Transformed or non-transformed cell lines from fish with stabilized responsiveness to CYP1A inducers have been described (Hahn et al., 1993; Lorenzen and Okey, 1990; Miranda et al., 1993). Fish-specific TEFs have recently been determined from the rainbow trout liver cell line RTL-W1 (Clemons et al., 1994). In order to facilitate the rapid analysis of large numbers of samples, an adapted version of a previously published multi-well plate method has recently been proposed to measure CYP1A induction in fish hepatoma cells by Hahn and co-workers (1995). Because of their low mitotic index, hepatocytes in primary cultures are not suitable for cytogenetic assays (e.g., with respect to induction of sister chromatid exchange and chromosome aberrations), mutagenicity assays (e.g., induction of ouabaine mutants), and DNA repair assays. So far, these approaches have been restricted to fibroblast-like fish cell lines (for review, see Babich and Borenfreund, 1991). In an attempt to develop a rapid and easily standardizable bioassay for the assessment of the estrogen receptor-dependent estrogenic potency of chemicals, a yeast strain transformed by plasmids containing the rainbow trout estrogen receptor cDNA and a reporter gene containing an estrogen responsive element has recently been developed (Petit et al., 1995).

Ecotoxicological perspectives require that the significance of pollutant effect observed in isolated cells is addressed at higher levels of biological organization. Unfortunately, up to now this has only been given very little attention in fish. Whereas extensive studies have been performed at the cellular level in rainbow trout, the population level is particularly difficult to investigate particularly in European ecosystems, where this species does normally not reproduce. Consequently, species allowing studies from the cellular level to the population would be of great interest for future prospects in fish ecotoxicology.

Acknowledgments
We thank F. Le Gac, M. Loir, J.-F. Masfaraud and J.P. Sumpter for providing us with unpublished data. We also thank L. Lagadic for his helpful comments.

56 G. Monod et al.

References

Anderson, R.S. and Brubacher, L.L. (1993) Inhibition by pentachlorophenol of production of reactive-oxygen intermediates by medaka phagocytic blood cells. *Mar. Environ. Res.* 35:125–129.

Anderson, M.J., Miller, M.R. and Hinton, D.E. (1996a) *In vitro* modulation of 17-β-estradiol-induced vitellogenin synthesis: effects of cytochrome P4501A1 inducing compounds on rainbow trout (*Oncorhynchus mykiss*) liver cells. *Aquat. Toxicol.* 34:327–350.

Anderson, M.J., Isen, H., Matsumura, F. and Hinton, D.E. (1996b) *In vivo* modulation of 17β-estradiol-induced vitellogenin synthesis and estrogen receptor in rainbow trout (*Oncorhynchus mykiss*) liver cells by β-naphthoflavone. *Toxicol. Appl. Pharmacol.* 1337:210–218.

Andersson, T. (1986) Cytochrome P-450-dependent metabolism in isolated liver cells from fed and starved rainbow trout, *Salmo gairdneri. Fish Physiol. Biochem.* 1:105–111.

Andersson, T. and Förlin, L. (1985) Spectral properties of substrate-cytochrome P-450 interaction and catalytic activity of xenobiotic metabolizing enzymes in isolated rainbow trout liver cells. *Biochem. Pharmacol.* 34:1407–1413.

Avella, M., Berhaut, J. and Payan, P. (1994) Primary culture of gill epithelial cells from the sea bass *Dicentrarchus labrax. In vitro Cell. Dev. Biol.* 30A:41–49.

Babich, H. and Borenfreund, E. (1987) Aquatic pollutants tested *in vitro* with early passage fish cells. *ATLA* 15:116–122.

Babich, H. and Borenfreund, F. (1991) Cytotoxicity and genotoxicity assays with cultured fish cells: a review. *Toxicol. in vitro* 5:91–100.

Bailey, G.S., Taylor, M.J. and Selivonchick, D.P. (1982) Aflatoxin B1 metabolism and DNA binding in isolated hepatocytes from rainbow trout (*Salmo gairdneri*). *Carcinogenesis* 3:511–518.

Bailey, G.S., Hendricks, J.D., Nixon, J E. and Pawlowski, N.E. (1984a) The sensitivity of rainbow trout and other fish to carcinogens. *Drug Metab. Rev.* 15:725–750.

Bailey, G.S., Taylor, M.J., Loveland, P.M., Wilcox, J.S., Sinnhuber, R.O. and Selivonchick, D.P. (1984b) Dietary modification of aflatoxin carcinogenesis: mechanism studies with isolated hepatocytes from rainbow trout. *Natl. Cancer Inst. Monogr.* 65:379–385.

Bailey, G.S., Williams, D.E., Wilcox, J.S., Loveland, P.M., Coulombe, R.A. and Hendricks J.D. (1988) Aflatoxin B1 carcinogenesis and its relation to DNA adduct formation and adduct persistence in sensitive and resistant salmonid fish. *Carcinogenesis* 9:1919–1926.

Baird, W.M., Smolarek, T.A., Plakunov, I., Hevizi, K., Kelley, J. and Klaunig, J. (1988) Formation of benzo[a]pyrene-7,8-dihydrodiol glucuronide is a major pathway of metabolism of benzo[a]pyrene in cell cultures from bluegill fry and brown bullhead. *Aquat. Toxicol.* 11:398 (abstract).

Baksi, S.M. and Frazier, J.M. (1990) Isolated fish hepatocytes-model systems for toxicology research. *Aquat. Toxicol.* 16:229–256.

Baldwin, L.A. and Calabrese, E.J. (1994) Gap junction-mediated intercellular communication in primary cultures of rainbow trout hepatocytes. *Ecotox. Environ. Safety* 28:201–207.

Beato, M. (1989) Gene regulation by steroid hormones. *Cell* 56:335–344.

Bories, G. and Cravedi, J.-P. (1994) Metabolism of chloramphenicol: a story of nearly 50 years. *Drug Metab. Rev.* 26:767–783.

Brambilla, G. and Martelli, A. (1990) Human hepatocytes in genotoxicity assays. *Pharmacol. Res.* 22:381–392.

Braunbeck, T. and Storch, V. (1992) Senescence of hepatocytes isolated from rainbow trout (*Oncorhynchus mykiss*) in primary culture. An ultrastructural study. *Protoplasma* 170:138–159.

Celander, M., Bremer, J., Hahn, M.E. and Stegeman, J.J. (1995) *CYP1A and CYP3A in the Poeciliopsis lucida hepatocellular carcinoma (PLHC-1) cell line treated with CYP1A and CYP3A inducers: corticosteroid potentiation of CYP1A induction.* 8th International Symposium on Pollutant Responses in Marine Organisms, California, abstract.

Clemons, J.H., van den Heuvel, M.R., Stegeman, J.J., Dixon, D.G. and Bols, N.C. (1994) Comparison of toxic equivalent factors for selected dioxin and furan congeners derived using fish and mammalian liver cell lines. *Can J. Fish. Aquat. Sci.* 51:1577–1584.

Coulombe R.A., Bailey, G.S. and Nixon, J. (1984) Comparative activation of aflatoxin B1 to mutagens by isolated hepatocytes from rainbow trout (*Salmo gairdneri*) and coho salmon (*Oncorhynchus kisutch*). *Carcinogenesis* 5:29–33.

Cravedi, J.-P. and Baradat, M. (1991) Comparative metabolic profiling of chloramphenicol by isolated hepatocytes from rat and trout (Oncorhynchus mykiss). Comp. Biochem. Physiol. 100C:649–652.

Cravedi, J.-P., Delous, G., Debrauwer, L. and Promé, D. (1993) Biotransformation and branchial excretion of 17 -methyltestosterone in trout. Drug Metab. Dispos. 21:377–385.

Cravedi, J.-P., Heuillet, G., Peleran, J.-C. and Wal, J.-M. (1985) Disposition and metabolism of chloramphenicol in trout. Xenobiotica 15:115–121.

Cravedi, J.-P., Perdu-Durand, E., Baradat, M. and Tulliez, J. (1989) Hydroxylation of pristane by isolated hepatocytes of rainbow trout: a comparison with in vivo metabolism and biotransformation by liver microsomes. Mar. Environ. Res. 28:15–18.

Cravedi, J.-P., Perdu-Durand, E., Baradat, M., Alary, J., Debrauwer, L. and Bories, G. (1995) Chloramphenicol oxamylethanolamine as an end product of chloramphenicol metabolism in rat and humans: evidence for the formation of a phospholipid adduct. Chem. Res. Toxicol. 8:642–648.

Cravedi, J.-P., Le Guen, I., Perdu-Durand, E., Baradat, M., Paris, A. and Prunet, P. (1996a) Xenobiotic and steroid biotransformation by trout gill epithelial cells in culture. 17th Annual conference of the European Society for Comparative Physiology and Biochemistry, Antwerp, Belgium, August 1996.

Cravedi, J.-P., Paris, A., Monod, G., Devaux, A., Flouriot, G. and Valotaire, Y. (1996b) Maintenance of cytochrome P450 content and phase I and phase II enzyme activities in trout hepatocytes cultured as spheroidal aggregates. Comp. Biochem. Physiol. 113C:241–246.

Devaux, A. (1990) Induction of rainbow trout cytochrome P-450 monooxygenases in primary hepatocyte culture and in vivo. 12th Annual Conference of the European Society for Comparative Physiology and Biochemistry, Utrecht, The Netherlands, abstract.

Devaux, A., Monod, G., Masfaraud, J.-F. and Vaillant, C. (1991) Use of a new coumarin analog in the study of cytochrome P-450 dependent monooxygenase activities in primary culture of rainbow trout hepatocytes (Oncorhynchus mykiss). C.R. Acad. Sci. 312(III):63–69.

Devaux, A., Pesonen, M., Monod, G. and Andersson, T. (1992) Glucocorticoid-mediated potentiation of P450 induction in primary culture of rainbow trout hepatocytes. Biochem. Pharmacol. 43:898–901.

Devaux, A., Pesonen, M., Monod, G. (1997) Alkaline comet assay in rainbow trout hepatocytes. Toxicol. in vitro 11:71–79.

Faisal, M. and Hugget, R.J. (1993) Effects of polycyclic aromatic hydrocarbons on the lymphocyte mitogenic response in spot, Leiostomus xanthurus. Mar. Environ. Res. 35:121–124.

Flouriot, G., Vaillant, C., Salbert, G., Pelissero, C., Guiraud, J.-M. and Valotaire, Y. (1993) Monolayer and aggregate cultures of rainbow trout hepatocytes: long term, stable liver specific expression in aggregates. J. Cell Sci. 105:407–416.

Flouriot, G., Monod, G., Valotaire, Y., Devaux, A. and Cravedi, J.-P. (1995a) Xenobiotic metabolizing enzyme activities in aggregate culture of rainbow trout hepatocytes. Mar. Environ. Res. 39:293–299.

Flouriot, G., Pakdel, F., Ducouret, B. and Valotaire, Y. (1995b) Influence of xenobiotics on rainbow trout liver estrogen receptor and vitellogenin gene expression. J. Mol. Endocrinol. 15:143–151.

Gagné, F. and Blaise, C. (1995) Evaluation of the genotoxicity of environmental contaminants in sediments to rainbow trout hepatocytes. Environ. Toxicol. Water Qual. 10:217–229.

Gagné, F., Marion, M. and Denizeau, F. (1990) Metal homeostasis and metallothionein induction in rainbow trout hepatocytes exposed to cadmium. Fund. Applied Toxicol. 14:429–437.

Gagné, F., Trottier, S., Blaise, C., Sproull, J. and Ernst, B. (1995) Genotoxicity of sediment extracts obtained in the vicinity of a creosote-treated wharf to rainbow trout hepatocytes. Toxicol. Lett. 78:175–182.

Gee, S.J., Green, C.E. and Tyson, C.A. (1984) Comparative metabolism of tolbutamide by isolated hepatocytes from rat, rabbit, dog and squirrel monkey. Drug Metab. Dispos. 12:174–178.

Gill, K.A. and Walsh P.J. (1990) Effects of temperature on metabolism of benzo[a]pyrene by toadfish (Opsanus beta) hepatocytes. Can. J. Fish. Aquat. Sci. 47:831–837.

Goksøyr, A. and Förlin, L. (1992) The cytochrome P-450 system in fish, aquatic toxicology and environmental monitoring. Aquat. Toxicol. 22:287–312.

Hahn, M.E., Lamb, T.M., Schultz, M.E., Smolowitz, R.M. and Stegeman, J.J. (1993) Cytochrome P4501A induction and inhibition by 3,3′,4,4′-tetrachlorobiphenyl in an Ah receptor-containing fish hepatoma cell ligne (PLHC-1). Aquat. Toxicol. 26:185–208.

Hahn, M.E., Poland, A., Glover, E. and Stegeman, J.J. (1994) Photoaffinity labeling of the Ah
    receptor: phylogenetic survey of diverse vertebrate and invertebrate species. *Arch. Biochem.
    Biophys.* 310:218–228.
Hahn, M.E., Patel, A.B. and Stegeman, J.J. (1995) Rapid assessment of cytochrome P4501A
    induction in fish hepatoma cells grown in multi-well plates. *Mar. Environ. Res.* 39:354
    (abstract).
Hightower, L.E. and Renfro, J.L. (1988) Recent applications of fish cell culture to biomedical
    research. *J. Exp. Zool.* 248:290–302.
Hyllner, S.J., Andersson, T., Haux, C. and Olsson, P.E. (1989) Cortisol induction of metallo-
    thionein in primary culture of rainbow trout hepatocytes. *J. Cell. Physiol.* 139:24–28.
Imlay, J.A., Chin, S.M. and Linn, S. (1988) Toxic DNA damage by hydrogen peroxide through
    the Fenton reaction *in vivo* and *in vitro*. *Science* 240:640–642.
Jobling, S., Reynold, T., White, R., Parker, M.G. and Sumpter, J.P. (1995) A variety of environ-
    mentally persistent chemicals, including some phthalate plasticizers are weakly estrogenic.
    Environ. Health Perspect. 103:582–587.
Kelly, J.J. and Maddock, M.B. (1985) *In vitro* induction of unscheduled DNA synthesis by geno-
    toxic carcinogens in the hepatocytes of the oyster toadfish. *Arch. Environ. Contam. Toxicol.*
    14:555–563.
Kennedy, C.J. and Walsh, P.J. (1994) The effects of temperature on the uptake and metabolism
    of benzo[a]pyrene in isolated gill cells of the gulf toadfish, *Opsanus beta*. *Fish Physiol. Bio-
    chem.* 13:93–103.
Kennedy, C.J., Gill, K.A. and Walsh, P.J. (1991) Temperature acclimation of xenobiotic meta-
    bolizing enzymes in cultured hepatocytes and whole liver of the gulf toadfish, *Opsanus beta*.
    Can. J. Fish. Aquat. Sci. 48:1212–1219.
Klaunig, J.E. (1984) Establishment of fish hepatocytes cultures for use in *in vitro* carcinogeni-
    city studies. *Natl. Cancer Inst. Monogr.* 65:163–173.
Leadon, S.A. (1990) Production and repair of DNA damage in mammalian cells. *Health Phys.*
    59:15–22.
Lee, L.E.J., Clemons, J.H., Bechtel, D.G., Caldwell, S.J., Han, K.B., Pasitschniak-Arts, M.,
    Mosser, D.D. and Bols, N. (1993) Development and characterization of a rainbow trout liver
    cell ligne expressing cytochrome P450-dependent monooxygenase activity. *Cell Biol. Toxi-
    col.* 9:279–294.
Lemaire, P. and Livingstone, D.R. (1993) Pro-oxidant/antioxidant processes and organic xeno-
    biotic interactions in marine organisms, in particular the flounder *Platichthys flesus* and the
    mussel *Mytilus edulis*. *Trends Comp. Biochem. Physiol.* 1:1119–1150.
Lorenzen, A. and Okey, A.B. (1990) Detection and characterization of [^{3}H]2,3,7,8-tetrachloro-
    dibenzo-*p*-dioxin binding to Ah receptor in a rainbow trout hepatoma cell line. *Toxicol.
    Applied Pharmacol.* 106:53–62.
Loveland, P.M., Wilcox, J.S., Pawlowski, N.E. and Bailey, G.S. (1987) Metabolism and DNA
    binding of aflatoxicol and aflatoxin B1 *in vivo* and in isolated hepatocytes from rainbow trout
    (*Salmo gairdneri*). *Carcinogenesis* 8:1065–1070.
Masfaraud, J.-F., Devaux, A., Pfhol-Leszkowicz, A., Malaveille, C. and Monod, G. (1992) DNA
    adduct formation and 7-ethoxyresorufin *O*-deethylase induction in primary culture of rain-
    bow trout hepatocytes exposed to benzo[a]pyrene. *Toxicol. in vitro* 6:523–531.
McQueen, C.A., Way, B.M. and Williams, G.M. (1988) Genotoxicity of carcinogens in human
    hepatocytes: application in hazard assessment. *Toxicol. Appl. Pharmacol.* 96:360–366.
Miller, M.R., Blair, J.B. and Hinton, D.E. (1989) DNA repair synthesis in isolated rainbow trout
    liver cells. *Carcinogenesis* 10:995–1001.
Miranda, C.L., Collodi, P., Zhao, X., Barnes, D.W., Buhler, D.R. (1993) Regulation of cyto-
    chrome P450 expression in a novel liver cell line from zebrafish (*Brachydanio rerio*). *Arch.
    Biochem Biophys.* 305:320–327.
Monod, G., De Mones, A. and Fostier, A. (1993) Inhibition of ovarian microsomal aromatase
    and follicular estradiol secretion by imidazole fungicides in rainbow trout. *Mar. Environ. Res.*
    35:153–157.
Mousavi, Y. and Adlercreutz, H. (1993) Genistein is an effective stimulator of sex hormones-
    binding globulin production in hepatocarcinoma human liver cancer cells and suppresses
    proliferation of these cells in culture. *Steroids* 58:301–304.
Nikinmaa, M. (1992) How does environmental pollution affect red cell function in fish? *Aquat.
    Toxicol.* 22:227–238.

Nishimoto, M. and Varanasi, U. (1986) Metabolism and DNA adduct formation of benzo-[a]pyrene and the 7,8-dihydrodiol of benzo[a]pyrene by fish liver enzymes. *In*: Cooke M. and Dennis A.J. (Eds) *Polynuclear Aromatic Hydrocarbons*. 9th International Symposium on Chemistry, Characterization and Carcinogenesis. Battelle Press, Columbus, Ohio, pp 685–699.

Nishimoto, M., Yanagida, G.K., Stein, J.E., Baird, W.M. and Varanasi, U. (1992) The metabolism of benzo[a]pyrene by English sole (*Parophrys vetulus*): comparison between isolated hepatocytes *in vitro* and *in vivo*. *Xenobiotica* 22:949–961.

Olive, P.L., Banath, J.P. and Durand, R.E. (1990) Detection of etoposide resistance by measuring DNA damage in individual Chinese hamster cells. *J. Natl. Cancer Inst.* 82:779–783.

Oppen-Berntsen D.O., Gram-Jensen E., Walther B.T. (1992) Zona radiata proteins are synthesized by rainbow trout (*Oncorhynchus mykiss*) hepatocytes in response to estradiol-17β. *J. Endocrinol.* 135:293–302.

Pakdel, F., Feon, S., Le Gac, F., Le Menn, F. and Valotaire, Y. (1991) *In vivo* estrogen induction of hepatic estrogen receptor mRNA and correlation with vitellogenin mRNA in rainbow trout. *Mol. Cell. Endocrinol.* 75:205–212.

Pärt, P. (1996) *Primary cultures of teleost branchial epithelial cells*. 17th Annual conference of the European Society for Comparative Physiology and Biochemistry, Antwerp, Belgium, August 1996.

Pärt, P. and Bergström, E. (1995) Primary cultures of teleost branchial epithelial cells. *In*: Wood, C.M. and Shuttleworth, T J. (Eds) *Fish Physiology – Cellular and Molecular Approaches to Fish Ionic Regulation*. Academic Press, San Francisco, pp 207–227.

Pärt, P., Norrgren, L., Bergström, E. and Sjöberg, P. (1993) Primary culture of epithelial cells from rainbow trout gills. *J. Exp. Biol.* 175:219–232.

Payne, J.F. (1976) Field evaluation of benzo[a]pyrene hydroxylase induction as a monitor for marine pollution. *Science* 191:945–946.

Pesonen, M. and Andersson, T. (1991) Characterization and induction of xenobiotic metabolizing enzyme activities in a primary culture of rainbow trout hepatocytes. *Xenobiotica* 21:461–471.

Pesonen, M. and Andersson, T. (1992) Toxic effects of bleached and unbleached paper mill effluents in primary cultures of rainbow trout hepatocytes. *Ecotox. Environ. Safety* 24:63–71.

Pesonen, M., Goksøyr, A. and Andersson, T. (1992) Expression of P450IA1 in a primary culture of rainbow trout hepatocytes exposed to β-naphthoflavone or 2,3,7,8-tetrachlorodibenzo-p-dioxin. *Arch. Biochem. Biophys.* 292:228–233.

Petit, F., Valotaire, Y. and Pakdel, F. (1995) Differential functional activities of rainbow trout and human estrogen receptors expressed in yeast *Saccharomyces cerevisiae*. *Eur. J. Biochem* 233:584–592.

Purdom, C.E., Hardiman, P.A., Bye, V.J., Eno, N.C., Tyler, C.R. and Sumpter, J.P. (1994) Estrogenic effects of effluents from sewage treatment works. *Chem. Ecol.* 8:275–285.

Roche, H. and Bogé, G. (1993) Effects of Cu, Zn and Cr salts on antioxidant enzyme activities *in vitro* of red blood cells of a marine fish *Dicentrarchus labrax*. *Toxicol. in vitro* 7:623–629.

Roesijadi, G. (1992) Metallothioneins in metal regulation and toxicity in aquatic animals. *Aquat. Toxicol.* 22:81–114.

Sadar, M.D., Ash, R., Sundqvist, J., Olsson, P.-E. and Andersson, T.B. (1996) Phenobarbital induction of *CYP1A1* gene expression in a primary culture of rainbow trout hepatocytes. *J. Biol. Chem.* 271:17635–17643.

Safe, S.H. (1990) Polychlorinated biphenyls (PCBs), dibenzo-p-dioxins (PCDDs), dibenzofurans (PCDFs) and related compounds: environmental and mechanistic considerations which support the development of toxic equivalency factors (TEFs). *Crit. Rev. Toxicol.* 21:51–88.

Safe, S.H. (1995) Modulation of gene expression and endocrine response pathways by 2,3,7,8-tetrachlorodibenzo-p-dioxin and related compounds. *Pharmac. Ther.* 67:247–281.

Seglen, P.O. (1975) Preparation of isolated rat liver cells. *Meth. Cell Biol.* 13:29–83.

Shen, A.L., Fahl, W.E. and Jefcoate, C.R. (1980) Metabolism of benzo[a]pyrene by isolated hepatocytes and factors affecting covalent binding of benzo[a]pyrene metabolites to DNA in hepatocytes and microsomal systems. *Arch. Biochem. Biophys.* 204:511–523.

Shugart, L.R. (1990) Biological monitoring: testing for genotoxicity. *In*: McCarthy, J.F. and Shugart, L.R. (Eds) *Biomarkers of Environmental Contamination*. CRC Press, Boca Raton, Florida, pp 205–216.

Sikka, H.C., Steward, A.R., Zaleski, J., Kandaswami, C., Rutkowski, J.P., Kumar, S. and Gupta, R.C. (1993) Comparative metabolism of benzo[a]pyrene by the liver microsomes and freshly isolated hepatocytes of brown bullhead and carp. *Polycyclic Aromat. Comp.* 3:1087–1094.

Singh, N.P., McCoy, M.T., Tice, R.R. and Schneider, E.L. (1988) A simple technique for quantitation of low levels of DNA damage in individual cells. *Exp. Cell Res.* 175:184–191.

Skett, P. (1994) Problems in using isolated and cultured hepatocytes for xenobiotic metabolism/metabolism-based toxicity testing; Solutions? *Toxicol. in vitro* 8:491–504.

Smolarek, T.A., Morgan, S.L., Moynihan, C.G., Lee, H., Harvey, R.G., Baird, W.M. (1987) Metabolism and DNA adduct formation of benzo[a]pyrene and 7,12-dimethylbenz[a]anthracene in fish cell lines in culture. *Carcinogenesis* 8:1501–1509.

Stegeman, J.J. and Hahn, M.E. (1994) Biochemistry and molecular biology of monooxygenases: current perspectives on forms, functions, and regulation of cytochrome P450 in aquatic species. *In:* Malins D.C. and Ostrander G.K. (Eds) *Aquatic toxicology. Molecular, biochemical, and cellular perspectives.* CRC Press, Boca Raton, Florida, pp 87–206.

Steward, A.R., Zaleski, J., Gupta, R.C. and Sikka, H.C. (1989) Comparative metabolism of benzo[a]pyrene and benzo[a]pyrene-7,8-dihydrodiol by hepatocytes isolated from two species of bottom-dwelling fish. *Mar. Environ. Res.* 28:137–140.

Steward, A.R., Kandaswami, C., Chidambaram, S., Ziper, C., Rutkowski, J.P. and Sikka, H.C. (1990a) Disposition and metabolic fate of benzo[a]pyrene in the brown bullhead. *Environ. Toxicol. Chem.* 9:1503–1512.

Steward, A.R., Zaleski, J. and Sikka, H.C. (1990b) Metabolism of benzo[a]pyrene and (–)-trans-benzo[a]pyrene-7,8-dihydrodiol by freshly isolated hepatocytes of brown bullheads. *Chem. Biol. Interact.* 74:119–138.

Thornton, S.C., Diamond, L. and Baird, W.M. (1982) Metabolism of benzo[a]pyrene by fish cells in culture. *J. Toxicol. Environ. Health* 10:157–167.

Vaillant, C., Monod, G., Valotaire, Y. and Rivière, J.-L. (1989) Measurement and induction of cytochrome P-450 and monooxygenase activities in a primary culture of rainbow trout (*Salmo gairdneri*) hepatocytes. *C.R. Acad. Sci.* 308(III):83–88.

Wahli, W., Dawid, I.B., Ryffel, G.U. and Weber, R. (1981) Vitellogenesis and vitellogenin gene family. *Science* 212:298–304.

Walton, D.G., Acton, A.B. and Stich, H.F. (1984) DNA repair synthesis following exposure to chemical mutagens in primary liver, stomach and intestinal cells isolated from rainbow trout. *Cancer Res.* 44:1120–1121.

Winston, G.W. and Di Giulio, R.T. (1991) Prooxidant and antioxidant mechanisms in aquatic organisms. *Aquat. Toxicol.* 19:137–161.

Witters, H., Berckmans, P. and Vangenechten, C. (1996) *Primary culture of fish gill epithelial cells: an in vitro model to screen the effects of pollutants on cell differentiation.* 17th Annual Conference of the European Society for Comparative Physiology and Biochemistry, Antwerp, Belgium, August 27–31.

Yamagami, K., Hamazaki, S., Yasumasu, S., Masuda, K. and Luchi, I. (1994) Molecular and cellular basis of formation, hardening, and breakdown of the egg envelope in fish. *Int. Rev. Cytol.* 136:51–92.

Zaleski, J., Steward, A.R. and Sikka, H.C. (1991) Metabolism of benzo[a]pyrene and (–)-trans-benzo[a]pyrene-7,8-dihydrodiol by freshly isolated hepatocytes from mirror carp. *Carcinogenesis* 12:167–174.

Fish Ecotoxicology
ed. by T. Braunbeck, D. E. Hinton and B. Streit
© 1998 Birkhäuser Verlag Basel/Switzerland

Cytological alterations in fish hepatocytes following *in vivo* and *in vitro* sublethal exposure to xenobiotics – structural biomarkers of environmental contamination

Thomas Braunbeck

*Department of Zoology I, Aquatic Ecology and Ecotoxicology Section,
University of Heidelberg, Im Neuenheimer Feld 230, D-69120 Heidelberg, Germany*

Summary. Cytopathological alterations in hepatocytes of fish following exposure to xenobiotic compounds represent a powerful tool to reveal sublethal effects of chemicals and to elucidate underlying modes of action. The present communication reviews the available information about ultrastructural changes in fish liver as well as isolated hepatocytes; whereas the discussion of *in vivo* effects is primarily focused on data from rainbow trout (*Oncorhynchus mykiss*) and zebrafish (*Danio rerio*), the presentation of *in vitro* data has been restricted to results from experiments with rainbow trout hepatocytes due to a lack of data from studies with hepatocytes from other species. Both *in vivo* and *in vitro* exposure to xenobiotics results in sensitive, selective, and, especially in *in vitro* experiments, extremely rapid responses of hepatocytes, which, however, may be confounded by internal parameters (species, sex, age, hormonal status) and external parameters (temperature, nutrition, duration of exposure). Thus, transfer of results and conclusions from one experiment to another is usually not possible. Likewise, *in vitro* results may not necessarily be extrapolated to the situation in intact fish, and effects by acute toxic exposure cannot be translated into sublethal effects.

Hepatocellular reactions consist of both unspecific and substance-specific effects; in any case, as a syndrome, the complex of all changes induced by a given xenobiotic, is specific. Especially in the lower exposure range, most, if not all, ultrastructural alterations appear to be fully reversible; upon cessation of exposure, restitution of hepatocellular integrity is usually accomplished within a few days. Most early reactions of hepatocytes apparently serve functions within the general adaptation syndrome, which is induced to compensate for the misbalance in organismic homeostasis. Most ultrastructural alterations after sublethal exposure have, therefore, to be classified as indicators of adaptive processes and may be contrasted to irreversible, i.e., degenerative and truly pathological phenomena. These adaptive processes should, by definition, not have consequences at higher levels of biological organization; yet, as biomarkers, they are of ecotoxicological relevance. Thus, with regard to their (eco)toxicological significance, components of this nonspecific "general toxicant adaptation syndrome" may serve as early and sensitive warning signals of chemical exposure, whereas more specific changes may be of advanced diagnostic value and may serve as indices for the identification of xenobiotics.

Integration of cytopathological alterations into routine aquatic toxicology requires quantification by means of stereological techniques, which make structural data accessible for statistical analysis and comparable with quantitative techniques such as biochemistry and molecular biological methods. Implementation of cytopathological techniques into routine long-term investigations with fish gives credit to the principles of animal welfare and protection, since more-in-depth analysis of internal mechanisms of sublethal chemical contamination in addition to the study of externally overt symptoms of intoxication adds to a *refinement* of fish experiments. One step further towards *reduction* of animal experiments may be achieved by translation of environmental cytopathology to primary hepatocyte cell cultures.

Introduction

Both the number of compounds and the scale of production of organic chemicals have increased tremendously over the past 20 years. On a worldwide scale, overall chemical production has raised from 7.5 million tons in 1950 to over 150 million tons in 1980 (Nagel, 1988) to more than 300 million tons in 1990 (Braunbeck, 1989), with a sales turnover of 14.3 billion dollars per year in 1985 (Umweltbundesamt, 1992). As early as 1970, 107 substances were sold at a rate of more than 50 000 tons per year; 28 were marketed at more than 1 million tons per year (Schenck, 1986). Since both deliberate release and accidental spillage has been reduced considerably by an increasing body of national and international regulations, at least in developed countries incidences of acute intoxication of organisms have become rare. In the view of ecotoxicology and aquatic toxicology, however, concealed but insidious contamination by low doses of a vast number of environmental contaminants has become a more important threat to living biota than spectacular accidents such as the Rhine chemical spill in November 1986 (Güttinger and Stumm, 1990; Segner and Braunbeck, 1998).

In an attempt to prevent harm to organisms by xenobiotic compounds, more emphasis has therefore to be put on sublethal rather than on acutely toxic effects. From a technical point of view, given the comparatively low concentrations but chronic nature of most environmental pollution, a shift has become necessary from the relatively crude methods used for the determination of acute toxicity, which hardly fulfill the requirements of modern legislation and most international guidelines, to the development of more refined methods, which are not only capable of monitoring acute toxicity, but also of giving hints to the underlying mechanisms of toxicity (Segner and Braunbeck, 1998).

The search for mechanistic principles in toxicology may be misinterpreted to be in some contrast to the definition of ecotoxicology, since research in mechanisms of toxicity can hardly focus on alterations at the population and ecosystem levels, but will rather emphasize sub-individual levels in the biological hierarchy of organization, i.e., the levels of molecules, organelles and cells (Hinton and Laurén, 1990; Huggett et al., 1992). Whereas, for the time being, the ecological relevance of biomarker reactions may appear limited, alterations at the level of molecules and cells will manifest themselves very rapidly and may be expected to be of extraordinary sensitivity (Fig. 1). With the advent of the concept of biomarkers, a theoretical basis has been provided for the integration of molecular, biochemical, physiological and histological markers of contamination into the basic concepts of ecotoxicology (Huggett et al., 1992; McCarty and Shugart, 1990; Peakall, 1992). Following the definition by Hinton and Laurén (1990), for the purpose of this review, biomarkers will be defined as any contaminant-induced physiological and/or biochemical change in a

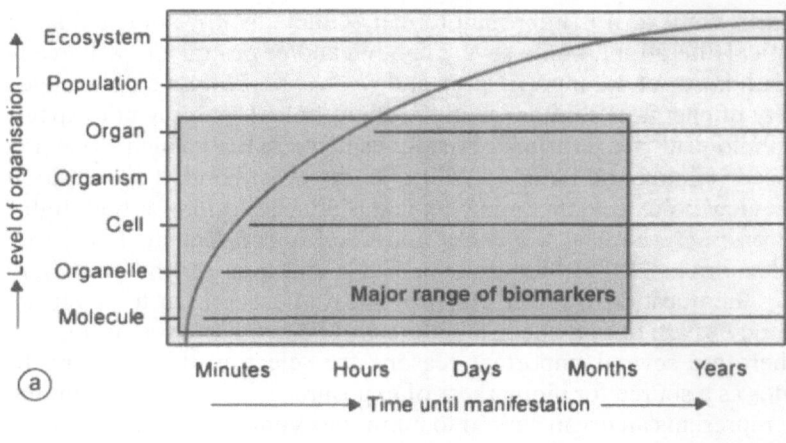

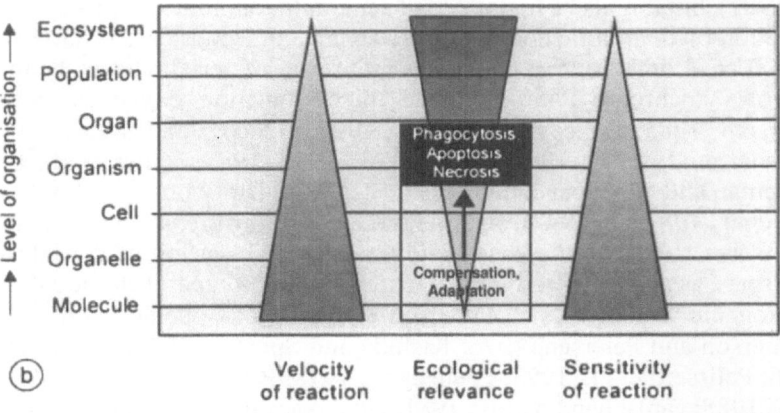

Figure 1. Levels of biological organization may be arranged in hierarchical order (a). The period until manifestation of an organism's reaction will increase with the organization level of the parameters investigated. While ecological relevance will increase with the level of organization, velocity and sensitivity of the reaction are likely to decrease (b). Whereas the range of biomarkers extends from single molecules over organelles and cells to the level of the individual organism, investigations centered around the cell as the smallest biological entity capable of expressing all characters of life are restricted to the sector between molecules and cells (a). Numerous reactions at the molecular levels are designed to compensate for the disturbance in homeostasis induced by xenobiotic exposure and, thus, have no or very little ecological relevance. However, once mechanisms of compensation are being overloaded, there will be a transition to deleterious processes eventually resulting in cell death including apoptosis, necrosis and phagocytosis by other cells (b; cf. Segner and Braunbeck, 1998).

not-too-sensitive organism, which leads to the formation of altered structure (a lesion) in the cells, tissues or organs. Thus, the major purpose of this review is a compilation of ultrastructural lesions in fish liver and hepatocytes following exposure to xenobiotics, i.e., morphological alterations at the level of the cell.

In this context, it is important to realize that any physiological and bio-chemical alteration, if only severe enough and/or protracted, will eventually result in structural modification and *vice versa*. Moreover, given the vast variety of chemical substances in the environment, we may not expect that morphological "fingerprints" exist for each toxicant to which a fish may be exposed (Hinton and Laurén, 1990). On the other hand, it seems doubtful if the view holds true that many toxicant-induced lesions at both light and electron microscopical levels are non-specific (cf. Couch, 1975; Meyers and Hendricks, 1985; Sindermann, 1984) and thus of limited diagnostic value; the question whether all cytological alterations in the liver and in hepatocytes are of non-specific nature will be a major issue of this review.

There are several important reasons for selecting the liver and hepatocytes as a source for biomarkers of exposure: The liver of vertebrates not only represents an organ central to numerous vital functions in basic metabolism (Arias et al., 1988; Gingerich, 1982; Moon et al., 1985; Phillips et al., 1987), but it is also a major site of accumulation, biotransformation and excretion of xenobiotic compounds (Meyers and Hendricks, 1985; Triebskorn et al., 1994a, b, 1997). Moreover, the liver of vertebrates is the major site of cytochrome P450-mediated mixed function oxygenase system (Goksøyr, 1985; Goksøyr and Förlin, 1992; Goksøyr and Solberg, 1987; Goksøyr et al., 1987, 1992; Lester et al., 1992, 1993; Miller et al., 1988; Stegeman and Kloepper-Sams, 1987; Stegeman and Woodin, 1984; Stegeman et al., 1987a, b, 1997), and bile produced within hepatocytes (Schmidt and Weber, 1973) and released into the proximal portion of the intestine may serve as carrier of conjugated toxicants. Since hepatocytes are the site of exogenic vitellogenesis (Anderson et al., 1996; Copeland et al., 1986; Emmerson and Petersen, 1976; Kishida and Specker, 1993; Maitre et al., 1986; Pelissero et al., 1993; Peute et al., 1978; Peyon et al., 1993; Vaillant et al., 1988; vanBohemen et al., 1981, 1982, Veranic and Pipan, 1995), toxicants may also be transferred to and stored in the yolk of the embryo. For these reasons, hepatocytes may be expected to be primary targets of toxic lesions, and selection of liver cells as appropriate targets should therefore provide an opportunity for the detection of suitable biomarkers of environmental pollution.

Light microscopical alterations of fish liver in consequence of exposure to xenobiotics have repeatedly been reviewed (e.g., Couch, 1975; Hinton and Laurén, 1990; Meyers and Hendricks, 1985; Sindermann, 1984). In contrast, data on liver ultrastructure of fish exposed to toxicants have not been compiled comprehensively so far, although a considerable body of information has been accumulated in recent years for *in vivo* effects of xenobiotics (Tabs. 1, 2). Data on morphological alterations in isolated hepatocytes after *in vitro* exposure to chemicals are scarce, but deserve particular attention, since these may give clues as to a distinction between intrinsic reactions of hepatocytes and alterations under the systemic control of the organism (Braunbeck, 1993, 1994b; Braunbeck et al., 1996, 1997a, b;

Table 1. Cytopathological studies on ultrastructural in fish liver after *in vivo* exposure to xenobiotic compounds

Species	Compound(s)	Reference
Brown trout (*Salmo trutta*)	Copper, zinc	Leland (1983)
Chinook salmon (*Oncorhynchus tshawytscha*)	PCB, petroleum hydrocarbon	Hawkes (1980)
Rainbow trout (*Oncorhynchus mykiss*)	Aflatoxin (AFB1)	Nunez et al. (1991), Wunder and Korn (1983)
	Allyl formate	Droy and Hinton (1988), Droy et al. (1989)
	Atrazine	Braunbeck (1994), Braunbeck et al. (1992a)
	Bis(tri-*n*-butyltin)oxide (TBTO)	Triebskorn et al. (1994)
	Cadmium	Förlin et al. (1986)
	4-Chloroaniline	Braunbeck (1994), Braunbeck et al. (1990b), Oulmi and Braunbeck (1996)
		Segner (1987)
	Copper	Braunbeck (1994)
	Diazinon	Arnold et al. (1995, 1996), Braunbeck (1994)
	Disulfoton	Arnold and Braunbeck (1994), Arnold and Braunbeck (1995), Braunbeck (1994)
	Endosulfan	Arnold and Braunbeck (1993), Arnold et al. (1995)
	Endosulfan and disulfoton	Moutou et al. (1996)
	Flumequine, oxolinic acid	Biagianti-Risbourg et al. (1996)
	Lindane (γ-hexachlorocyclohexane)	Oulmi et al. (1993), Oulmi et al. (1995)
	Linuron	Khan and Semalu (1995), Kontir et al. (1984)
	3-Methylcholanthrene	Braunbeck (1994)
	Ochratoxin	Scarpelli (1977)
	Phenylbutazone	Hacking et al. (1978), Norrgren et al. (1993)
	PCBs, polychlorinated naphthalenes, polychlorinated paraffines, polybrominated diphenyl ethers	
Atlantic tomcod (*Microgadus tomcod*)	Solvent-refined coal heavy distillate	Cormier (1986), Stoker et al. (1985)

Table 1 (continued)

Species	Compound(s)	Reference
Eel (*Anguilla anguilla*)	Cadmium	Biagianti et al. (1986), Gony et al. (1988)
	Dinotro-*o*-cresol	Braunbeck et al. (1987), Braunbeck and Völkl (1993)
	Cadmim and benzo[a]pyrene	Lemaire-Gony and Lemaire (1992)
Carp (*Cyprinus carpio*)	2,4-Dichlorophenoxy acetic acid	Benedeczky et al. (1984)
	Endosulfan	Braunbeck and Appelbaum (1998)
	Microcystin-LR	Raberg et al. (1991)
	Paraquat, Ultracid 40, copper sulphate	Benedeczky et al. (1986), Rojik et al. (1983)
	Phenol	Benedeczky and Nemcsok (1990)
Silver carp (*Hypophthalmichthys molitrix*)	Paraquat, Ultracid 40, copper sulphate	Rojik et al. (1983)
Goldfish (*Carassius auratus*)	Lead	Franchini et al. (1991)
Roach (*Rutilus rutilus*)	Copper	Segner (1987)
Zebrafish (*Danio rerio*)	4-Nitrophenol	Braunbeck et al. (1989)
	Atrazine	Braunbeck et al. (1992a)
	DDT	Weis (1974)
	Lindane (γ-hexachlorocyclohexane)	Braunbeck et al. (1990a)
	2,4-Dichlorophenol	Oulmi et al. (1994)
	Bis(tri-*n*-phenyltin)acetate (TPTA)	Oulmi et al. (1995), Strmac and Braunbeck (1996, 1998)
Golden ide (*Leuciscus idus melanotus*)	Dinotro-*o*-cresol	Braunbeck and Völkl (1993)
	4-Chloroaniline	Braunbeck and Segner (1991)
Channel catfish (*Ictalurus punctatus*)	PCB	Hinton et al. (1978), Lipsky et al. (1978)
European catfish (*Silurus glanis*)	Paraquat, Ultracid 40, copper sulphate	Rojik et al. (1983)

Species	Compound(s)	Reference
Brown bullhead (*Ictalurus nebulosus*)	Diethylnitrosamine	Hampton et al. (1988)
	Solvent-refined coal heavy distillate	Stoker et al. (1985)
South American catfish (*Pimelodus maculatus*)	Cadmium	Bonates and Ferri (1980), Ferri (1980), Ferri and Macha (1980)
Sheepshead minnow (*Cyprinodon variegatus*)	N-Nitrosodiethylamine	Couch (1991), Couch (1993), Couch and Courtney (1985), Couch and Courtney (1987)
Green sunfish (*Lepomis cyanellus*)	Sodium arsenate	Sorensen (1976), Sorensen et al. (1980, 1985)
Milkfish (*Chanos chanos*)	Copper	Segner and Braunbeck (1990)
Medaka (*Oryzias latipes*)	Adriamycin	Harada et al. (1992)
	Diethylnitrosamine	Braunbeck et al. (1992b), Bunton (1990), Hinton and Laurén (1990), Hinton et al. (1984, 1985b, 1988), Ishikawa et al. (1975), Laurén et al. (1990)
	Methylazoxymethanol acetate	Harada et al. (1988), Hawkins et al. (1986) Hinton et al. (1984)
Guppy (*Poecilia reticulata*)	DDT	Weis (1974)
	Methylazoxymethanol acetate	Hawkins et al. (1986)
	Iron	Segner and Storch (1985)
Grey mullet (*Liza aurata, L. ramada*)	Atrazine	Affandi and Biagianti (1987), Biagianti (1986), Biagianti (1987), Biagianti-Risbourg and Bastide (1995) Biagianti-Risbourg (1991, 1992)
Mullet (*Mugil cephalus*)	2-Methylcholanthrene	Schoor and Couch (1979)
See bas (*Dicentrarchus labrax*)	Benzo[a]pyrene	Lemaire et al. (1992)
Snake-headed fish (*Channa puntatus*)	Copper	Khangarot (1992)

Zahn et al., 1996a). Thus, a further focus of this communication will be put on the comparison of cytopathological effects in hepatocytes after *in vivo* and *in vitro* exposure.

The primarily qualitative nature of morphological studies imposes serious restrictions on their potential integration with more quantitatively oriented disciplines in environmental sciences (Braunbeck, 1994a; Braunbeck and Segner, 1991; Hinton et al., 1987; Peute et al., 1985; Segner and Braunbeck, 1990b, c; Veranic and Pipan, 1995). Whereas in mammalian histology and pathology, quantitative stereological methods have already been introduced from the late 1960s (e.g., Blouin, 1977; Blouin et al., 1977; Bolender, 1979; Bolender and Weibel, 1973; Hess et al., 1973; Loud, 1968; Pfeifer, 1973; Rohr and Riede, 1973; Stäubli et al., 1969; Weibel, 1979; Weibel et al., 1969), quantitative approaches have been scant in teleost histology (Braunbeck, 1994a; Braunbeck and Storch, 1992; Hampton et al., 1989; Hinton et al., 1985a; Peute et al., 1985; Segner and Braunbeck, 1990a; Rocha et al., 1995, 1997; Sorensen et al., 1985; Veranic and Pipan, 1995). Thus, this communication will also put emphasis on the quantification of ultrastructural data, which are available for toxicant-induced cytopathology in fish liver and hepatocytes.

Electron microscopical alterations in the liver of fish following *in vivo* exposure to xenobiotic compounds

So far, only a limited number of little more than 20 fish species has been used as model systems in laboratory studies on ultrastructural effects in fish liver after exposure to xenobiotics (Tab. 1). With about 30 experiments published so far, rainbow trout (*Oncorhynchus mykiss*) is by far the best-investigated fish species not only with respect to liver cytopathology, but also with regard to confounding internal factors such as hormones (Ng et al., 1984; Saez et al., 1984a; Vera et al., 1993), age (Biagianti-Risbourg et al., 1996; Barni et al., 1985; Braunbeck et al., 1990c, 1992a Leatherland and Sonstegard, 1988; Kalashnikova and Kadilov, 1991; Ng et al., 1986; Robertson and Bradley, 1991, 1992; Vernier, 1975; Yamamoto and Egami, 1974) or sexual maturation (vanBohemen et al., 1981), as well as external parameters such as temperature (e.g., Berlin and Dean, 1967; Hodson and Blunt, 1981; Howe et al., 1994; Storch et al., 1984b) and nutrition (e.g., Andersson et al., 1985; Gas, 1973; Hilton, 1982; Langer, 1978; Leatherland, 1982; Mehrle et al., 1974; Moon, 1983; Segner, 1985; Storch et al., 1984a). Whereas in most studies with rainbow trout juvenile fish were used in order to avoid hepatic sexual dimorphism as a confounding factor, most long-term exposures with, e.g., zebrafish were carried out with either sex, so that a distinction between female and male individuals was necessary (Braunbeck et al., 1989, 1990b, c, 1992a; Oulmi and Braunbeck, 1996; Strmac and Braunbeck, 1998). In field studies (see below), the number of studies using electron microscopy as a tool to identify liver lesions is even more restricted (cf. Tab. 12); however, predominance of rainbow trout is not as evident as in laboratory studies.

Hepatic cytopathology of rainbow trout (*Oncorhynchus mykiss*) after *in vivo* exposure to xenobiotics: the salmonid model

Ultrastructure of control rainbow trout liver

The ultrastructure of rainbow trout liver has repeatedly been described in detail (Berlin and Dean, 1967; Braunbeck, 1994a, 1995, 1996; Braunbeck et al., 1990c, 1992a; Chapman, 1981; Hacking et al., 1978; Hampton et al., 1985, 1988a, b, 1989; Leatherland and Sonstegard, 1983; Scarpelli et al., 1963). Rainbow trout hepatocytes are typical epithelial cells with a size range from 11 to 25 µm characterized by distinct polarity, with the apical pole facing the biliary system and the basal pole located *vis-a-vis* to the endothelial lining of sinusoids (Fig. 2). Cells are regularly arranged in tubules with 3–5 hepatocytes contributing to form the central bile canaliculus (*tubular liver architecture*; Hampton et al., 1985, 1988a) and characterized by a conspicuous separation into extended peripheral glycogen fields with few lipid inclusions and perinuclear organelle-containing portions of cytoplasm (*intracellular compartmentation*; Braunbeck et al., 1990c). Thus, hepatocytes of untreated rainbow trout display an exceptionally regular arrangement of cellular components.

The ovoid nucleus bears little randomly scattered heterochromatin and a distinct slightly eccentrically located nucleolus of considerable electron density and usually honeycomb-like appearance displaying a clear-cut separation into a pars granulosa and a pars fibrosa. Extensive stacks of up to 40 highly ordered non-fenestrated cisternae of the rough endoplasmic reticulum (RER) interspersed with few mitochondria form an almost continuous sheath around the centrally located nucleus occasionally with a single layer of spherical to ovoid mitochondria intervening between nucleus and ER (Fig. 3); additional RER cisternae connected to the central RER stacks in a spoke-like fashion may be found immediately underneath the plasma membrane. In contrast, smooth endoplasmic reticulum (SER) in its typical development as an irregular undulating and anastomosing network of tubular and vesicular profiles is usually restricted to minute peribiliary areas.

The RER piles are regularly bordered by large amounts of spherical peroxisomes, and the RER stacks are regularly separated from glycogen fields by large amounts of peroxisomes. Up to 45 peroxisomes with a diameter ranging from 0.15 to 1.2 µm can be counted on a single cell section. The staining intensity of the peroxisomes after incubation with the alkaline diaminobenzidine medium (LeHir et al., 1979) is of considerable variabili-

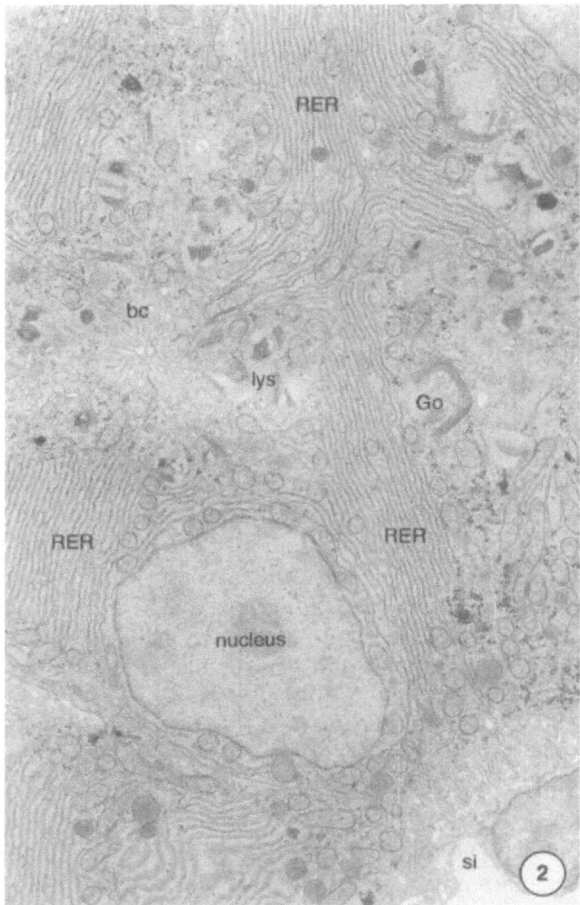

Figure 2. In the liver of non-contaminated rainbow trout (*Oncorhynchus mykiss*), hepatocytes are characterized by a very regular arrangement of large piles of non-fenestrated cisternae of the rough endoplasmic reticulum (RER) around a centrally located nucleus. Normally, the RER stacks are separated from the nucleus by a layer of mitochondria and peroxisomes, which have been visualized by incubation in alkaline diaminobenzidine. At the apical cell pole, the RER sheath is interrupted by the peribiliary complex including Golgi fields (Go), which give rise to Golgi vesicles migrating towards the pluricellular bile canaliculus (bc), and a limited number of lysosomes (lys). The basal pole of the hepatocytes facing the blood sinusoids (si) extends a variable number of microvilli into the space of Disse between hepatocytes and endothelial cells (en). Hepatocyte of control rainbow trout; magnification × 6000.

ty, suggesting variable catalase activities in adjacent hepatocytes. If compared to control zebrafish (*Danio rerio*; see below), staining for catalase activity in rainbow trout is low. As in other fish species, a peroxisomal core, which is typical of peroxisomes in mammals except some species (Angermüller and Fahimi, 1981, 1983, 1986; LeHir et al., 1979; Litwin et al., 1988; Völkl et al., 1988), could not be revealed so far. The uniform array of

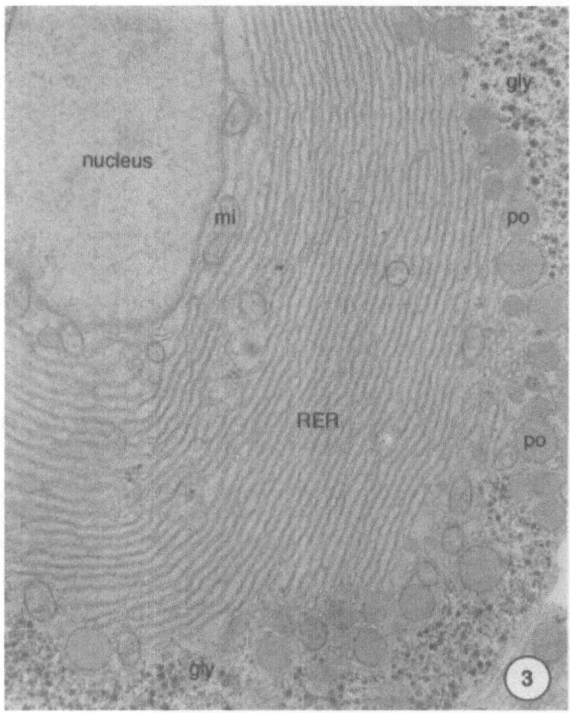

Figure 3. At higher magnifications, rainbow trout (*Oncorhynchus mykiss*) hepatocytes display their characteristic separation into the central organelle-containing portion containing mitochondria (mi), peroxisomes (po) and RER cisternae around the nucleus, and a peripheral field mainly comprising glycogen (gly) as the major storing product in rainbow trout liver. Typically, this "cytoplasmic compartmentation" (Braunbeck et al., 1989) is even more pronounced by a layer of peroxisomes separating the two major areas. Hepatocyte of control rainbow trout; reprinted from Braunbeck et al. (1992b); magnification × 9200.

peroxisomes may be interspersed with small spherical to ovoid mitochondria with an average diameter of 0.5–1 μm. In rainbow trout hepatocytes, only occasionally elongated or irregularly shaped mitochondrial profiles can be detected.

Between the nucleus and the bile canaliculus, the RER sheath is interrupted by strongly developed Golgi fields consisting of several stacks of 3–5 Golgi cisternae displaying a marked polarity and budding off many vacuoles of variable size containing numerous VLDL (very low density lipoprotein; cf. Vernier, 1975) granules. Especially in older individuals, the peribiliary area may exhibit abundant lysosomes of considerable diversity with regard to their size and contents, frequently arranged in clusters between the Golgi complexes and the bile canaliculi. In controls, the cell periphery is occupied by extensive storage fields, mainly glycogen deposits; lipid inclusions are scant.

Ultrastructure of rainbow trout liver after exposure to xenobiotics

Table 2 provides a compilation of data on ultrastructural alterations in hepatocytes from a selection of long-term experiments, in which rainbow trout were exposed to reference compounds (4-chloroaniline, an intermediate in the manufacture of dyestuffs, pigments, antioxidants, various pharmaceuticals and agricultural chemicals), different pesticides (endosulfan, atrazine, diazinon, disulfoton, linuron, and bis(tri-*n*-butyltin)oxide), as well as the mycotoxin ochratoxin. Since in an earlier report, a detailed comparative analysis of the hepatocellular effects in these experiments was given (Braunbeck, 1994a), only a brief outline of the major cytological effects will be provided.

Liver parenchyma. From the data in Table 2 it is evident that increased intra- and interindividual variability of the liver parenchyma as well as disturbance of the highly regular intracellular compartmentation as defined by Braunbeck et al. (1990c) are characteristic of contamination by many toxicants and, thus, have to be classified as unspecific markers of chemical exposure. It should be noted, however, that especially the deterioration of the regular compartmentation into perinuclear organelle-containing portions of the cytoplasm and peripheral storage fields is a very early signal of disturbance of hepatocellular homeostasis, since it not only occurs as a consequence of increased chemical burden, but also following changes in ambient temperature (Berlin and Dean, 1967; Braunbeck et al., 1987; Storch et al., 1984b), malnutrition (Bac et al., 1983; Langer and Storch, 1978; Leatherland, 1982; Mehrle et al., 1974; Moon, 1983; Segner and Braunbeck, 1988; Segner and Juario, 1986; Segner and Möller, 1984; Segner et al., 1984, 1993; Storch and Juario, 1983; Storch et al., 1984a, b) as well as during adaption to seasonal variation (Braunbeck and Segner, 1991; March and Reisman, 1995; Quaglia, 1976a, b; Saez et al., 1984a, b; Segner and Braunbeck, 1990a; vanBohemen et al., 1981), aging and sexual maturation (Barni et al., 1985; Takahashi et al., 1977; Yamamoto and Egami, 1974). Peribiliary as well as perisinusoidal accumulation of cytoplasmic vacuoles, however, could only be documented for diazinon intoxication and should thus – at least on the basis of data available so far – be classified as specific for this compound (Braunbeck, 1994a).

Nuclei. Whereas control animals are almost free of binucleate cells, stimulation of mitosis (Fig. 4) and subsequent augmentation of binucleate cells were observed for all toxicants except for endosulfan, bis(tri-*n*-butyltin)oxide (TBTO) and disulfoton, and, thus, appear to be an unspecific feature typical of intoxication by xenobiotic compounds (Tab. 2). As further non-specific symptoms, multiple nucleoli are frequent, and, in parallel to corresponding alterations in the cytoplasm, the nuclear envelope is often distorted. In contrast, free macrovesicular (not membrane-bound) lipid

Table 2. Cytological alterations in the liver of rainbow trout (*Oncorhynchus mykiss*) after prolonged sublethal *in vivo* exposure to 4-chloroaniline, endosulfan, atrazine, diazinon, disulfoton, linuron, tributyltin oxide (TBTO), and ochratoxin

	4-Chloroaniline	Endosulfan	Atrazine	Diazinon	Disulfoton	Linuron	TBTO	Ochratoxin
Variability of liver parenchyma								
Increased intraindividual parenchyma	–	0.1	40	40	–	30	–	0.1
Increased intraindividual parenchyma	–	0.05	40	–	–	30	–	0.1
Intracellular compartmentation								
Disturbance of compartmentation	200	0.1	40	–	–	30	2	0.1
Perisinusoidal vacuolation	–	–	–	*40*	–	–	–	–
Peribiliary vacuolation	–	–	–	*40*	–	–	–	–
Nuclei								
Stimulation of mitosis	200	–	160	40	–	30	–	1
Augmentation of binucleate cells	200	–	160	40	–	–	–	1
Free nuclear lipid inclusions	1000	–	–	–	–	120	–	0.1
Intranuclear steatosis	–	–	–	–	–	–	–	*1*
Intranuclear myelin formation	*200*	–	–	–	–	–	–	–
Deformation of nuclear envelope	200	0.01	80	–	–	30	0.5	–
Inflation of nuclear envelope	200	–	–	–	–	–	2	–
Augmentation of nucleoli	200	0.01	160	40	–	120	0.5	–
Rough endoplasmic reticulum								
Increased heterogeneity	200	0.01	40	–	–	120	0.5	0.1
Reduction	200	–	160	–	0.1*	–	2	0.1
Fragmentation, vesiculation	200	0.05	40	–	–	120	0.5	0.1
Dilation of cisternae	200	0.01	40	–	–	120	–	0.1
Fenestration of cisternae	–	–	–	–	–	–	–	*1*
Collapse of cisternae	–	–	–	–	–	–	–	*0.1*
Steatosis	–	–	–	–	–	120	–	0.1
Formation of RER whorls	–	–	–	–	–	30	–	1
Transformation into myelin bodies	–	–	–	–	–	–	–	*1*
Smooth endoplasmic reticulum								
Proliferation	–	0.05	160	40	5	30?	2	1

Table 2 (continued)

	4-Chloroaniline	Endosulfan	Atrazine	Diazinon	Disulfoton	Linuron	TBTO	Ochratoxin
Golgi apparatus								
Increased heterogeneity	200	0.05	–	200	–	120	–	–
Hypertrophy	–	–	–	200	5	120	–	–
Fenestration of cisternae	–	*0.05*	–	–	–	–	–	–
Dilation of cisternae	–	–	–	200	–	*120*	–	–
Stimulation of VLDL synthesis	–	–	–	–	–	–	–	1
Decrease of VLDL synthesis	200	0.05	–	–	–	–	–	*1*
Reduction of Golgi vesicles	–	–	–	200	–	–	–	1
Augmentation of Golgi vesicles	–	–	–	–	–	–	–	–
Collapse of cisternae	–	–	–	–	–	–	–	–
Mitochondria								
Proliferation	200	–	–	–	0.1*	120	–	–
Increased heterogeneity	200	0.05	40	40	–	120	–	0.1
Increase in volume	–	–	–	*40*	–	–	–	–
Dilation of intermembranous space	–	0.05	–	–	–	30	–	–
Induction of longitudinal cristae	–	–	*40*	–	–	–	–	*1*
Collapse of mitochondria	–	–	–	–	–	–	–	–
Intramitochondrial myelin formation	–	0.05	160	–	–	–	–	–
Partial lysis	–	*0.1*	–	–	–	–	–	–
Peroxisomes								
Proliferation	*200*	0.01	–	200	–	120	–	–
Reduction	–	–	–	–	–	–	–	–
Increased heterogeneity	–	0.05	–	200	–	120	–	–
Increased catalase activity	–	*0.05*	–	–	–	–	–	–
Reduction in catalase activity	*200*	–	–	–	–	–	–	–
Heterogeneity in catalase distribution	–	*0.05*	–	40?	–	–	–	–
Cluster formation	–	0.01	–	–	–	*120*	–	–
Tail formation (division?)	–	–	–	–	–	*120*	–	–
Myelin formation in matrix	–	*0.05*	–	–	–	–	–	–
Accumulation around lipid	–	–	–	–	–	*120*	–	–

	4-Chloroaniline	Endosulfan	Atrazine	Diazinon	Disulfoton	Linuron	TBTO	Ochratoxin
Lysosomal elements								
Proliferation	–	0.05	80	200	0.1	30	2	–
Reduction	*200*	–	160	–	–	–	–	–
Induction of new lysosome types	–	–	–	–	–	120	–	–
Crystal formation in matrix	*200*	–	–	200	–	–	–	–
Phospholipidosis	–	–	–	200	–	120	0.5	1
Myelinated bodies	–	0.05	40	–	–	30	–	–
Induction of autophagosomes	–	0.01	40	40	–	30	–	–
Induction of multivesicular bodies	–	0.05	160	–	–	120	–	–
Induction of glycogenosomes	*200?*	–	–	40	–	–	–	–
Increase of lipid deposits	–	–	–	–	–	30	0.5	–
Perisinusoidal polarisation of lipids	–	–	–	–	–	*30*	–	–
Steatosis	–	–	–	–	–	120	–	0.1
Increased density of lipid droplets	–	–	–	–	–	–	–	*0.1*
Formation of cholesterol crystals	*200*	–	–	–	–	–	–	–
Storage materials								
Formation of lipid clusters	–	–	–	–	–	*120*	–	–
Nuclear lipid inclusions	–	–	–	–	–	*120*	–	–
Decrease in glycogen stores	*200*	0.1	40	–	5	120	2	0.1
Glycogen condensation	*200*	–	–	–	–	–	–	–
Non-parenchymal cells								
Immigration of macrophages	*200*	0.05	40	200	–	30	–	1
Formation of macrophage centres	*200*	0.1	80	–	–	120	–	1
Immigration of granulocytes	–	–	*40*	–	–	–	–	–
Increased glycogen phagocytosis	*200*	0.05	80	–	–	120	–	1
Phagocytosis of entire cells	–	0.1	80	–	–	30	–	–
Proliferation of Ito cells	–	–	–	*40*	–	–	–	–

Data presented as lowest concentration (in µg/L) inducing the respective alteration. Specific alterations within the given set of experiments are printed in italics. * Transient reaction at 0.1 and 1 µg/L only.
Literature: 4-chloroaniline: Braunbeck et al. (1990); endosulfan: Arnold et al. (1996); atrazine: Braunbeck et al. (1992); Diazinon: Braunbeck (1994); Disulfoton: Arnold and Braunbeck (1995), Arnold et al. (1996); Linuron: Oulmi et al. (1995); TBTO: Braunbeck (1994); Ochratoxin: Braunbeck (1994).

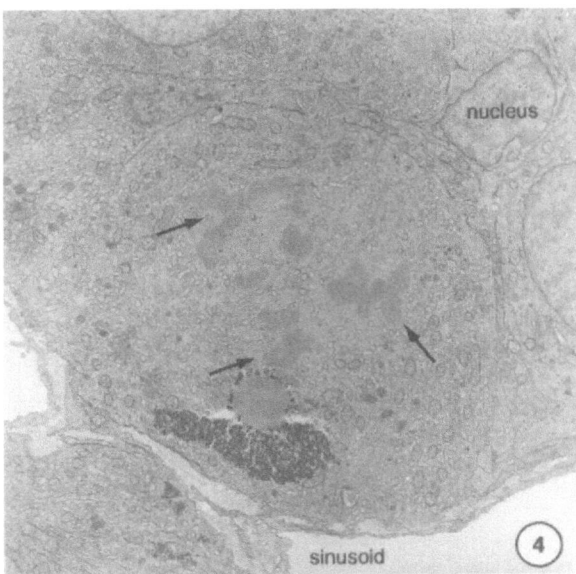

Figure 4. Whereas hepatocytes in control rainbow trout (*Oncorhynchus mykiss*) are almost free of binucleate cells, stimulation of mitosis and subsequent augmentation of binucleate cells can regularly be observed and can, thus, be classified as an unspecific feature typical of intoxication by xenobiotic compounds. At arrows, note heavy clumps of highly condensed chromatin in chromosomes. Hepatocyte of rainbow trout exposed to 1 mg/L 4-chloroaniline; reprinted from Braunbeck et al. (1990c); magnification × 3000.

inclusions could only be observed after exposure to 4-chloroaniline, linuron, and ochratoxin; intranuclear steatosis (i.e., membrane-bound lipid accumulation; Baglio and Farber, 1965) is restricted to ochratoxin, and intranuclear myelin formation is unique to 4-chloroaniline intoxication.

Rough endoplasmic reticulum. Although mostly not specific of particular compounds, changes in the ultrastructural appearance of the rough endoplasmic reticulum (RER) represent an extremely sensitive indicator of xenobiotic exposure in rainbow trout. Reduction in the overall volume as well as fragmentation, vesiculation and dilation of RER cisternae as symptoms of increased morphological heterogeneity range among the most conspicuous cytopathological features in rainbow trout hepatocytes (Fig. 5). Notable exceptions are diazinon, which fails to induce any RER alteration (Tab. 2), and ochratoxin, which stimulates additional modifications such as extreme fenestration and fusion of opposite cisternal membranes (Fig. 6), complete degranulation of the RER system (Fig. 7), as well as transformation of parts or the entire RER into huge myelin bodies (Figs. 8, 9). Steatosis in RER is observed with linuron and ochratoxin. As a consequence, the syndrome of cytological modifications in the RER system consists of both specific and non-specific symptoms; this situation is representative of

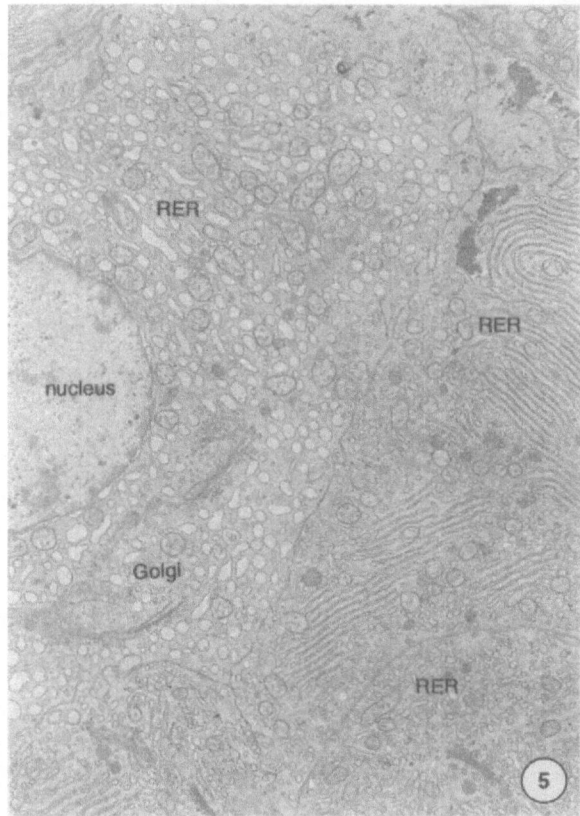

Figure 5. In contrast to the almost geometric array of RER in hepatocytes of control rainbow trout (*Oncorhynchus mykiss*), exposure to organic chemicals regularly results in dilation, fractionation and vesiculation of the RER cisternae, an unspecific alteration, which could be observed following exposure to, for example, atrazine, 4-chloroaniline, endosulfan, diazinon, and linuron. As a rule, changes in the nuclear envelope correspond to parallel modifications in the RER system. Note the increased heterogeneity between single hepatocytes, i.e., the increase in intraparenchymal heterogeneity. Hepatocyte of rainbow trout exposed to 1 mg/L 4-chloroaniline; reprinted from Braunbeck et al. (1990c); magnification × 5000.

most cytopathological alterations observed in rainbow trout hepatocytes (Braunbeck, 1994a).

Smooth endoplasmic reticulum. Except for 4-chloroaniline, all substances induce an SER augmentation; the degree of proliferation, however, varies from low in the linuron experiment to appreciable with diazinon. It should further be noted that in many cases (especially ochratoxin) differentiation between degranulated RER (cf. Fig. 7) and true (i.e., functionally active) SER is practically impossible by means of electron microscopy.

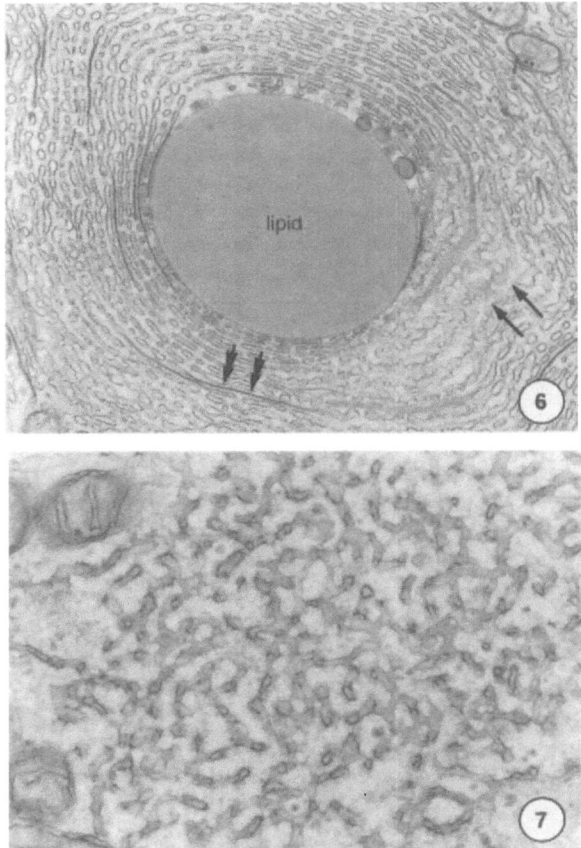

Figures 6 and 7.   Following exposure of rainbow trout (*Oncorhynchus mykiss*) to very low
doses of food-borne ochratoxin, a mycotoxin synthesized by *Aspergillus ochraceus*, the originally non-fenestrated cisternae of the rough endoplasmic reticulum (RER) show extreme
fenestration, which can only be discriminated from vesiculation in sections tangential to the
cisternae (Fig. 6, at arrows), fusion of opposite cisternal membranes (Fig. 6, double arrows) and
complete degranulation of the RER system, which may appear transformed into an irregular
network of short anastomosing cisternae and vesicles strongly resembling smooth endoplasmic
reticulum (Fig. 7). Hepatocyte of rainbow trout exposed to 0.5 mg ochratoxin per kg food.
Figure 6: magnification × 14 300; Fig. 7: × 39 700.

Golgi fields. A typical, but unspecific character of contaminated rainbow
trout liver is increased Golgi heterogeneity, which, however, in contrast to
the RER reaction, is of diverse nature: 4-chloroaniline and endosulfan
induce reduced VLDL secretion, whereas linuron stimulates VLDL production in hypertrophic Golgi fields; diazinon provokes an increased number of Golgi vesicles without stimulation of VLDL production; endosulfan
prompts fenestration, linuron induces dilation of Golgi cisternae; ochratoxin, finally, apparently causes complete morphological disintegration of

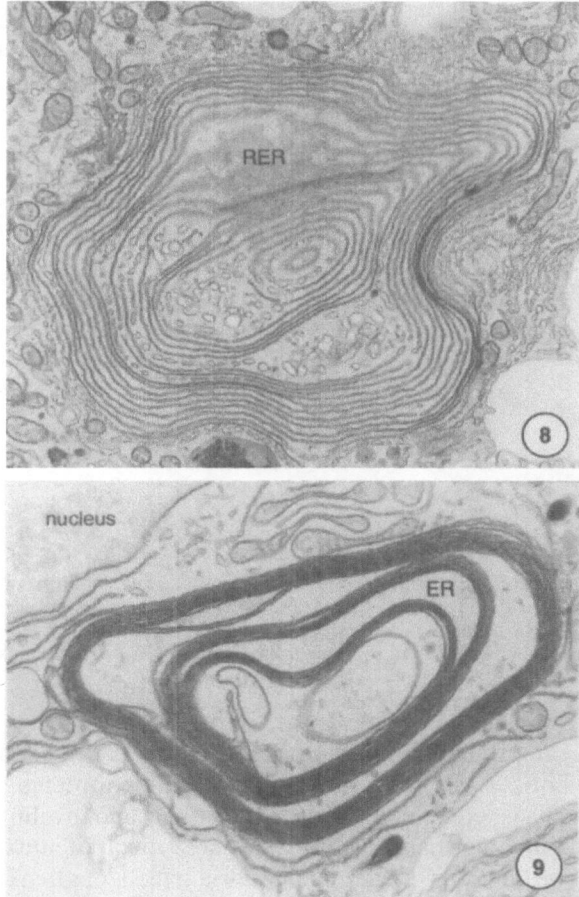

Figures 8 and 9: Additional alterations in the rough endoplasmic reticulum of rainbow trout (*Oncorhynchus mykiss*) exposed to the mycotoxin ochratoxin include transformation into huge concentric membrane whorls (Fig. 8) characterized by subsequent condensation into myelinated bodies (Fig. 9). In hepatocytes with progressive damage, the cytoplasm usually appears translucent. Hepatocyte of rainbow trout exposed to 1 mg ochratoxin per kg food. Figure 8: magnification × 11 200; Fig. 9: × 10 800.

dictyosomes and cessation of normal Golgi functions. Both atrazine and TBTO fail to induce any changes in the Golgi fields.

Mitochondria. In healthy fish, mitochondria are predominantly located as spherical to only slightly elongated profiles in intimate association with RER lamellae and peroxisomes close to the nucleus or the edges of RER stacks. Again, generally increased heterogeneity is the sole feature common to all treatments (Tab. 2). Both linuron and 4-chloroaniline evoke conspicuous proliferation of mitochondria; endosulfan exposure results in

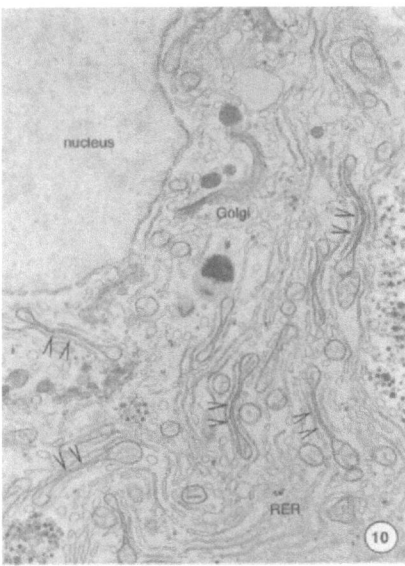

Figure 10. An unusual change in mitochondria of hepatocytes in rainbow trout (*Oncorhynchus mykiss*) exposed to low doses of food-borne ochratoxin is the acquisition of an extremely dumb-bell-like shape (at arrowheads). In the central portion of the dumbbell, mitochondrial cristae are no longer discernible. Hepatocyte of rainbow trout exposed to 0.1 mg ochratoxin per kg food; magnification × 9800.

dilation of the intermembranous space, formation of extended myelin-like membrane whorls in different mitochondrial compartments and eventual mitochondrial lysis; elongated longitudinal cristae and myelin formation in the matrix and the intermembranous space are typical of atrazine contamination; diazinon results in megamitochondria with a length of up to 12 μm; and, with ochratoxin, mitochondria appear to collapse, i.e., the matrix volume is reduced to zero (Fig. 10). This observation indicates that mitochondrial reactions to xenobiotics are – in contrast to those of the endoplasmic reticulum – much more diverse and are, thus, of higher diagnostic value with respect to attempts to identify the effects of specific compounds in complex chemical mixtures.

Peroxisomes. Peroxisomal proliferation is weak after diazinon and linuron treatment, but most conspicuous after endosulfan exposure. In most cases, proliferation of peroxisomes is accompanied by the formation of slender tail-like projections, which might be interpreted as the morphological expression of peroxisome division. In hepatocytes of rainbow trout subjected to endosulfan, peroxisomes also form aggregates of up to 20 profiles and display intensely staining matrix portions strongly reminding of the core typical of mammalian peroxisomes. In contrast, the number of peroxisomes is reduced following 4-chloroaniline exposure of rainbow trout.

Lysosomes. With respect to the lysosomal compartment, rainbow trout hepatocytes usually show quite stereotype reactions: Except for 4-chloroaniline (reduction) and ochratoxin (no reaction), all chemicals tested entail a proliferation of lysosomes accompanied by an augmentation of other lysosomal elements (myelinated and multivesicular bodies, autophagosomes and glycogenosomes). Following exposure to 4-chloroaniline, lysosomes display long, slender crystalline inclusions in their matrix, and, with both diazinon and linuron, lysosomes contain stacks of obliterating membrane fragments, a phenomenon strongly reminiscent of what has been described as phospholipidosis in mammals (cf. Phillips et al., 1987). Glycogenosomes, i.e., organelles degrading glycogen within membrane-bound vacuoles most likely derived from lysosomes (Gas and Pequignot, 1972; Gas and Serfaty, 1972), can be observed after rainbow trout exposure to diazinon and, much more conspicuously, to 4-chloroaniline (Braunbeck et al., 1990c). Since exactly the same cytopathological phenomena can be observed after metabolic disturbance by other internal or external factors, hepatocellular alterations in the lysosomal compartment can only be used as a general indicator of metabolic misbalance.

Storage products. Likewise, since hepatic storage products underlie the influence of a multitude of internal and external parameters, chemical-specific effects can hardly be expected. Except for linuron and ochratoxin, which stimulate an augmentation of cytosolic lipid deposits (macrovesicular lipid accumulation, i.e., aggregation of lipoid materials to cytoplasmic vacuoles without a limiting membrane) and steatosis (microvesicular lipid accumulation, i.e., within cisternae of the rough and smooth endoplasmic reticulum; Baglio and Farber, 1965), hardly any change can be observed with respect to lipid reserves. Following exposure to linuron, hepatocytes display a conspicuously inhomogeneous distribution of lipid droplets, which accumulate at the basal (sinusoidal) face of the hepatocytes. In contrast, glycogen stores are depleted after exposure to 4-chloroaniline, endosulfan, atrazine, linuron and ochratoxin. 4-Chloroaniline is further marked by accumulation of cholesterol-like crystals and conspicuous condensation of glycogen into very dense masses, which eventually end up in glycogenosome-like particles.

Non-parenchymal liver elements. In contrast to hepatocytes, non-parenchymal cells display little change in consequence of exposure to xenobiotic compounds. Whereas the liver of unimpaired rainbow trout is almost clear of migrating phagocytic cells, in all experiments except for disulfoton and TBTO, toxicant exposure results in varying degrees of macrophage immigration and subsequent formation of macrophage centers. Elevated phagocytic activity, both of glycogen, cellular debris and entire hepatic cells, is evident. Exceptional, however, are the immigration of granulocytes following atrazine exposure, and the proliferation of Ito cells in rainbow trout exposed to diazinon.

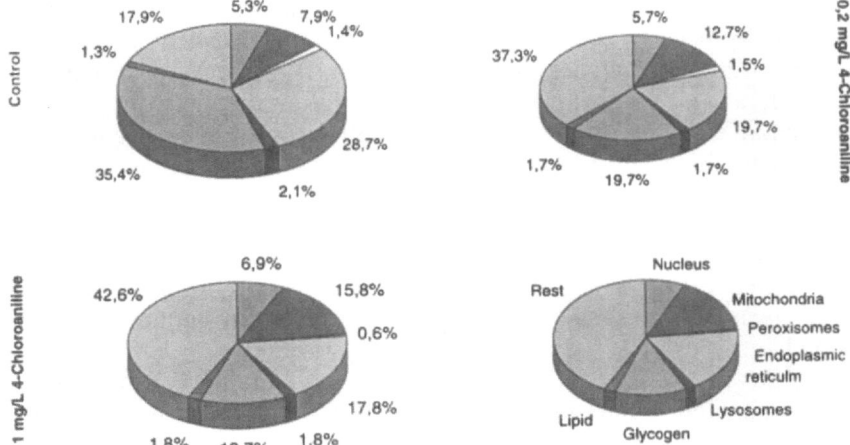

Figure 11. Volumetric alterations in hepatocytes of rainbow trout (*Oncorhynchus mykiss*) exposed to sublethal concentrations of 4-chloroaniline for 30 days. Note the dose-dependent decline in rough endoplasmic reticulum and glycogen in contrast to an augmentation of mitochondrial and lipid volume. Data as quantified from results published by Braunbeck et al. (1990c).

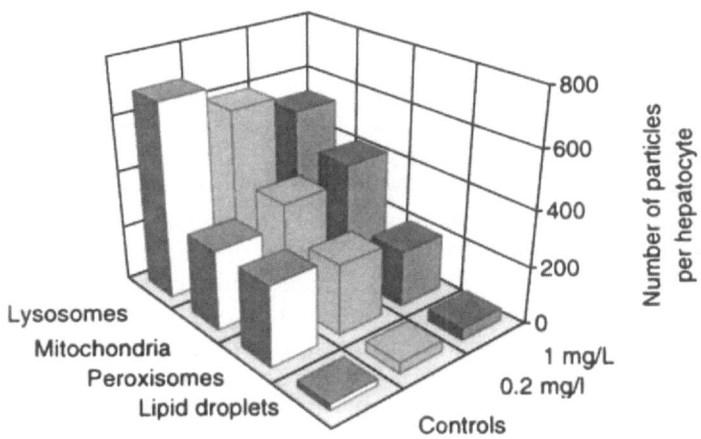

Figure 12. Numerical changes in selected particulate organelles of hepatocytes from rainbow trout (*Oncorhynchus mykiss*) exposed to sublethal concentrations of 4-chloroaniline for 30 days. There is a dose-dependent decline in lysosomes and lipid droplets in contrast to an increase in the number of mitochondria (cf. volumetric increase of mitochondria as shown in Fig. 11). Likewise, note the parallel development of volume (Fig. 11) and number of peroxisomes. Data as quantified from results published by Braunbeck et al. (1990c).

Quantitative changes in the liver of rainbow trout after chemical exposure.
In order to make the primarily qualitative ultrastructural modifications by
xenobiotic compounds directly comparable to results from more quantitatively oriented disciplines such as biochemistry and molecular biology,
changes may be analyzed by means of stereological techniques (Weibel,
1979). As shown in Figures 11 and 12 for the reference compound 4-
chloroaniline, semiquantitative data (Tab. 2) can thus be corroborated by
strictly numerical data, which are accessible to statistical evaluation.

Hepatic cytopathology in zebrafish (*Danio rerio*) after *in vivo* exposure to xenobiotics: the cyprinid model

With regard to the body of information, which has been made available in
the last two decades, zebrafish is probably the second-best investigated fish
species (Ekker and Akimenko, 1991; Laale, 1977), and, especially in the
course of growing importance of vertebrate neuro- and developmental biology, zebrafish is about to pass rainbow trout in its popularity (Brown,
1997; Eisen, 1996; Ekker and Akimenko, 1991; Granato and Nüsslein-Volhard, 1996; Haffter and Nüsslein-Volhard, 1996; Ingham, 1997; Mizell and
Romig, 1997; Postlethwait and Talbot, 1997).

Ultrastructure of control zebrafish liver

The liver of zebrafish exhibits a conspicuous sexual dimorphism, especially with regard to the organization of the rough endoplasmic reticulum
(RER) and the distribution of storage materials. In either sex, there is a
distinct separation of a perinuclear sheath of organelle-containing cytoplasm from peripheral portions mainly containing glycogen and lipid deposits, i.e., a cytoplasmic compartmentation comparable to that in rainbow
trout.

Liver of female control zebrafish. The polyhedral hepatocytes of female
zebrafish measure 15–19 µm in diameter and are dominated by an extensive development of the protein- and lipoprotein-synthesizing apparatus as
a morphological expression of exogenic vitellogenesis in the liver of
sexually mature female zebrafish and other fish species (Heesen and
Engels, 1973; Ng and Idler, 1983; Ng et al., 1984; Peute et al., 1978, 1985;
Takahashi et al. 1977; van der Gaag et al., 1977; Veranic and Pipan, 1995;
Weis, 1972; Wiegand, 1996). The RER forms elaborate stacks of up to
more than 20 parallel, non-fenestrated lamellae enveloping the centrally
located nucleus, which is about 5 µm in diameter and normally exhibits
only very little heterochromatin (Fig. 13). Three to five Golgi fields per
hepatocyte are regularly located between the nucleus and the unicellular

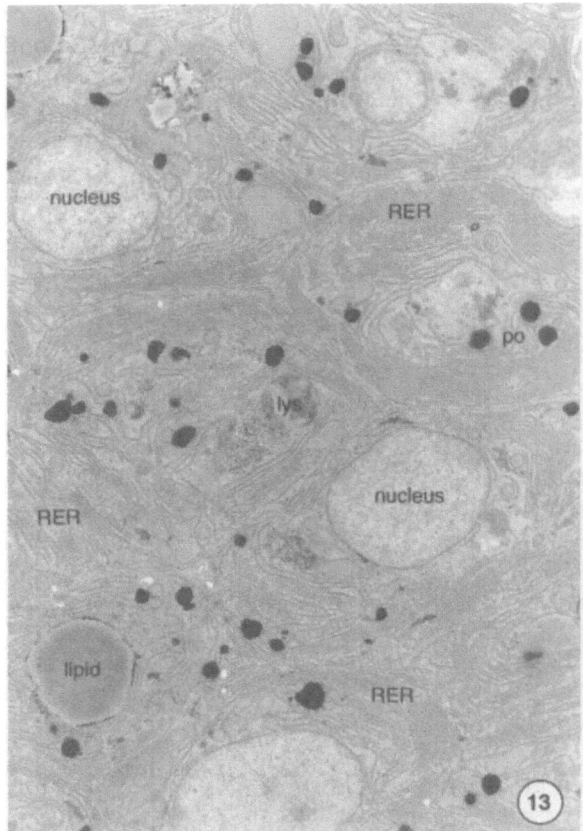

Figure 13. Hepatocytes of non-treated female zebrafish (*Danio rerio*) is characterized by an abundance of rough endoplasmic reticulum (RER) arranged in regular stacks of up to 20 parallel, non-fenestrated cisternae and numerous peroxisomes (po) intensively stained with diaminobenzidine. In contrast to males, the amount of glycogen stores is limited. Reprinted from Braunbeck et al. (1990b); magnification × 3600.

bile canaliculus typical of cyprinid fish species (Braunbeck et al., 1987, 1990a, b, 1992; Braunbeck and Segner, 1991; Byczkowska-Smyk, 1968, 1970, 1971; David, 1961; Gilloteaux et al., 1996; Langer, 1979a, b; Nopanitaya et al., 1979; Rutschke and Brozio, 1974; Saez et al., 1984; Tanuma, 1980; Yamamoto, 1962, 1964, 1965) and composed of 2 to 5 cisternae budding off numerous vesicles containing VLDL as well as material of moderate electron density representing yolk materials. Mitochondria appear as spherical or elongated profiles (up to 4.5 μm in length and up to 0.8 μm in width) adjacent to RER stacks. After incubation in the alkaline diaminobenzidine medium (LeHir et al., 1979), abundant peroxisomes can easily be distinguished as intensely staining roundish or ellipsoid particles of 0.2 to 1.8 μm in diameter (Fig. 13). Peroxisomes, which are frequently arranged in groups in close spatial relationship to mitochondria and RER cisternae, consistently lack a crystalline core.

Lipid represents the major storage product in the liver of female zebra fish; amount and size of hepatocellular lipid droplets, however, are subject to appreciable variation due to interindividual variability as well as individual position within the gonadal cycle (Bresch, 1982, 1991; Bresch et al., 1986; Eaton and Farley, 1974a, b; Heesen, 1977; Heesen and Engels, 1973; Hisaoka and Battle, 1958; Hisaoka and Firlit, 1960, 1962; Nagel, 1986; Peute et al., 1978; van der Gaag et al., 1977).

Liver of male control zebrafish. In contrast to females, the organelle-containing portion of the hepatocytes of male zebrafish, which are of a size comparable to that of female specimens, is typically small: One to three discontinuous layers of short RER cisternae, few mitochondria (up to 3.5 µm in length and 0.5 µm in width) of roundish to elongated shape and a small number of moderately sized (0.3–1 µm) spherical peroxisomes surround the large central nucleus, which contains little heterochromatin, but a conspicuous nucleolus. Golgi fields are small when compared to females. The cell periphery comprises extended glycogen areas (Fig. 14) free of organelles except for few lysosomes. In male control zebrafish, there is only a minor amount of lipid inclusions. In either sex, smooth endo-

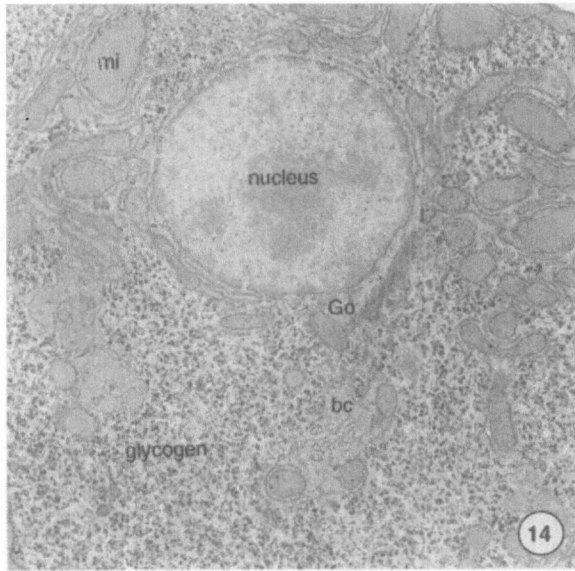

Figure 14. The liver of untreated male zebrafish (*Danio rerio*) typically contains comparatively few organelles (cf. Fig. 13), but immense amounts of glycogen deposits. Rough endoplasmic reticulum is explicitly scarce. Note the characteristic arrangement of nucleus, Golgi complex (Go) and bile canaliculus (bc); mi – mitochondria; magnification × 6700.

plasmic reticulum and lysosomes are rare, the latter being restricted to the peribiliary complex in control fish. Myelinated bodies, multivesicular bodies and autophagic vacuoles cannot be observed. For further details of the ultrastructure of normal zebrafish see Braunbeck et al. (1989, 1990b, c), Peute et al. (1978, 1985), Phromkunthong and Braunbeck (1994) and van der Gaag et al. (1977).

Ultrastructure of zebrafish liver after exposure to xenobiotics

So far, in zebrafish five experiments using the reference compounds 4-nitrophenol, 4-chloroaniline and 3,4-dichloroaniline as well as the pesticides atrazine and lindane as model compounds have been conducted in our laboratory to investigate sublethal effects of xenobiotic compounds on the ultrastructure of the liver. Major results are listed in Table 3.

Nuclei. In contrast to control animals (Figs. 13, 14), exposure of zebrafish to 4-nitrophenol, 3,4-dichloroaniline and lindane result in a conspicuous increase in the number of binucleate hepatocytes. Whereas following lindane exposure predominantly male fish display stimulation of nuclear division, 4-nitrophenol and 3,4-dichloroaniline induce karyokinesis in both sexes at comparable rates (up to 10% binucleate cells in single individuals). In female zebrafish exposed to 3,4-dichloroaniline, the nuclei frequently show two nucleoli. In parallel to analogous changes in the endoplasmic reticulum (see below), the nuclear envelope appears distended following exposure to 4-nitrophenol and 3,4-dichloroaniline. As a further analogy, both micro-(membrane-bound) and macrovesicular (not membrane-delineated) lipid inclusions may be encountered in the nuclei of steatotic hepatocytes of fish exposed to lindane.

Rough endoplasmic reticulum. The most remarkable ultrastructural alteration in the hepatocytes of female zebrafish induced by 4-nitrophenol, 4-chloroaniline and 3,4-dichloroaniline is the fenestration of the RER lamellae (Fig. 15), with the stacking of the cisternae being preserved. In tangential sections, fenestrated RER cisternae display perforations with a diameter of 70–100 nm at a fairly constant spacing. The degree of fenestration is highest in animals exposed to 3,4-dichloroaniline; occasionally, concentric whorls of fenestrated RER lamellae can be documented. In all experiments, additional fractionation and vesiculation as well as a considerably less regular array of remaining RER cisternae result in a less uniform appearance of hepatocytes than in controls. Fractionation, vesiculation and dilation of the RER represent the only alteration detectable after exposure to atrazine.

In either sex, but more pronounced in female specimens, 40 µg/L lindane induce steatosis (Baglio and Farber, 1965), i.e., a microvesicular (mem-

Table 3. Cytopathological alterations incuded in heptacytes of zebrafish (*Danio rerio*) by long-term *in vivo* exposure to sublethal concentrations of selected xenobiotics

	4-Nitro-phenol	4-Chloro-aniline	3,4-Dichloro-aniline	Atrazine	Lindane
Cytoplasmic compartmentation					
Disturbance of compartmentation	1000	40	2	1000	40
Nuclei					
Stimulation of karyokinesis	1000	–	20	–	80
Deformation of nuclear envelope	5000	–	20	–	
Nuclear lipid inclusions	–	–	–	–	40
Endoplasmic reticulum					
SER – proliferation in males	1000	–	20	–	80
SER – proliferation in females	5000	–	–	–	–
RER – fenestration in females	1000	200	2	–	–
RER – fragmentation in females	–	200	20	1000	–
RER collapse in females	–	–	20	–	–
RER dilation in females	–	200	20	–	40
RER decrease in females	–	–	–	1000	40
RER decrease in males	5000	–	–	1000	80
Steatosis	–	–	2	–	40
Golgi fields					
VLDL decrease in females	–	–	20	–	–
Golgi hypertrophy in females	–	200	20	–	–
Golgi fenestration in females	–	–	20	–	–
Mitochondria					
Increase in number	–	–	20	–	–
Increase in size	–	200	20	–	–
Club-shaped deformation	–	–	20	1000	40
Longitudinal crystalloids	–	–	20	–	–
Stratified inclusions	–	1000	–	–	–
Peroxisomes					
Decrease in catalase activity	–	1000	–	–	80
Decrease in number	–	200	–	–	–
Peroxisomal proliferation	–	–	20	–	–
Peroxisome aggregation	–	–	20	–	–
Lysosomal inclusions					
Lysosomal proliferation	100	40	20	1000	–
Crystal formation	–	200	20	–	–
Myelin formation	1000	200	20	1000	80
Autophagic vacuoles	1000	–	20	–	80
Storage products					
Lipid increase	–	–	–	–	40
Stetosis	–	–	2	–	40
Cholesterol deposition	–	2000	–	–	–
Glycogen decrease	1000	–	–	–	80
Macrophages					
Immigration of macrophages	1000	200	2	1000	40

Data presented as lowest concentration of the toxicants (μg/L), at which effect could be observed. For reference, see Table 1.

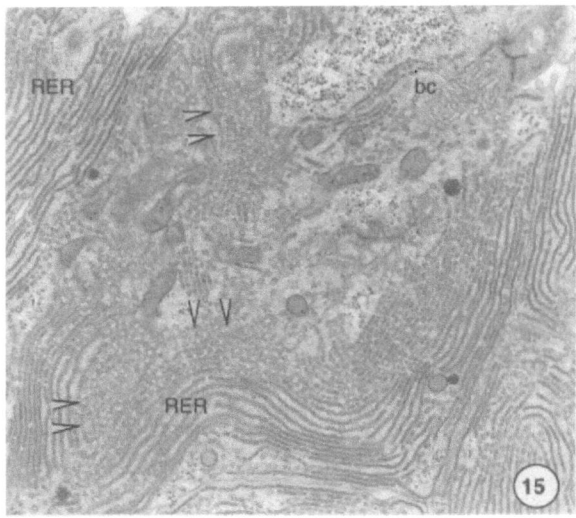

Figure 15. A typical reaction of hepatocytes from female zebrafish (*Danio rerio*) exposed to xenobiotic compounds (in this example, 3,4-dichloroaniline) is the fenestration of the cisternal system of the rough endoplasmic reticulum (RER). In tangential sections, fenestration may be readily distinguished by the appearance of a high number of apertures spaced at fairly constant intervals in an otherwise planar area from vesiculation (arrowheads), which is characterized by a dissolution of the continuous cisternae into bleb-like structures (cf. e.g., Fig. 5). bc – bile canaliculus. Hepatocyte of female zebrafish exposed to 20 μg/L 3,4-dichloroaniline for 5 weeks; magnification × 7500.

brane-bound) fatty transformation of the RER cisternae (Fig. 16). Starting from isolated cisternae, the lumen of the RER is increasingly distended, and small lipid inclusions can be discerned within the cisternal lumen. During progressive fatty transformation, the extent of RER dilation increases and finally most of the hepatocytic cytoplasm is occupied by steatotic RER. This microvesicular fatty change induced by lindane is accompanied by a concomitant increase in size and number of normal cytoplasmic, macro-vesicular lipid inclusions. Steatotic hepatocytes never display RER fenestration and lack the cytoplasmic segregation into storage fields and organelle-containing areas typical of control specimens. At higher lindane concentrations, the overall amount of RER is conspicuously reduced. Steatosis can also be discerned in fish exposed to 3,4-dichloroaniline, however, limited to primordial stages without progression at higher toxicant concentrations.

The "collapse" of the opposite membranes of RER cisternae in females is an alteration unique for 3,4-dichloroaniline exposure: The membranes lining RER segments up to 4 μm in length appear to fuse resulting in complete obstruction of the RER lumen. The remaining RER in these hepatocytes consist of heavily dilated vesicular profiles.

Due to the completely different amount and organization of the RER, and except for the steatosis induced by lindane, none of the alterations de-

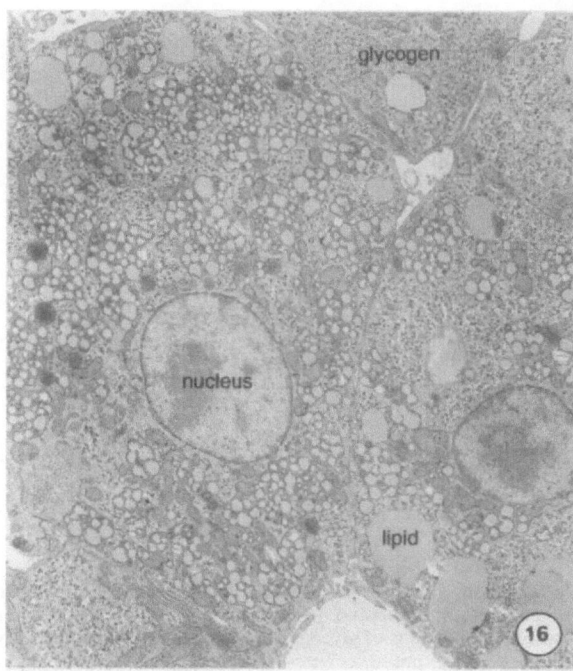

Figure 16. A feature typical of exposure to lindane (γ-hexachlorocyclohexane) in hepatocytes of female zebrafish (*Danio rerio*) is the occurrence of microvesicular lipid, i.e., membrane-bound accumulations of small, but occasionally confluent lipid droplets derived from cisternae of the rough endoplasmic reticulum (RER; "steatosis" according to Baglio and Farber, 1964). In males, steatosis can be demonstrated, however, to a much lower extent than in female zebrafish. Note the entire transformation of the RER system into a steatotic reticulum. Hepatocyte of female zebrafish exposed to 40 µg/L γ-hexachlorocyclohexane; reprinted from Braunbeck et al. (1990b); magnification × 3800.

scribed for female specimens can be found in male zebrafish. Instead, treatment with 4-nitrophenol and lindane only results in a minor decrease in the amount of RER cisternae.

Smooth endoplasmic reticulum. Whereas in female zebrafish SER of limited extent can only be found after treatment with 4-nitrophenol, there is a pronounced proliferation of SER in male animals exposed to 4-nitrophenol, 3,4-dichloroaniline and lindane (Fig. 17). SER induction is strongest in zebrafish exposed to 4-nitrophenol and in juveniles kept under exposure to 200 µg/L 3,4-dichloroaniline. In contrast, atrazine fails to stimulate any change in the SER compartment.

Golgi fields. A stimulation of Golgi activity or disturbance in secretory processes following exposure to 4-chloroaniline is indicated by elevated numbers of secretory vacuoles as well as by inflation of Golgi cisternae. After 3,4-

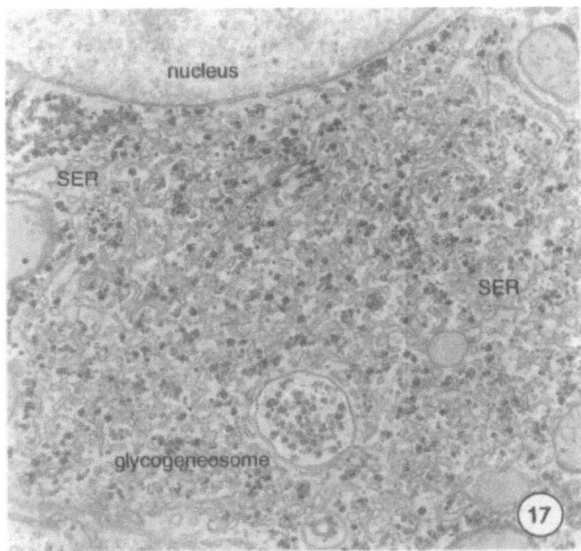

Figure 17. In male zebrafish (*Danio rerio*) exposed to 4-nitrophenol, 3,4-dichloroaniline or lindane, the smooth endoplasmic reticulum (SER) forms massive accumulations within peripheral glycogen areas. Small portions of glycogen are enclosed in lysosomes (glycogenosome). Hepatocyte of male zebrafish exposed to 1 mg/L 4-nitrophenol for 4 weeks; reprinted from Braunbeck et al. (1989); magnification × 12 800.

dichloroaniline exposure, in analogy to the perforation of the RER lamellae, Golgi cisternae in female zebrafish display numerous fenestrations with a mean diameter of 30 nm. The gradient in electron density from *cis*- to *trans*-face typical of controls is preserved. The only alteration in the Golgi fields of male specimens is a reduction in the amount of VLDL particles after exposure to 3,4-dichloroaniline, which is most striking in juvenile fish.

Mitochondria. Club-shaped deformation of mitochondria encountered after exposure to 3,4-dichloroaniline, atrazine and lindane most likely represents an unspecific symptom of stress, whereas, so far, other morphological modifications must be regarded specific of certain xenobiotics: 4-chloroaniline exposure resulted in ovoid giant mitochondria up to 8 µm in diameter displaying stratified, electron-dense inclusions with a maximum diameter of 350 nm in addition to normal, unaltered granula intramitochondrialia. In these mitochondria, cristae are restricted to the periphery of the organelles. In contrast, especially juvenile and male adult zebrafish exposed 3,4-dichloroaniline frequently display conspicuous accumulations of mitochondria with increased numbers of slender, parallel cristae penetrating deeply into the mitochondrial matrix at the sinusoidal pole of the hepatocytes, as well as large mitochondria with rod-like crystalloid inclusions arranged in parallel to the longitudinal axis of the mitochondria (Fig. 18).

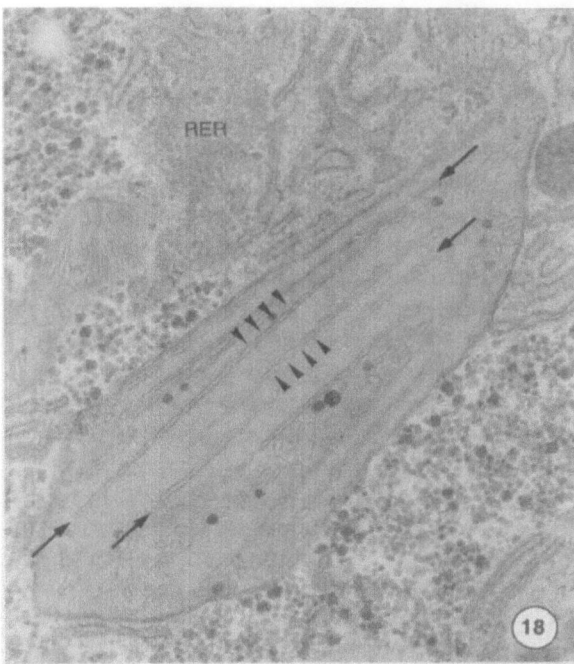

Figure 18. An ultrastructural alteration in hepatocytes from zebrafish (*Danio rerio*) of either sex unique for exposure to 3,4-dichloroaniline is the occurrence of conspicuous accumulations of mitochondria with increased numbers of slender, parallel cristae penetrating deeply into the mitochondrial matrix at the sinusoidal pole of the hepatocytes, as well as large mitochondria with rod-like crystalloid inclusions strictly arranged in parallel to the longitudinal axis of the mitochondria (at arrows). The crystalloid structure of the mitochondrial matrix is characterized by an inconspicuous cross-striation spaced at 13–16 nm (at arrowheads). Hepatocyte of female zebrafish exposed to 20 µg/L 3,4-dichloroaniline for 5 weeks RER – rough endoplasmic reticulum; magnification × 34 200.

Peroxisomes. Both number and staining activity of peroxisomes are reduced in consequence of exposure to 4-chloroaniline and lindane, thus indicating lower catalase activities. In parallel, the peroxisomal matrix of 4-chloroaniline-exposed hepatocytes takes a coarsely granular appearance. In contrast, exposure to 3,4-dichloroaniline induces a proliferation of peroxisomes accompanied by an increased number of peroxisomes displaying tail-like protrusions most likely indicative of the budding-off of smaller peroxisomal profiles from larger ones. A further alteration observed in fish exposed to 3,4-dichloroaniline is the accumulation of peroxisomes along crystal-shaped, membrane-bound extensions of the RER (Fig. 19).

Lysosomes. Except for lindane, proliferations of lysosomes and myelinated bodies are induced to varying degrees after exposure to all toxicants tested so far. Especially at higher concentrations, their occurrence is no longer restrict-

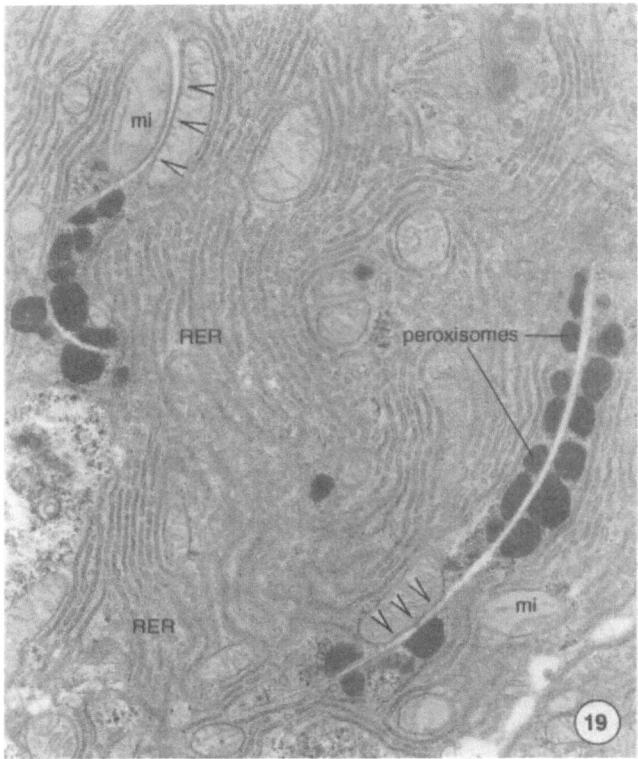

Figure 19. In hepatocytes of zebrafish (*Danio rerio*) exposed to 3,4-dichloroaniline only, there is a conspicuous accumulation of peroxisomes staining intensively for catalase activity after incubation with alkaline diaminobenzidine along crystal-shaped, membrane-bound extensions (arrowheads) of the rough endoplasmic reticulum (RER), which, however, are devoid of any ribosomes. Most likely, the inner space of these peculiarly-shaped ER cisternae originally contained some crystal-like materials, which were extracted during the dehydration process. Hepatocyte of female zebrafish exposed to 20 μg/L 3,4-dichloroaniline for 5 weeks; magnification × 16 800.

ed to the peribiliary area. Except for atrazine, all xenobiotics studied also stimulate the formation of autophagic vacuoles, which is indicative of an increased turnover of cellular components. In 4-chloroaniline- and 3,4-dichloroaniline-exposed zebrafish only, crystal-like inclusions with an extension of up to 4 μm cause distortion of the otherwise spherical lysosomal profiles. In lysosomes of hepatocytes from adult zebrafish exposed to 3,4-dichloroaniline, transverse sections reveal the crystalline lattice to be composed of 5 nm particles of moderate electron density spaced at distances of 5 nm and arranged in a rhombic pattern at angles of 60° and 120°, respectively.

Hepatic storage products. Whereas glycogen contents are not influenced by 4-chloroaniline and atrazine, there is a progressive depletion of hepatocytic

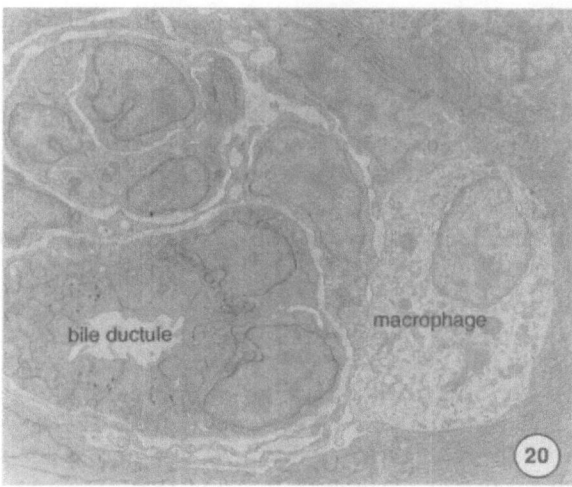

Figure 20. An unspecific, yet undoubtedly chemical-induced alteration in the liver not only of zebrafish (*Danio rerio*), but most likely all fish species is the immigration of macrophages into the liver parenchyma. As a rule, the routes of invasion are along the biliary system as well as via the space of Disse, from where macrophages penetrate into the recessus hepatis. In fish contaminated more severely, macrophages frequently aggregate to form clusters ([melano]macrophage centers), which usually no longer display closer spatial relationships to the original routes of invasion. Liver of female zebrafish exposed to 160 µg/L atrazine for 5 weeks; reprinted from Braunbeck et al. (1992b); magnification × 4300.

carbohydrate deposits with increasing concentrations of 4-nitrophenol and lindane. In turn, a drastic increase in the amount of lipid stored can be observed in conjunction with steatosis after exposure to lindane. 4-Chloroaniline exposure induces an accumulation of cholesterol crystals within glycogen fields of the hepatocytes of female zebrafish.

Macrophages. With regard to changes in non-parenchymal elements in zebrafish liver, an increase in the number of macrophages can be documented after exposure to all chemicals tested so far. At lower toxicant levels, free phagocytic cells are confined to their routes of immigration, i.e., the space of Disse and along the biliary ducts and ductules (Fig. 20; cf. Rocha et al., 1994, 1996); at higher concentrations, however, the occurrence of macrophages is no longer restricted to the perisinusoidal space and biliary system. Especially in males exposed to lindane, large accumulations of macrophages containing cellular debris or even entire hepatocytes can be detected in the parenchyma without apparent relation to the perisinusoidal space of Disse.

Species-specificity of cytopathological reactions of fish hepatocytes

A comparison of data on cytopathological effects by 4-chloroaniline and atrazine on rainbow trout (Tab. 2) and zebrafish (Tab. 3) reveals that, in addition to sex, the systematic position of the species investigated has to be taken into consideration as a confounding factor: Except for some basic, unspecific cellular reactions (see below), the cytopathological syndromes in rainbow trout and zebrafish are not only qualitatively different; the two species also show a contrast in sensitivity to 4-chloroaniline (Braunbeck et al., 1992b, c). Similar results were obtained from further comparative studies on the effects of dinitro-*o*-cresol in the livers of European eel (*Anguilla anguilla*) and golden ide (*Leuciscus idus melanotus*; Braunbeck and Völkl, 1993), of disulfoton in rainbow trout and European eel (Arnold, 1995; Arnold and Braunbeck, 1994a; Arnold et al., 1996), as well as of endosulfan in the livers of rainbow trout (Arnold and Braunbeck, 1994b, 1996) and carp (*Cyprinus carpio*; Braunbeck, 1994c). As a consequence, transfer of data from one species to another in both qualitative and quantitative terms is not allowed (Braunbeck, 1994c; Braunbeck et al., 1992b, c).

Cellular changes in the liver of fish exposed to combinations of xenobiotic compounds

Similar conclusions can be drawn from an analysis of hepatocellular effects by isolated compounds with those induced by combinations of xenobiotics: The syndromes of cytopathological changes by 3,4-dichloroaniline and lindane are not simply reflected in the syndrome induced by simultaneous exposure to both substances (Tab. 4). Effects by isolated compounds cannot be easily translated into a plain addition of effects in the liver of fish exposed concurrently, although a number of alterations can be traced back to single compounds. Similar results were obtained from a combination experiment with rainbow trout simultaneously exposed to disulfoton and endosulfan (Arnold and Braunbeck, 1993; Arnold et al., 1995). As a conclusion, identification of the effects of single compounds in the complex syndrome of hepatocellular changes induced by chemical mixtures sampled, for example, after a chemical spill, can be difficult (Arnold and Braunbeck, 1994a; Braunbeck et al., 1990a, 1994).

Dose-response relationships in the reaction of fish hepatocytes to xenobiotics

From a summary of toxicological data for the xenobiotics investigated with zebrafish, appreciable differences with regard to the molar toxicant concentration (nmol/g fish) killing 50% of the experimental fish (LD_{50} values)

Table 4. Cytopathological alterations induced in hepatocytes of zebrafish (*Danio rerio*) by long-term exposure to sublethal concentrations of pure 3,4-dichloroaniline and lindane, as well as combinations of 3,4-dichloroaniline and lindane

	3,4-Dichloro-aniline [3,4-DCA] →	Combination 3,4-DCA + HCH ←	Lindane [HCH]
Cytoplasmic compartmentation			
Disturbance of compartmentation	2	[100 + 40]	40
Nuclei			
Stimulation of karyokinesis	20	–	80
Increase in heterochromatin	–	100 + 40	–
Deformation of nuclear envelope	20	100 + 40	–
Nuclear lipid inclusions	–	40 + 40	40
Endoplasmic reticulum			
SER proliferation in males	20		80
RER fenestration in females	2	–	–
RER fragmentation in females	20	–	–
Collapse of RER cisternae in females	20	–	–
Dilation of RER cisternae in females	20	[40 + 40]	40
RER decrease in females	–	–	40
RER decrease in males	–	–	80
Induction of steatosis	[2]	[40 + 40]	40
Golgi fields			
VLDL decrease in females	20	–	–
Golgi hypertrophy in females	20	–	–
Golgi fenestration in females	20	100 + 40	–
Mitochondria			
Increase in number and size	20	–	–
Club-shaped deformation	20	100 + 40	40
Formation of longitudinal cristae	–	100 + 40	–
Longitudinal intramitochondrial crystalloids	20	100 + 40	–
Stratified intramitochondrial inclusions	20	–	–
Peroxisomes			
Decrease in catalase activity	–	–	80
Peroxisomal proliferation	20	40 + 40	–
Peroxisome aggregation	20	40 + 40	–
Lysosomal inclusions			
Lysosomal proliferation	20	100 + 40	–
Intralysosomal crystal formation	20	–	–
Myelin formation	20	–	80
Formation of electron-lucent vacuoles	–	100 + 40	80
Formation of autophagic vacuoles	20	100 + 40	80
Storage products			
Lipid increase	–	100 + 40	40
Induction of steatosis	[2]	[40 + 40]	40
Cholesterol deposition	–	100 + 40	–
Glycogen decrease	–	–	80
Macrophages and granulocytes			
Immigration of macrophages	2	40 + 40	40
Immigration of granulocytes	–	100 + 40	–

Data presented as the lowest concentration of the toxicants (μg/L), at which the effect could be observed. Similarities between 3,4-DCA/HCH-combination inverted. Weak reactions are given in brackets. Data for lindane from Braunbeck et al. (1990).

and to the bioconcentration factors of both the total animal and the liver become apparent. Considering the actual toxicant concentrations in the liver at the level of the cytological lowest observed effect concentration (LOEC) level (nmol/g liver), these substance-specific differences become even more apparent.

Likewise, the ratios between acute and chronic toxicity (ACR[1], in the present context defined as the ratio of the molar toxicant concentrations in (a) total fish in the range of LD_{50} and (b) the liver in the range of earliest cytological sublethal effects; Tab. 5) are different between atrazine as well as lindane (ACR: 1–4), 4-nitrophenol (ACR: 41) and 3,4-dichloroaniline (ACR: 1317). The finding of the exceptional characteristics of chronic toxicity of 3,4-dichloroaniline based on cytological data are substantiated by the conclusions drawn by Kenaga (1982) and Call et al. (1983) founded on ACR calculations from organismic acute and chronic toxicity data: Chronic toxicity cannot be extrapolated from acute toxicity (see also the chapter by R. Nagel, this volume).

Relative sensitivity of cytological techniques

Comparing the potential value of the cytological approach in environmental sciences with that of other methods, we have to consider the hierarchical organization of biological systems (e.g., Bartholomew, 1964; Sheehan, 1984; Stebbing, 1985; Vogt, 1987). The impact of toxicity as an environmental factor becomes first apparent at the lowest level of biological organization – i.e., the molecular level – and subsequently on the organelle, cellular, tissue, organismic and population level. Histology and cytology covering the organelle, cellular and tissue levels are more integrative, but less selective and sensitive than techniques working at the molecular level, but they are less integrative and more selective and sensitive than approaches using the organism or population level (cf. Sheehan, 1984).

Loss of selectivity at a higher level of biological organization, for example, may be illustrated by volume density changes of the rough endoplasmic reticulum (RER): Both under conditions of cold acclimation and acclimation to natural food there is an RER proliferation in the liver of golden ide (*Leuciscus idus melanotus*; Rafael and Braunbeck, 1988). However, only in the case of natural food versus artificial food can this phenomenon be correlated with increased fish growth (Rafael and Braunbeck, 1988), indicating that, with regard to cold acclimation, RER proliferation serves a function other than growth promotion. In female zebrafish exposed to organic xenobiotics, RER proliferation appears to be a

---

[1] Acute-chronic ratio (ACR) according to Kenaga (1982): ACR = 1/application factor AF = LC_{50} value/MATC, with MATC = maximum acceptable toxicant concentration.

Table 5. Toxicological data of organic xenobiotics tested in zebrafish (*Brachydanio rerio*) in relation to hepatocellular cytolopathology

	4-Nitrophenol	4-Chloroaniline	3,4-Dichloroaniline	Atrazine	Lindane
Water toxicant concentration (μg/L)	100	40	2	100	40
	1000	200	20	1000	80
	5000	1000	100		
Molecular weight (g/mol)	139	128	162	216	291
LC$_{50}$ (96 h; nmol/L)	13.76	34.43	8.42	36.94	0.09
LC$_{50}$ (96 h; nmol/L)	99	269	52	171	0.3
Bioconcentration factor (BCF)	41	8	30	5–7	798
LD$_{50}$ (96 h, nmol/g fish)	4671	2156	1581	856	239
Hepatic accumulation factor	4	n.d.	3	8	2
Liver toxicant concentration (nmol/g liver)	115	n.d.	1.2	23	219
	1150	n.d.	12.0	230	438
	5750	n.d.			
Cytological LOEC (nmol/g liver)	115	n.d.	1.2	230	219
Cytological acute: chronic ratio	41	n.d.	1318	4	1

BCF – Bioconcentration factor as ratio between concentrations in water and total body (Nagel, 1988). Hepatic accumulation factor calculated as the ratio between concentrations in total body and liver (Nagel, 1988). LOEC – Lowest observed toxicant concentration with cytological effects. Acute: chronic ratio as the ratio between acute toxicity (nmol/g fish) and lowest toxicant concentration in the liver (nmol/g liver), at which cytological alterations could be observed. n.d. – not determined.

common component of the general toxicant adaptation syndrome (see below). Thus, the observation of RER proliferation may yield a general hint about elevated hepatocellular protein synthesis capacity, but it fails to provide conclusive information about both cause and consequences of the change.

Gain of sensitivity at a lower level of organization can be exemplified by the fact that cytological alterations in the 4-nitrophenol and 4-chloroaniline experiments precede effects in subsequent generations: There is morphological evidence of a decrease in hepatic capacity of protein synthesis after zebrafish exposure to 4-nitrophenol (Braunbeck et al., 1989), thus anticipating impairment of reproductive success, which, in turn, could only be documented in the survival rate of the following generation (Nagel, 1988). In zebrafish subjected to 4-chloroaniline exposure, ultrastructural alterations of hepatocytes could be correlated to a dose-dependent reduction in egg production and fertilization rate as well as to macroscopical deformations in the subsequent generation (Bresch et al., 1990).

An analysis of different parameters measured within the experiments described with respect to liver effects in zebrafish reveals that, except for atrazine, cytopathology was the most sensitive endpoint to reveal sublethal toxicity (Tab. 6). Similar conclusions can be drawn for toxicant-induced cytological alterations in rainbow trout liver (Tab. 7); again, for atrazine, the liver is most likely only of minor importance as an target organ. The sensitivity of the cytological approach in terms of LOEC data may even be enhanced by the introduction of quantitative morphological techniques. Particularly in cases when toxicants without prominent structural effects are investigated, minor changes, which may be present nonetheless, can be

Table 6. Lowest toxicant concentrations for effects on hepatic ultrastructure, survival, reproduction (egg release and fertilization), growth and behaviour in zebrafish (*Danio rerio*)

	4-Nitro-phenol	4-Chloro-aniline	3,4-Di-chloro-aniline	Atrazine	Lindane
Lowest concentration with cytological effects (µg/L)	100	40	2	1000	40
Lowest concentration for effects on survival (µg/L)*	100 (F_1)	> 1000	20 (F1)	> 1000 (F_1)	> 80 (F_1)
Lowest concentration for effects on reproduction (µg/L) *	> 5000	40 (F1)	> 20	> 1000	80
Lowest concentration for effects on growth (µg/L) *	not determined	> 1000 **	200	100	> 80 ***
Lowest concentration for effects on behaviour (µg/L) *	> 5000	> 1000	> 100	> 1000	40

* Data from Bresch et al. (1990), Görge and Nagel (1990) as well as Nagel et al. (1991).
** Increase in weight at lowest test concentration.
*** Transitional growth reduction in larval stages at 40 and 80 µg/L.

Table 7. Toxicological data for selected organic toxicants tested with reference to cytological alterations in rainbow trout (*Oncorhynchus mykiss*) liver after prolonged *in vivo* exposure

	LC$_{50}$ value	"Conventional" NOEC	"Conventional" LOEC	Cytological LOEC
4-Chloroaniline	Rainbow trout: 14 mg/L Zebrafish: 46 mg/L	Zebrafish: 1.8 mg/L	Zebrafish: 3.2 mg/L	**Zebrafish: 40 µg/L** **Rainbow trout: 200 µg/L**
Endosulfan	Striped bass: 0.1 µg/L Rainbow trout: 0.3–1.9 µg/L Carp: 0.9–7 µg/L	Carp: 0.14 µg/L Guppy: 0.6 µg/L Tilapia: 0.63 µg/L	Carp: 0.33 µg/L Tilapia: 0.44 µg/L	**Rainbow trout: 10 ng/L** **Carp: 1 ng/L**
Atrazine	Catfish: 0.2–0.3 µg/L Rainbow trout: 1–9 mg/L	Zebrafish: 80 µg/L Brook trout: 60–120 µg/L	Zebrafish: 100 µg/L Carp: 100 µg/L	**Rainbow trout: 40 µg/L** Zebrafish: 1 mg/L
Diazinon	Brook trout: 1.6 mg/L Bluegill sunfish: 0.46 mg/L	Fathead minnow: < 3.2 µg/L Brook trout: < 0.55 µg/L	**Brook trout: 0.55 µg/L** Fathead minnow: 3.2 µg/L	Rainbow trout: 40 µg/L
Disulfoton	Bluegill sunfish: 0.3 mg/L Rainbow trout: 1.85 mg/L			**Rainbow trout: 0.1 µg/L**
Linuron	Harlekin fish: 0.6–4.6 mg/L Rainbow trout: 5–16 mg/L		**Rainbow trout: 30 µg/L***	**Rainbow trout: 30 µg/L**
Ochratoxin	Rainbow trout: 4.67 mg/kg **			**Rainbow trout: 100 µg/L**

Abbreviations: * Hematological alterations in same experiment. ** LD$_{50}$.
Fields in bold indicate presently vaild LOEC data.

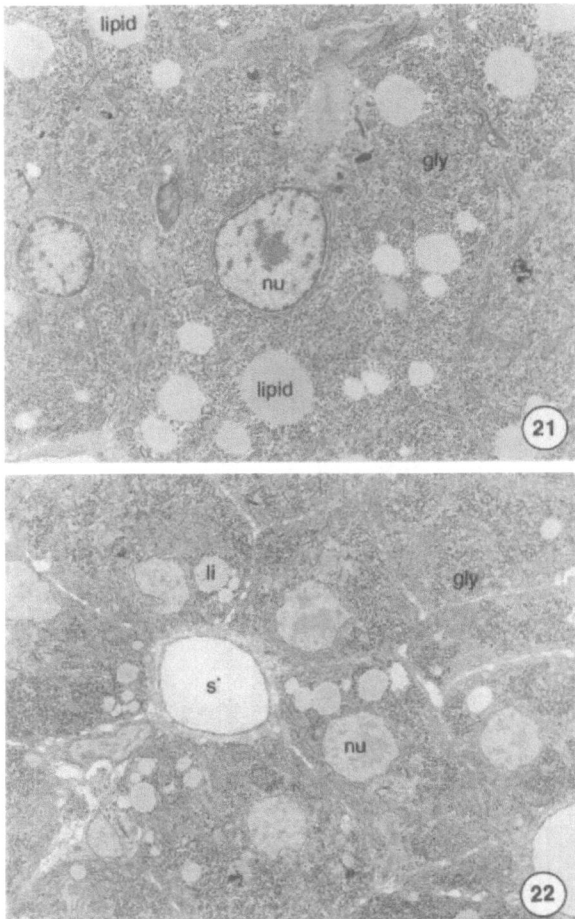

Figures 21 and 22. In low-power electron micrographs of hepatocytes of juvenile control carp (*Cyprinus carpio*; Fig. 21) and fingerling carp exposed to 0.5 µg endosulfan per kg food (Fig. 22), chemical-dependent differences momentarily appear minor. However, a closer inspection reveals numerous alterations by endosulfan. Whereas in controls hepatocytes are dominated by vast accumulations of glycogen in the cell periphery, lipid (li) inclusions are scattered over the entire peripheral storage areas, and rough endoplasmic reticulum (RER) around the spherical nucleus (nu) is scarce, in hepatocytes of carp exposed to 0.5 µg/kg endosulfan, cell size is reduced, the nuclear outline appears less regular, and the RER forms extensive stacks. The volume of Golgi fields is increased, the amount of reserve materials is significantly lower than in controls, and lipid droplets accumulate primarily at the vascular cell pole. Volumetric changes are detailed in Figure 23. Reprinted from Braunbeck and Appelbaum (1998); Figure 21: controls, magnification × 2800; Figure 22: 0.5 µg/kg endosulfan, magnification × 1600.

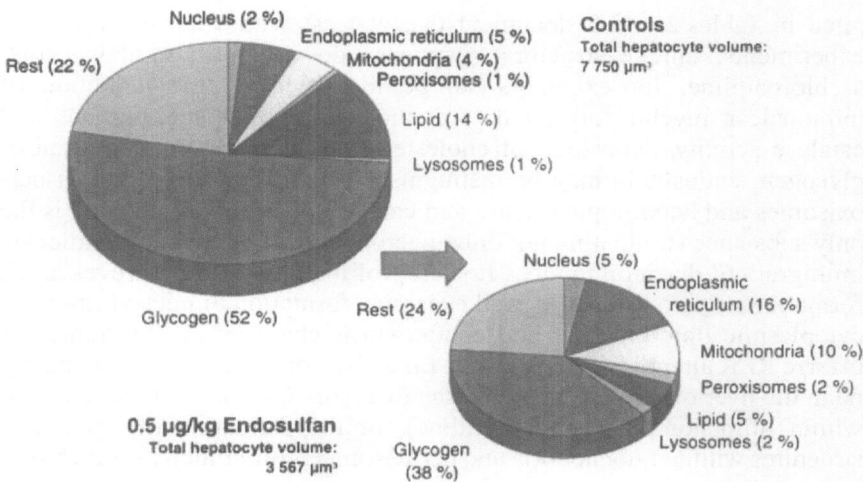

Figure 23. Volumetric changes induced in hepatocytes of juvenile carp (*Cyprinus carpio*) by oral administration of 0.5 µg endosulfan per kg food. Diameters of graphs represent relative volume. Data as quantified according to morphometric techniques following the principles by Weibel (1979) from results published by Braunbeck and Appelbaum (1998).

revealed by quantitative analysis. For instance, exposure of carp to traces of dietary endosulfan results (at least at lower magnifications) in modest cellular alterations in the liver (Figs. 21, 22). Stereological analysis, however, reveals that not only is hepatocyte volume reduced by 50%, but that relative volumes of almost all cellular organelles are doubled, whereas cellular storage products are drastically reduced (Fig. 23).

Substance-specificity of cytological alterations: the "general toxicant adaptation syndrome"

As a general rule, structural and ultrastructural acclimation to environmental conditions is only rarely based on the imposition of entirely new, different structures, but rather consists of qualitative and quantitative alterations of pre-existing characters (Forbus, 1952). The adaptive response itself may be of general nature or may be specific to a particular type of stressor (Moore, 1980).

When considering the rather limited set of structural variables of a hepatocyte, it is evident that specificity is likely to be relative. Mainly based on light microscopical investigations, this led several authors to the conclusion that chemical exposure exclusively results in unspecific structural alterations (Couch, 1975; Gingerich, 1982; Sindermann, 1980, 1984; Sindermann et al., 1980; Walsh and Ribelin, 1975). However, the data compiled in Tables 2 and 3 document that at least within the given sets of experiments, unique structural responses do occur. In rainbow trout, 4-chloroaniline, for example, can be identified by the induction of intranuclear myelin formation, reduction of peroxisome number and catalase activity, deposition of cholesterol and extreme condensation of glycogen; endosulfan may be distinguished by myelin formation in peroxisomes and heterotopic uricase and catalase distribution; atrazine is the only substance stimulating not only macrophage invasion, but granulocyte immigration; diazinon induces Ito cell proliferation; linuron provokes tail formation of peroxisomes as well excessive formation of nuclear lipid and cytoplasmic lipid clusters; and ochratoxin is characterized by numerous bizarre RER alterations (Figs. 6–9). Likewise, toxicant-specific phenomena in the liver of zebrafish include steatosis (lindane), stratified inclusions within mitochondria (4-chloroaniline), or the formation of crystalloid structures within mitochondria and peroxisomal proliferation (3,4-dichloroaniline).

In contrast, typical unspecific ultrastructural reactions of fish hepatocytes apparently comprise changes such as disturbance of compartmentation, proliferation or reduction and fenestration and/or vesiculation within the endoplasmic reticulum, club-shaped deformation of mitochondria, lysosomal proliferation, myelin formation and invasion of macrophages. However, primarily unspecific changes such as lysosomal proliferation may become specific by the presence of additional qualitative alterations, e.g., the 4-chloroaniline- or 3,4-dichloroaniline-induced occurrence of crystalloid inclusions within the lysosomal matrix. For these changes, the term "general toxicant adaptation syndrome" is proposed, since this set of ultrastructural modifications of fish hepatocytes may be expected following exposure to *any* toxicant and many other disturbing environmental factors other than chemical substances (the term "syndrome" has already been used by other authors in a similar context; cf. Couch, 1993).

This illustrates that only the simultaneous consideration of all aspects of a given organelle reflects the true adaptive response to the exogenous stimulant (Braunbeck, 1994a). Likewise, at the next higher level of biological organization, the acclimation of the hepatocyte is characterized by the integration of the reaction of all organelles. In fact, for any of the experiments reported, the syndrome, i.e., the combination of qualitative and quantitative, specific and unspecific alterations, proves to be factor-specific.

Even compounds which are chemically closely related, such as 4-chloroaniline and 3,4-dichloroaniline, do not reveal any more similarities in zebrafish hepatocytes than a combination of any two other of the substances tested (Tab. 3). These examples re-emphasize that extreme care should be taken in the extrapolation of experimental data from one experiment to another.

Time-pattern and reversibility of cytological alterations in fish liver: Adaptive *versus* degenerative responses

With respect to the significance of structural alterations for the next higher levels of biological organization, a classification has to be made into regenerative or adaptive reactions, i.e., processes to be regarded as expression of defense mechanisms, or, alternatively, into symptoms of irreversible damage, i.e., plainly degenerative phenomena (cf. Altmann, 1980; Braunbeck et al., 1990b; Sindermann, 1984). Since a major character of adaptive changes is their disappearance following cessation of the impact of a disturbing factor, most electron microscopical changes should be reversible. In order to elucidate time-pattern and reversibility of cytological alterations in fish liver, a long-term study on ultrastructural alterations by oral application of the fish therapeutic agent malachite green for up to 3 months was carried out with five ornamental fish species (*Paracheirodon axelrodi*, *Megalomphodus sweglesi*, *Barbus conchonius*, *Xiphophorus helleri*, and *Gymnogeophagus daffodil*). Fish were investigated after 0, 7, 15 and 22 days as well as 3 months of treatment; in addition, recovery was studied after 1 and 4 weeks following a 2-week application and 4 weeks after a 3-month dosage (Braunbeck, 1994c). Early, but feeble effects by malachite green could already be observed after 7 days. Although partly dependent on sex and age, reactions of hepatocytes were basically identical in all species, indicating that alterations were actually induced by malachite green. Again, hepatocytes displayed a clear time-response relationship, and modifications due to species-dependent peculiarities appeared to be of minor importance. Apparently, the primary cellular target of malachite green were mitochondria, which is in line with the decoupling of oxidative phosphorylation repeatedly documented for malachite green (Alderman, 1985, 1991; Gerundo et al., 1991; Werth and Boiteux, 1968); further important alterations were found in nuclei and the endoplasmic reticulum. As secondary consequences of malachite green exposure, increased heterogeneity of the hepatic parenchyma and time-dependent glycogen depletion were recorded. However, in none of the fish species investigated were effects in mitochondria and nuclei as severe as those depicted by Gerundo and co-workers (1991), which most likely represent artifacts by inadequate immersion fixation. There was no indication of mutagenic, teratogenic or cancerogenic action by malachite green, which was concluded from the intercalating properties of malachite green (Alderman, 1991; Gerundo et al., 1991). Except for some degenerative alterations in mitochondria, most ultrastructural effects, even those by a 4 months' exposure to malachite green, had to be classified as unspecific stress phenomena (Braunbeck, 1994c).

After a 14-day medication period, 1 week of feeding with a malachite green-free diet was sufficient for far-reaching recovery of hepatocellular ultrastructure. After 14 days, reconvalescence was complete except for an increased number of tertiary lysosomes indicative of former exposure.

Except for persisting high numbers of lysosomal elements, effects observed after 3 months of malachite green medication also proved to be fully reversible.

This indicates that especially the unspecific changes of fish hepatocytes represent adaptive rather than degenerative responses. For instance, proliferation of the smooth endoplasmic reticulum and lysosomes as well as infiltration of migrating cells of the unspecific cellular immune response most likely indicate stimulation of defense and regenerative processes in the liver, respectively, and have, thus, to be classified as adaptive.

On the other hand, reactions such as fractionation, dilation and vesiculation of the endoplasmic reticulum and the nuclear envelope are assumed to express the onset of degenerative processes, i.e., to be truly pathological symptoms related to functional disturbance or even dysfunction. In European eel and golden ide exposed to sublethal concentrations of dinitro-o-cresol, for example, the transition from adaptive to degenerative hepatocellular changes could be identified between 50 and 250 µg/L dinitro-o-cresol by both electron microscopical and biochemical techniques (Braunbeck and Völkl, 1991a, b, 1993).

Functional significance and diagnostic value of cytological alterations

According to the principle of the relationship and mutual interference between structure and function, an alteration of liver structure may be expected to imply an alteration of liver function (Hinton et al., 1987; Weibel, 1981). A well-known example of structure-function relationship in toxicology is the correlation between the proliferation of SER and the induction of xenobiotic biotransformation processes (mammals: Ghadially, 1988; Miayi, 1980; Phillips et al., 1987; Stäubli et al., 1969; fish: Braunbeck and Völkl, 1991a, b, 1993; Hacking et al., 1978; Hinton et al., 1978, 1985b; Klaunig et al., 1979; Lipsky et al., 1978; Schoor and Couch, 1979). 4-Nitrophenol and lindane, which are known to induce profound alterations in the hepatocellular ER compartment and to stimulate enzyme induction in mammals (for references, see Braunbeck et al. 1989, 1990b), were found to evoke similar effects in the liver of zebrafish. In males, the degree of SER induction by 80 µg/L lindane, however, is comparatively low when compared to 1 mg/L 4-nitrophenol, which, in turn, is lower than after 3,4-dichloroaniline exposure (≥20 µg/L). Again, sexual dimorphism has to be considered, since, except for exposure to 4-nitrophenol (≥5 mg/L), female specimens of zebrafish do not display SER proliferation, but a progressive fenestration of the RER. The latter structural response may be interpreted as a similar functional acclimation to intoxication, i.e., the morphological expression of the onset of biotransformation processes, since in fish cytochrome P4501A is not exclusively located on the surface of SER cisternae, but also on cisternae of the RER (Lester et al., 1992, 1993).

The functional interpretation of hepatic structural alterations in fish with regard to their functional implications is handicapped by our still very limited knowledge of the underlying acclimation processes. Particularly in the field of aquatic toxicology, there is still a fundamental lack of appropriate data on metabolic pathways of xenobiotics in fish. This situation has two consequences: On the one hand, those structural reactions gain highest diagnostic value, for which functional interpretation is available. Therefore, comparatively unspecific responses such as proliferation of the SER, which is functionally well characterized, may be a more important diagnostic tool than any factor- or toxicant-specific structural change. On the other hand, specific morphological changes provide at least a hint to the possible location of the toxic attack, and (maybe) even to the mode of action of the toxicant investigated (Braunbeck, 1994a; Nagel, 1988). Moreover, specific structural modifications of hepatocytes are, no doubt, of increased diagnostic value with respect to the identification of the substance inducing them (for other examples of specific cytological alterations in the hepatocytes of further fish species by organic chemicals, see Braunbeck and Völkl, 1991b).

With regard to the functional interpretation of toxicant-induced cytological modifications of fish liver, it is essential to realize that adaptive changes designed to compensate for disturbances in the homeostasis of the organism cannot – by definition – be verified at higher levels of biological organization, unless cellular and organismic defense mechanisms are overloaded (Braunbeck et al., 1990a; Braunbeck and Völkl, 1991b, 1993). In terms of ecotoxicology, the fact that many ultrastructural changes seemingly lack consequences at the level of individuals and populations may not be misinterpreted that they are not of ecotoxicological significance. Rather, they are indicative of the onset of chemical intoxication, which, however, is still within the range of compensation. The fact that histological and cytological investigations are capable of simultaneously providing evidence of both adaptive reactions, truly pathological responses *and* the transition from adaptation to degeneration is certainly one of the major advantages of cytopathological investigations.

Isolated fish hepatocytes as a tool for the *in vitro* diagnosis of toxicity

After 1969, when Berry and Friend were the first to describe the isolation of rat liver parenchymal cells by collagenase perfusion, the use of isolated hepatocytes was soon established as an ideal system for the study of many aspects of hepatic metabolism (Moon et al., 1985). Since 1976, when Birnbaum and colleagues studied hormone-stimulated glycogenolysis in goldfish hepatocytes, isolated fish hepatocytes have mainly been used intensively in fresh suspensions and in primary culture as a powerful, yet simple tool for the investigation of piscine hepatic functions. Cultured hepatocytes are particularly well suited for applications not only in basic fish physiology, but also in the fields of pharmacology and environmental toxicology (for reviews, see Baksi and Frazier, 1990; Hightower and Renfro, 1988; Moon et al., 1985). In environmental toxicology studies, the application of non-mammalian systems is rapidly expanding (Blair et al., 1990; Braunbeck and Storch, 1992), and, for aquatic systems, fish have become an indispensable model system for the evaluation of the effects of noxious compounds. In fulfillment of the need for new bioassay development, and in an attempt to further reveal basic modes of toxic action, as well as to reduce the number of animals used for toxicological studies, a particularly fruitful approach has been to find and to develop *in vitro* model systems which retain the basic characteristics of the more complex *in vivo* condition, yet can easily be manipulated experimentally (Baksi and Frazier, 1990).

In mammalian toxicology, isolated liver cells in primary culture have proved a most valuable tool to assess toxic effects of chemicals, since hepatocytes are not only central to systemic metabolic regulation, but are also primarily involved in xenobiotic metabolism and are frequently the specific target for chemicals (Baksi and Frazier, 1990; Fry and Bridges, 1977, 1979; Rauckman and Padilla, 1987). These properties may also be regarded as the major advantages of isolated hepatocytes over permanent (fibroblastic) fish cell lines (Braunbeck, 1993; Zahn et al., 1996b). The fundamental advantages of cell culture systems for the study of hepatic function in pharmacological and toxicological research have been repeatedly reviewed (Baksi and Frazier, 1990; Fry and Bridges, 1979; Hightower and Renfro, 1988; Klaassen and Stacey, 1982; Moon et al., 1985; Stammati et al., 1981): (1) control of experimental conditions without the elaborate control systems typical of whole animal tests; (2) elimination of interactive systemic effects and multiple defense mechanisms in the intact organism, and, thus, simplification of interpretation; (3) reduced variability between experiments; (4) simultaneous and multiple sampling over time (screening); (5) smaller quantities of test chemicals needed for complete dose-response studies resulting in reduced amounts of toxic wastes to be disposed off; and, above all, (6) reduction of experimental time and financial expense (for further reference, see also the chapter by G. Monod and co-authors, this volume).

In aquatic toxicology, however, the use of hepatocyte cell culture systems has lagged behind that of their medical (mammalian) counterpart (Kocan et al., 1985), but recently a whole set of potential physiological endpoints for toxicological studies with isolated fish hepatocytes has been evaluated including lactate dehydrogenase leakage and inhibition of protein synthesis rate (Gagné et al., 1990; Klaunig et al., 1985; Kocal et al., 1988b; Walton and Cowey, 1979), induction of metallothioneins (Baksi and Frazier, 1988), plasma membrane permeability, cell volume regulation and cellular ATP levels (Ballatori et al., 1988), formation of DNA adducts (Bailey et al., 1982; Coulombe et al., 1984; Goeger et al., 1986; Smolarek et al., 1987), as well as induction of unscheduled DNA synthesis (Klaunig et al., 1985; Walton et al., 1983, 1984). Morphological criteria, however, have only rarely been applied to assess the viability of piscine cell culture systems. Although the ultrastructure of isolated hepatocytes has been recommended as a suitable viability control in the fundamental study by Berry and Friend (1969) and has repeatedly been successfully utilized in several mammalian studies (Chapman et al., 1973; Drochmans et al., 1975; Frimmer et al., 1976; Guillouzo et al., 1978; Isom et al., 1985; Jeejeebhoy and Phillips, 1976; Phillips et al., 1974; Roux et al., 1978; Schreiber and Schreiber, 1973; Stacey and Fanning, 1981; Wanson et al., 1977), in studies with fish hepatocytes only Andersson and Förlin (1985) have used the cytological integrity of their preparations to diagnose cellular viability. Likewise, ultrastructural alterations in isolated hepatocytes of fish only rarely have been considered in long-term culture studies (Blair et al., 1990, 1995; Bouche et al., 1979; Klaunig, 1984; Klaunig et al., 1985; Miller et al., 1993; Ostrander et al., 1995; Saez et al., 1982), before our group began to systematically investigate the suitability of cytopathological alterations in isolated hepatocytes from rainbow trout and other species (for references, see Tab. 8).

As for *in vivo* experiments, rainbow trout was chosen by most researchers as the preferred donor for the cells, since rainbow trout hepatocytes are by far the best-investigated fish cell culture system not only in physiological studies (for reviews, see Moon et al. (1985), but also in toxicological research (Baksi and Frazier, 1990). To date, only primary cultures of normal hepatocytes from eel (Zahn et al., 1995), carp (Bouche et al., 1979; Saez et al., 1982), channel catfish (*Ictalurus punctatus*; Klaunig, 1984) and rainbow trout (Blair et al., 1990, 1995; Braunbeck and Storch, 1992; Klaunig et al., 1985; Segner et al., 1994) have been at least partly characterized cytologically. Many reports are available on specific culture conditions for rainbow trout hepatocytes (Blair et al., 1990; Braunbeck, 1992; Braunbeck

Table 8. Cytopathological studies on ultrastructural alterations in isolated fish hepatocytes after *in vitro* exposure to xenobiotic compounds

Species	Compound(s)	Reference
Rainbow trout (*Oncorhynchus mykiss*)	Acetaminophen	Miller et al. (1993)
	4-Chloroaniline	Braunbeck et al. (1993)
	2,4-Dichloroaniline	Zahn (1994)
	Dinotro-*o*-cresol	Zahn (1994)
	Disulfoton	Arnold (1995), Zahn et al. (1995, 1996a, b)
	Malchite green	Zahn and Braunbeck (1995b)
	4-Nitrophenol	Zahn and Braunbeck (1998)
	Seepage waters from garbage dumps	Zahn and Braunbeck (1995a)
Eel (*Anguilla anguilla*)	Disulfoton	Arnold (1995), Arnold et al. (1995)
Carp (*Cyprinus carpio*)	Mimosine	Vogt et al. (1993, 1994)

and Storch, 1992; Klaunig, 1984; Klaunig et al., 1985; Kocal et al., 1988a, b; Lipsky et al., 1986) as well as on cytopathological reactions under the influence of various toxicants such as carcinogens (Bailey et al., 1982; Coulombe et al., 1984; Goeger et al., 1986; Loveland et al., 1987; Perdew et al., 1986; Walton et al., 1983, 1984; Whitham et al., 1982) and heavy metals (Denizeau and Marion, 1990; Gagné et al., 1990; Marion and Denizeau, 1983a, b). Likewise, among fish, our knowledge on the role of isolated hepatocytes in xenobiotic metabolism is best in rainbow trout (Andersson, 1984, 1986; Andersson and Förlin, 1985; Andersson and Koivusaari, 1986; Bailey et al., 1984; Coulombe et al., 1984; Devaux et al., 1992; Loveland et al., 1987, 1988; Parker et al., 1981; Pesonen and Andersson, 1991; Walton et al., 1984).

Ultrastructural alterations in isolated fish hepatocytes following *in vitro* exposure to xenobiotic compounds

Ultrastructure of isolated rainbow trout hepatocytes – an example for cell senescence

In order to evaluate the suitability of morphological criteria in isolated fish hepatocytes during prolonged primary culture for toxicological studies, hepatocytes were isolated from rainbow trout by liver collagenase perfusion (Braunbeck and Storch, 1992). Isolated hepatocytes were investigated for up to 10 days in primary culture on uncoated Petri dishes by means of light and electron microscopy. Viability of isolated hepatocytes as estimated from trypan blue exclusion declined from > 90% at the beginning of the incubation to ≤ 80% after 8 days of primary culture. Survival of hepatocytes was best at incubation temperatures of 14°C, and addition of fetal calf serum failed to improve cell performance. The ability of hepatocytes to attach to the substratum and to form a monolayer is certainly a criterion to assess their integrity (Berry and Friend, 1969). In serum- and hormone-free medium, Koban (1986) was able to study temperature acclimatization of catfish hepatocytes in monolayer culture on plates coated with catfish liver biomatrix for 20 days, and Lipsky and co-workers (1986) described 93% attachment of isolated rainbow trout hepatocytes to culture plates coated with a commercially available extracellular matrix. Only Maitre and colleagues (1986) were successful in maintaining trout hepatocytes in monolayer culture on dishes without pretreatment for 10 to 12 days; however, cell attachment was slow and remained fairly weak. In contrast to Mommsen and Lazier (1986), who described the cultivation of salmon hepatocytes on Falcon Primaria plates (Becton & Dickinson Labware) for 9 days, Blair et al. (1990) reported only transient attachment of isolated rainbow trout hepatocytes to Falcon plates for 24 h with subsequent detachment, but found good adhesion for up to 6 days on culture plates coated with an extract of rainbow trout skin. In fact, the quality of the isolated hepatocyte suspension seems to be mainly dependent on the amount of stress the cells are exposed to during the isolation procedure; thus, rainbow trout hepatocytes isolated as rapidly as possible at adequate temperature with a suitable mixture of dissociating enzymes will, albeit loosely, adhere to the dishes without additional coating. Since the undefined matrix of collagen and other coatings is liable to reduce the standardization of the assay conditions in subsequent toxicity experiments, coating should, if ever possible, not be used in order to avoid ill-defined experimental conditions.

Freshly isolated hepatocytes appear as solitary spherical cells with numerous microvilli at the outer surface. Except for a 30% reduction in cell size due to stress-induced glycogen reduction, the ultrastructure of freshly isolated hepatocytes closely resembles that of rainbow trout hepatocytes *in vivo* (for comparison, see Tabs. 3 and 9). Within 1 day of culture, about 60–80% of the isolated hepatocytes sediment to form a monolayer attached to the culture dishes, whereas up to 20% remain in suspension forming hepatocyte aggregates. Cell adhesion is generally weak, and during prolonged culture increasing amounts of cells detach, whereas the floating cell accumulations grow to aggregates of up to more than 100 cells. Cell viability and ultrastructure is similar in monolayers and spheroids, and only from the fifth day in culture, hepatocytes in the center of floating aggregates become necrotic.

From days 1 to 5, rainbow trout hepatocytes in primary culture display only minor cytological alterations (Fig. 24; Tab. 9). Excellent cytoplasmic compartmentation, restoration of hepatocytic glycogen stores (cf. Foster and Moon, 1987; French et al., 1981; Hayashi and Ooshiro, 1985, 1986; Mommsen et al., 1988), high secretion rates of VLDL by dictyosomes, establishment of cell-to-cell contacts, restitution of cellular polarity and the epithelial character of the cells, as well as formation of bile canaliculi document recovery of the hepatocytes in primary culture. From day 5 in culture, an increasing number of cells detach from the substrate, and cell senescence is indicated by a marked increase in ultrastructural heterogeneity, progressive vesiculation and fractionation of the RER, transformation of RER stacks into huge membrane whorls, aggregation and proliferation of peroxisomes and SER, lack of dictyosomal VLDL production, drastic accumulation of lysosomes, myelinated bodies and autophagic vacuoles, as well as successive exhaustion of cellular glycogen deposits (for quantitative changes, see Tab. 10). Whereas with conventional methods for the assessment of cell viability most hepatocytes appear intact for up to 10 days in culture, cytological investigations reveal severe deterioration of cellular integrity from day 7. Since almost identical results were obtained for eel hepatocytes (Arnold, 1995; Zahn et al., 1995), both rainbow trout and eel hepatocytes can be recommended as a model for toxicological studies, provided incubation periods do not exceed 5 days.

Table 9. Semiquantitative evaluation of cytological alterations in isolated hepatocytes of rainbow trout (*Oncorhynchus mykiss*) during prolonged culture

	Day 1	Day 2	Day 3	Day 4	Day 5	Day 6	Day 7
Plasma membrane							
Microvilli	+++	++	++	++	++	++	++
Bile canaliculi	+	++	+++	+++	+++	+++	++
Nucleus							
Spherical transformation	–	–	–	–	+	+	++
Increased heterochromatin	–	–	–	–	–	+	++
Irregular outline	–	–	–	–	–	(+)	(+)
Rough endoplasmic reticulum							
Arrangement in stacks	++++	++++	+++	++++	+++	++	++
Association with other organelles	+++	+++	+++	+++	+++	+++	++
Vesiculation/fragmentation	(+)	(+)	(+)	+	+	++	++
Dilation of cisternae				+	+	++	+
Mitochondria							
Number	++++	++++	++++	++++	+++	+++	+++
Spherical outline	+++	+++	+++	++	++	+	+
Elongated outline	–	–	–	–	+	++	++
Increased polymorphism	–	–	–	+	+	++	++
Alterations in matrix	–	–	–	–	–	(+)	+
Peroxisomes							
Number	++	++	++	++	+++	+++	+++
Increase in size	–	–	–	–	+	+	+
Golgi fields							
Active secretion	+++	+++	+++	+++	+++	++	++
Size increase of single stacks	–	–	–	–	+	–	–
Dilation of cisternae	–	–	–	–	(+)	++	++
Lysosomes							
Proliferation	–	–	+	++	+++	+++	+++
Myelinated bodies	–	–	–	++	+	++	++
Autophagic vacuoles	–	–	–	+	++	+++	+++
Formation of clusters	–	–	–	–	+	+	+
Storage materials							
Lipid	++	++	+	+++	++	+	–
Glycogen	(+)	+	+	+	+	+	(+)

Abbreviations: + = little developed; ++ = moderately developed; +++ = strongly developed; ++++ = very strongly developed.

Ultrastructure of isolated rainbow trout hepatocytes following exposure to xenobiotic compounds

Table 11 gives a compilation of results from a selection of *in vitro* experiments with isolated rainbow trout hepatocytes exposed to sublethal concentrations of the reference toxicants 4-chloroaniline, 2,4-dichlorophenol and malachite green as well as the pesticides triphenyltin acetate, atrazine, dinitro-*o*-cresol and disulfoton for up to 8 days. With any of these sub-

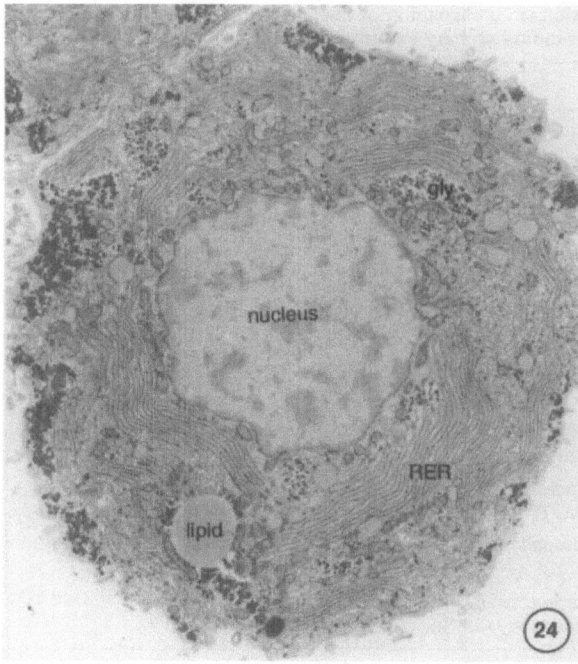

Figure 24. Except for abundant slender microvilli protruding from the plasmalemma, the cytological appearance of hepatocytes isolated from rainbow trout (*Oncorhynchus mykiss*) liver immediately after isolation resembles that of hepatocytes *in vivo* (cf. Figs. 2, 3): Freshly isolated hepatocytes are rich in glycogen (gly) deposits and display a clear-cut cytoplasmic compartmentation into large glycogen areas and organelle-containing cytoplasm. Rough endoplasmic reticulum (RER) lamellae show intensive stacking, mitochondria are in intimate relationship to cisternae of RER and accumulate around the nucleus and along the edges of RER piles. Most peroxisomes aggregate between RER stacks and glycogen fields, and part of the Golgi cisternae appear slightly fenestrated, but are clearly active in secreting VLDL particles. Control isolated hepatocyte from rainbow trout (*Oncorhynchus mykiss*); reprinted from Braunbeck (1993); magnification × 8000.

stances, first cytological modifications in isolated hepatocytes could be detected as early as 1 day at the lowest toxicant concentration tested. In fact, as a rule, no observed effect concentrations (NOEC) values could be established with the concentrations selected for the tests. For instance, whereas trypan blue exclusion failed to reveal any effect of 4-chloroaniline exposure up to concentrations of 72.2 µM, cytological modifications could be detected as early as 1 day at 7.8 µM.

As in the *in vivo* experiments, ultrastructural analysis of *in vitro* cytopathology in hepatocytes unequivocally documented a complex set of sublethal effects in a dose- and time-dependent manner. For 4-chloroaniline, e.g., changes include: modifications in nuclear morphology (Fig. 25); reduction and marginalization of heterochromatin; displacement and even-

Table 10. Alterations in composition of hepatocytes isolated from rainbow trout (*Oncorhynchus mykiss*) liver during primary culture for up to 7 days

	Day 1	Day 3	Day 5	Day 7
Mean volume of hepatocyte (μm^3)[a]	1231 ± 173	1346 ± 221	882 ± 107*	796 ± 89*
Mean volume of nucleus (μm^3)[a]	70.2 ± 8.4	74.0 ± 9.7	52.9 ± 8.6*	55.0 ± 9.8
Nuclear-cytoplasmic ratio	0.057	0.055	0.060	0.069
Mitochondrial volume per hepatocyte (μm^3)[b]	96.1 ± 13.2	101.0 ± 12.4	69.7 ± 6.0*	48.6 ± 6.9*
Peroxisomal volume per hepatocyte (μm^3)[b]	16.0 ± 2.7	16.2 ± 1.3	14.8 ± 1.6	21.5 ± 3.9*
RER volume per hepatocyte (μm^3)[b]	291.9 ± 34.5	272.0 ± 38.6	135.9 ± 22.8*	101.9 ± 14.5*
Lysosomal volume per hepatocyte (μm^3)[b]	22.2 ± 4.2	22.9 ± 3.9	25.6 ± 3.4	29.5 ± 4.2*
Volume of glycogen fields per hepatocyte (μm^3)[b]	70.2 ± 10.1	172.3 ± 30.6*	66.2 ± 8.9	31.1 ± 3.2*
Volume of lipid droplets per hepatocyte (μm^3)[b]	8.6 ± 1.4	8.1 ± 2.1	23.8 ± 3.9*	7.2 ± 0.9
Volume of cell parameters not measured individually (μm^3)[c]	656.4 ± 100.7	680.0 ± 99.7	499.4 ± 103.5	503.3 ± 82.7

Data are given as means ± S.D; n = 4. Differences from 1 day values were evaluated by the Wilcoxon-Mann-Whitney U-test: * p < 0.05.
[a] Volume of hepatocytes and nuclei as calculated for spherical profiles from direct measurement of cellular and nuclear diameter.
[b] Volumes as calculated from volume densities of respective structures.
[c] Parameter comprising Golgi fields, smooth endoplasmic reticulum, bile canaliculi and cytosol. Calculated as hepatocytic volume – (nuclear + mitochondrial + peroxisomal + RER + lysosomal + glycogen + lipid volume).

tual loss of the nucleolus (Fig. 25); proliferation and increased pleomorphism of mitochondria and peroxisomes; matrix dilation and loss of cristae in mitochondria; progressive degranulation, fenestration, vesiculation, fragmentation and transformation into huge concentric membrane whorls of the RER (Figs. 25, 26); proliferation of the SER (Fig. 26); accumulation of lipid within ER cisternae (steatosis); formation of glycogen bodies; stagnation of Golgi secretory activity; induction of myelinated bodies and autophagic vacuoles, but only transient induction of lysosomes followed by a suppression of lysosomal proliferation.

Table 11. Lowest observed effect concentration of *in vitro* exposure to selected organic xenobiotic compounds with respect to cytological alterations in isolated rainbow trout (*Oncorhynchus mykiss*) hepatocytes

	Dinitro-o-cresol	2,4-Dichlorophenol	4-Chloroaniline	Atrazine	Disulfoton	Triphenyltin acetate	Malachite green
Nucleus							
Irregular shape	7.2 (2)	—	7.8 (1)	—	7.3 (3)	0.24 (1)	—
Dilation of nuclear envelope	0.7 (2)	6.1 (4)	—	2.32 (5)	—	—	—
Rupture of nuclear envelope	0.7 (5)	—	—	—	72.9 (1)	2.44 (1)	—
Condensation of heterochromatin	0.7 (5)	—	7.8 (5)	23.2 (5)	—	0.24 (3)	0.029 (1)
Marginalization of heterochromatin	—	—	—	—	—	0.24 (5)	—
Proliferation of heterochromatin	—	—	—	—	—	2.44 (3)	—
Dense heterochromatin aggregates	—	—	—	—	—	—	—
Reduction of heterochromatin	—	—	7.8 (5)	2.32 (5)	72.9 (1)	—	0.029 (1)
Altered heterochromatin pattern	—	61.3 (4)	—	—	—	—	—
Loss of nucleolus	—	—	78.4 (5)	—	—	24.4 (3)	—
Mitochondria							
Heterogeneity in size	72.2 (4)	613 (2)	7.8 (3)	—	72.9 (1)	0.24 (1)	0.29 (1)
Proliferation of mitochondria	—	—	23.5 (1)	—	—	—	—
Reduction of mitochondria	—	—	—	—	—	0.24 (5)	—
Formation of mitochondria clusters	0.7 (2)	61.3 (3)	—	0.2 (1)	—	0.24 (1)	0.029 (3)
Formation of myelin-like whorls	—	—	7.8 (1)	23.2 (5)	—	—	—
Loss of mitochondrial cristae	—	—	23.5 (5)	—	—	—	—
Association with lipid droplets	—	—	—	0.2 (1)	—	—	0.029 (5)
Peroxisomes							
Proliferation	—	—	23.5 (5)	0.2 (1)	—	0.24 (1)	0.29 (1)
Formation of peroxisome clusters	—	—	78.4 (1)	—	—	0.24 (1)	—

Table 11 (continued)

	Dinitro-o-cresol	2,4-Dichlorophenol	4-Chloroaniline	Atrazine	Disulfoton	Triphenyltin acetate	Malachite green
Rough endoplasmic reticulum							
Reduction	7.2 (3)	6.1 (4)	7.8 (3)	0.2 (3)	–	0.24 (1)	0.29 (3)
Loss of cisternal stacking	0.7 (3)	6.1 (1)	7.8 (3)	0.2 (1)	7.3 (3)	0.24 (1)	0.29 (3)
Proliferation of cisternae	–	–	–	–	72.9 (1)	–	–
Fenestration of cisternae	–	–	23.5 (2)	–	–	–	–
Vesiculation and/or fragmentation	0.7 (4)	6.1 (1)	7.8 (1)	0.2 (5)	7.3 (3)	0.24 (1)	0.029 (1)
Degranulation	–	–	7.8 (1)	–	–	–	–
Concentric arrangement (whorls)	0.7 (3)	6.1 (1)	7.8 (1)	0.2 (1)	7.3 (3)	0.24 (1)	0.029 (1)
Dilation of cisternae	7.2 (1)	6.1 (1)	–	0.2 (3)	7.3 (3)	0.24 (1)	0.29 (1)
Smooth endoplasmic reticulum							
Proliferation	–	6.1 (2)	7.8 (1)	23.2 (5)	–	0.24 (1)	0.029 (5)
Steatosis (microvesicular lipid)	–	–	7.8 (1)	23.2 (5)	72.9 (1)	–	–
Golgi fields							
Inactivation of VLDL secretion	–	–	7.8 (5)	2.3 (3)	–	–	0.029 (5)
Dilation of cisternae	7.2 (1)	6.1 (2)	–	23.2 (3)	–	0.24 (1)	–
Concentric arrangement of cisternae	7.2 (3)	–	–	–	–	–	–
Lysosomal elements							
Proliferation of lysosomes	7.2 (1)	6.1 (2)	7.8 (1)	0.2 (3)	7.3 (3)	0.24 (1)	0.29 (1)
Formation of lysosome clusters	–	–	7.8 (3)	–	–	0.24 (1)	–
Induction of myelinated bodies	72.2 (2)	61.3 (4)	7.8 (3)	2.3 (3)	0.7 (1)	0.24 (1)	0.029 (5)
Induction of cytoplasmic vacuoles	7.2 (1)	6.1 (3)	23.5 (3)	23.2 (5)	0.7 (1)	0.24 (1)	0.029 (3)
Storage products							
Decrease in glycogen contents	7.2 (59	61.3 (4)	–	23.2 (3)	–	0.24 (1)	0.029 (5)
Formation of glycogenosomes	0.7 (2)	6.1 (19	–	2.3 (3)	–	–	0.029 (3)
Formation of glycogen bodies	–	–	7.8 (3)	0.2 (5)	–	–	0.029 (3)
Increase in lipid droplets	0.7 (4)	6.1 (4)	–	2.3 (3)	72.9 (1)	2.44 (3)	0.29 (1)

Data are presented as LOEC (lowest observed effect concentration) in µM plus earliest experimental period in days, after which the alteration was recorded (in brackets). For reference, see Table 8.

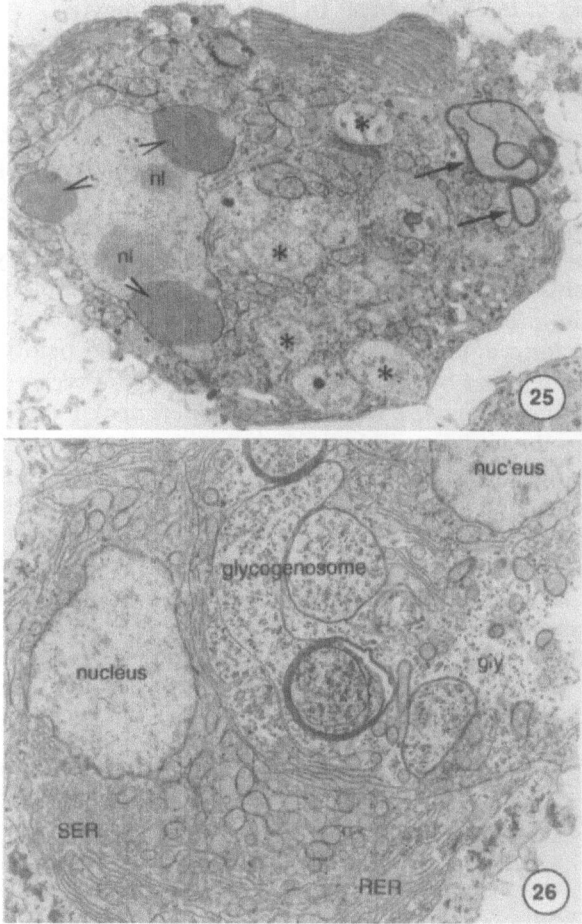

Figures 25 and 26. Nuclear changes in primary cultures of hepatocytes isolated from rainbow trout (*Oncorhynchus mykiss*) following *in vitro* exposure to 4-chloroaniline include a more ir-regular outline, reduction and marginalization of the heterochromatin (arrowheads), dislocation of the nucleoli to the nuclear periphery as well as induction of multiple nucleoli (nl; Fig. 25). Note the circular arrangement and partial transformation of the rough endoplasmic reticulum (RER) into myelinated bodies (arrows), as well as the high number of autophagic vacuoles (Fig. 25: *). Further alterations of the endoplasmic reticulum induced by 4-chloroaniline comprise the inclusion of considerable glycogen areas (gly) into myelin-like structures (glycogenosomes; Fig. 26) as well as the proliferation of the smooth endoplasmic reticulum (SER; Fig. 26). Reprinted from Braunbeck (1993); Figure 25: 3 mg/L 4-chloroaniline *in vitro* for 5 days, magnification × 6700; Figure 26: 10 mg/L 45-chloroaniline *in vitro* for 3 days; × 6800.

Comparability of *in vitro* changes in isolated hepatocytes and *in vivo* alterations in the liver of intact rainbow trout: the question of transferability

A detailed evaluation and comparison of hepatocellular *in vitro* alterations by 4-chloroaniline with effects in intact rainbow trout liver suggest major correspondence in some fundamental hepatocellular functions such as intracellular compartmentation, number of mitochondria and lysosomes, ultrastructural organization of the RER, as well as the secretory activity of the Golgi fields (Fig. 27; Braunbeck, 1993). Other symptoms, e.g., ER transformation into myelinated structures, can be interpreted as toxicant-induced premature aging of isolated hepatocytes (see below). However, since a similar number of ultrastructural changes, however, could only be observed in either the *in vitro* or the *in vivo* experiment, a direct transfer of *in vitro* data to the situation in intact fish does not appear justified, which is not surprising, since any systemic effect in the intact animal is – by definition – excluded by the *in vitro* approach. The ultrastructural evaluation of cytopathology in isolated hepatocytes represents a test system different from the corresponding *in vivo* approach.

In contrast, a comparison of the *in vitro* and *in vivo* approaches with respect to sensitivity in terms of the lowest observed effect concentrations reveals that the *in vitro* assay is just as sensitive as cytopathology in the intact fish, if not more sensitive; it is, thus, extremely sensitive, especially

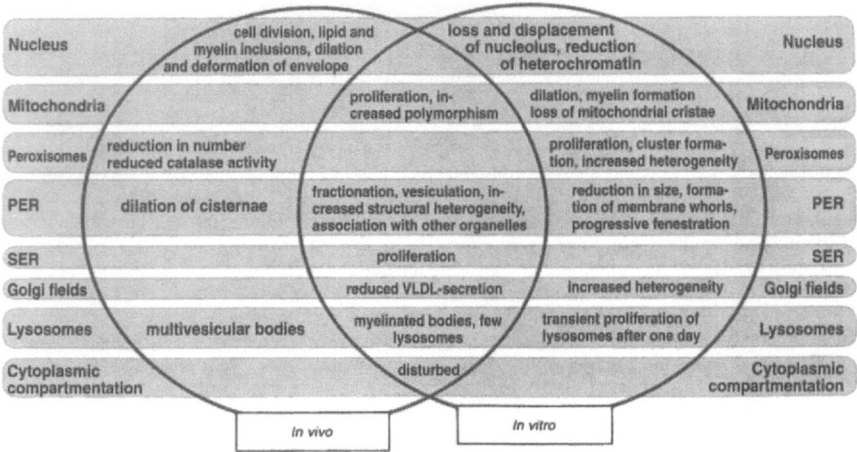

Figure 27. Comparison of cytopathological effects induced in rainbow trout (*Oncorhynchus mykiss*) hepatocytes after *in vivo* exposure of rainbow trout as well as after *in vitro* exposure of primary hepatocyte cultures to selected sublethal concentrations of the reference compound 4-chloroaniline. RER – rough endoplasmic reticulum; SER – smooth endoplasmic reticulum; reprinted from Braunbeck (1994c). Data as compiled from *in vitro* results published by Braunbeck (1993) as well as *in vivo* findings published by Braunbeck et al. (1990c).

when exposure time is taken into account: whereas *in vivo* experiments usually take 4 to 5 weeks, *in vitro* results can be revealed as early as 1 to 3 days after the onset of exposure. If compared to permanent fish cell cultures (cf. the chapter by H. Segner, this volume), isolated hepatocytes are undoubtedly more sensitive (Braunbeck, 1995; Braunbeck et al., 1996; Hollert and Braunbeck, 1997; Zahn and Braunbeck, 1995a; Zahn et al., 1993, 1996a, b). On the basis of these considerations, the *in vitro* approach with isolated fish hepatocytes can be particularly recommended for purposes of rapid and sensitive screening for basic toxic effects of xenobiotic compounds.

Specificity of *in vitro* reactions in the ultrastructure of isolated hepatocytes

A closer evaluation of effects by 4-chloroaniline, dinitro-*o*-cresol, 2,4-dichlorophenol, triphenyltin acetate and malachite green reveals that – as in corresponding *in vivo* experiments – each cytopathological syndrome consists of both unspecific modifications as listed for 4-chloroaniline and substance-specific changes. Transformation of Golgi cisternae into concentric membrane whorls, e.g., is specific of dinitro-*o*-cresol, whereas reactions to 2,4-dichlorophenol are characterized by unusual heterochromatin distribution patterns and formation of mitochondria aggregations (Zahn and Braunbeck, 1993). Comparison of results with changes induced by 4-chloroaniline reveals loss of mitochondrial cristae, proliferation of mitochondria, and cessation of VLDL secretion as typical of 4-chloroaniline exposure. Uncommon reactions to triphenyltin acetate include a suppression of mitochondria, whereas atrazine and malachite green apparently failed to induce specific cytopathological alterations (Zahn and Braunbeck, 1995b). The differences in the reaction to the phenol derivatives 2,4-dichlorophenol and dinitro-*o*-cresol again lead to the conclusion that even within a given class of chemical compounds transfer of results on sublethal toxicity from one compound to another may be extremely delicate.

Time-course of cytopathological changes in isolated hepatocytes: senescence as a major mechanism of toxicity?

Since, if compared to corresponding *in vivo* assays, the expenditure of time and labor is certainly lower in the *in vitro* approach, the latter appears particularly suitable for the systematic analysis of the time-course of cytopathological changes in hepatocytes. A more-in-depth analysis of changes induced by the reference compound 4-chloroaniline reveals that most of the unspecific changes listed in Table 11 can also be found in control hepatocytes (Fig. 28), however, after a longer period of exposure. This led to the

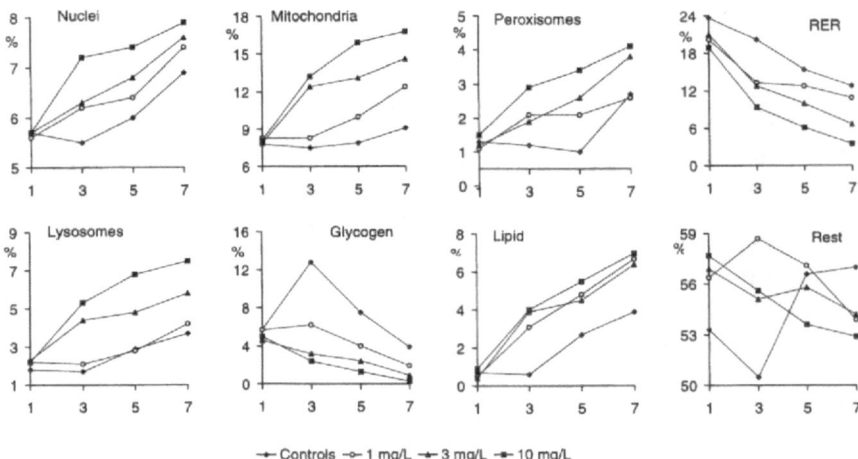

Figure 28. A closer analysis of time-dependent volumetric alterations in primary hepatocytes isolated from rainbow trout (*Oncorhynchus mykiss*) following *in vitro* exposure for up to 5 days to selected sublethal concentrations of the reference compound 4-chloroaniline reveals that chemical-induced changes (temporally) precede volumetric changes observed in control cells during senescence in the course of prolonged primary culture and thus give rise to the hypothesis of protracted senescence as a fundamental toxicological process (cf. Braunbeck, 1994b, c, 1997). Data as quantified according to morphometric techniques following the principles by Weibel (1979) from results published by Braunbeck (1993).

conclusion that especially the stereotype reactions observed both in *in vivo* and in *in vitro* experiments with rainbow trout liver and hepatocytes, respectively, might be symptoms of protracted senescence (Braunbeck, 1994b, 1997). In fact, comparison of corresponding *in vivo* and *in vitro* changes in eel hepatocytes by the phosphorodithioate disulfoton strongly corroborate this view (Arnold, 1995; Zahn et al., 1996a).

Field studies: Liver cytopathology as a source of biomarkers of environmental contamination

If compared to laboratory studies (Tab. 2), the number of field studies using changes in hepatocellular ultrastructure as markers of exposure to an appreciable extent is restricted to about 25 communications (Tab. 12). In all of these, however, numerous alterations have been described. In contrast to laboratory experiments, which usually produce a considerable number of stereotype reactions, effects described in the liver of fish sampled in the field are usually much more diverse. For instance, the most prominent effect in hepatocytes of Arctic char (*Salvelinus alpinus*) sampled in high-alpine lakes with occasionally extreme living conditions (Hofer and Medgyesy, 1997; Hofer et al., 1997; Köck et al., 1996a, b) is accumulation of

Table 12. Selected cytopathological investigations on ultrastructural alterations in fish liver in field studies

Species	Case study	Reference
Rainbow trout (*Oncorhynchus mykiss*)	Tumors in hatchery-reared trout (USA)	Halver (1976), Scarpelli et al. (1963, 1967)
	Sediments from Black river (Ohio, USA)	Stoker et al. (1985)
	Prudhoe Bay crude oil (p.o.)	Hawkes (1977)
Arctic charr (*Salvelinus alpinus*)	Acidic high alpine lakes (Austria)	Braunbeck et al. (1993), Hofer et al. (1997)
Common barbel (*Barbus barbus*)	River Meuse (Netherlands)	Hugla et al. (1995)
Rock sole (*Lepdisopsetta bilineata*)	U.S. west coast	Stehr et al. (1998)
English sole (*Parophrys vetulus*)	Puget Sound	Stehr (1986), Stehr and Myers (1990), Stehr et al. (1988)
	Experimentally oiled sediments	McCain et al. (1978)
Winter flounder (*Pseudopleuronectes americanus*)	Boston harbor	Bodammer and Murchelano (1990), Moore et al. (1989), Murchelano and Wolke (1985)
European flounder (*Platichthys flesus*)	Regeneration of flounder from the Elbe estuary (German Bight)	Köhler (1989)
Starry flounder (*Platichthys stellatus*)	U.S. west coast	Stehr et al. (1998)
Brown bullhead (*Ictalurus nebulosus*)	Sediments from Black river (Ohio, USA)	Stoker et al. (1985)
Eel (*Anguilla anguilla*)	Sandoz chemical spill (Switzerland)	Braunbeck et al. (1994)
	Bages-Segean, Gulf of Lyons, France	Biagianti-Risbourg et al. (1993)
Mummichog (killifish; *Fundulus heteroclitus*)	Regeneration after oil spill exposure	Sabo and Stegeman (1977)
	Naphthalene	DiMichele and Taylor (1978)
Bullhead (*Cottus gobio*)	Paper mill effluents (Austria)	Bucher et al. (1992)
White croaker (*Genyonemus lineatus*)	U.S. west coast	Stehr et al. (1998)
Ruffe (*Gymnocephalus cernua*)	Lower Elbe River (Germany)	Kranz and Peters (1984)

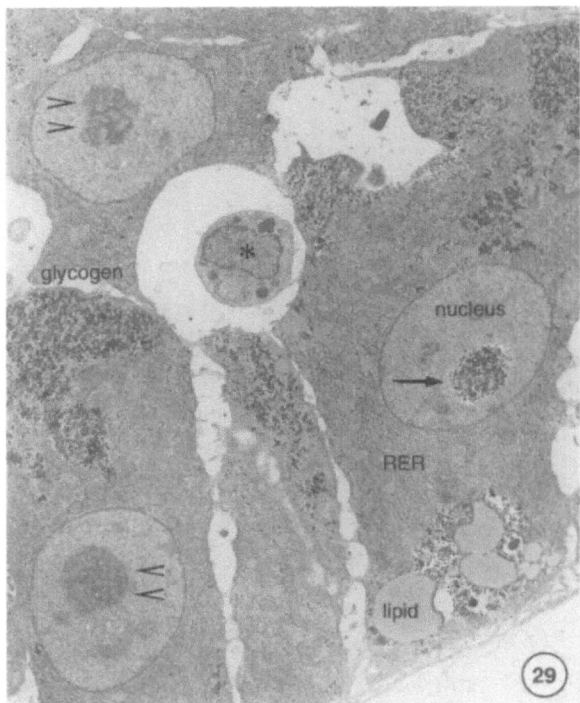

Figure 29. In contrast to laboratory experiments, which usually produce a considerable number of unspecific cytopathological alterations, fish from the field frequently show highly unusual effects. The most prominent effect in hepatocytes of Arctic char (*Salvelinus alpinus*) sampled in high-alpine lakes with occasionally extreme living conditions, for example, is accumulation of glycogen within nuclei (at arrows) as well as bizarre deformations of the nucleolus (at arrowheads), which have, so far, not been reported from any laboratory study. At asterisk, note macrophage in dilated intercellular space. Reprinted from Hofer et al. (1997); magnification × 2800.

glycogen within nuclei resembling the inclusions described by Thiyagarajah and Grizzle (1985, 1986) in neoplastic liver of *Rivulus ocellatus marmoratus* exposed to diethylnitrosamine as well as bizarre deformations of the nucleolus, which have, so far, not been reported from any laboratory study (Fig. 29; Hofer et al., 1997).

Likewise, in addition to a number of unspecific changes, hepatocytes from European eel exposed to the chemical spill into the Rhine river system at Basel in November 1986 displayed extraordinary features such as annulate lamellae (Fig. 30), crystallization of the peroxisomal matrix (Fig. 31) and accumulation of protein-like crystals within distended cisternae of the RER (Fig. 32). There is increasing evidence that in liver annulate lamellae, which have been intensively reviewed by Kessel (1968, 1983, 1989) as well as Ghadially (1988), are restricted to pathological conditions and hepatoma cells (Hoshino, 1963; Hruban et al., 1965a, b; Svoboda, 1964),

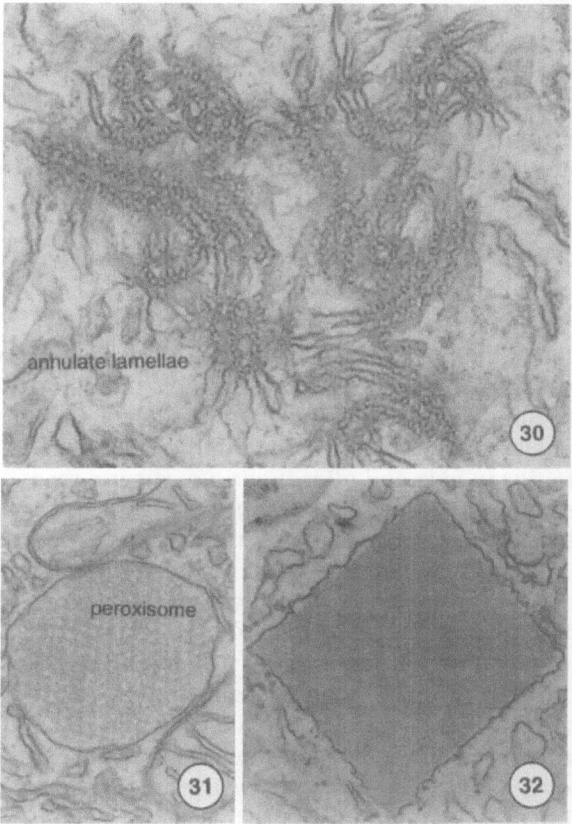

Figures 30–32. In addition to a number of unspecific changes, hepatocytes from European eel (*Anguilla anguilla*) exposed to the chemical spill into the Rhine river system at Basel in November 1986 displayed extraordinary features such as annulate lamellae (Fig. 30), crystallization of the peroxisomal matrix (Fig. 31) and accumulation of protein-like crystals within distended cisternae of the rough endoplasmic reticulum (RER). Reprinted from Braunbeck et al. (1990a); Figure 30: magnification × 64 800; Figure 31: × 30 200; Figure 32: × 32 400.

except for their normal occurrence in embryonic liver (Benzo, 1974). In fish liver, only Mugnaini and Harboe (1966) reported annulate lamellae in the hagfish (*Myxine glutinosa*).

Transformation even of parts of the peroxisomal matrix into a huge crystalloid core (Fig. 31) is highly unusual in fish. Although fish liver peroxisomes lack a crystalline core (Braunbeck et al., 1987; Goldenberg et al., 1978; Heusequin, 1973; Kramar et al., 1974), uricase, the major component of the crystalloid in mammals (Angermüller and Fahimi, 1986; Hruban and Swift, 1964), was found to be present in a soluble form (Braunbeck et al., 1987; Fujiwara et al., 1987; Goldenberg, 1977; Hayashi et al., 1989). However, soluble amphibian uricase purified from isolated peroxisomes can be precipitated by raising pH, applying detergents or transferring to hyper- or hypoosmotic media (Fujiwara et al., 1987). Although the latter authors failed to reveal the same properties in fish liver uricase, it appears reasonable to consider the crystalline lattice overlying the peroxisomal matrix in the liver of exposed eels to be caused by precipitation of uricase (and/or other peroxisomal matrix proteins) under such modified intracellular conditions.

Conclusions

Cytopathological alterations in hepatocytes of fish as a consequence of sublethal exposure to xenobiotic compounds undoubtedly represent a powerful tool to not only reveal the effects of chemicals, but also to give hints as to the site of damage and – potentially – to the underlying mode of action. Both after *in vivo* and after *in vitro* exposure, reactions of hepatocytes to even traces of chemical compounds are very sensitive, highly selective, and, especially in *in vitro* experiments, extremely rapid: Whereas *in vivo* several days may be required to elucidate the effect of chemicals, early *in vitro* reactions may be revealed within a few hours.

Hepatocellular reactions are relative to internal parameters such as species, sex, age and hormonal status of the fish investigated; moreover, external (environmental) parameters such as temperature, nutrition and duration of exposure need to be considered as confounding factors. As a consequence, transfer of results and conclusions from one experiment to another is usually not possible. Likewise, *in vitro* results may not necessarily be extrapolated to the situation in intact fish. In any case, alterations in hepatocytes (from fish) exposed to acutely toxic concentrations do not allow predictions as to potential reactions at the sublethal level of contamination. In particular, due to the complex exposure situation in the field, results from field studies will most likely be unique for the given situation.

Although many changes do not appear to be specific of particular compounds, there are examples of ultrastructural modifications, which seem unique for particular chemicals or at least a class of chemical. In any case, as a syndrome, the complex of all changes induced by a given xenobiotic, is specific. Especially in the lower exposure range, most, if not all, ultrastructural alterations appear to be fully reversible; upon cessation of exposure, except for deposition of defect cell components in secondary and tertiary lysosomes (residual bodies), restitution of hepatocellular integrity is usually rapid and accomplished within a few days.

Moreover, most early reactions of hepatocytes apparently serve functions within the general toxicant adaptation syndrome of fish to chemical exposure, which is induced to compensate for the misbalance in organismic homeostasis. Most ultrastructural alterations observed after sublethal exposure to xenobiotic compounds have, therefore, to be classified as indicators of adaptive processes and may be contrasted to degenerative, i.e., truly pathological phenomena, with most components of the latter category being irreversible. Once operative and not overloaded, these adaptive processes should, by definition, not have consequences and become overt at higher levels of biological organization; yet, as markers of the induction of compensative processes, they are of ecotoxicological relevance. Thus, with regard to their (eco)toxicological significance, components of the unspecific general toxicant adaptation syndrome may serve as early and sensitive warning signals of chemical exposure, whereas more specific changes may be of advanced diagnostic value and thus may even serve as indices for the identification of certain chemicals (chemical classes). Thus, the diagnostic value of cytopathological changes is relative to their specific quality.

In order to avoid the potentially subjective nature of primarily qualitative morphological data, integration of cytopathological alterations in fish liver and hepatocytes into routine aquatic toxicology requires quantification by means of stereological techniques. Morphometric quantification makes structural data accessible for statistical analysis and, thus, comparable with primarily (semi)quantitative techniques such as biochemistry and molecular biological techniques. Especially since the functional interpretation of morphological observations and, thus, the assessment of their ecological and ecotoxicological relevance are frequently difficult, statistical evaluation may at least assist in objectively defining thresholds of effect.

Last, but not least, implementation of cytopathological techniques into routine long-term investigations with fish gives credit to the principles of animal welfare and protection, since more-in-depth analysis of internal mechanisms of sublethal chemical contamination in addition to the study of externally overt symptoms of intoxication undoubtedly adds to a *refinement* of animal experiments with fish. One step further towards *reduction* of animal experiments is the translation of the principles of environmental cytopathology to cell cultures: Even with primary cultures of hepatocytes, donor fish are no longer exposed to chemical stress, and, since isolation of hepatocytes from the liver of, for example, rainbow trout gives an average yield of 10^8 cells per g liver with maximum cell recoveries of $5-6 \times 10^8$/g liver, hepatocytes from one individual fish may be sufficient to test $3-4$ concentrations of a given chemical compound or field samples in replicate.

References

Affandi, R. and Biagianti, S. (1987) A study of the liver of eels kept in captivity: disturbances induced in hepatocytes by artificial diets. *Aquaculture* 67:226–228.

Alderman, D.J. (1985) Malachite green: a review. *J. Fish Dis.* 8:289–298.

Alderman, D.J. (1991) Malachite green and alternatives. *EAS Spec. Publ., Gent, Belgium*, 14:17–24.

Altmann, H.W. (1980) Drug-induced liver reactions: a morphological approach. *Curr. Top. Pathol.* 69:69–142.

Anderson, M.J., Olsen, H., Matsumara, F. and Hinton, D.E. (1996) *In vivo* modulation of 17β-estradiol-induced vitellogenin synthesis and estrogen receptor in rainbow trout (*Oncorhynchus mykiss*) liver cells by β-naphthoflavone. *Toxicol. Appl. Pharmacol.* 137:210–218.

Andersson, T. (1984) Oxidative and conjugative metabolism in isolated rainbow trout liver cells. *Mar. Environ. Res.* 14:442–443.

Andersson, T. (1986) Cytochrome P-450-dependent metabolism in isolated liver cells from fed and starved rainbow trout, *Salmo gairdneri. Fish. Physiol. Biochem.* 1:105–111.

Andersson, T. and Förlin, L. (1985) Spectral properties of substrate-cytochrome P-450 interaction and catalytic activity of xenobiotic metabolizing enzymes in isolated rainbow trout liver cells. *Biochem. Pharmacol.* 34:1407–1413.

Andersson, T. and Koivusaari, U. (1986) Oxidative and conjugative metabolism of xenobiotics in isolated liver cells from thermally acclimated rainbow trout. *Aquat. Toxicol.* 8:85–92.

Andersson, T., Koivusaari, U. and Förlin, L. (1985) Xenobiotic transformation in the rainbow trout liver and kidney during starvation. *Comp. Biochem. Physiol.* 82C:221–225.

Angermüller, S. and Fahimi, H.D. (1981) Selective cytochemical localization of peroxidase, cytochrome oxidase and catalase in rat liver with 3,3'-diaminobenzidine. *Histochemistry* 71:33–44.

Angermüller, S. and Fahimi, H.D. (1983) Selective staining of cell organelles in rat liver with 3,3'-diaminobenzidine. *J. Histochem. Cytochem.* 31:230–232.

Angermüller, S. and Fahimi, H.D. (1986) Ultrastructural cytochemical localization of uricase in peroxisomes of rat liver. *J. Histochem. Cytochem.* 34:159–165.

Arias, I.M., Popper, H., Schachter, D. and Schafritz, D.A. (1988) *The liver: biology and pathobiology*. Raven Press, New York., 664 pp.

Arnold, H. (1995) *Die Wirkung von Pestiziden auf Monitororgane und -zellen bei Fischen. Ein Vergleich von subletalen Effekten in Einzelstudien und im Kombinationsexperiment*. Ph.D. Thesis, Dept. of Zoology, University of Heidelberg, 194 pp.

Arnold, H. and Braunbeck, T. (1993) Subletale Toxizität von Schadstoffgemischen: Die Wirkung von Endosulfan und Disulfoton auf die Leber der Regenbogenforelle (*Oncorhynchus mykiss*). *Verh. Dtsch. Zool. Ges.* 86:189.

Arnold, H. and Braunbeck, T. (1994a) Disulfoton as a major toxicant in the chemical spill at Basel in November 1986: Acute and chronic studies with eel and rainbow trout. *In:* Müller, R. and Lloyd, R. (eds) *Sublethal and chronic effects of pollutants on freshwater fish*. Blackwell, Oxford, pp 79–86.

Arnold, H. and Braunbeck, T. (1994b) Qualitative und quantitative Effekte in Leberzellen der Regenbogenforelle (*Oncorhynchus mykiss*) infolge einer subletalen Endosulfanexposition: Bedingt das eine das andere? *Verh. Dtsch. Zool. Ges.* 87:70.

Arnold, H., Pluta, H.J. and Braunbeck, T. (1995) Simultaneous exposure of fish to endosulfan and disulfoton *in vivo*: ultrastructural stereological and biochemical reactions in hepatocytes of male rainbow trout (*Oncorhynchus mykiss*). *Aquat. Toxicol.* 33:17–43.

Arnold, H., Pluta, H.J. and Braunbeck, T. (1996a) Sublethal effects of prolonged exposure to disulfoton in rainbow trout (*Oncorhynchus mykiss*): cytological alterations in the liver by a potent acetylcholine esterase inhibitor. *Ecotox. Environ. Safety* 34:43–55.

Arnold, H, Pluta, H.J. and Braunbeck, T. (1996b) Cytological alterations in the liver of rainbow trout (*Oncorhynchus mykiss*) after prolonged exposure to water-borne endosulfan. *Dis. Aquat. Org.* 25:39–52.

Bac, N., Biagianti, S. and Bruslé, J. (1983) Etude cytologique ultrastructurale des anomalies hépatiques du loup, de la daurade et de l'anguille induites par une alimentation artificielle. *IFREMER Actes Coll.* 1:473–484.

Baglio, C.M. and Farber, E. (1965) Reversal by adenine of the ethionine-induced lipid-accumulation in the endoplasmic reticulum of the rat liver. *J. Cell Biol.* 27:591–601.

Bailey, G.S., Taylor, M.J. and Selivonchick, D.P. (1982) Aflatoxin B1 metabolism and DNA binding in isolated hepatocytes from rainbow trout (*Salmo gairdneri*). *Carcinogenesis* 3:511–518.

Bailey, G.S., Taylor, M.J., Loveland, P.M., Wilcox, J.S., Sinnhuber, R.O. et al. (1984) Dietary modification of aflatoxin B1 carcinogenesis: mechanism studies with isolated hepatocytes from rainbow trout. *Natl. Cancer Inst. Monogr.* 65:379–385.

Baksi, S.M. and Frazier, J.M. (1988) A fish hepatocyte model for the investigation of the effects of environmental contaminants. *Mar. Environ. Res.* 243:141–145.

Baksi, S.M. and Frazier, J.M. (1990) Review. Isolated fish hepatocytes – model systems for toxicology research. *Aquat. Toxicol.* 16:229–256.

Ballatori, N., Shi, C. and Boyer, J.L. (1988) Altered plasma membrane ion permeability in mercury-induced cell injury: studies in hepatocytes of elasmobranch *Raja erinacea*. *Toxicol. Appl. Pharmacol.* 95:279–291.

Barni, S., Bernocchi, G. and Gerzeli, G. (1985) Morphohistochemical changes in hepatocytes during the life cycle of the European eel. *Tissue & Cell* 17:97–109.

Bartholomew, G.A. (1964) The roles of physiology and behavior in the maintenance of homeostasis in the desert environment. *SEB Symposia Series* 18:7–21.

Benedeczky, I. and Nemcsok, J. (1990) Detection of phenol-induced subcellular alteration by electron microscopy in the liver and pancreas of carp. *Environ. Monit. Assess.* 14:385–394.

Benedeczky, I., Biro, P. and Schaff, Z. (1984) The effect of 2,4-D-containing herbicide (Dikonirt) on the ultrastructure of carp (*Cyprinus carpio*) liver cells. *Acta Biol. Hung.* 30:107–125.

Benedeczky, I., Nemcsok, J. and Halasy, K. (1986) Electron microscopic analysis of the cytopathological effect of pesticides in the liver, kidney and gill tissues of carp. *Acta Biol. Szeged* 32:69–91.

Benzo, C.A. (1974) Annulate lamellae in hepatic and pancreatic cells of the chick. *Am. J. Anat.* 140:139–143.

Berlin, J.D. and Dean, J.M. (1967) Temperature-induced alterations in hepatocyte structure of rainbow trout. *J. Exp. Zool.* 164:117–132.

Berry, M.N. and Friend, D.S. (1969) High yield preparation of isolated rat liver parenchymal cells. *J. Cell Biol.* 43:506–520.

Biagianti, S. (1986) Cytopathological effects of a herbicide (a S-triazine: atrazine) on the liver of grey mullet juveniles. *In:* Vivarès, C.P., Bonami, J.R. and Jaspers, E. (eds) *Pathology in Marine Aquaculture. European Aquaculture Society Spec. Publ.*, pp 91–92.

Biagianti, S. (1987) Perturbations structurales et métaboliques induites par une dose sublétale d'atrazine (s-triazine, herbicide), dans les hépatocytes d'un téléostéen, le muge (*Liza ramada*). Symposium International sur les Impacts Cellulaires en Ecotoxicologie, Lyon.

Biagianti-Risbourg, S. (1991) Fine structure of hepatocytes in juvenile grey mullets: *Liza saliens* Risso, L. *ramada* Risso and *L. aurata* Risso (Teleostei, Mugilidae). *J. Fish Biol.* 39:687–703.

Biagianti-Risbourg, S. (1992) Intérêt (éco)toxicologique d'une approche hysto-cytologique pour la compréhension des reponses hépatiques de mugilides (Téléostéens) à des contaminations subaigues par l'atrazine. *Ichthyophys. Acta* 15:205–223.

Biagianti-Risbourg, S. and Bastide, J. (1995) Hepatic perturbations induced by a herbicide (atrazine) in juvenile grey mullet *Liza ramada* (Mugilidae, Teleostei): an ultrastructural study. *Aquat. Toxicol.* 31:217–229.

Biagianti, S., Gony, S. and Lecomte, R. (1986) Experimental effects of cadmium on glass eel (*Anguilla anguilla*). *Vie Milieu* 36:317.

Biagianti-Risbourg, S., Brutel, A., Lemaire, C., Hildebrand, C. and Lecomte-Finiger, R. (1993) Liver pertubations in *Anguilla anguilla* from a polluted mediterranean lagoon (Bages-Segean, Gulf of Lyons, France). 8th EIFAC Working Party on Eel, Olsztyn, Poland.

Biagianti-Risbourg, S., Pairault, C., Vernet, G. and Boulekbache, H. (1996) Effect of lindane on the ultrastructure of the liver of the rainbow trout, *Oncorhynchus mykiss*, sac-fry. *Chemosphere* 33:2065–2079.

Birnbaum, M.J., Schultz, J. and Fain, J.N. (1976) Hormone-stimulated glycogenolysis in isolated goldfish hepatocytes. *Am. J. Physiol.* 231:191–197.

Blair, J.B., Miller, M.R., Pack, D., Barnes, R., Teh, S.J. and Hinton, D.E. (1990) Isolated trout liver cells: establishing short-term primary cultures exhibiting cell-to-cell interactions. *In Vitro Cell Dev. Biol.* 26:237–249.

Blair, M.G., Ostrander, G.K., Miller, M.R. and Hinton, D.E. (1995) Isolation and characterization of biliary epithelial cells from rainbow trout liver. *In Vitro Cell. Dev. Biol. Anim.* 31:780–789.

Blouin, A. (1977) Morphometry of liver sinusoidal cells. *In:* Wisse, E. and Knook, D.L. (eds) *Kupffer Cells and other Liver Sinusoidal Cells.* Elsevier, Amsterdam, pp 66–71.

Blouin, A., Bolender, R.P. and Weibel, E.R. (1977) Distribution of organelles and membranes between hepatocytes and non-hepatocytes in the rat liver parenchyma. *J. Cell Biol.* 72:441–455.

Bodammer, J.E. and Murchelano, R.A. (1990) Cytological study of vacuolated cells and other aberrant hepatocytes in winter flounder from Boston Harbor. *Cancer Res.* 50:6744–6756.

Bolender, R.P. (1979) Morphometric analysis in the assessment of the response of the liver to drugs. *Pharm. Rev.* 30:429–443.

Bolender, R.P. and Weibel, E.R. (1973) A morphometric study of the removal of phenobarbital-induced membranes from hepatocytes after cessation of treatment. *J. Cell Biol.* 56:746–761.

Bonates, A.B.I. and Ferri, S. (1980) Fine structure of a freshwater teleost (*Pimelodus maculatus*) hepatocytes revealed by ultrathin sections. *Anat. Anz.* 148:132–144.

Bouche, G., Gas, N. and Paris, H. (1979) Isolation of carp hepatocytes by centrifugation on a discontinuous ficoll gradient. A biochemical and ultrastructural study. *Biol. Cellul.* 36:17–24.

Braunbeck, T. (1992) Isolated fish hepatocytes and permanent fish cell lines in cytotoxicity tests – an alternative to fish hepatocytes *in vivo*? *In:* Schoeffl, H., Schulte-Hermann, R. and Tritthart, H.A. (eds) *Möglichkeiten und Grenzen der Reduktion von Tierversuchen*. Springer, Wien & New York, pp 153–154.

Braunbeck, T. (1993) Cytological alterations in isolated hepatocytes from rainbow trout (*Oncorhynchus mykiss*) exposed to 4-chloroaniline. *Aquat. Toxicol.* 25:83–110.

Braunbeck, T. (1994a) Detection of environmental relevant pesticide concentrations using cytological parameters: Pesticide specificity in the reaction of rainbow trout liver? *In:* Müller, R. and Lloyd, R. (eds) *Sublethal and Chronical Effects of Pollutants on Freshwater Fish*. Blackwell, Oxford., pp 15–29.

Braunbeck, T. (1994b) Protracted senescence as a mode of toxic action in isolated hepatocytes of rainbow trout? Cytological investigations on the *in vitro* toxicity of environmental contaminants. Workshop on Physiological Aspects of Cell Culture. Annual Meeting of the Society of Experimental Biology, *Society of Experimental Biology*, Swansea.

Braunbeck, T. (1994c) *Strukturelle und funktionelle Veränderungen in Hepatocyten von Fischen als Biomarker für die Belastung durch Umweltchemikalien*. Habilitation Thesis, Dept. of Zoology, University of Heidelberg, 213 pp.

Braunbeck, T. (1995) *Zelltests in der Ökotoxikologie – Cytotoxizitätstests mit Zellkulturen aus Fischen als Alternative und Ergänzung zu konventionellen Fischtests*. Veröff. PAÖ 11: 204 pp.

Braunbeck, T. (1997) Protracted senescence – a fundamental process of toxicity in isolated fish hepatocytes? *Verh. Dtsch. Zool. Ges.* 90.1:345.

Braunbeck, T. and Appelbaum, S. (1998) Ultrastructural alterations in the liver and intestine of carp (*Cyprinus carpio*) induced by ultra-low peroral doses of endosulfan. *Dis. Aquat. Org.*, in press.

Braunbeck, T. and Segner, H. (1991) Preexposure temperature acclimation and diet as modifying factors for the tolerance of golden ide (*Leuciscus idus melanotus*) to short-term exposure to 4-chloroaniline. *Ecotox. Environ. Safety*: 24:72–94.

Braunbeck, T. and Storch, V. (1989) Zelle und Umwelt – Wie wirken sich Umweltgifte auf Zellen aus? *Biologie in unserer Zeit* 19:127–132.

Braunbeck, T. and Storch, V. (1992) Senescence of hepatocytes isolated from rainbow trout (*Oncorhynchus mykiss*) in primary culture: an ultrastructural study. *Protoplasma* 170:138–159.

Braunbeck, T. and Völkl, A. (1991a) Applications of experimental histology and cytology in the diagnosis of fish kills after chemical spills. *Verh. Intern. Verein. Limnol.* 24:2477.

Braunbeck, T. and Völkl, A. (1991b) Induction of biotransformation in the liver of eel (*Anguilla anguilla* L.) by sublethal exposure to dinitro-*o*-kresol: an ultrastructural and biochemical study. *Ecotox. Environ. Safety* 21:109–127.

Braunbeck, T. and Völkl, A. (1993) Cytological alterations in the livers of golden ide (*Leuciscus idus melanotus*) and eel (*Anguilla anguilla*) induced by sublethal doses of dinitro-*o*-cresol. *In:* Braunbeck, T., Hanke, W. and Segner, H. (eds) *Fish Ecotoxicology and Ecophysiology*,. VCH, Weinheim, pp 55–80.

Braunbeck, T., Gorgas, K., Storch, V. and Völkl, A. (1987) Ultrastructure of hepatocytes in golden ide (*Leuciscus idus melanotus* L.; Cyprinidae: Teleostei) during thermal adaptation. *Anat. Embryol.* 175:303–313.

Braunbeck, T., Storch, V. and Nagel, R. (1989) Sex-specific reaction of liver ultrastructure in zebrafish (*Brachydanio rerio*) after prolonged sublethal exposure to 4-nitrophenol. *Aquat. Toxicol.* 14:185–202.

Braunbeck, T., Burkhardt-Holm, P. and Storch, V. (1990a) Liver pathology in eels (*Anguilla anguilla* L.) from the Rhine river exposed to the chemical spill at Basle in November 1986. *In:* Kinzelbach, R. and Friedrich, G. (eds) *Biologie des Rheins*. G. Fischer, Stuttgart, pp 371–392.

Braunbeck, T., Görge, G., Storch, V. and Nagel, R. (1990b) Hepatic steatosis in zebrafish (*Brachydanio rerio*) induced by long-term exposure to γ-hexachlorocyclohexane. *Ecotox. Environ. Safety* 19:355–374.

Braunbeck, T., Storch, V. and Bresch, H. (1990c) Species-specific reaction of liver ultrastructure of zebra fish (*Brachydanio rerio*) and trout (*Salmo gairdneri*) after prolonged exposure to 4- chloroaniline. *Arch. Environ. Contam. Toxicol.* 19:405–418.

Braunbeck, T., Teh, S.J., Lester, S.M. and Hinton, D.E. (1992a) Ultrastructural alterations in hepatocytes of medaka (*Oryzias latipes*) during the cytotoxic phase of diethylnitrosamine. *Toxicol. Pathol.* 20:179–196.

Braunbeck, T., Burkhardt-Holm, P., Görge, G., Nagel, R., Negele, R.D., Storch, V. (1992b) Regenbogenforelle und Zebrabärbling, zwei Modelle für verlängerte Toxizitätstests: relative Empfindlichkeit, Art- und Organspezifität in der cytopathologischen Reaktion von Leber und Darm auf Atrazin. *Schriftenr. Ver. Wasser-, Boden-, Lufthygiene* 89:109–145.

Braunbeck, T., Oulmi, Y., Sordyl, H. and Hofer, R. (1993) Seesaiblinge (*Salvelinus alpinus*) in einem Tiroler Hochgebirgssee: Auswirkungen extremer Umweltbedingungen auf Organe und Gewebe. Teil II – Cytologie von Leber und Niere. *Verh. Dtsch. Zool. Ges.* 86:5.

Braunbeck, T., Arnold, H. and Burkhardt-Holm, P. (1994) Subletale Veränderungen bei Fischen durch Pestizide. *Proceedings des Abschlusssymposiums zum Sandoz-Rheinfonds*, pp 60–67.

Braunbeck, T., Hauck, C., Scholz, S. and Segner, H. (1996) Mixed function oxygenases in cultured fish cells: contributions of *in vitro* studies to the understanding of MFO induction. *Z. Angew. Zool.* 81:55–72.

Braunbeck, T., Arnold, H., Oulmi, Y., Zahn, T. and Storch, V. (1997a) Neue Ansätze zur Bewertung der Ökotoxizität von Xenobiotika. Teil 1: *In vitro*-Tests mit Fischzellkulturen. *Bioforum* 1/97:12–16.

Braunbeck, T., Arnold, H., Oulmi, Y., Zahn, T. and Storch, V. (1997b) Neue Ansätze zur Bewertung der Ökotoxizität von Xenobiotika. Teil 2: Mikroinjektionsexperimente mit Fischeiern. *Bioforum* 4/97:149–151.

Bresch, H. (1982) Investigation of the long-term action of xenobiotics on fish with special regard to reproduction. *Ecotox. Environ. Safety* 6:102–112.

Bresch, H. (1991) Early life stages in zebrafish versus a growth test in rainbow trout to evaluate toxic effects. *Bull. Environ. Contam. Toxicol.* 46:641–648.

Bresch, H., Markert, M., Munk, R. and Spieser, O.H. (1986) Long-term fish toxicity test using the zebrafish: effect of group formation and temporary separation by sex on spawning. *Bull. Environ. Contam. Toxicol.* 37:606–614.

Bresch, H., Beck, H., Ehlermann, D., Schlaszus, H. and Urbanek, M. (1990) A long-term toxicity test comprising reproduction and growth of zebrafish with 4-chloroaniline. *Arch. Environ. Contam. Toxicol.* 19:419–527.

Brown, K.S. (1997) Making a splash with zebrafish. *Bioscience* 47:68–71.

Bucher, F., Hofer, R. and Salvenmoser, W. (1992) Effects of treated paper mill effluents on hepatic morphology in male bullhead (*Cottus gobio* L.). *Arch. Environ. Contam. Toxicol.* 23:410–419.

Bunton, T.E. (1990) Hepatopathology of diethylnitrosamine in the medaka (*Oryzias latipes*) following short-term exposure. *Toxicol. Pathol.* 18:313–323.

Byczkowska-Smyk, W. (1968) Bile canaliculi in hepatic cells of the crucian carp (*Carassius carassius* L.) and tench (*Tinca tinca* L.). *Acta Biol. Crac. ser. Zool.* 9:93–99.

Byczkowska-Smyk, W. (1970) The ultrastructure of the hepatic cells of the carp (*Cyprinus carpio* L.) and the godgeon (*Gobio gobio* L.). *Acta Biol. Crac., Krakow* 13:105–110.

Byczkowska-Smyk, W. (1971) Observations on the ultrastructure of hepatic cells of the chub (*Leuciscus cephalus* L.) and roach (*Rutilus rutilus* L.). *Acta Biol. Crac., Krakow* 14:279–284.

Call, D.J., Brooke, L.T., Kent, R.J., Knuth, M.L., Anderson, C., Moriarty, C. (1983) Toxicity, bioconcentration, and metabolism of the herbicide propanil (3',4'-dichloropriopionanilide) in freshwater fish. *Arch. Environ. Contam. Toxicol.* 12:175–182.

Chapman, G.B. (1981) Ultrastructure of the liver of the fingerling rainbow trout *Salmo gairdneri. J. Fish Biol.* 18:553–567.

Chapman, G.S., Jones, A.L., Meyer, U.A. and Bissell, D.M. (1973) Parenchymal cells from adult rat liver in non-proliferating monolayer culture. II. Ultrastructural studies. *J. Cell Biol.* 59:735–747.

Copeland, P.A., Sumpter, J.P., T.K., Walker; and Croft, M. (1986) Vitellogenin levels in male and female rainbow trout (*Salmo gairdneri* Richardson) at variuos stages of the reproductive cycle. *Comp. Biochem. Physiol.* 83B/2:487–493.

Cormier, S.M. (1986) Fine structure of hepatocytes and hepatocellular carcinoma of the Atlantic tomcod *Microgadus tomcod* Walbaum. *J. Fish Dis.* 9:179–194.

Couch, J.A. (1975) Histopathological effects of pesticides and related chemicals on the liver of fishes. *In*: Ribelin, W.E. and Migaki, G. (eds) *The Pathology of Fishes*. Univ. Wisconsin Press, Madison, pp 559–584.

Couch, J.A. (1991) Spongiosis hepatis: chemical induction, pathogenesis, and possible neoplastic fate in a teleost fish model. *Toxicol. Pathol.* 19:237–250.

Couch, J.A. (1993) Light and electron microscopic comparisons of normal hepatocytes and neoplastic hepatocytes of well-differentiated hepatocellular carcinomas in a teleost fish. *Dis. Aquat. Org.* 16:1–14.

Couch, J.A. and Courtney, L.A. (1985) Attempts to abbreviate time to endpoint in fish hepatocarcinogenesis assays. *Water Chlorination* 5:22–43.

Couch, J.A. and Courtney, L.A. (1987) N-Nitrosodiethylamine-induced hepatocarcinogenesis in estuarine shipshead minnow (*Cyprinodon variegatus*): neoplasms and related lesions compared with mammalian lesions. *J. Natl. Cancer Inst.* 79:297–321.

Coulombe, R.A., Bailey, G.S. and Nixon, J.E. (1984) Comparative activation of aflatoxin B1 to mutagens by isolated hepatocytes from rainbow trout (*Salmo gairdneri*) and coho salmon (*Oncorhynchus kisutch*). *Carcinogenesis* 5:29–33.

David, H. (1961) Zur submikroskopischen Morphologie intrazellulärer Gallenkapillaren. *Acta anat.* 47:216–224.

Denizeau, F. and Marion, M. (1990) Toxicity of cadmium, copper, and mercury to isolated trout hepatocytes. *Can. J. Fish. Aquat. Sci.* 47:1038–1042.

Devaux, A., Pesonen, M., Monod, G. and Andersson, T. (1992) Glucocorticoid-mediated potentiation of P-450 induction in primary culture of rainbow trout hepatocytes. *Biochem. Pharmacol.* 43:898–901.

DiMichele, L. and Taylor, M.H. (1978) Histopathological and physiological responses of *Fundulus heteroclitus* to naphthalene exposure. *J. Fish. Res. Bd. Can.* 35:1061–1066.

Drochmans, P., Wanson, J.C. and Mosselmans, R. (1975) Isolation and subfractionation on Ficoll gradients of adult rat hepatocytes. *J. Cell Biol.* 66:1–22.

Droy, B. and Hinton, D.E. (1988) Allyl formate-induced hepatotoxicity in rainbow trout. *Mar. Environ. Res.* 24:259–264.

Droy, B.F., Davis, M.E. and Hinton, D.E. (1989) Mechanism of allyl formate-induced hepatotoxicity in rainbow trout. *Toxicol. Appl. Pharmacol.* 98:313–324.

Eaton, R.C. and Farley, R.D. (1974a) Spawning cycle and egg production of zebrafish, *Brachydanio rerio*, in the laboratory. *Copeia* 1:195–204.

Eaton, R.C. and Farley, R.D. (1974b) Growth and the reduction of depensation of zebrafish, *Brachydanio rerio*, reared in the laboratory. *Copeia* 1:204–209.

Eisen, J.S. (1996) Zebrafish make a big splash. *Cell* 87:969–977.

Ekker, M. and Akimenko, M.A. (1991) Le poisson zèbre (*Danio rerio*), un modèle en biologie du développement. *Médicine/Sciences.* 7:663–560.

Emmerson, K.M. and Petersen, I.M. (1976) Natural occurrence and experimental induction by estradiol-17α of a lipophosphoprotein (vitellogenin) in flounder (*Platichthys flesus* L.). *Comp. Biochem. Physiol.* 54B:443–446.

Ferri, S. (1980) Effect of cadmium on the Golgi complex of a freshwater teleost (*Pimelodus maculatus*). *Protoplasma* 103:99–103.

Ferri, S. and Macha, N. (1980) Lysosomal enhancement in hepatic cells of a teleost fish induced by cadmium. *Cell Biol. Internatl. Rep.* 4:357–363.

Forbus, W.D. (1952) *Reaction to injury; pathology for students of disease*. Williams and Wilkins, Baltimore, 257 pp.

Förlin, L., Haux, C., Karlson-Norrgren, L., Runn, P. and Larsson, A. (1986) Biotransformation enzyme activities and histopathology in rainbow trout, *Salmo gairdneri*, treated with cadmium. *Aquat. Toxicol.* 8:51–64.

Foster, G.D. and Moon, T.W. (1987) Metabolism in sea raven (*Hemitripterus americanus*) hepatocytes: the effects of insulin and glucagon. *Gen. Comp. Endocrin.* 66:102–115.

Franchini, A., Barbanti, E. and Fantin, A.M. (1991) Effects of lead on hepatocyte ultrastructure in *Carassius carassius* var. *auratus*. *Tissue & Cell* 23:893–901.

French, C.J., Mommsen, T.P. and Hochachka, P.W. (1981) Amino acid utilization in isolated hepatocytes from rainbow trout. *Europ. J. Biochem.* 113:311–317.

Frimmer, M., Homann, J., Petzinger, E., Rufeger, U. and Scharmann, W. (1976) Comparative studies on isolated rat hepatocytes and AS-30D hepatoma cells with leucoidin from *Pseudomonas aeruginosa*. *Naunyn Schmiedbergs Arch. Pharmacol.* 295:63–69.

Fry, J.R. and Bridges, J.W. (1977) A novel mixed hepatocyte-fibroblast culture system and its use as a test for metabolism-mediated cytotoxicity. *Biochem. Pharmacol.* 26:969–973.

Fry, J.R. and Bridges, J.W. (1979) Use of primary hepatocyte cultures in biochemical toxicology. *In:* Hodgson, E., Bend, J.R. and Philpot, R.M. (eds) *Review in Biochemical Toxicology*. Elsevier, North Holland, pp 201–246.

Fujiwara, S., Ohashi, H. and Noguchi, T. (1987) Comparison of intraperoxisomal localization form and properties of amphibian (*Rana catesbiana*) uricase from those found in other animal uricases. *Comp. Biochem. Physiol.* 86B:23–26.

Gagné, F., Marion, M. and Denizeau, F. (1990) Metallothionein induction and metal homeostasis in rainbow trout hepatocytes exposed to mercury. *Tox. Letters* 51:99–107.

Gas, N. (1973) Cytophysiologie du foie du carpe (Cyprinus carpio L.) II. Modalités d'alterations des ultrastructures au cours du jeûne prolongé. *J. Physiol.*, Paris, 66:283–302.

Gas, N. and Pequignot, J. (1972) Restoration of the structures of the liver cell of carps renourished by two synthetic regimes after a prolonged starvation. *C. r. Seanc. Soc. Biol.* 166: 446–453.

Gas, N. and Serfaty, A. (1972) Cytophysiologie du foie du carpe (*Cyprinus carpio* L.). Modifications ultrastructurales consecutives au maintien dans les conditions de jeûne hivernal. *J. Physiol.*, Paris 64:57–67.

Gerundo, N., Alderman, D.J., Clifton-Hadley, R.S. and Feist, S.W. (1991) Pathological effects of repeated doses of malachite green: a preliminary study. *J. Fish Dis.* 14:521–532.

Ghadially, F.N. (1988) *Ultrastructural pathology of the cell and matrix. Vols. I + II.* Butterworths, London, Boston, Singapore, Sidney, Toronto, Wellington, 1340 pp.

Gilloteaux, J., Oldham, C.K. and Biagianti-Risbourg, S. (1996) Ultrastructural diversity of the biliary tract and the gallbladder in fish. *In:* Munshi, J.S., Dutta, H.M. (eds) *Fish Morphology. Horizon of New Research*. Oxford & IBH Publishing, New Delhi, Calcutta, pp 95–110.

Gingerich, W.H. (1982) Hepatic toxicology in fish. *In:* L.J., Weber (ed) *Aquatic Toxicology*. Raven Press, New York, pp 55–105.

Goeger, D.E., Shelton, D.W., Hendricks, J.D. and Bailey, G.S. (1986) Mechanisms of anticarcinogenesis by indole-3-carbinol: effect on the distribution and metabolism of aflatoxin B1 in rainbow trout. *Carcinogenesis* 7:2025–2031.

Goksøyr, A. (1985) Purification of hepatic microsomal cytochromes P-450 from β-naphthoflavone-treated Atlantic cod (*Gadus morhua*), a marine teleost fish. *Biochim. Biophys. Acta* 840:409–417.

Goksøyr, A. and Förlin, L. (1992) The cytochrome P-450 system in fish, aquatic toxicology and environmental monitoring. *Aquat. Toxicol.* 22:287–312.

Goksøyr, A. and Solberg, T.S. (1987) Cytochromes P-450 in fish larvae: immunochemical detection of responses to oil pollution. *Sarsia* 72:405–407.

Goksøyr, A., Andersson, T., Hansson, T., Klungsoyr, J., Zhang, Y. and Förling, L. (1987) Species characteristics of the hepatic xenobiotic and steroid biotransformation systems in two teleost fish, Atlantic cod (*Gadus morhua*) and rainbow trout (*Salmo gairdneri*). *Toxicol. Appl. Pharmacol.* 89:347–360.

Goksøyr, A., Larsen, H.E., Blom, S. and Förlin, L. (1992) Detection of cytochrome P450 1A1 in North Sea dab liver and kidney. *Mar. Ecol. Progr. Ser.* 91:83–88.

Goldenberg, H. (1977) Organzation of purine degradation in the liver of a teleost (carp; *Cyprinus carpio* L.). A study of its subcellular distribution. *Mol. Cell. Biochem.* 16:17–21.

Goldenberg, H., Huettinger, M., Kampfer, P. and Kramar, R. (1978) Preparation of peroxisomes from carp liver by zonal rotor density gradient centrifugation. *Histochem. J.* 10:103–113.

Gony, S., Lecomte-Finiger, R., Faguet, D., Biagianti, S. and Bruslé, J. (1988) An experimental study on the action of cadmium on juvenile eels: biology of development and cytopathology. *Océanis* 14:141–148.

Görge, G. and Nagel, R. (1990) Toxicity of lindane, atrazine, and deltamethrin to early life stages of zebrafish (*Brachydanio rerio*). *Ecotox. Environ. Safety* 20:246–255.

Granato, M. and Nüsslein-Volhard, C. (1996) Fishing for genes controlling development. *Curr. Opin. Genet. Dev.* 6:461–468.

Güttinger, H. and Stumm, W. (1990) Ökotoxikologie am Beispiel der Rheinverschmutzung durch den Chemie-Unfall bei Sandoz in Basel. *Naturwissenschaften* 77:253–261.

Guillouzo, A., Guguen-Guillouzo, C., Boisnard, M., Bourel, M. and Benhamou, J.P. (1978) Smooth endoplasmic reticulum proliferation and increased cell multiplication in cultured hepatocytes of the newborn rat in the presence of phenobarbital. *Exp. Mol. Pathol.* 28:1–9.

Hacking, M.A., Budd, J. and Hodson, K. (1978) The ultrastructure of the liver of the rainbow trout: normal structure and modifications after chronic administration of a polychlorinated biphenyl Aroclor 1254. *Can. J. Zool.* 56:477–491.

Haffter, P. and Nüsslein-Volhard, C. (1996) Large scale genetics in a small vertebrate, the zebra-fish. *Int. J. Dev. Biol.* 40:221–227.

Halver, J.E. (1976) Alflatoxicosis and adventitious toxins in fish. *Fish Path.* 10:199–219.

Hampton, J.A., McCuskey, P.A., McCuskey, R.S. and Hinton, D.E. (1985) Functional units in rainbow trout (*Salmo gairdneri*) liver: I. Arrangement and histochemical properties of hepatocytes. *Anat. Rec.* 213:166–175.

Hampton, J.A., Lantz, R.C., Goldblatt, P.J., Laurén, D.J. and Hinton, D.E. (1988a) Functional units in rainbow trout (*Salmo gairdneri*, Richardson) liver: II. The biliary system. *Anat. Rec.* 221:619–634.

Hampton, J.A., Klaunig, J.E. and Goldblatt, P.J. (1988b) Ultrastructure of hepatic tumors in teleosts. *In*: Motta, P.M. (ed) *Biopathology of the Liver: An Ultrastructural Approach.* Kluwer Academic Publ., Boston, pp 167–175.

Hampton, J.A., Lantz, R.C. and Hinton, D.E. (1989) Functional units in rainbow trout (*Salmo gairdneri*, Richardson) liver: III. Morphometric analysis of parenchyma, stroma, and component cell types. *Am. J. Anat.* 185:58–73.

Harada, T., Hatanaka, J. and Enomoto, M. (1988) Liver cell carcinomas in the medaka (*Oryzias latipes*) induced by methylazoxymethanol-acetate. *J. Comp. Pathol.* 98:441–452.

Harada, T., Okazaki, N., Sato, Y. and Hatanaka, J. (1992) Influence of the carcinostatic drug, adriamycin, on the medaka (*Oryzias latipes*). *Bull. Nipp. Vet. Anim. Sci. Univ.* 17:106–113.

Hawkes, J.W. (1977) The effects of petroleum-hydrocarbon exposure on the structure of fish tissues. *In*: Wolfe, D.A. (ed.) *Fate and Effects of Petroleum Hydrocarbons in Marine Ecosystems and Organisms.* Pergamon, New York., pp 115–128.

Hawkes, J.W. (1980) The effects of xenobiotics on fish tissues: morphological studies. *Fed. Proc.* 39:3230–3236.

Hawkins, W.E., Overstreet, R.M. and Walker, W.W: (1986) Ultrastructural analysis of hepatic carcinogenesis in two small fish species. *Proc. 44th Ann. Meeting EMSA*:124–125.

Hayashi, S. and Ooshiro, Z. (1985) Primary culture of the freshly isolated liver cells of the eel. *Bull. Jpn. Soc. Sci. Fish.* 51:765–771.

Hayashi, S. and Ooshiro, Z. (1986) Primary culture of eel hepatocytes in serum-free medium. *Bull. Jap. Soc. Sci. Fish.* 52:1641–1651.

Hayashi, S., Fujiwara, S. and Noguchi, T. (1989) Degradation of uric acid in fish liver peroxisomes *J. Biol. Chem.* 264:3211–3215.

Heesen, D.T. (1977) Immunologische Untersuchungen an exo- und endogenen Dotterproteinen von *Brachydanio rerio* (Teleostei, Cyprinidae) und verwandten Arten. *Zool. Jb. Anat.* 97:566–582.

Heesen, D.T. and Engels, W. (1973) Elektrophoretische Untersuchungen zur Vitellogenese von *Brachydanio rerio* (Cyprinidae, Teleostei). *Roux's Arch. Entwicklungsmech. Organ* 173:46–59.

Hess, F.A., Weibel, E.R. and Preisig, R. (1973) Morphometry of dog liver: normal base-line data. *Virch. Arch.* 12B:303–317.

Heusequin, E. (1973) Sur les éléments mis en évidence par la 3,3′-diaminobenzidine dans le foie de *Lebistes reticulatus. Arch. Biol.* 84:243–279.

Hightower, L.E. and Renfro, J.L. (1988) Recent applications of fish cell culture to biomedical research. *J. Exp. Zool.* 248:290–302.

Hilton, J.W. (1982) The effect of pre-fasting diet and water temperature on liver glycogen and liver weight in rainbow trout, *Salmo gairdneri* Richardson during fasting. *J. Fish. Biol.* 20:69–78.

Hinton, D.E. and Laurén, D.J. (1990) Liver structural alterations accompanying chronic toxicity in fishes: potential biomarkers of exposure. *In:* McCarty, J.F and Shugart, L.R. (eds) *Biomarkers of environmental contamination.* Lewis Publ., Boca Raton., pp 15–57.

Hinton, D.E., Klaunig, J.E. and Lipsky, M.M. (1978) PCB-induced alterations in teleost liver: a model for environmental disease in fish. *Mar. Fish. Rev.* 40:47–50.

Hinton, D.E., Lantz, R.C. and Hampton, J.A. (1984) Effect of age and exposure to a carcinogen on the structure of the medaka liver: a morphometric study. *Natl. Cancer Inst. Monogr.* 65:239–249.

Hinton, D.E., Hampton, J.A. and Lantz, R.C. (1985a) Morphometric analysis of liver in rainbow trout quantitatively defining an organ of xenobiotic metabolism. *Mar. Environ. Res.* 238–239.

Hinton, D.E., Hampton, J.A. and McCuskey, P.A. (1985b) Japanese medaka liver tumor model: review of literature and new findings. *In:* Jolley, R.L., Bull, R.J., Davis, W.P., Katz, S., Roberts, M.H. and Jacobs, V.A. (eds) *Water Chlorination.* Lewis Publ., Chelsea.

Hinton, D.E., Lantz, R.C., Hampton, J.A., McCuskey, P.R. and McCuskey, R.S. (1987) Normal versus abnormal structure: considerations in morphologic responses of teleost to pollutants. *Environ. Health Perspect.* 71:139–146.

Hinton, D.E., Laurén, D.J. and Teh, S.J. (1988) Cellular composition and ultrastructure of hepatic neoplasms induced by diethylnitrosamine in *Oryzias latipes. Mar. Environ. Res.* 24:307–310.

Hisaoka, K.K. and Battle, H.I. (1958) The normal developmental stages of the zebrafish, *Brachydanio rerio* (Hamilton-Buchanan). *J. Morph.* 102:302–328.

Hisaoka, K.K. and Firlit, C.F. (1960) Further studies on the embryonic development of the zebrafish, *Brachydanio rerio (Hamilton-Buchanan). J. Morph.* 107:205–225.

Hisaoka, K.K. and Firlit, C.C. (1962) Ovarian cycle and egg production in the zebrafish, *Brachydanio rerio. Copeia* 4:788–792.

Hodson, P.V. and Blunt, B.R. (1981) Temperature-induced changes in pentachlorophenol chronic toxicity to early life stages of rainbow trout. *Aquat. Toxicol.* 1:113–127.

Hofer, R. and Medgyesy, N. (1997) Growth, reproduction and feeding of dwarf arctic char, Salvelinus alpinus, from an Alpine high mountain lake. *Arch. Hydrobiol.* 138:509– 524.

Hofer, R., Köck, G. and Braunbeck, T. (1997) Nuclear histo- and cytopathology as indicators of disturbed liver metabolism in Arctic char (*Salvelinus alpinus*) from acidic high alpine lakes. *Dis. aquat. Org.* 28:139–150.

Hollert, H and Braunbeck, T. (1997) *Ökotoxikologie in vitro – Gefährdungspotential in Wasser, Sediment und Schwebstoffen.* Veröff. PAÖ, 189 pp.

Hoshino, M. (1963) Submicroscopic characteristics of four strains of *Yoshida ascites* hepatoma of rats: a comparative study. *Cancer Res.* 23:209–632.

Howe, G.E., Marking, L.L., Bills, T.D., Boogaard, M.A. and Mayer, F.L. (1994) Effects of water temperature on the toxicity of 4-nitrophenol and 2,4-dinitrophenol to developing rainbow trout (*Oncorhynchus mykiss*). *Environ. Toxicol. Chem.* 13:79–84.

Hruban, Z. and Swift, H. (1964) Uricase: localization in hepatic microbodies. *Science* 146:1316–1317.

Hruban, Z., Swift, H., Dunn, F.W. and Lewis, D.E. (1965a) Effect of α-3-furylaniline on the ultrastructure of the hepatocytes and pancreatic acinar cells. *Lab. Invest.* 14:70–80.

Hruban, Z., Swift, H. and Rechzigl, M. (1965b) Fine structure of transplantable hepatomas of the rat. *J. Nat. Cancer Inst.* 35:459–495.

Huggett, R.J., Kimerle, R.A., Mehrle, P.M. and Bergman, H.L. (1992) *Biomarkers. Biochemical, physiological, and histological markers of anthropogenic stress.* Lewis Publ., Boca Raton, 347 pp.

Hugla, J.L., Philippart, J.C., Kremers, P. , Goffinet, G. and Thome, J.P. (1995) PCB contamination of the common barbel, *Barbus barbus* (Pisces, Cyprinidae), in the River Meuse in relation to hepatic monooxygenase activity and ultrastructural liver change. *Neth. J. Aquat. Ecol.* 29:135–145.

Ingham, P.W. (1997) Zebrafish genetics and its implications for understanding vertebrate development. *Human Molec. Genet.* 6:1755–1760.

Ishikawa, T., Shimamine, T. and Takayama, S. (1975) Histologic and electron microscopy observations on diethylnitrosamine-induced hepatomas in small aquarium fish (*Oryzias latipes*). *J. Natl. Canc. Inst.* 55:909–916.

Isom, H.C., Secott, T., Georgoff, I., Woodworth, C. and Mummaw, J. (1985) Maintenance of differentiated rat hepatocytes in primary culture. *Proc. Nat. Acad. Sci.* 82:3252–3256.

Jeejeebhoy, K.N. and Phillips, M.J. (1976) Isolated mammalian hepatocytes in culture. *Gastroenterology* 71:1086–1096.

Kalashnikova, M.M. and Kadilov, A.E. (1991) The heaptocyte ultrastructure of the Sevan trout (*Salmo ischchan gegarkuni* Kessel) in ontogeny. *Bull. Eksp. Biol. Med.* 111:208–211.

Kenaga, E.E. (1982) Predictability of chronic toxicity from acute toxicity of chemicals in fish and aquatic invertebrates. *Environ. Tox. Chem.* 1:347–358.

Kessel, R.G. (1968) Annulate lamellae. *J. Ultrastr. Res., Suppl.* 10:1–82.

Kessel, R.G. (1983) The structure and function of annulate lamellae: porous cytoplasmic and intranuclear membranes. *Intern. Rev. Cytol.* 82:181–302.

Kessel, R.G. (1989) The annulate lamellae – from obscurity to spotlight. *Electron Microsc. Rev.* 2:257–348.

Khan, A.A. and Semalu, S.S. (1995) Ultrastructural and biochemical effects of 3-methyl-cholanthrene in rainbow trout. *Bull. Environ. Contam. Toxicol.* 54:731–736.

Khangarot, B.S. (1992) Copper-induced hepatic ultrastructural alterations in the snake-headed fish, *Channa punctatus. Ecotox. Environ. Safety* 23:282–293.

Kishida, M. and Specker, J.L. (1993) Vitellogenin in tilapia (*Oreochromis mossambicus*): Induction of two forms by estradiol, quantification in plasma and characterization in oocyte extract. *Fish. Physiol. Biochem.* 12:171–182.

Klaassen, C.D. and Stacey, N.H. (1982) Use of isolated heaptocytes in toxicity assessment. *In:* Plaa, G. and Hewitt, W.R. (eds) *Toxicology of the Liver.* Raven Press, New York, pp 147–179.

Klaunig, J.E. (1984) Establishment of fish hepatocyte cultures for use in *in vitro* carcinogenicity studies. *Natl. Cancer Inst. Monogr.* 65:163–173.

Klaunig, J.E., Lipsky, M.M., Trump, B.F. and Hinton, D.E. (1979) Biochemical and ultrastructural changes in teleost liver following subacute exposure to PCB. *J. Environ. Path. Toxicol.* 2:953–963.

Klaunig, J.E., Ruch, R.J. and Goldblatt, J.P. (1985) Trout hepatocyte culture: Isolation and primary culture. *In Vitro Cell Dev. Biol.* 21:221–228.

Koban, M. (1986) Can cultures of teleost hepatocytes show temperature acclimation? *Am. J. Physiol.* 250:R211–R220.

Kocal, T., Crane, T.L. and Quinn, B.A. et al. (1988a) Degradation of extracellular thymidine by cultured hepatocytes from rainbow trout (*Salmo gairdneri*). *Comp. Biochem. Physiol.* 91B:557–561.

Kocal, T., Quinn, B.A., Smith, I.R., Ferguson, H.W. and Hayes, H.A. (1988b) Use of trout serum to prepare primary attached monolayer cultures of hepatocytes from rainbow trout (*Salmo gairdneri*). *In Vitro Cell Dev. Biol.* 24:304–308.

Kocan, R.M., Sabo, K.M. and Landolt, M.L. (1985) Cytotoxicity/genotoxicity: The application of cell culture techniques to the measurement of marine sediment pollution. *Aquat. Toxicol.* 6:165–177.

Köck, G., Noggler, M. and Hofer, R. (1996a) Pb in otoliths and opercula of Arctic char (*Salvelinus alpinus*) from oligotrophic lakes. *Water Res.* 30:1919–1923.

Köck, G., Triendl, M. and Hofer, R. (1996b) Seasonal patterns of metal accumulation in Arctic char (*Salvelinus alpinus*) from an oligotrophic Alpine lake related to temperature. *Can. J. Fish. Aquat. Sci.* 53:780–786.

Köhler, A. (1989) Regeneration of contaminant-induced liver lesions in flounder – experimental studies towards the identification of cause-effect relationships. *Aquat. Toxicol.* 14:203–232.

Kontir, D.M., Lantz, R.C. and Hinton, D.E. (1984) Ultrastructural morphometry of rainbow trout (*Salmo gairdneri*) hepatocytes during 3-methylcholanthrene (3-MC) induction of MFOs. *Mar. Environ. Res.* 14:420–421.

Kramar, R., Goldenberg, H., Böck, P. and Klobucar, N. (1974) Peroxisomes in the liver of the carp (*Cyprinus carpio* L.). Electron microscopic cytochemical and biochemical studies. *Histochem.* 40:137–154.

Kranz, H. and Peters, N. (1984) Melano-macrophage centers in liver and spleen of ruffe (*Gymnocephalus cernua*) from the Elbe estuary. *Helgol. Meeresunters.* 37:415–424.

Laale, H.W. (1977) The biology and use of zebrafish, *Brachydanio rerio*, in fisheries research. A literature review. *J. Fish Biol.* 10:121–173.

Langer, M (1979a) Histologische Untersuchungen an der Teleosteerleber. I. Der Aufbau des Leberparenchyms. *Z. mikr.-anat. Forsch.* 93:829–848.

Langer, M. (1979b) Histologische Untersuchungen an der Teleosteerleber. III. Das galleausführende System. *Z. mikr.-anat. Forsch.* 93:1105–1136.

Langer, M. and Storch, V. (1978) Zur Ultrastruktur der Hepatocyten von Teleosteern bei Nahrungsentzug. Z. mikr.-anat. Forsch. 92:641–654.

Laurén, D.J., Teh, S.T. and Hinton, D.E. (1990) Cytotoxicity phase of diethylnitrosamine-induced hepatic neoplasia in medaka. Cancer Res. 50:5504–5514.

Leatherland, J.F (1982) Effect of commercial trout diet on liver ultrastructure of fed and fasted coho salmon, Oncorhynchus kisutch Walbaum. J. Fish Biol. 21:311–319.

Leatherland, J.F. and Sonstegard, R.A. (1983) Interlake comparison of liver morphology and in vitro hepatic monodeiodination of L-thyroxine in sexually mature coho salmon, Oncorhynchus kisutch Walbaum, from Lakes Erie, Ontario, Michigan and Superior. J. Fish Biol. 22:519–536.

Leatherland, J.F. and Sonstegard, R.A. (1988) Ultrastructure of the liver of Lake Erie coho salmon from post-hatching until spawning. Cytobios 54:195–208.

LeHir, M., Herzog, V. and Fahimi, H.D. (1979) Cytochemical detection of catalase with 3,3′-diaminobenzidine. A quantitative reinvestigation of the optimal assay conditions. Histochem. 64:51–66.

Leland, H.V. (1983) Ultrastructural changes in the hepatocytes of juvenile rainbow trout and mature brown trout exposed to copper and zinc. Environ. Toxicol. Chem. 2:353–368.

Lemaire, P., Berhaut, J., Lemaire-Gony, S. and Lafaurie, M. (1992) Ultrastructural changes induced by benzo[a]pyrene in sea bass (Dicentrarchus labrax) liver and intestine: Importance of the intoxication route. Environ. Res. 57:59–72.

Lemaire-Gony, S. and Lemaire, P. (1992) Interactive effects of cadmium and benzo[a]pyrene on cellular structure and biotransformation enzymes of the liver of the European eel Anguilla anguilla. Aquat. Toxicol. 22:145–160.

Lester, S.M., Braunbeck, T., Tehg, S.J., Stegeman, J.J., Miller, M.R. et al. (1992) Immunocytochemical localization of cytochrome P450 IA1 in liver of rainbow trout (Oncorhynchus mykiss). Mar. Environ. Res. 34:117–122.

Lester, S.M., Braunbeck, T., Teh, S.J., Stegeman, J.J., Miller, M.R. et al. (1993) Hepatic cellular distribution of cytochrome CYP 1A in rainbow trout (Oncorhynchus mykiss): an immunohisto- and cytochemical study. Cancer Res. 53:3700–3706.

Lipsky, M.M., Klauing, J.E. and Hinton, D.E. (1978) Comparison of acute response to PCB in livers of rat and channel catfish. A biochemical and morphologic study. J. Toxicol. Environ. Health 4:107–121.

Lipsky, M.M., Sheridan, T.R., Bennett, R.O. and May, E.B. (1986) Comparison of trout hepatocyte culture on different substrates. In Vitro Cell. Dev. Biol. 22:360–362.

Litwin, J.A., Völkl, A., Mueller-Hoecker, J. and Fahimi, H.D. (1988) Immunocytochemical demonstration of peroxisomal enzymes in human kidney biopsies. Virchows Arch. B Cell Pathol. 128:207–213.

Loud, A.V. (1968) A quantitative stereological description of the ultrastructure of normal rat liver parenchymal cells. J. Cell Biol. 37:27–46.

Loveland, P.M., Wilcox, J.S., Pawlowski, N.E. and Bailey, G.S. (1987) Metabolism and DNA binding of aflatoxicol and aflatoxin B1 in vivo and in isolated hepatocytes from rainbow trout (Salmo gairdneri). Carcinogenesis 8:1065–1070.

Loveland, P.M., Wilcox, J.S., Hendricks, J.D. and Bailey, G.S. (1988) Comparative metabolism and DNA binding of aflatoxin B1, aflatoxin M1, aflatoxicol and aflatoxicol-M1 in hepatocytes from rainbow trout (Salmo gairdneri). Carcinogenesis. 9:441–446.

Maitre, J.L., Valotaire, Y. and Guguen-Guillouzo, C. (1986) Estradiol-17α stimulation of vitellogenin synthesis in primary culture of male trout hepatocytes. In Vitro Cell. Dev. Biol. 22:337–343.

March, P.E. and Reisman, H.M. (1995) Seasonal changes in hepatocyte ultrastructure correlated with the cyclic synthesis of secretory proteins in the winter flounder (Pleuronectes americanus). Cell Tissue Res. 281:153–161.

Marion, M. and Denizeau, F. (1983a) Rainbow trout and human cells in culture for the evaluation of the toxicity of aquatic pollutants: a study with cadmium. Aquat. Toxicol. 3:329–343.

Marion, M. and Denizeau, F. (1983b) Rainbow trout and human cells in culture for the evaluation of the toxicity of aquatic pollutants: a study with lead. Aquat. Toxicol. 3:47–60.

McCain, B.B., Hodgins, H.O., Gronlund, W.D., Hawkes, J.W., Brown, D.W. et al. (1978) Bioavailability of crude oil from experimentally oiled sediments to English sole (Parophrys vetulus), and pathological consequences. J. Fish. Res. Bd. Can. 35:657–664.

McCarty, J.F. and Shugart, L.R. (1990) *Biomarkers of environmental contamination*. Lewis Publishers, CRC Press, Boca Raton, Florida., 457 pp.

Mehrle, P.M., Johnson, W.W. and Mayer, F.L. (1974) Nutritional effects on chlordane toxicity in rainbow trout. *Bull. Environ. Contam. Toxicol.* 12:513–517.

Meyers, T.R. and Hendricks, J.D. (1985) Histopathology. *In:* Rand, G.M. and Petrocelli, S.R. (eds) *Fundamentals of aquatic toxicology*. Hemisphere Publ. Corp., Washington, pp 283–334.

Miayi, K. (1980) Ultrastructural basis for toxic liver injury. *In:* Farber, E. and Fisher, M.M. (eds) *Toxic injury of the liver*. Marcel Dekker, New York, pp 59–154.

Miller, M.R., Hinton, D.E., Blair, J.J. and Stegeman, J.J. (1988) Immunohistochemical localization of cytochrome P-450 E in liver, gill and heart of scup (*Stenotomus chrysops*) and rainbow trout (*Salmo gairdneri*). *Mar. Environ. Res.* 24:37–39.

Miller, M.R., Wentz, E., Blair, J.B., Pack, D. and Hinton, D.E. (1993) Acetaminophen toxicity in cultured trout liver cells. I. Morphological alterations and effects on cytochrome P450 1A1. *Exp. Mol. Pathol.* 58:114–126.

Mizell, M. and Romig, E.S. (1997) The aquatic vertebrate embryo as a sentinel for toxins: Zebrafish embryo dechorionation and perivitelline space microinjection. *Int. J. Develop. Biol.* 41:411–423.

Mommsen, T.P. and Lazier, C.B. (1986) Stimulation of estrogen receptor accumulation by estradiol in primary cultures of salmon hepatocytes. *FEBS Lett.* 195:269–271.

Mommsen, T.P., Walsh, P.J., Perry, S.F. and Moon, T.W. (1988) Interactive effects of catecholamines and hypercapnia on glucose production in isolated trout hepatocytes. *Gen. Comp. Endocrin.* 69:1–11.

Moon, T.W. (1983) Changes in tissue ion contents and ultrastructure of food-deprived immature American eel, *Anguilla rostrata* (Le Sueur). *Can. J. Zool.* 61:812–821.

Moon, T.W., Walsh, P.J. and Mommsen, T.P. (1985) Fish hepatocytes: a model metabolic system. *Can. J. Fish. Aquat. Sci.* 42:1772–1782.

Moore, M.N. (1980) Cytochemical determination of cellular responses to environmental stressors in marine organisms. *Rapp. P.-v. Réun. Cons. int. Explor. Mer* 179:7–15.

Moore, M.J., Smolowitz, R. and Stegeman, J.J. (1989) Cellular alteration preceding neoplasia in *Pseudopleuronectes americanus* from Boston Harbor. *Mar. Environ. Res.* 28:425–429.

Moutou, K., Braunbeck, T. and Houlihan, D. (1996) Quantitative analysis of alterations in liver ultrastructure of rainbow trout (*Oncorhynchus mykiss*) after administration of the aquaculture antibacterials oxolinic acid and flumequine. *Dis. aquat. Org.* 29:21–34.

Mugnaini, E. and Harboe, S.B. (1966) The liver of *Myxine glutinosa*: a true tubular gland. *Z. Zellforsch.* 78:342–369.

Murchelano, R.A. and Wolke, R.E. (1985) Epizootic carcinoma in the winter flounder, *Pseudopleuronectes americanus*. *Science* 228:587–589.

Nagel, R. (1986) Untersuchungen zur Eiproduktion beim Zebrabärbling (*Brachydanio rerio*, Ham.-Buch.). *J. Appl. Ichthyol.* 2:173–181.

Nagel, R. (1988) *Umweltchemikalien und Fische – Beiträge zu einer Bewertung*. Habilitation thesis, University of Mainz, 256 pp.

Nagel, R., Bresch, H., Caspers, N., Hansen, P.D., Markert, M. et al. (1991) Effects of 3,4-dichloroaniline on the early life stages of the zebrafish (*Brachydanio rerio*): results of a comparative laboratory study. *Ecotox. Environ. Safety* 21:157–164.

Ng, T.B. and Idler, D.R. (1983) Yolk formation and differentiation in teleosts. *In:* Hoar, W.S., Randall, D.J. and Donalson, E.M. (eds) *Fish Physiology, Vol. IX. Reproduction*. Academic Press, New York, London, pp 373–404.

Ng, B.T., Woo, N.Y.S., Tam, P.P. and Au, C.Y.W. (1984) Changes in metabolism and hepatic ultrastructure induced by estradiol and testosterone in immature female *Epinephalus akaara* (Teleostei, Serranidae). *Cell Tissue Res.* 236:651–659.

Ng, T.P., Tam, P.P.L. and Woo, N.Y.S. (1986) Sexual maturation in the black seabream, *Mylio macrocephalus* Teleostei, Sparidae: changes in pituitary gonadotropes, hepatocytes and related biochemical constituents in liver and serum. *Cell Tissue Res.* 245:207–213.

Nopanitaya, W., Carson, J.L., Grisham, J.W. and Aghajanian, G. (1979) New observations on the fine structure of the liver in goldfish (*Carassius auratus*). *Cell Tissue Res.* 196:249–261.

Norrgren, L., Andersson, T. and Bjoerk, M. (1993) Liver morphology and cytochrome P450 activity in fry of rainbow trout after microinjection of lipid-soluble xenobiotics in the yolk-sac embryos *Aquat. Toxicol.* 26:307–316.

Nunez, O., Hendricks, J.D. and Duimstra, J.R. (1991) Ultrastructure of hepatocellular neoplasms in aflatoxin B1 (AFB1)-initiated rainbow trout (*Oncorhynchus mykiss*). *Toxicol. Pathol.* 19:11–23.

Ostrander, G.K., Blair, J.B., Stark, B.A., Marley, G.M. and Bales, W.D. (1995) Long-term primary culture of epithelial cells from rainbow trout (*Oncorhynchus mykiss*) liver. *In Vitro Cell. Dev. Biol. Anim.* 31:367–378.

Oulmi, Y. and Braunbeck, T. (1996) Toxicity of 4-chloroaniline in early life-stages of zebrafish (*Brachydanio rerio*): I. Cytopathology of liver and kidney after microinjection. *Arch. Environ. Contam. Toxicol.* 30:390–402.

Oulmi, Y., Negele, R.D. and Braunbeck, T. (1993) Die Fischniere als Biomonitor für subletale Schadstoffbelastung: Ein Vergleich cytologischer Effekte durch die Herbizide Atrazin und Linuron bei der Regenbogenforelle (*Oncorhynchus mykiss*). *Verh. Dtsch. Zool. Ges.* 86:193.

Oulmi, Y., Schroeder, A. and Braunbeck, T. (1994) Einfluss einer chronischen Wasserkontamination durch subletale Konzentrationen von 2,4-Dichlorphenol auf larvale Zebrabärblinge (*Brachydanio rerio*): Schadstoffbedingte cytologische Veränderungen in Leber und Niere. *Verh. Dtsch. Zool. Ges.* 87:330.

Oulmi, Y., Storch, V. and Braunbeck, T. (1995a) Cytopathology of liver and kidney in rainbow trout (*Oncorhynchus mykiss*) after long-term exposure to sublethal concentrations of linuron. *Dis. Aquat. Org.* 21:35–52.

Oulmi, Y., Strmac, M. and Braunbeck, T. (1995b) Ein toxikologischer Mikroinjektions-Bioassay mit Embryonen des Zebrabärblings (*Brachydanio rerio*): Cytopathologische Veränderungen durch Triphenylzinnacetat in Leber und Nierentubulus. *Verh. Dtsch. Zool. Ges.* 88:262.

Parker, R.S., Morrissey, M.T., Moldeus, P. and Selivonchick, D.P. (1981) The use of isolated hepatocytes from rainbow trout (*Salmo gairdneri*) in the metabolism of acetaminophen. *Comp. Biochem. Physiol.* 70B:631–633.

Peakall, D. (1992) *Animal biomarkers as pollution indicators.* Chapman & Hall, London, New York, Tokyo, Melbourne, Madras., 291 pp.

Pelissero, C., Flouriot, G., Foucher, J.L., Bennetau, B, Dunoguès, J., LeGac, F. et al. (1993) Vitellogenin synthesis in cultured hepatocytes; an *in vitro* test for the estrogenic potency of chemicals *J. Steroid Biochem. Mol. Biol.* 44:263–272.

Perdew, G.H., Schaup, H.W., Becker, J.L., Williams, J.L. and Selivonchick, D.P. (1986) Alterations in the synthesis of proteins in the hepatocytes of rainbow trout fed cyclopropenoid acids. *Biochim. Biophys. Acta* 877:9–19.

Pesonen, M. and Andersson, T. (1991) Characterization and induction of xenobiotic metabolizing enzyme activities in a primary culture of rainbow trout hepatocytes. *Xenobiotica* 21:461–471.

Peute, J., van der Gaag, M.A. and Lambert, J.G.D. (1978) Ultrastructure and lipid content of the liver of the zebrafish, *Brachydanio rerio*, related to vitellogenin synthesis. *Cell Tissue Res.* 186:297–308.

Peute, J., Huiskamp, R. and Van Oordt, P.G.W.J. (1985) Quantitative analysis of estradiol.17α-induced changes in the ultrastructure of the liver of the male zebrafish, *Brachydanio rerio*. *Cell Tissue Res.* 242:377–382.

Peyon, P., Baloche, S. and Burzawa and Gerard, E. (1993) Synthesis of vitellogenin by eel (*Anguilla anguilla* L.) hepatocytes in primary culture: requirement of 17 β-estradiol-priming. *Gen. Comp. Endocrinol.* 91:318–329.

Pfeifer, U. (1973) Cellular autophagy and cell atrophy in the rat liver during long-term starvation. *Virch. Arch. Abt. B Zellpathol.* 12:195–211.

Phillips, M.J., Oda, M., Edwards, V.D., Greenberg, G.R. and Jeejeebhoy, K.N. (1974) Ultrastructural and functional studies of cultured hepatocytes. *Lab. Invest.* 31:533–542.

Phillips, M.J., Poucell, S., Patterson, J. and Valencia, P. (1987) *The liver: an atlas and text of ultrastructural pathology.* Raven Press, New York., 585 pp.

Phromkunthong, W. and Braunbeck, T. (1994) Sexual dimorphism in the reaction of zebrafish (*Brachydanio rerio*) to ascorbic acid deficiency: Induction of steatosis in hepatocytes of male fish. *J. Appl. Ichthyol.* 10:146–153.

Postlethwait, J.H. and Talbot, W.S. (1997) Zebrafish genomics: From mutants to genes. *Trends in Genetics* 13:183–190.

Roux, F., Guillam, C. and Bescol-Liversac, J. (1978) Pathologie toxique de l'hépatocyte en culture histiotypique. II-Action d'un organochlore: le lindane. Effet de substance étrangère métabolisable et effet toxique. *Ann. Anat. Pathol.* 23:253–275.

Quaglia, A. (1976a) Observations on the fine structure of the hepatocytic cells in the sprat (*Sprattus sprattus* L.). *Boll. Pesca Piscic. Idrobiol.* 31:221–226.

Quaglia, A. (1976b) Seasonal variations in the ultrastructur of the liver of the pilchard *Sardinia pilchardus* Walb. *Arch. Oceanogr. Limn.* 18:525–530.

Rabergh, C.M.I., Bylund, G. and Eriksson, J.E. (1991) Histopathological effects of microcystin-LR, a cyclic peptide toxin from the cyanobacterium (blue-green alga) *Microcystis aeroginosa*, on common carp (*Cyprinus carpio* L.). *Aquat. Toxicol.* 20:131–146.

Rafael, J. and Braunbeck, T. (1988) Interacting effects of diet and environmental temperature on biochemical parameters in the liver of *Leuciscus idus melanotus* (Cyprinidae, Teleostei). *Fish Physiol. Biochem.* 5:9–19.

Rauckman, E.J. and Padilla, G.M. (1987) *The isolated hepatocyte: use in toxicology and xenobiotic biotransformations.* Academic Press, New York, 292 pp.

Robertson, J.C. and Bradley, T.M. (1991) Hepatic ultrastructure changes associated with the parr-smolt transformation on Atlantic salmon (*Salmo salar*). *J. Exp. Zool.* 260:135–148.

Robertson, J.C. and Bradley, T.M. (1992) Liver ultrastructure of juvenile Atlantic Salmon (*Salmo salar*). *J. Morph.* 211:41–54.

Rocha, E., Monteiro, R.A. and Pereira, C.A. (1994) The liver of the brown trout, *Salmo trutta fario*: a light and electron microscopic study. *J. Anat.* 185:241–249.

Rocha, E., Monteiro, R.A.F. and Pereira, C.A. (1995) Microanatomical organization of hepatic stroma of the brown trout, *Salmo trutta fario* (Teleostei, Salmonidae): A qualitative and quantitative approach. *J. Morph.* 223:1–11.

Rocha, E., Monteiro, R.A.F. and Pereira, C.A. (1996) The pale-grey interhepatic cells of brown trout (*Salmo trutta*) are a subpopulation of liver resident macrophages or do they establish a different cellular type? *J. Submicr. Cytol. Pathol.* 28:357–368.

Rocha, E., Monteiro, R.A. and Pereira, C.A. (1997) Liver of the brown trout, *Salmo trutta* (Teleostei, Salmonidae): A stereological study at light and electron microscopic levels. *Anat. Rec.* 247:317–328.

Rohr, H.P. and Riede, U.N. (1973) Experimental metabolic disorders and the subcellular reaction pattern. *Curr. Top. in Pathology* 58:1–48.

Rojik, I., Nemcsok, J. and Boross, L. (1983) Morphological and biochemical studies on liver, kidney and gill of fishes affected by pesticides. *Acta Biol. Hung.* 34:81–92.

Rutschke, E. and Brozio, F. (1974) Bemerkungen zur Substruktur der Leber des Karpfens (*Cyprinus carpio* L.). *Z. mikr. anat. F.* 88:745–758.

Sabo, D.J. and Stegeman, J.J. (1977) Some metabolic effects of petroleum hydrocarbons in marine fish. *In:* Vernberg, F.J., Calabrese, A., Thurberg, F.P. and Vernberg, W.B. (eds) *Physiological responses of marine biota to pollutants.* Academic Press, New York, pp 279–287.

Saez, L., Goicoechea, O., Amthauer, R. and Krauskopf, M. (1982) Behaviour of RNA and protein synthesis during the acclimatization of the carp. Studies with isolated hepatocytes. *Comp. Biochem. Physiol.* 72B:31–38.

Saez, L., Amthauer, R., Rodriguez, E. and Krauskopf, M. (1984a) Effects of insulin on the fine structure of hepatocytes from winter-acclimatized carps: studies on protein synthesis. *J. Exp. Zool.* 230:187–197.

Saez, L., Zuvic, T., Amthauer, R., Rodriguez, E. and Krauskopf, M. (1984b) Fish liver protein synthesis during cold acclimation: seasonal changes of the ultrastructure of the carp hepatocyte. *J. Exp. Zool.* 230:175–286.

Scarpelli, D.G. (1967) Ultrastructural and biochemical observations in trout hepatoma. *Fish. Wild. Serv. Res. Rep.* 70:60–71.

Scarpelli, D.G. (1977) General discussion to part IV. Biological effects on marine animals: Health implications for man. Aquatic pollutants and biological effects with emphasis on neoplasia. *Ann. N. Y. Acad. Sci.* 298:463–481.

Scarpelli, D.G., Greider, M.H. and Frajola, W.J. (1963) Observations on hepatic cell hyperplasia, adenoma and hepatoma of rainbow trout (*Salmo gairdneri*). *Cancer Res.* 23:848–856.

Schenck, H.P. (1986) Tabellierte Produktionsmengen organischer Chemikalien. Umweltbundesamt, Berlin, Forschungsbericht 106 01 023/03.

Schmidt, D.C. and Weber, L.J. (1973) Metabolism and biliary excretion of sulfobromophthalein in rainbow trout (*Salmo gairdneri*). *J. Fish. Res. Bd. Can.* 30:1301–1308.

Schoor, W.P. and Couch, J.A. (1979) Correlation of mixed-function activity with ultrastructural changes in the liver of a marine fish. *Cancer Biochem. Biophys.* 4:95–103.

Schramm, M., Müller, E. and Triebskorn, R. (1998) Brown trout (*Salmo trutta f. fario*) liver ultrastructure as biomarker of small stream pollution. *Biomarkers* 3:93–108.

Schreiber, G. and Schreiber, M. (1973) The preparation of single cell suspensions from liver and their use for the study of protein synthesis. *Subcell. Biochem.* 2:307–353.

Segner, H. (1985) Influence of starvation and refeeding with different diets on the hepatocyte ultrastructure of juvenile *Siganus guttatus* Bloch (Teleostei: Siganidae). *Zool. Anz.* 214:81–90.

Segner, H. (1987) Response of fed and starved roach, *Rutilus rutilus*, to sublethal copper contamination *J. Fish Biol.* 30:423–437.

Segner, H. and Braunbeck, T. (1988) Hepatocellular adaptation to extreme nutritional conditions in ide, *Leuciscus idus melanotus* L. (Cyprinidae). A morphofunctional analysis. *Fish Physiol. Biochem.* 5:79–97.

Segner, H. and Braunbeck, T. (1990a) Adaptive changes of liver composition and structure in golden ide during winter acclimatization. *J. Exp. Zool.* 255:171–185.

Segner, H. and Braunbeck, T. (1990b) Environmental adaptation of the cyprinid teleost *Leuciscus idus melanotus*: Changes in liver composition and structure during the winter season. *J. Exp. Zool.* 255:171–185.

Segner, H. and Braunbeck, T. (1990c) Qualitative and quantitative assessment of the response of milkfish, *Chanos chanos*, fry to low-level copper exposure. *In:* Perkins, F.O. and Cheng, T.C. (eds) *Pathology in Marine Science.* Academic Press, San Diego, pp 347–368.

Segner, H. and Braunbeck, T. (1998) Cellular response profile to chemical stress. *In:* Schüürmann, G. and Markert, B. (eds) *Ecotoxicology.* Wiley and Spektrum Akadem. Verlag, New York and Heidelberg, pp 521–569.

Segner, H. and Juario, J.V. (1986) Histological observations on the rearing of milkfish, *Chanos chanos*, fry using different diets. *J. Appl. Ichthyol.* 2:162–173.

Segner, H. and Möller, H. (1984) Electron microscopical investigations on starvation – induced liver pathology in flounders *Platichthys flesus. Mar. Ecol. Progr. Ser.* 19:193–196.

Segner, H. and Storch, V. (1985) Influence of water-borne iron on the liver of *Poecilia reticulata. J. Appl. Ichthyol.* 1:39–47.

Segner, H., Orejana-Acosta, B. and Juario, J.V. (1984) The effect of *Brachionus plicatilis* grown on three different species of phytoplancton on the ultrastructure of the hepatocytes of *Chanos chanos* (Forskal) fry. *Aquaculture* 42:109–115.

Segner, H., Rösch, R., Verreth, J. and Witt, U. (1993) Larval nutrition physiology: studies with *Clarias gariepinus, Coregonus lavaretus* and *Scophthalmus maximus. J. World Aquac. Soc.* 24:121–134.

Segner, H., Blair, J. B., Wirtz, G. and Miller, M. R. (1994) Cultured trout liver cells: utilization of substrates and response to hormones. *In Vitro Cell. Dev. Biol.* 30A:306–311.

Sheehan, P.J. (1984) Effects on individuals and populations. *In:* Sheehan, P.J., Miller, D.R., Butler, G.C. and Bourdeau, P. (eds) *Effects of pollutants at the ecosystems level.* Wiley, New York, pp 23–50.

Sindermann, C.J. (1980) The use of pathological effects of pollutants in marine environmental monitoring programs. *Rapp. P.-v. Reun. Cons. int. Explor. Mer* 179:129–134.

Sindermann, C.J. (1984) Fish and environmental impact. *Arch. Fisch. Wiss.* 35:125–160.

Sindermann, C.J., Bang, F.B., Christensen, N.O., Dethlefsen, V., Harshbarger, J.C., Mitchell, J.R. (1980) The role and value of pathobiology in pollution effects monitoring programs. *Rapp. P.-v. Réun. Cons. int. Explor. Mer* 179:135–151.

Smolarek, T.A., Morgan, S.L., Moynihan, C.G., Lee, H., Harvey, R.G., Baird, W.M. (1987) Metabolism and DNA adduct formation of benzo[a]pyrene and 7,12-dimethylbenz(a)anthracene in fish cell lines in culture. *Carcinogenesis* 8:1501–1508.

Sorensen, E.M.B. (1976) Ultrastructural changes in the hepatocytes of the green sunfish, *Lepomis cyanellus* Rafinesque, exposed to solutions of sodium arsenate. *J. Fish Biol.* 8:229–240.

Sorensen, E.M.B., Ramirez-Mitchell, R., Harlan, C.W. and Bell, J.S. (1980) Cytological changes in the fish liver following chronic environmental arsenic exposure. *Bull. Environ. Contam. Toxicol.* 25:93–99.

Sorensen, E.M.B., Ramirez-Mitchell, R., Pradzynsk, A., Bayer, T.L. and Wenz, L.L. (1985) Stereological analyses of hepatocyte changes parallel arsenic accumulation in the livers of green sunfish. *J. Environ. Pathol. Toxicol.* 6:195–210.

Stacey, N.H. and Fanning, J.C. (1981) Ultrastructural changes in isolated rat hepatocytes after incubation with carbon tetrachloride. *Toxicology* 22:69–77.

Stammati, A.P., Silano, V. and Zucco, F. (1981) Toxicology investigations with cell culture systems. *Toxicology* 20:91–153.

Stäubli, W., Hess, R. and Weibel, E.R. (1969) Correlated morphometric and biochemical studies on the liver cell. II. Effects of phenobarbital on rat hepatocytes. *J. Cell Biol.* 42:92–112.

Stebbing, A.R.D. (1985) A possible synthesis. *In:* Bayne, B.L. (ed.) *The effects of stress and pollution on marine animals.* Praeger Publishers, New York, pp 301–314.

Stegeman, J.J. and Kloepper-Sams, P.J. (1987) Cytochrome P-450 isozymes and monooxygenase activity in aquatic animals. *Env. Health Persp.* 71:87–95.

Stegeman, J.J. and Woodin, B.R. (1984) Differential regulation of hepatic xenobiotic and steroid metabolism in marine teleost species. *Mar. Environ. Res.* 14:422–425.

Stegeman, J.J., Miller, M., Singh, H. and Hinton, D. (1987) Cytochrome P-450E induction and localization in liver and endothelial tissue of extrahepatic organs of scup (*Stenotomus chrysops*). *Fed. Proc.* 46:379.

Stegeman, J.J., Woodin, B.R., Singh, H., Oleksiak, M.F. and Celander, M. (1997) Cytochromes P450 (CYP) in tropical fishes: Catalytic activities, expression of multiple CYP proteins and high levels of microsomal P450 in liver of fishes from Bermuda. *Comp. Biochem. Physiol.* 116C/1:61–75.

Stehr, C. (1986) The ultrastructure of hepatocellular carcinomas in English sole (*Parophrys vetulus*) from Puget Sound. *Proc. 44th Ann. Meeting Electr. Micr. Soc. Am.* 330–33.

Stehr, C.M. and Myers, M.S. (1990) The ultrastructure and histology of cholangiocellular carcinomas in English sole (*Parophrys vetulus*) from Puget Sound, Washington. *Toxicol. Pathol.* 18:362–372.

Stehr, C.M., Rhodes, L.D. and Myers, M.S. (1988) The ultrastructure and histology of hepatocellular carcinomas of English sole (*Parophrys vetulus*) from Puget Sound, Washington. *Toxicol. Pathol.* 16:418–431.

Stehr, C. M., L. L. Johnson and Myers, M.S. (1998) Hydropic vacuolation in the liver of three species of fish from the U.S. west coast: lesion description and risk assessment associated with contaminant exposure. *Dis. aquat. Org.* 32:119–135.

Stoker, P.W., Larsen, J.R., Booth, G.M. and Lee, M.L. (1985) Pathology of liver and gill tissues from two genera of fishes exposed to two coal-derived materials. *J. Fish. Biol.* 27:31–46.

Storch, V. and Juario, J.V. (1983) The effect of starvation and subsequent feeding on the hepatocytes of *Chanos chanos* (Forsskal) fingerlings and fry. *J. Fish Biol.* 23:95–103.

Storch, V., Segner, H., Juario, J. and Duray, M.N. (1984a) Influence of nutrition on the hepatocytes of *Chanos chanos* (Chanidae: Teleostei). *Zool. Anz., Jena* 213/3–4:151–160.

Storch, V., Welsch, U., Schünke, M. and Wodtke, E. (1984b) Einfluss von Temperatur und Nahrungsentzug auf die Hepatocyten von *Cyprinus carpio* (Cyprinidae, Teleostei). *Zool. Beitr. NF.* 28:253–269.

Strmac, M. and Braunbeck, T. (1996) Toxikologische Untersuchungen zur Wirkung von Triphenylzinnacetat auf frühe Lebensstadien des Zerbabärblings (*Brachydanio rerio*). *Verh. Dtsch. Zool. Ges.* 89:331.

Strmac, M. and Braunbeck, T. (1998) Effects of triphenyltin acetate on early life-stages of zebrafish (*Brachydanio rerio*). *Ecotox. Environ. Safety; submitted.*

Svoboda, D.J. (1964) Fine structure of hepatomas induced in rats induced with *p*-dimethylaminobenzene. *J. Natl. Cancer Inst.* 33:315–339.

Takahashi, K., Sugawara, Y. and Sato, R. (1977) Fine structure of the liver cells in maturing and spawning female chum salmon, *Oncorhynchus keta* (Walbaum). *Tohoku J. Agric. Sci.* 28:103–110.

Tanuma, Y. (1980) Electron microscope observations on the intrahepatic bile canaliculus and subsequent bile ductules of the crucian, *Carassius carassius*. *Arch. Histol. Jap.* 43:1–21.

Thiyagarajah, A. and Grizzle, J.M. (1985) Pathology of diethylnitrosamine toxicity in the fish *Rivulus marmoratus*. *Pathol. Toxicol. Oncol.* 6:219–232.

Thiyagarajah, A. and Grizzle, J.M. (1986) Pathology of diethylnitrosamine toxicity in the fish *Rivulus ocellatus marmoratus*. *Can. J. Zool.* 64:2868–2870.

Triebskorn, R., Koehler, H.R., Flemming, J., Braunbeck, T., Negele, R.D. et al. (1994a) Evaluation of bis(tri-*n*-butyltin)oxide (TBTO) neurotoxicity in rainbow trout (*Oncorhynchus mykiss*). I. Behaviour, weight increase, and tin contents. *Aquat. Toxicol.* 30:189–197.

Triebskorn, R., Köhler, H.R., Koertje, K.H., Negele, R.D., Rahmann, H. et al. (1994b) Evaluation of bis(tri-n-butyltin)oxide (TBTO neurotoxicity in rainbow trout (*Oncorhnychus mykiss*). II. Ultrastructural diagnosis and tin localization by energy filtering transmission electron microscopy (EFTEM). *Aquat. Toxicol.* 30:199–213.

Triebskorn, R., Köhler, H.-R., Honnen, W., Schramm, M., Adams, S.M. et al. (1997) Induction of heat shock proteins, changes in liver ultrastructure, and alterations of fish behavior: are these biomarkers related and are they useful to reflect the state of pollution in the field? *J. Aquat. Ecos. Stress Recov.* 6:57–73.

Umweltbundesamt (1992) *Daten zur Umwelt* 1990/1991. Erich Schmidt Verlag, Bonn.

Vaillant, C., Le Guellec, C. and Pakdal, F. and Valotaire, Y. (1988) Vitellogenin gene expression in primary culture of male rainbow trout hepatocytes. *Gen. Comp. Endocrinol.* 70:284–290.

van der Gaag, M., Lambert, J.G.D., Peute, J. and Van Oordt, P.G.W.J. (1977) Ultrastructural aspects of the liver of the female zebrafish, *Brachydanio rerio*, during the reproductive cycle. *J. Endocrin.* 72:50P–51M.

vanBohemen, C.G., Lambert, J.G.D. and Peute, J. (1981) Annual changes in plasma and liver in relation to vitellogenesis in the female rainbow trout, *Salmo gairdneri. Gen. Comp. Endocrin.* 44:94–107.

vanBohemen, C.G., Lambert, J.G.D., Goos, H.J.T. and Van Oordt, P.G.W.J. (1982) Estrone and estradiol participation during exogeneous vitellogenesis in the female rainbow trout, *Salmo gairdneri. Gen. Comp. Endocrin.* 46:81–92.

Vera, M.J., Romero, F., Figueroa, J., Amthauer, R., Leon, G. Villanueva, J. and Krauskopf, M. (1993) Oral administration of insulin in winter-acclimatized carp (*Cyprinus carpio*) induces hepatic ultrastructural changes. *Comp. Biochem. Physiol.* 106A:677–682.

Veranic, P. and Pipan, N. (1995) Two degradation pathways of endoplasmic reticulum in fish hepatocytes. *Biol. Cell* 83:69–75.

Vernier, J.M. (1975) Etude ultrastructurelle des lipoprotéines hépatiques de très basse densité au cours du développement de la triute arc-en-ciel, *Salmo gairdneri* Rich. *J. Microsc. Biol. Cell.* 23:39–50.

Vogt, G. (1987) Monitoring of environmental pollutants such as pesticides in prawn aquaculture by histological diagnosis. *Aquaculture* 67:157–164.

Völkl, A., Baumgart, E. and Fahimi, H.D. (1988) Localization of urate oxidase in the crystalline core of rat liver peroxisomes by immunocytochemistry and immunoblotting. *J. Histochem. Cytochem.* 36:329–336.

Walsh, A.H. and Ribelin, W.E. (1975) The pathology of pesticide poisoning. *In*: Ribelin, W.E. and Migaki, G. (eds) *Pathology of fishes*. Univ. Wisconsin Press, Madison, pp 515–557.

Walton, M.J. and Cowey, C.B. (1979) Gluconeogenesis by isolated hepatocytes from rainbow trout, *Salmo gairdneri. Comp. Biochem. Physiol.* 62:75–79.

Walton, D.B., Acton, A.B. and Stich, H.F. (1983) DNA repair synthesis in cultured mammalian and fish cells following exposure to chemical mutagens. *Mutat. Res.* 124:153–161.

Walton, D.B., Acton, A.B. and Stich, H.F. (1984) DNA repair synthesis following exposure to chemical mutagens in primary liver, stomach, and intestinal cells isolated from rainbow trout. *Cancer Res.* 44:1120–1121.

Wanson, J.C., Drochmans, P., Mosselmans, R. and Ronveaux, M.F. (1977) Adult rat hepatocytes in primary monolayer culture. Ultrastructural characteristics of intercellular contacts and cell membrane differentiations. *J. Cell Biol.* 74:858–873.

Weibel, E.R. (1979) *Stereological Methods*. Academic Press, New York, London, 384 pp.

Weibel, E.R. (1981) Stereological methods in cell biology: where are we – where are we going? *J. Histochem. Cytochem.* 29:1043–1052.

Weibel, E.R., Stäubli, W., Gnägi, H.R. and Hess, F.A. (1969) Correlated morphometric and biochemical studies on the liver cell. I. Morphometric model, stereologic methods, and normal morphometric data for rat liver. *J. Cell Biol.* 42:68–91.

Weis, P. (1972) Hepatic ultrastructure in two species of normal, fasted and gravid teleost fishes. *Am. J. Anat.* 133:317–332.

Weis, P. (1974) Ultrastructural changes induced by low concentrations of DDT in the livers of the zebrafish and the guppy. *Chem. Biol. Interact.* 8:25–30.

Werth, G. and Boiteux, A. (1968) Zur Toxikologie der Triphenylmethanharnstoffe. Malachitgrün als Entkoppler der oxidativen Phosphorylierung *in vivo* und *in vitro. Arch. Toxikol.* 23:82–103.

Whitham, M., Nixon, J.E. and Sinnhuber, R.O. (1982) Liver DNA bound *in vivo* with aflatoxin B1 as a measure of hepatocarcinoma initiation in rainbow trout. *J. Natl. Cancer Inst.* 68:623–628.

Wiegand, M.D. (1996) Composition, accumulation and utilization of yolk lipids in teleost fish *Rev. Fish Biol. Fisheries* 6:259–286.

Wunder, W. and Korn, H. (1983) Aflatoxin cancer (hepatoma) in the liver of the rainbow trout (*Salmo irideus*). *Zool. Beitr. NF* 28:99–109.

Yamamoto, T. (1962) Some observations on the fine structure of the terminal biliary passages in the goldfish liver. *Anat. Rec.* 142:293.

Yamamoto, T. (1964) Some observations on the fine structure of liver cells in the starved goldfish (*Carassius auratus*), with special reference to the morphology of fat mobilization during starvation to the liver. *Arch. Histol. Jap.* 24:335–345.

Yamamoto, T. (1965) Some observations on the fine structure of the intrahepatic biliary passage in goldfish (*Carassius auratus*). *Z. Zellf.* 65:319–330.

Yamamoto, M. and Egami, N. (1974) Sexual differences and age changes in the fine structure of hepatocytes in the medaka, *Oryzias latipes. J. Fac. Sci. Univ. Tokio, Sec. IV* 13:199–210.

Zahn, T. and Braunbeck, T. (1993) Isolated fish hepatocytes as a tool in aquatic toxicology – sublethal effects of dinitro-*o*-cresol and 2,4-dichlorophenol. *Sci. Total Environ., Suppl.* 721–734.

Zahn, T. and Braunbeck, T. (1995a) Acute and sublethal toxicity of seepage water from garbage dumps to permanent cell cultures and primary cultures of hepatocytes from rainbow trout (*Oncorhynchus mykiss*): a novel approach to environmental risk assessment *Zbl. Hyg.* 196:455–479.

Zahn, T. and Braunbeck, T. (1995b) Cytologic evidence for malachite green toxicity in isolated rainbow trout (*Oncorhynchus mykiss*) hepatocytes. *Toxicol. in Vitro* 9:729–741.

Zahn, T., Hauck, C. and Braunbeck, T. (1993) Cytological alterations in fish fibrocytic R1 cells as an alternative test system for the detection of sublethal effects of environmental pollutants: a case-study with 4-chloroaniline. *In:* Braunbeck, T., Hanke, W. and Segner, H. (eds) *Fish in Ecotoxicology and Ecophysiology.* VCH, Weinheim, pp 103–126.

Zahn, T., Arnold, H. and Braunbeck, T. (1995) Ultrastrukturelle Veränderungen in isolierten Hepatocyten aus Regenbogenforelle (*Oncorhynchus mykiss*) und Aal (*Anguilla anguilla*) nach Belastung mit Disulfoton. *Verh. Dtsch. Zool. Ges.* 88:268.

Zahn, T., Arnold, H. and Braunbeck, T. (1996a) Cytological and biochemical response of R1 cells and isolated hepatocytes from rainbow trout (*Oncorhynchus mykiss*) to subacute *in vitro* exposure to disulfoton. *Exp. Toxicol. Pathol.* 48:47–64.

Zahn, T., Hollert, H. and Braunbeck, T. (1996b) Vergleich der Zytotoxizität von umweltrelevanten Chemikalien auf isolierte Hepatocyten aus der Regenbogenforelle und S9-supplementierte RTG-2-Zellen. *Verh. Dtsch. Zool. Ges.* 89:332.

Fish Ecotoxicology
ed. by T. Braunbeck, D. E. Hinton and B. Streit
© 1998 Birkhäuser Verlag Basel/Switzerland

Architectural pattern, tissue and cellular morphology in livers of fishes: Relationship to experimentally-induced neoplastic responses

David E. Hinton[1] and John A. Couch[2]

[1] Department of Anatomy, Physiology and Cell Biology, School of Veterinary Medicine, University of California at Davis, Davis, California 95916, USA
[2] 4703 Soule Place, Gulf Breeze, Florida 32561, USA

Summary. The teleost liver is one of the most sensitive organs to show alteration in biochemistry, physiology and structure following exposure to various types of environmental pollutants. Despite the importance of this organ to environmental toxicology and to ecotoxicology where biomarkers of exposure and of deleterious effect are found, the architectural pattern is not well known. This chapter reviews an architectural plan for teleost liver and compares that to the often cited mammalian pattern. Hepatic tubules composed principally of hepatocytes and biliary epithelial cells are in close proximity to lacunae which are of mesodermal origin. As is described, the tubule and lacunae concepts provide a means to better interpret morphologic alterations following exposure. These concepts are used to illustrate features of the chronic toxicity following exposure to proven carcinogens.

Introduction

Teleost models are important vertebrate *in vivo* systems for the analysis of environmental contaminants including detection of the role(s) of these xenobiotics as initiating carcinogens or as modulators of the carcinogenic process (Bailey et al., 1996; Boorman et al., 1997; Bunton, 1996). As endpoints in field investigations, neoplastic lesions have an important role in coastal toxicology programs in the USA (Malins et al., 1985; Myers et al., 1994), Great Britain (Moore and Evans, 1992), Germany (Kranz and Dethlefsen, 1990), Canada (Balch et al., 1995), and the Netherlands (Vethaak and Jol, 1996; Vethaak and Wester, 1996). We feel that these investigations contain far more information beyond simply suggesting causal relationship of a given compound to neoplasia; they reveal much about adaptive strategies that populations of cells make to withstand a toxic milieu and to be able to respond to growth stimuli.

Of the target organs of fishes in which tumors develop, the liver has emerged as one very important site. During the progression of hepatic neoplasia, events/processes occur which may be detected with our experimental pathology toolkit and include: cellular toxicity and necrosis, regeneration, apoptosis, mitosis, differential growth kinetics of focal lesions *versus* the non-focal liver and changes in cytology (immunochemical and enzymic) and tissue architecture.

Since certain of the above events/processes are also seen in endocrine-regulated normal growth; in development and in the acute toxic responses of cells, a detailed description of the hepatic tissues and cell types, their architecture, and resultant lesions is important. Livers of fishes and other oviparous vertebrates have similarities to those of mammalian species; but important differences are seen both in health and in toxicity. For example, most species of fishes lack fixed macrophages (Küpffer cells) within sinusoidal lumina (McCuskey et al., 1986), although these are known to be essential for clearance of endotoxins from portal blood of mammals. Fish have unique macrophage aggregates in perisinusoidal locations, a feature not common to mammalian species. These differences may influence the metabolism, the site of action of potential toxicants and the clearance of foreign compounds. As will be presented in detail in this chapter, position, arrangement, and number of biliary epithelial cells differ in livers of fishes versus their mammalian counterparts. It is possible that the relative abundance of mixed biliary and hepatic neoplasms following exposure to hepatocarcinogens reflects this difference in normal anatomy. Finally, vitellogenin and choriogenin are liver-derived products which are synthesized under stimulation of hepatocytes by estradiol (Mommsen and Walsh, 1988; Murata et al., 1997) or estrogen-like xenobiotics (Arukwe et al., 1997; Sumpter, 1995; Sumpter et al., 1996) and are special features of oviparous vertebrates.

The tubular architecture of fishes

The comparative hepatic histologic literature of the past century, reviewed by Shore and Jones (1889), revealed a widely held view that the architectural unit most consistent with the arrangement of hepatocytes to each other, and with the intrahepatic vascular and biliary systems, was the tubule. Tubules were more or less cylindrical arrangements of hepatocytes, 3 to 7 cells in cross-section, the apices of which surrounded a biliary passageway (the lumen). Bases of hepatocytes faced the perisinusoidal space (space of Disse, currently). Figures 1−4 illustrate features of the hepatic tubule in two species commonly employed in laboratory carcinogen exposures. In Figure 1, from rainbow trout (*Oncorhynchus mykiss*) a routine survey preparation processed in paraffin and stained by hematoxylin and eosin shows tubules. The increased resolution of a semithin section (0.5−1.0 μm in thickness), stained with toluidine blue, readily reveals hepatocytes and biliary epithelial cells within the curvilinear, cylindrical hepatic tubules (Fig. 2). A hematoxylin and eosin-stained paraffin section of liver from medaka (*Oryzias latipes*) is shown in Figure 4. Tubules are apparent in this species as well. Higher resolution (Fig. 4) in this species shows both biliary epithelial cells and hepatocytes in hepatic tubules.

After reviewing mammalian liver literature, Elias (1949) compiled a complete list of the various synonymous terms for the above tubular

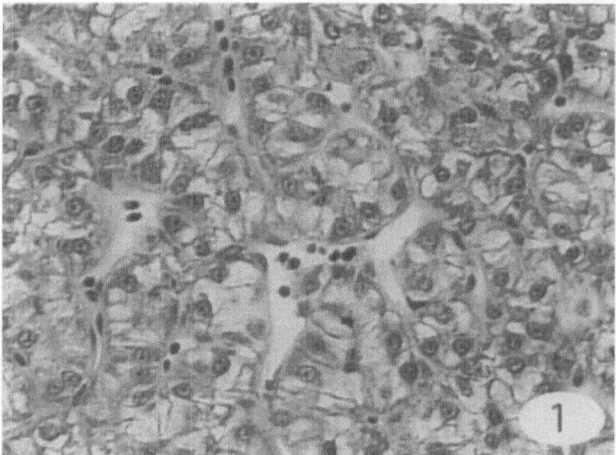

Figure 1. Section of liver from young rainbow trout (*Oncorhynchus mykiss*) showing tubules in longitudinal and transectional array. Lacunae are represented by sinusoidal lumina containing nucleated red blood cells. Hematoxylin and eosin stain, magnification × 400.

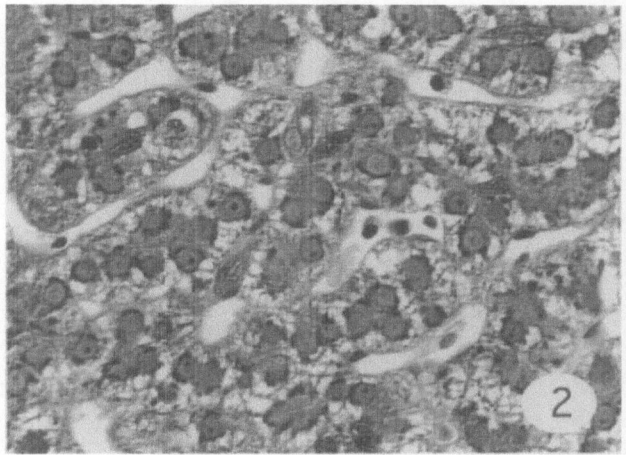

Figure 2. Higher magnification view of section from liver of trout in Figure 1. Note oval shaped nuclei of the centrotubular bile preductular epithelial cells surrounded by hepatocytes of the tubules. Hematoxylin and eosin stain, magnification × 540.

arrangement. These included: cords, trabeculae, columns, Stränge, or cordone, depending on the book and the author's language. Interestingly, the impetus for the re-evaluation by Elias (1949) was his attempt to produce a filmstrip for medical students on the normal histology of the mammalian liver. For maximum teaching impact, he used stereographic (i.e., three-dimensional) methods, and, analysis of profiles from the stereograms,

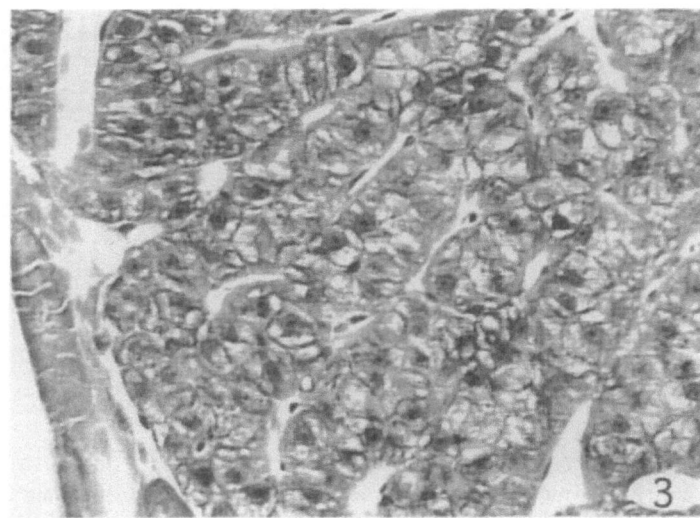

Figure 3. Section of liver from young medaka (*Oryzias latipes*) showing hepatic tubules and lacunae. Hematoxylin and eosin stain, magnification × 400.

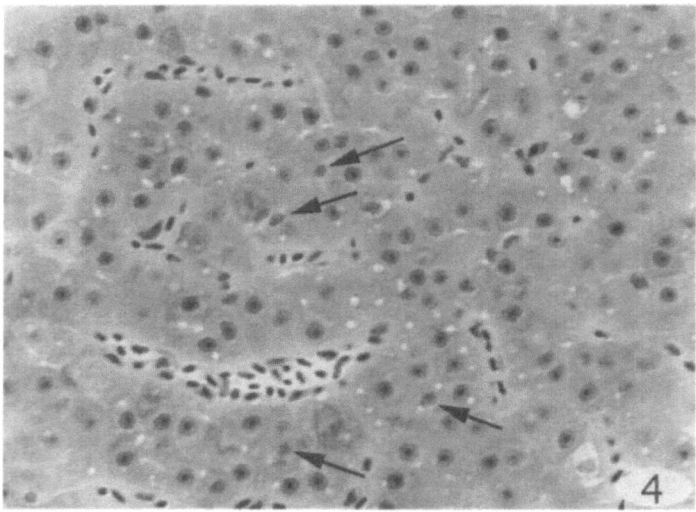

Figure 4. Toluidine blue stained semithin resin section of medaka liver. The resolution is sufficient to resolve hepatocytes and biliary epithelial cells (arrows) within tubules, magnification × 400.

based on drawings from the earlier literature, were simply not compatible with profiles of sectioned liver from human, rabbit, cat and dog. Rather, his stereograms revealed "long rows of cells, one cell wide". From this, Elias (1949) concluded that the liver architecture was comprised of plates, or laminae, one cell thick and that plates were frequently perforated by lacunae (a collective term including perisinusoidal spaces and sinusoids). Interlaminary bridges of parenchymal cells frequently connected adjacent laminae. Later, Elias referred to the composite lacunae as the hepatic labyrinth and the composite of parenchymal cells as the hepatic muralium or muralium simplex (Elias and Sherrick, 1969). This architectural analysis holds true for mammalian liver and is frequently figured in various texts.

Based strictly on diameters of lacunae, Elias (1949) proposed a classification of livers into two types; one, the saccular liver, possessed saccular lacunae and was illustrated by sections from liver of cat and man. The other, the tubular liver, possessed cylindrical lacunae and was illustrated by sections through liver of horse and rabbit. In the 1949 papers, Elias did not refer to the review by Shore and Jones (1889), which had already employed the term "tubular" to define arrangement of hepatic parenchymal cells. Later (1952), Elias, writing with Bengelsdorf, recognized his lack of awareness of the comparative aspects of liver morphology in non-mammalian vertebrates, as reviewed by Shore and Jones (1889). He changed his previous term "tubular liver" to "tubulosinusoidal" liver, thereby, in our opinion, removing perhaps the most useful term that we would prefer for describing the architecture of livers of Osteichthyes. When Elias and Bengelsdorf (1952) extended their analysis to lower vertebrates, they used 5 μm sections of paraffin-embedded liver from various vertebrate classes including: Cyclostomata, Selachii, Teleostei, Ganoidei, Dipnoi, Amphibia, Reptilia, and Aves. They concluded that the livers of non-mammalian vertebrates were solid masses of hepatic cells perforated by cylindrical lacunae containing sinusoids. The walls separating neighboring sinusoids were regarded as predominantly two cells thick. This gave rise to the term dual-plated muralium (muralium duplex).

Several features of the dual-plated muralium model can be questioned, when higher resolution methods including toluidine blue stained 0.5–1.0 μm sections (high resolution light microscopy; Figs. 1 and 3) or transmission electron microscopy (TEM) are used. TEM studies showed tubules in hepatic parenchyma in eel (Ito et al., 1962), lamprey (Bertolini, 1965) and hagfish (Mugnaini and Harboe, 1967). These findings led Elias and Sherrick (1969) to accept the tubular architectural model in these vertebrates. Subsequently, extensive investigations of the lamprey liver (Youson, 1981) and rainbow trout liver (Hampton et al., 1988, 1989) confirmed the tubular arrangement.

Figures 5 and 6 are drawings of the curvilinear hepatic tubules and labyrinth (lacunae) based on analysis of liver sections from a variety of teleosts. In Figure 5, the hepatic tubules which are made up of hepatocytes and biliary epithelial cells (bile preductular cells; Hinton and Pool, 1976) are in white. In addition to hepatocytes and biliary epithelial cells, we have figured interhepatocytic, perisinusoidal macrophages (McCuskey et al., 1986; Rocha et al., 1996) and included them in the tubular space. Between the

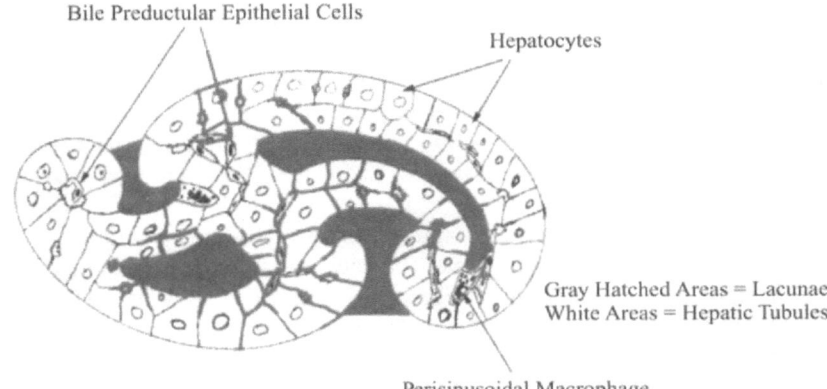

Figure 5. Drawing of hepatic tubules and lacunae (gray areas). Cell types as indicated by labels.

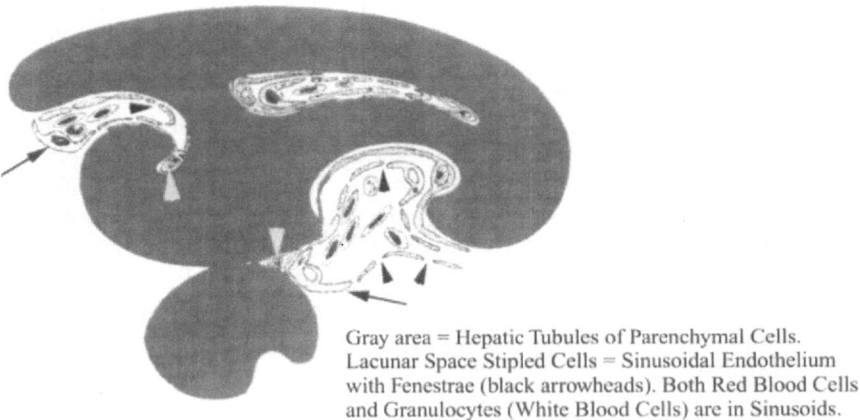

Gray area = Hepatic Tubules of Parenchymal Cells.
Lacunar Space Stipled Cells = Sinusoidal Endothelium
with Fenestrae (black arrowheads). Both Red Blood Cells
and Granulocytes (White Blood Cells) are in Sinusoids.

Figure 6. Drawing of hepatic tubules (gray areas) and lacunae. Cells of lacunae include: sinusoidal endothelial cells (arrows) with gaps indicating fenestrae (black arrow heads). Cells between endothelial and darkened tubule profiles are perisinusoidal cells of Ito (vertical arrow heads). Nucleated red and granulocytic white blood cells are shown in sinusoidal lumina.

tubules, lacunae shown in gray (Fig. 5) fill the space. Cellular components of lacunae are emphasized in Figure 6. These include: sinusoidal endothelial cells, various blood cells within the vascular lumina and perisinusoidal cells of Ito within the space of Disse and sending long extensions between endothelial cells and finger-like processes of hepatocytes (not shown). When present, melanomacrophage aggregates (Wolke, 1992) are also in lacunae. Hepatic tubules of Figure 6 are darkened to emphasize lacunae.

The collective hepatic tubules form the parenchymal compartment of the liver, while the stromal compartment includes all lacunae, the portal vein and its branches, the paired hepatic veins, the intrahepatic artery and its composite branches including arterioles, and finally the large bile ducts. In addition, exocrine pancreas, the abundance of which is strongly species dependent, may be present as acini within the adventitia of the portal vein and its larger branches.

Important features of teleost liver different from mammalian liver

The mammalian architectural pattern for liver is the one that is by far the most well-known and, while overall similarity exists in livers of various vertebrates, the mammalian pattern has been incorrectly applied to certain descriptions of hepatic structure of fishes (cf. discussion in Hampton et al., 1985). There are seven points of digression which are of particular relevance to our consideration in this chapter.

First, livers of many fish are a single lobe enabling sampling of the entire organ thereby avoiding problems inherent with distribution and size of focal lesions common to two-dimensional studies (Harada et al., 1989; Morris, 1989; Pitot et al., 1989).

Second, the lobular pattern common to mammalian liver (Banks, 1993a; Ham, 1974) is not as distinct or is absent in conventionally processed liver preparations from fishes (Hampton et al., 1985; Simon et al., 1967). Compare figures in Ham (1974) or any more recent mammalian text (Banks, 1993a) with Figures 7 and 8 of the present communication and with those of Hampton et al. (1985). As is shown in Figure 7, perilobular

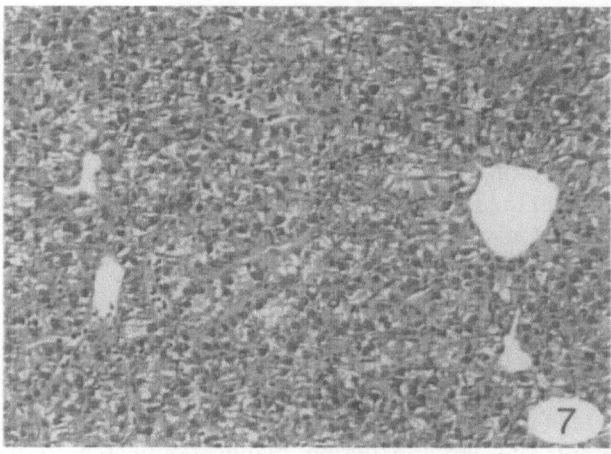

Figure 7. Section of liver showing portion of hepatic lobule. Venous profile at right of field was transected. Venules at left of field were tangentially sectioned. Note absence of identifying features to designate these profiles as portal *versus* hepatic venules. Hematoxylin and eosin stain, magnification × 150.

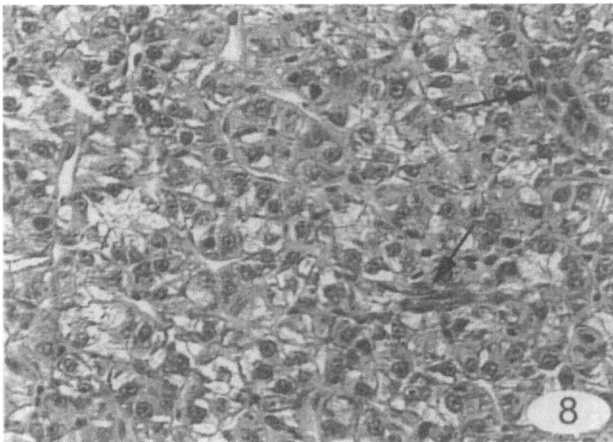

Figure 8. Section of liver showing parenchymal compartment of trout liver. Tubules are tightly packed and bile duct (arrows) without accompanying venule is present. Hematoxylin and eosin stain, magnification × 300.

connective tissue is absent. In addition, very few of the venular profiles, to which sinusoids are connected, have adjacent bile ducts, arterioles or lymphatics which enable us to differentiate the peripheries of hepatic lobules. Rather, as is shown in Figure 8, small bile ducts are often seen within parenchyma and not in the adventitia of venules as is the case in mammalian liver (portal venules and bile ducts – see below). Using high resolution *in vivo* microscopic methods in trout which were anaesthetized and maintained by gill suffusion of MS-222, livers were epi-illuminated at various wavelengths from 350–700 nm and such illumination permitted good definition of tubules, sinusoids, venules, and some cellular detail (Hinton et al., 1987). Despite following individual erythrocytes through sinusoids into hepatic veins, this method only examines the most superficial aspects of the liver and relationships initially visualized using this technique, are lost when followed by conventional processing. In summary, there is no current concept of the microcirculation in fish liver. Hinton and co-workers (1987) stated that if we are to use aquatic species in toxicity testing it is imperative that we understand how tubular liver of teleosts responds to toxicants. Questions included: (1) Do subpopulations (functional units) of hepatocytes exist in the tubular liver? (2) Based on the location of the hepatocytes adjacent to defined portions of the hepatic vasculature, will some cells show toxic effects, while others will not?

Third, livers of mammals (Macchiarelli et al., 1990) and fishes (Elias and Sherrick, 1969) are large exocrine glands connected to the intestine through the common bile duct and to the remainder of the splanchnic circulation through the portal vein and the hepatic artery. Both contain extensive capillary-like structures, the hepatic sinusoids. In the latter, however,

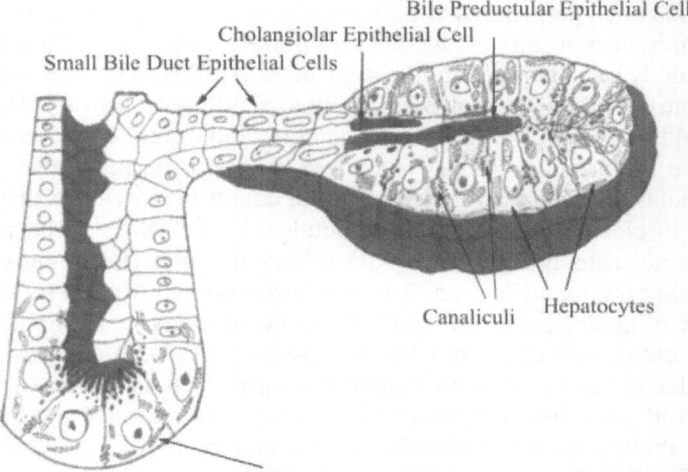

Figure 9. Drawing of components of hepatic tubule and its exocrine duct. Tubule (right) and its exocrine duct, bile duct, have been shortened to reveal components. Length of individual tubules and their relationships in space are not known. Specific cell types are labeled.

the parenchyma retains a glandular structure, the adenomere (Banks, 1993b) of which is a tubule of hepatocytes arranged about a lumen of biliary space delimited by hepatocytes (canaliculi) and by transitional biliary epithelial cells (bile preductular epithelial cells; Hinton and Pool, 1976). Figure 9 is a drawing of the hepatic tubule illustrating hepatocytes and their relationship to various portions of the intrahepatic biliary system. In a two-dimensional view, the hepatic tubule of teleosts appears as a muralium with a double row of hepatocytes intervening between adjacent sinusoidal profiles. However, as is presented herein (Figs. 1–6, and 9) and as described in the papers by Hampton et al. (1985, 1988, 1989) from this laboratory, and reviewed in detail by Hinton (1997), biliary epithelial cells intervene between the layers of hepatocytes. High resolution light microscopy (Okihiro, 1996), enzyme histochemistry (Okihiro, 1996) and transmission electron microscopic analysis of freeze-fractured material (Hampton et al., 1988) all support the concept of a tubule with biliary space as the lumen.

Fourth, spatial relationships between biliary ductules, portal venules, and hepatic venules (central veins of lobule) differ in teleost fishes as compared with mammals. Mammalian livers have portal tracts, and though differing among species in the amount of connective tissue, portal tracts contain portal venule, hepatic arteriole, bile ductule and lymphatic space (Banks, 1986; Ham, 1974). This co-distribution of biliary ductule and portal venule permits the microscopist to distinguish portal from hepatic venules (initial tributaries of the hepatic venous system) and therefore to distinguish hepatocytes of portal, or afferent, regions of the lobule from those of perihepatovenular (also called perivenous or pericentral), or efferent regions. Because of these differences in biliary ductule and portal venule distributions, the teleost lobule is less distinct, and the associated venules cannot be unequivocally identified. By observing relationships of individual profiles, Hinton (1998) found that 48–52% of the venules in rat and mouse liver were portal (in close proximity to biliary ductules), whereas the remainder were hepatic venules. In rainbow trout liver, similar counts recorded less than 7% of venules in close proximity to bile ductules. In the teleost liver, portal venules are not usually distributed with biliary ductules (Hampton et al., 1985), making the identification of portal (afferent) and hepatic (efferent) venules and of associated efferent or afferent hepatocytes problematic.

Fifth, heterogeneity of mammalian hepatocytes has been demonstrated with regard to oxygen saturation, cell volume, glycogen, isozymes of cytochrome P-450 and other enzymes, and cell shape. This has led to the concept of metabolic zonation within the mammalian liver lobule. Less (Schär et al., 1985), little (Segner and Braunbeck, 1988), or no (Hampton et al., 1985), evidence exists, however, for preferential localization of enzymes analogous to the zonation demonstrated in mammalian livers. Hepatotoxicants like carbon tetrachloride have been classified by the zone in which they produce alteration in rodent liver. Use of this reference toxicant in trout has produced no consistent pattern of alteration (Gingerich et al., 1978; Pfeifer et al., 1980; Statham et al., 1978) despite the fact that recent investigations with isolated hepatocytes of trout (Råbergh and Lipsky, 1997) show depletion of glutathione and appearance of the enzyme, lactic dehydrogenase, in culture medium following exposure.

Sixth, canaliculi, specializations of the hepatocyte plasma membrane (Macchiarelli et al., 1988), are for practical purposes the sole intralobular biliary passageway within the mammalian lobule. Only short paraportal ductules (canals of Hering; Elias and Pauly, 1966), extend brief distances from portal tracts to penetrate adjacent parenchyma. These are the only mammalian intralobular biliary epithelial cells (Grisham, 1980) and are regarded as the source of "oval cell" proliferation (Grisham, 1980). In comparison, canaliculi are fewer in lobules of fish liver, where they extend for only short distances within the hepatic parenchyma before they empty into transitional passageways (bile preductules) lined by hepatocytes and biliary epithelial cells. These aspects are illustrated in the drawing (Fig. 9). Bile preductules and ductules are numerous and are extensively located *within* the hepatic parenchyma (Hampton et al., 1985, 1988, 1989; Hinton and Pool, 1976). When morphometric procedures were used to estimate the numerical densities (e.g., number per volume of liver) for biliary epithelial cells *within* the rainbow trout parenchyma (Hampton et al., 1989), a ratio of one biliary epithelial cell to seven (male) or ten (female) hepatocytes was obtained.

Seventh, sexual dimorphism is particularly evident during vitellogenin production in fishes. It is possible to identify the sex of an adult fish by analysis of electron micrographs of hepatocytes. Hepatocytes from males contain abundant glycogen, while those from females are enriched in protein synthetic organelles (granular endoplasmic reticulum and Golgi apparatus). Glycogen is scarcer and lipid is generally more abundant in the female (Braunbeck et al., 1989). Male and female differences in hepatocyte volume are common to fish (Hampton et al., 1989) and rodents (Pitot et al., 1989), but the hepatocyte contribution to vitellogenin production, seen in Golgi-derived vesicles as coarse granular material of medium electron density, is the strongest hepatic indicator of sex (female). Under the light microscope, actively vitellogenic livers show increased hepatocyte lipid and cytoplasmic basophilia (Ng et al., 1984; Olivereau and Olivereau, 1979; Selman and Wallace, 1983; Wallace, 1978; Wester and Canton, 1986), features not generally associated with male fish or with females not in vitellogenic phase, but certainly within range of normality under pre-spawning conditions.

Cellular relationships in neoplastic lesions of fish liver

The following section describes features of teleost hepatic resident cell types as they are related to the hepatic tubule and to the hepatic lacunae (Figs. 5 and 6). Next, the known and proposed role(s) of these cells in lineages of specific hepatic neoplasms are presented. These are organized around the framework presented in Figure 10.

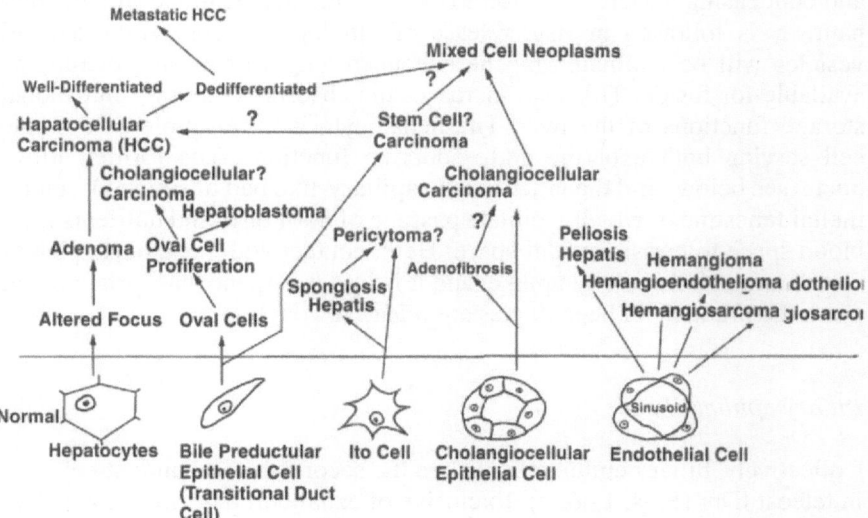

Figure 10. Known and proposed cellular lineages in hepatic carcinogenesis of teleosts. Normal cell types are at bottom of figure and various known and/or proposed steps are shown. From the hepatic tubules, hepatocellular, cholangiolar and/or mixed cell neoplasms may arise. Pericytomas and sarcomas involving endothelial cells are from lacunae.

Cells of the hepatic tubule

These are the epithelial cells of the hepatic parenchyma and are responsible for the major functions associated with this organ.

Hepatocytes

The hepatocyte is the most conspicuous cell and the one occupying the greatest percentage of the organ volume. Total hepatocyte volume has been estimated at 80% (rainbow trout; Hampton et al., 1989), 83% (brown trout; Rocha et al., 1997) and 88% (medaka; Hinton et al., 1984a) of total liver volume. As is shown in Figures 1–6, and 9, hepatocytes form almost all of the hepatic tubules. This arrangement is appropriate for the exocrine functions of liver. In fact, surficial modification of the hepatocyte plasma membranes forms canaliculi, the initial portion of the hierarchy of biliary passageways within the liver. Hormones such as estradiol stimulate hepatocyte production of vitellogenin, an endocrine-related function. Structural hepatocyte features which facilitate this function include the numerous finger-like processes within the perisinusoidal space of Disse. Herbener et al. (1983, 1984) used immunocytochemistry to visualize and quantify the organelles used in the hepatocyte production, transport and release of vitellogenin by the frog (*Rana pipiens*). Immunogold localization of an antibody to frog vitellogenin showed that a pathway involving endoplasmic reticulum, Golgi apparatus, and condensing vesicles was used. If we assume that a similar intracellular pathway is followed in fish, release of vitellogenin from Golgi-derived vesicles will be facilitated by the enhanced length of plasma membrane available for fusion. This also increases the absorptive area for nutritional storage functions of the liver. The hepatocyte is an example of a single cell serving both exocrine and endocrine functions. The former utilize ducts (see below) and the latter a rich capillary-like bed of sinusoids, endothelial fenestrae of which facilitate passage of hormones and nutrients from blood space to perisinusoidal space. Here, contact with hepatocyte plasma membrane is followed by uptake, and it is into this space that certain of the synthetic products of hepatocytes are released.

Biliary epithelial cells

Collectively, biliary epithelial cells are the second most abundant cell type in teleost liver (Figs. 1–6, 9). Exclusive of canaliculi described above (see hepatocytes), these cells are the entire epithelial lining for the intrahepatic hierarchy of biliary passageways. Biliary epithelial cells have been categorized by the portion of the intrahepatic biliary tree that they are associated with, i.e., bile preductular epithelial cells, ductular or cholangiolar epithelial cells, small – and large – bile ductal epithelial cells, or, by morphology, as oval, cuboidal, short- and tall columnar epithelial cells.

The bile preductular epithelial cell of fishes was first described in channel catfish (*Ictalurus punctatus*) by Hinton and Pool (1976). This term was selected based on the flow of bile, viz., manufactured by the hepatocyte and then passed through the hierarchy of biliary passageways. The bile preductular cell shares junctional complexes with hepatocytes and is located in the central portion of the hepatic tubule, and, together with hepatocytes, forms a portion of the wall of the biliary system between the canaliculus and the bile ductule or cholangiole (Braunbeck et al., 1992; Hampton et al., 1988; Hinton and Pool, 1976). Based on its transitional position between the secretomere and the duct, the bile preductular epithelial cell is to the teleost liver what the centroacinar cell of the mammalian exocrine pancreas is to the acinus or alveolus (Banks, 1993a) of that organ.

The next level in the intrahepatic hierarchy of biliary passageways is the cholangiole or ductule. Epithelial cells of these structures are similar in morphology to that of the bile preductular epithelial cells. The difference is that the ductule lumen is completely lined by biliary epithelial cells. These cells are oval to cuboidal and continue to lie at the center of the hepatic tubule (Figs. 4 and 5). This location of biliary epithelial cells in the hepatic tubule was first suggested after comparison of localization patterns for magnesium-dependent adenosine triphosphatase in enzyme histochemical preparations (Hampton et al., 1988) of rat and rainbow trout liver. The lead phosphate marked the wall of the biliary passageways within the liver. In the former, a pattern resembling chicken wire was seen throughout the hepatic parenchyma. All parenchymal biliary passageways were uniform, black lines and represented canaliculi in the rat. By contrast, the pattern in rainbow trout did not resemble the chicken wire pattern of the rat, rather, four gradations of thickness of lead phosphate reaction product were seen within the hepatic parenchyma of the rainbow trout. Companion transmission electron microscopic studies (Hampton et al., 1988) showed that the larger diameters were associated with ductules, small and intermediate bile ducts. Biliary epithelial cells of preductules, ductules, and small bile ducts lacked basal laminae. When bile duct epithelium became more columnar in appearance, a basal lamina was apparent, and this was surrounded by a capillary plexus containing nerve processes, often closely associated with hepatic arterioles, forming the biliary arterial tract (Hampton et al., 1988). Our findings suggested that bile preductules, ductules, and small bile ducts were found inside and were a component of hepatic tubules. Larger ducts, possessing columnar epithelial cells, with their adnexia were analogous to ducts within portal tracts of mammalian liver. Large bile ducts of fish liver are surrounded by connective tissue and separated from the parenchymal compartment (Hampton et al., 1988). In our experience, these large intrahepatic bile ducts in medaka liver are restricted to locations near the hilus (D.E. Hinton, unpublished observations).

In addition to hepatocytes and biliary epithelial cells described above, a perisinusoidal, interhepatocytic macrophage has been described (McCuskey et al., 1986; Rocha et al., 1996). The latter report, in brown trout (*Salmo trutta*), showed close relationship of this cell to the intratubular bile ductules. There are no reports of this macrophage in the lineage of hepatic neoplasms to date. Rather, it is included here simply due to its presence within the hepatic tubule. The report by Rocha and colleagues (1996) forms an excellent descriptive platform from which more definitive investigations as to the role of this cell in health and disease could be launched. Of particular importance is their demonstration of the macrophage in transit from sinusoidal lumen into perisinusoidal space suggesting that a portion of the life history of this cell is within the vascular system. Definitive studies on the origin and life history of teleost hepatic macrophages are needed.

Cells of the hepatic lacunae

To understand the concept of hepatic lacunae, it is first necessary to understand aspects of the development of tubules and spaces within liver. Elias (1955) studied the embryology of the liver in 30 species from 16 orders of eight classes including two teleost genera, *Salmo* (species not given) and guppy (*Lebistes reticulatus*). In the former, the median ventral entoderm of the foregut became thicker and the developing liver came to lie immediately on the yolk sphere. In the latter, a mass of entodermal cells was found in intimate contact with the yolk. In both, the entoderm was regarded as the sole contributor to liver development, and internal rearrangements of liver cells and yolk-associated blood islands occurred. Blood vessels from the margin of the yolk sac entered the hepatic tissue thereby forming an organ whose hepatic tubules were an entodermal contribution and whose lacunae were provided by mesoderm. From the above, it is apparent that the hepatic lacunae contain sinusoids and the perisinusoidal cells and that these components arose from a different germ layer, the mesoderm, and retain their distinction from the entodermally-derived hepatic tubules.

Perisinusoidal or fat-storing cells of Ito

Location, shape, contents (or lack thereof) and the name of the original Japanese investigator have found usage in the nomenclature of this resident liver cell. The drawing (Fig. 6) shows relationship of the perisinusoidal or fat-storing cell of Ito to the hepatic tubules and to the endothelial cell lining hepatic sinusoids. Morphology and functional properties including storage of vitamin A by this cell in various vertebrates have been reviewed (Wake, 1980). Perisinusoidal cells contain an elongated and frequently indented (by lipid droplets, the solvent of vitamin A) nucleus. "Empty" variants of this cell (Ito and Shibasaki, 1968) are fibroblast-like cells with well-developed granular endoplasmic reticulum. First described in human adults and embryos (Yamamoto and Enzan, 1975), these cells have subsequently been described in various fishes and reptiles. In teleosts, injection of vitamin A causes the number of fibroblast-like cells to decrease while the number of lipid-laden perisinusoidal fat-storing cells increases (Fujita et al., 1980; Takahashi et al., 1978).

Sinusoidal endothelial cells

The sinusoidal endothelial cell is the final cell type considered in this review as having potential to give rise to neoplasms in the liver of fishes. The relationship of sinusoidal endothelial cells to perisinusoidal space and to vascular lumen is shown (Fig. 6). Few papers have presented information on the fine structure of sinusoidal endothelial cells (McCuskey et al., 1986; Nopanitaya et al., 1979) and these were restricted to goldfish (*Carassius auratus*) and rainbow trout (*Oncorhynchus mykiss*), respectively. Freeze-etch replicas revealed fenestrae on endothelial cells of goldfish. Collections of approximately 15–20 fenestrae appeared similar to sieve plates described in other vertebrates (Wisse, 1974). Abundant cytofilaments characterize fat-storing cells, biliary epithelial cells, and endothelial cells. Identification of individual cells in lesions of liver will likely have to be done on an immunohistochemical basis since common cytologic features (filaments in all three cell types, and vesicles in fat-storing cells and endothelial cells) exist.

Specific neoplastic lesions

Hepatocyte derived neoplastic lesions

That fish mature hepatocytes are targets of both the toxic injurious and carcinogenic effects of known carcinogens is well supported by published studies to date. Following the hepatotoxic changes dictated by concentration and/or duration of exposure agent, the earliest possible preneoplastic or neoplastic change appears to involve fully differentiated, mature hepatocytes. These early changes consist of the formation of foci of altered hepatocytes detected by tinctorial or staining differences between the focus and surrounding, extrafocal hepatocytes that constitute a normal or regenerated hepatic parenchyma. Teh and Hinton (1993), using freeze-dried, vacuum-embedded fresh tissue have studied enzyme histochemistry of foci in medaka induced by brief exposure to diethylnitrosamine. Serial section analysis permitted correlation of histochemistry with conventional staining in the same lesions. Foci detected by conventional staining also showed enzyme alteration and in that model, foci appeared first and preceded the other lesions shown in Figure 10 and discussed below.

While a number of phenotypes can be demonstrated using conventional stains (Okihiro, 1996) and while it is important to distinguish each in thoroughly addressing histogenesis of hepatic neoplasia (Bannasch et al., 1996), three are more commonly described (Boorman et al., 1997; Couch and Courtney, 1987). These are basophilic, eosinophilic, and clear cell foci (Couch and Courtney, 1987; Hinton, 1998). Biochemical, cytochemical and ultrastructural properties of foci in rat hepatocarcinogenesis (Bannasch et al., 1985, 1996) and to a lesser extent in mouse (Ruebner et al., 1996) have been reported. Enzyme histochemical features of foci of altered hepatocytes have been established in serial sections processed for various enzymes and for conventional staining (Teh, 1996; Teh and Hinton, 1993). Janis Cooke (unpublished observations, this laboratory) examined the three focal phenotypes under a promotional assay which included feeding estradiol to previously initiated medaka (diethylnitrosamine bath exposure for 48 hours at 21 days after hatch). Her findings show that basophilic foci underwent preferential growth during estradiol feeding and that eosinophilic foci were more commonly encountered in males given the initiating carcinogen alone. Clear cell foci appeared early but were not promoted by estradiol and eventual neoplasms lacked cells with this phenotype.

Adenomas appear to be bridging lesions between certain foci and eventual hepatocellular carcinomas (described below). When serial sampling follows stop exposure, adenomas typically overlap with the occurrence of foci of altered hepatocytes but are distinguished from them by their more distinct borders, thickened hepatic tubules, cellular pleomorphism, and variable overall shape. Adenomas are typically larger than foci (Couch and Courtney, 1987; Hinton, 1993).

Characteristics of hepatocellular carcinoma in fishes were described (Hinton, 1993). Overall size of these lesions varies from focal to very large, often occupying a major portion of the overall organ. Masses of cells show numerous mitotic figures and they may be bizarre. In addition, the masses are often solid and resemble engorged tubules. The margin of the tumor is less well-defined and invasion into otherwise normal surrounding parenchyma is common. As is shown in Figure 10, both poorly- and well-differentiated hepatocellular carcinomas are seen as subsets of hepatocellular carcinomas. It is possible that well-differentiated carcinomas develop from cells in some of the adenomas that, as a group, begin to decrease in numbers coincident with the first appearance of the well-differentiated carcinomas (Couch and Courtney, 1987; Hinton et al., 1992; Okihiro, 1996). Although metastasis occurs, we simply lack sufficient data to state whether this process involves only dedifferentiated or poorly differentiated carcinoma and not well differentiated or both.

Neoplastic lesions derived from bile preductular epithelial cells – a proposed lineage

Neoplastic lesions derived from bile preductular epithelial cells, the second cell in the lineage scheme (Fig. 10) could putatively give rise to hepatocellular, biliary and perhaps mixed cell neoplasms. As is shown in Figure 9, the bile preductular cells and the centroacinar cells of the pancreas have analogous positions within the tubule (liver) and acinus with its duct (pancreas). Okihiro (1996) has provided evidence that demonstrates proliferation of bile preductular cells after partial hepatectomy in both medaka and rainbow trout. Further serial analysis of the repair/regeneration showed the appearance of cells with transitional features between biliary epithelium and hepatocytes, and, finally with time, the hepatocyte phenotype characterized these cell populations. Taken together, this suggests that the bile preductular epithelial cell may be a stem cell capable of giving rise to both adult cell types within the hepatic tubule. Further similarity between descriptions of fine structure of oval cells in rodents and bile preductular cells in rainbow trout were demonstrated by Hampton et al. (1988). More detailed serial progression between initiation and first appearance of neoplasms is needed and this is under investigation in medaka in which the interval of interest is 16 weeks. Based on the sheer number of the bile preductular epithelial cells (Hampton et al., 1989), rainbow trout and medaka as well as other teleosts are excellent models for analysis of the role(s) of this cell type. Okihiro (1996) added new markers to the list of those reviewed by Hinton (1993) including lectins and cytofilaments.

The lineage suggested by the line from bile preductular epithelial cells in Figure 10 is proposed as a possibility; one which currently lacks experimental evidence. In addition, the database with various carcinogens is limited and studies with additional compounds are needed. Interestingly, livers of many species of teleosts normally include exocrine pancreatic tissues that follow the course of the larger intrahepatic branches of portal veins (Couch et al., 1974; Hinton et al., 1984b). Therefore, the liver of certain fishes might be called a hepatopancreas. Sell (1990) reported that occasionally hepatocytes appear in the pancreas of a variety of mammals spontaneously or after experimental injury or treatment with carcinogens. He concluded that the pancreas, like the liver, possesses a stem cell that is at least bipotent, with the capacity of differentiating into adenomere or ductal epithelial cell. Thus, the "hepatopancreas" of fishes may provide a reservoir that harbors a multipotent precursor or stem cell that could be transformed by carcinogenic agents and then partially differentiate into one of several neoplastic phenotypes directly or *via* the oval cell route. The rare occurrence of metaplastic pancreas in fish liver (rainbow trout) neoplasms induced by diethylnitrosamine (Lee et al., 1989), and the findings by Groff et al. (1992) as reviewed in Laurén and Hinton (1990) of relatively massive pancreatic metaplasia (20% of liver) in the non-neoplastic liver of striped bass (*Morone saxatilis*) that were dying in a commercial pond might reflect the induction of a stem cell in their liver to differentiate into pancreatic cells.

Neoplastic lesions derived from perisinusoidal cells (stellate cells, Ito cells, fat-storing cells)

Bannasch et al. (1981) first described a pathological liver condition involving fat-storing cells in rats experimentally exposed to N-nitrosomorpholine, spongiosis hepatis. These mesh-like lesions in toxicant injured hepatic tissue were made up of attenuated, empty perisinusoidal cells, bound to one another by desmosomes, and framed larger to smaller spaces previously occupied by hepatocytes. In fishes, the spongiosis hepatis lesions were first reported by Hinton et al. (1984) and studied in altered, pathological stages by Hinton (1984), Couch and Courtney (1987, 1991) and Bunton (1990). Fish spongiosis hepatis lesions appear to be derived from perisinusoidal cells similarly to that in mammals (Fig. 10). Couch (1991) was able to illustrate clearly, with ultrastructural, morphological evidence, that the early formation of spongiotic structures by perisinusoidal cells began while the cells were still in the perisinusoidal space adjacent to endothelium, following induction of liver toxic degeneration by exposure of the fish to N-nitrosodiethylamine. The proliferation of perisinusoidal cells and their extended growth beyond the perisinusoidal space, resulted further in variably sized, but larger spongiosis hepatis lesions in injured liver. Morphogenetic observations (electron and light microscopic levels) suggest that spongiosis hepatis in rats (Bannasch et al., 1981, 1986) and in fish (Couch, 1991), following exposure to certain nitrosamines, has as its cell of origin the perisinusoidal cell (Fig. 10).

In addition, the normal perisinusoidal cell and the reticular framework cell of spongiosis hepatis share certain cytoplasmic features to such a degree as to indicate that they are probably the same cell type. Hinton (1984) and Couch (1991) demonstrated intermediate filaments in normal perisinusoidal cells as well as those of spongiosis hepatis. This finding was also made by Bannasch et al. (1981). The lipid droplets in normal perisinusoidal cells largely disappear in the attenuated perisinusoidal cells of spongiosis hepatis because of the distorted shapes they assume during spongiosis hepatis formation and because of their loss of cytoplasmic volume.

It is possible that highly advanced, chemically-induced spongiosis hepatis cells, in both rats and certain fishes, may transform into neoplasms. In these cases, the spongiosis hepatis may form pericytomas in rats (Bannasch et al., 1986) or polymorphic cell neoplasms in fishes (Couch, 1991), probably a form of pericytoma, since perisinusoidal cells are probably pericytes or are in the pericyte lineage in the liver and other splanchnic organs (Couch, 1991; Wake, 1980). The appropriate names for liver perisinusoidal cell neoplasms, therefore, would be pericytomas or pericytic sarcomas, depending upon their degree of dedifferentiation, histological patterns and invasiveness. Indeed, Couch and Courtney (1987) and Grizzle and Thiyagarajah (1988) described classical appearing hemangiopericytomas near or in the livers of the fishes, sheepshead minnows (*Cyprinodon varie-gatus*) and *Rivulus*, that had been exposed to diethylnitrosamine. These neoplasms consisted of apparent concentric whorls of pericytes around small vessels attached to the liver or in the liver, and could have arisen from pericytes around hepatic arterioles or veins (see Couch, 1991).

Neoplastic lesions derived from endothelial cells

Experimentally-induced, endothelial cell neoplasms in the livers of fishes have been reported only once. Grizzle and Thiyagarajah (1988) described cavernous hemangiomas in *Rivulus marmoratus* exposed continuously for 8 months to diethylnitrosamine (beginning as larval fish). These lesions began as peliosis as early as 4 weeks of exposure in the larvae and pro-gressed (four cases) to cavernous hemangioma characterized as large blood-filled cavities lined by single layers of swollen endothelial cells. There is a question as to whether these lesions are neoplastic or represen-tative of advanced peliosis hepatis. It is apparent that the sinusoids and their endothelia, however, are the structures involved in their origin (Fig. 10).

Endothelium-derived neoplasms of spontaneous origins have been reported in other tissues such as skin, heart, and gills of fishes (Couch, 1995). They include: hemangiomas, hemangioendotheliomas and heman-gioendotheliosarcomas (Fournie et al., 1998). As more work is done using small teleost fishes in carcinogenesis studies, additional cases will likely be described.

Concluding summary

The normal teleost liver is a tubular gland and likely a compound tubular gland, the functional unit or adenomere of which is a tubule of hepatocytes surrounding a biliary structure, the lumen. Exocrine functions of the liver include production of bile by the hepatocytes and its transport within a hierarchy of biliary passageways. In a manner analogous to the exocrine pancreas, transitional biliary cells, "bile preductular epithelial" cells, serve – under certain conditions – as progenitors for hepatocytes and bile duc-tular and ductal cells. Based on current evidence presented herein, hepato-cellular carcinomas arise from adult, fully differentiated hepatocytes, and lineage includes: foci of altered hepatocytes, adenomas and carcinomas. A proposed lineage, presently lacking experimental evidence, uses bile pre-ductular epithelial cells which resemble oval cells of mammalian liver as sources of teleost oval cells which may then proceed toward cholangiocellular, hepatocellular, or mixed neoplasms. The other lineage involving cells of the hepatic tubule is that leading from altered cholangiolar epithelial cell to cholangiocellular neoplasms. Cells of hepatic lacunae which may give rise to neoplasms involve the perisinusoidal cell of Ito and the endothelial cell of the hepatic sinusoids. By considering the teleost liver as a collection of entoderm-derived tubules and mesoderm-derived vascular and perivascular components, the lacunae, interpretation of specific lesions is enhanced. This architectural pattern is also useful in interpretation of non-neoplastic, toxic alterations. What we need to determine are the three dimensional features of individual tubules and whether recurring patterns exist between the tubule, or a portion of it, and specific vessels of the hepatic microcirculation.

Acknowledgements
Funded in part by US Public Health Service Grants CA-45131 from the National Cancer Institute, ES-04699 from the Superfund Basic Science Research Project, CR8191658-010 from the U.C. Davis, US EPA Center for Ecological Health Research and US EPA #R823297.

References

Arukwe, A., Knudsen, F.R. and Goksøyr, A. (1997) Fish zona radiata (egg shell) protein: a sensitive biomarker for environmental estrogens. *Environ. Health Perspect.* 105:418–422.

Bailey, G.S., Williams, D.E. and Hendricks, J.D. (1996) Fish models for environmental carcinogenesis: The rainbow trout. *Environ. Health. Perspec.* 104:5–21.

Balch, G.C., Metcalfe, C.D. and Huestis, S.Y. (1995) Identification of potential fish carcinogens in sediment from Hamilton Harbour, Ontario, Canada. *Environ. Toxicol. Chem.* 14:79–91.

Banks, W.J. (1986) *Applied Veterinary Histology.* 2nd edition, Williams and Wilkins, Baltimore.

Banks, W.J. (1993a) Digestive system II – extramural organs. *In:* Banks, W.J. (Ed.) *Applied veterinary histology*, third edition,. Mosby-Year Book Inc., St. Louis, pp 360–373.

Banks, W.J. (1993b) Epithelia. *In:* Banks, W.J. (Ed.) *Applied veterinary histology.* 3rd ed. Mosby-Year Book Inc., St. Louis, pp 48–67.

Bannasch, P., Bloch, M. and Zerban, H. (1981) Spongiosis hepatis, specific changes of the perisinusoidal liver cells induced in rats by N-nitrosomorpholine. *Lab. Invest.* 44:252–264.

Bannasch, P., Moore, M.A., Hacker, H.J., Klimek, F., Mayer, D., Enzmann, H. and Zerban, H. (1985) Potential significance of phenotypic instability in focal and nodular liver lesions induced by hepatocarcinogens. *In:* Brunner, H. and Thaler, H. (Eds) *Hepatology.* Raven Press, New York. pp 191–209.

Bannasch, P., Griesemer, R.A., Anders, F., Becker, R., Cabral, J.R., Porta, G.D., Feron, V.J., Henschler, D., Ito, N., Kroes, R., Magee, P.N., Montesano, R., Napalkov, N.P., Nesnow, S., Rao, G.N. et al. (1986) Early preneoplastic lesions. *In:* Montesano, R., Bartsch, H., Vainaio, H., Wilbourn, J., Yamasaki, H. (Eds) *Long-term and short-term assays for carcinogens: a critical appraisal.* International Agency for Research on Cancer/International Programme on Chemical Safety Commission of the European Communities, Lyon. IARC Scientific Publications 83:85–101.

Bannasch, P., Zerban, H. and Hacker, H.J. (1996) Foci of altered hepatocytes, rat. *In:* Jones, T.C., Popp, J.A. and Mohr, U. (Eds) *Monographs on pathology of laboratory animals. Digestive system.* 2nd Edition, Springer Verlag, Berlin, Heidelberg, New York, pp 3–37.

Bertolini, B. (1965) The structure of the liver cells during the life cycle of a brook-lamprey (*Lampetra zanandreai*). *Z. Zellforsch.* 67:297–318.

Boorman, G.A., Botts, S., Bunton, T.E., Fournie, J.W., Harshbarger, J.C., Hawkins, W.E., Hinton, D.E., Jokinen, M.P., Okihiro, M.S. and Wolfe, M.J. (1997) Diagnostic criteria for degenerative, inflammatory, proliferative non-neoplasia and neoplastic liver lesions in medaka (*Oryzias latipes*): Consensus of a National Toxicology Program Pathology Working Group. *Toxicol. Pathol.* 25:202–210.

Braunbeck, T., Storch, V. and Nagel, R. (1989) Sex-specific reaction of liver ultrastructure in zebra fish (*Brachydanio rerio*) after prolonged sublethal exposure to 4-nitrophenol. *Aquat. Toxicol.* 14:185–202.

Braunbeck, T.A., Teh, S., Lester, S.M. and Hinton, D.E. (1992) Ultrastructural alterations in hepatocytes of medaka (*Oryzias latipes*) exposed to diethylnitrosamine. *Toxicol. Pathol.* 20:179–196.

Bunton, T.E. (1990) Hepatopathology of diethylnitrosamine in the medaka (*Oryzias latipes*) following short-term exposure. *Toxicol. Pathol.* 18:313–323.

Bunton, T.E. (1996) Experimental chemical carcinogenesis in fish. *Toxicol. Pathol.* 24:603–618.

Couch, J.A. (1991) Spongiosis hepatis: Chemical induction, pathogenesis, and possible neoplastic fate in a teleost fish model. *Toxicol. Pathol.* 19:237–250.

Couch, J.A. (1995) Invading and metastasizing cardiac hemangioendothelial neoplasms in a cohort of the fish *Rivulus marmoratus*: unusually high prevalence, histopathology, and possible etiologies. *Cancer Res.* 55:2438–2447.

Couch, J.A. and Courtney, L.A. (1987) N-Nitrosodiethylamine-induced hepatocarcinogenesis in estuarine sheepshead minnow (*Cyprinodon variegatus*): neoplasms and related lesions compared with mammalian lesions. *J. Natl. Cancer Inst.* 79:297–321.

Couch, J.A., Gardner, G., Harsbarger, J.C., Tripp, M.R. and Yevich, P.P. (1974) Histological and physiological evaluations in some marine fauna. *In:* Laroche, G. (Ed.) *Marine bioassays*. American Petroleum Institute, U.S. Environmental Protection Agency and Marine Technology Society, Washington, D.C. pp 156–172.

Elias, H. (1949a) A re-examination of the structure of the mammalian liver. I. Parenchymal architecture. *Am. J. Anat.* 84:311–333.

Elias, H. (1949b) A re-examination of the structure of the mammalian liver. II. The hepatic lobule and its relation to the vascular and biliary systems. *Am. J. Anat.* 85:379–456.

Elias, H. (1955) Origin and early development of the liver in various vertebrates. *Acta. Hepatol.* 3:1–56.

Elias, H. and Bengelsdorf, H. (1952) The structure of the liver of vertebrates. *Acta. Anat.* 14:297–337.

Elias, H. and Pauly, J. (1966) Human microanatomy. 3rd edition, F.A. Davis Co., Philadelphia. pp 214–223.

Elias, H. and Sherrick, J.C. (1969) *Morphology of the Liver.* Academic Press, New York, London. pp 6–9.

Fournie, J.W., Herman, R.L., Couch, J.A. and Howse, H. (1998) Tumors of heart and endothelial cells. *In:* Dawe, C.J., Harshbarger, J.C., Wellings, S.R. and Strandberg, J.D. (Eds) *Pathobiology of spontaneous and induced neoplasms in fishes: comparative characterization, nomenclature and literature*, Academic Press, New York, London, in press.

Fujita, J., Tamaru, T. and Miyagawa, J. (1980) Fine structural characteristics of the hepatic sinusoidal walls of the goldfish (*Carassius auratus*) *Arch. Histol. Jpn.* 43:265–273.

Gingerich, W.H., Weber, L.J. and Larson, R.E. (1978) Carbon tetrachloride-induced retention of sulfobromophthalein in the plasma of rainbow trout. *Toxicol. Appl. Pharmacol.* 43:147–158.

Grisham, J.W. (1980) Cell types in long-term propagable cultures of rat liver. *Ann. N.Y. Acad. Sci.* 349:128–137.

Grizzle, J.M. and Thiyagarajah, A. (1988) Diethylnitrosamine-induced hepatic neoplasms in the fish *Rivulus ocellatus marmoratus. Dis. Aquat. Org.* 5:39–50.

Groff, J.M., Hinton, D.E., McDowell, T.S. and Hedrick, R.P. (1992) Progression and resolution of megalocytic hepatopathy with exocrine pancreatic metaplasia in a population of cultured juvenile striped bass *Morone saxatilis. Dis. Aquat. Org.* 13:189–202.

Ham, A.W. (1974) *Histology.* 7th edition, J.B. Lippincott Co, Philadelphia. pp 686–708.

Hampton, J.A., McCuskey, P.A., McCuskey, R.S. and Hinton, D.E. (1985) Functional units in rainbow trout (*Salmo gairdneri,* Richardson) liver. I. Histochemical properties and arrangement of hepatocytes. *Anat. Rec.* 213:166–175.

Hampton, J.A., Lantz, R.C., Goldblatt, P.J., Laurén, D.J. and Hinton, D.E. (1988) Functional units in rainbow trout (*Salmo gairdneri*, Richardson) Liver: II. The biliary system. *Anat. Rec.* 221:619–634.

Hampton, J.A., Lantz, R.C. and Hinton, D.E. (1989) Functional units in rainbow trout (*Salmo gairdneri*, Richardson) liver: III. Morphometric analysis of parenchyma, stroma, and component cell types. *Am. J. Anat.* 185:58–73.

Harada, T., Maronpot, R.R., Morris, R.W., Stitzel, K.A. and Boorman, G.A. (1989) Morphological and stereological characterization of hepatic foci of cellular alteration in control Fisher 344 rats. *Toxicol. Pathol.* 17:579–593.

Herbener, G.H., Bendayan, M. and Feldhoff, R.C. (1984) The intracellular pathway of vitellogenin secretion in the frog hepatocyte as revealed by protein A-gold immunocytochemistry. *J. Histochem. Cytochem.* 32:697–704.

Hinton, D.E. (1993) Cells, cellular responses, and their markers in chronic toxicity of fishes. *In:* Malins, D.C. and Ostrander, G.K. (Eds) *Aquatic toxicology: molecular, biochemical and cellular perspectives.* Lewis Publishers, Boca Raton, FL, pp 207–239.

Hinton, D.E. (1998) Structural considerations in teleost hepatocarcinogenesis: Gross and microscopic features including architecture, specific cell types and focal lesions. *In:* Dawe, C.J., Harshbarger, J.C., Wellings, S.R. and Strandberg, J.D. (Eds) *Pathobiology of spontaneous and induced neoplasms in fishes: comparative characterization, nomenclature and literature,* in press. Academic Press, New York.

Hinton, D.E. and Pool, C.R. (1976) Ultrastructure of the liver in channel catfish *Ictalurus punctatus* (Rafinesque). *J. Fish Biol.* 8:209–219.

Hinton, D.E., Lantz, R.C. and Hampton, J.A. (1984a) Effect of age and exposure to a carcinogen on the structure of the medaka liver: A morphometric study. *Natl. Cancer Inst. Monogr.* 65:239–249.

Hinton, D.E., Walker, E.R., Pinkstaff, C.A. and Zuchelkowski, E.M. (1984b) Morphological survey of teleost organs important in carcinogenesis with attention to fixation. *Natl. Cancer Inst. Monogr.* 65:291–320.

Hinton, D.E., Lantz, R.C., Hampton, J.A., McCuskey, P.R. and McCuskey, R.S. (1987) Normal versus abnormal structure: considerations in morphologic responses of teleosts to pollutants. *Environ. Health Perspect.* 71:139–146.

Hinton, D.E., Couch, J.A., Teh, S.J. and Courtney, L.A. (1988) Cytological changes during progression of neoplasia in selected fish species. *Aquat. Toxicol.* 11:77–112.

Hinton, D.E., Teh, S.J., Okihiro, M.S., Cooke, J.B. and Parker, L.M. (1992) Phenotypically altered hepatocyte populations in diethylnitrosamine-induced medaka liver carcinogenesis: resistance, growth and fate. *Mar. Environ. Res.* 34:1–5.

Ito, T. and Shibasaki, S. (1968) Electron microscopic study on the hepatic sinusoidal wall and the fat-storing cells in the normal human liver. *Arch. Histol. Jpn.* 29:137–192.

Ito, T., Watanabe, A. and Takahashi, Y. (1962) Histologische und cytologische Untersuchungen der Leber bei Fischen und Cyclostomata, nebst Bemerkungen über die Fettspeicherungszellen. *Arch. Histol. Jpn.* 22:429–463.

Kranz, H. and Dethlefsen, V. (1990) Liver anomalies in dab (*Limanda limanda*) from the southern North Sea with special consideration given to neoplastic lesions. *Dis. Aquat. Org.* 9:171–185.

Laurén, D.J., Teh, S.J. and Hinton, D.E. (1990) Cytotoxicity phase of diethylnitrosamine-induced hepatic neoplasia in medaka. *Cancer Res.* 50:5504–5514.

Lee, B.C., Hendricks, J.D. and Bailey, G.S. (1989) Iron resistance of hepatic lesions and nephroblastoma in rainbow trout (*Salmo gairdneri*) exposed to MNNG. *Toxicol. Pathol.* 17:474–482.

Macchiarelli, G., Motta, P.M. and Fujita, T. (1988) Scanning electron microscopy of the liver cells. *In:* Motta, P.M. (Ed.) *Biopathology of the liver. An ultrastructural approach.* Kluwer Academic Publishers, Lancaster, UK., pp 37–57.

Macchiarelli, G., Makabe, S. and Motta, P.M. (1990) The structural basis of mammalian liver function. *In:* Riva, A. and Motta, P.M. (Eds) *Ultrastructure of the extraparietal glands of the digestive tract.* Kluwer Academic Publishers, Norwell, MA., pp 185–211.

Malins, D.C., Krahn, M.M., Brown, D.W., Rhodes, L.D., Myers, M.S., McCain, B.B. and Chan, S.-L. (1985) Toxic chemicals in marine sediment and biota from Mukilteo, Washington: relationships with hepatic neoplasms and other hepatic lesions in English sole (*Parophrys vetulus*). *J. Natl. Cancer Inst.* 74:487–494.

McCuskey, P.A., McCuskey, R.S. and Hinton, D.E. (1986) Electron microscopy of cells of the hepatic sinusoids in rainbow trout (*Salmo gairdneri*) *In:* Kirn, A., Knook, D.L. and Wisse, E. (Eds) *Cells of the hepatic sinusoid*, Vol. 1. Kupffer Cell Foundation, Leiden, The Netherlands, pp 489–494.

Mommsen, T.P. and Walsh, P.J. (1988) Vitellogenesis and oocyte assembly. *In:* Hoar, W.S. and Randall, D.J. (Eds) *Fish physiology*, Vol. XIA. Academic Press, San Diego, pp 347–406.

Moore, M.N. and Evans, B. (1992) Detection of *ras* oncoprotein in liver cells of flatfish (dab) from a contaminated site in the North Sea. *Mar. Environ. Res.* 34:33–38.

Morris, R.W. (1989) Testing statistical hypotheses about rat liver foci. *Toxicol. Pathol.* 17:569–578.

Mugnaini, E. and Harboe, S.B. (1967) The liver of *Myxine glutinosa*: a true tubular gland. *Z. Zellforsch.* 78:341–369.

Myers, M.S., Stehr, C.M., Olson, O.P., Johnson, L.L., McCain, B.B., Chan, S.-L. and Varanasi, U. (1994) Relationship between toxicopathic hepatic lesions and exposure to chemical contaminants in English sole (*Pleuronectes vetulus*), starry flounder (*Platichthys stellatus*), and white croaker (*Genyonemus lineatus*) from selected marine sites on the Pacific Coast, USA. *Environ. Health Perspect.* 102:200–215.

Ng, T.B., Woo, N.Y.S., Tam, P.P.L. and Au, C.Y.W. (1984) Changes in metabolism and hepatic ultrastructure induced by estradiol and testosterone in immature female *Epinephelus akaara* (Teleostei, Serranidae). *Cell Tiss. Res.* 236:651–659.

Nopanitaya, W., Aghajanian, J., Grisham, J.W. and Carson, J.L. (1979) An ultrastructural study on a new type of hepatic perisinusoidal cell in fish. *Cell Tissue Res.* 198:35–42.

Okihiro, M.S. (1996) *Regenerative, hyperplastic, and neoplastic-hepatic growth in medaka (Oryzias latipes) and rainbow trout (Oncorhynchus mykiss): an investigation of the role of the biopolar hepatic stem cell.* PhD Thesis, Comparative Pathology. Department of Anatomy, Physiology & Cell Biology, School of Veterinary Medicine, University of California, Davis.

Olivereau, M. and Olivereau, J. (1979) Effect of estradiol-17β on the cytology of the liver, gonads and pituitary, and on plasma electrolytes in the female freshwater eel. *Cell Tissue Res.* 199:431–454.

Pfeifer, K.F., Weber, L.J. and Larson, R.E. (1980) Carbon tetrachloride-induced hepatotoxic response in rainbow trout, *Salmo gairdneri*, as influenced by two commercial fish diets. *Comp. Biochem. Physiol.* 67C:91–96.

Pitot, H.C., Campbell, H.A., Maronpot, R., Bawa, N., Rizvi, T.A., Xu, Y., Sargent, L., Dragan, Y. and Pyron, M. (1989) Critical parameters in the quantitation of the states of initiation, promotion, and progression in one model of hepatocarcinogenesis in the rat. *Toxicol. Pathol.* 17:594–612.

Råbergh, C.M.I. and Lipsky, M.M. (1997) Toxicity of chloroform and carbon tetrachloride in primary cultures of rainbow trout hepatocytes. *Aquat. Tox.* 37:169–182.

Rocha, E., Monteiro, R.A.F. and Pereira, C.A. (1996) The pale-grey interhepatocytic cells of brown trout (*Salmo trutta*) are a subpopulation of liver resident macrophages or do they establish a different cellular type? *J. Submicrosc. Cytol. Pathol* 28:357–368.

Rocha, E., Monteiro, R.F. and Pereira, C.A. (1997) Liver of the brown trout, *Salmo trutta* (Teleostei, Salmonidae): a stereological study at light and electron microscopic levels. *Anat. Rec.* 247:317–328.

Ruebner, B.H., Bannasch, P., Hinton, D.E., Cullen, J.M. and Ward, J.M. (1996) Foci of altered hepatocytes, mouse. *In:* Jones, T.C., Popp, J.A. and Mohr, U. (Eds) *Monographs of pathology of laboratory animals, digestive system.* 2nd edition. Springer-Verlag, Berlin, Heidelberg, New York, pp 38–49.

Schär, M., Maly, I.P. and Sasse, D. (1985) Histochemical studies on metabolic zonation of the liver in the trout (*Salmo gairdneri*). *Histochem.* 83:147–151.

Segner, H. and Braunbeck, T. (1988) Hepatocellular adaptation to extreme nutritional conditions in ide, *Leuciscus idus melanotus* L. (Cyprinidae). A morphofunctional analysis. *Fish Physiol. Biochem.* 5:79–97.

Sell, S. (1990) Is there a liver stem cell? *Cancer Res.* 50:3811–3815.

Selman, K. and Wallace, R.A. (1983) Oogenesis in *Fundulus heteroclitus* III. Vitellogenesis. *J. Exp. Zool.* 226:441–457.

Shore, T.W. and Jones, H.L. (1889) On the structure of the vertebrate livers. *J. Physiol.* (London) 10:408–428.

Simon, R.C., Dollar, A.M. and Smuckler, E.A. (1967) Descriptive classification of normal and altered histology of trout livers. *In:* Halver, J.E. and Mitchell, I.A. (Eds) *Trout hepatoma research: conference papers.* Research Report No. 70, US Fish and Wildlife Service, Washington, pp 18–28.

Statham, C.N., Croft, W.A. and Lech, J.J. (1978) Uptake, distribution, and effects of carbon tetrachloride in rainbow trout (*Salmo gairdneri*). *Toxicol. Appl. Pharmacol.* 45:131–140.

Sumpter, J.P. (1995) Feminized responses in fish to environmental estrogens. *Toxicol. Lett.* 82/83:737–742.

Sumpter, J.P., Jobling, S. and Tyler, C.R. (1996) Estrogenic substances in the aquatic environment and their potential impact on animals, particular fish. *In:* Tayler, E.W. (Ed.) *Toxicology of aquatic pollution. Physiological, molecular and cellular approaches.* Cambridge University Press, Cambridge, pp 205–224.

Takahashi, Y., Tsubouchi, H. and Kobayashi, K. (1978) Effects of vitamin A administration upon Ito's fat-storing cells of the liver in the carp. *Arch. Histol. Jpn.* 41:339–349.

Teh, S.J. (1996) *Cellular aspects of hepatocarcinogenesis in medaka (Oryzias latipes): dynamics of histogenesis and gender-related sensitivity.* PhD thesis, Comparative Pathology. Department of Anatomy, Physiology and Cell Biology, School of Veterinary Medicine, University of California, Davis.

Teh, S.J. and Hinton, D.E. (1993) Detection of enzyme histochemical markers of hepatic preneoplasia and neoplasia in medaka (*Oryzias latipes*). *Aquat. Toxicol.* 24:163–182.

Vethaak, A.D. and Jol, J.G. (1996) Diseases of flounder *Platichthys flesus* in Dutch coastal and estuarine waters, with particular reference to environmental stress factors. I. Epizootiology of gross lesions. *Dis. Aquat. Org.* 26:81–97.

Vethaak, A.D. and Wester, P.W. (1996) Diseases of flounder *Platichthys flesus* in Dutch coastal and estuarine waters, with particular reference to environmental stress factors. II. Liver histopathology. *Dis. Aquat. Org.* 26:99–116.

Wake, K. (1980) Perisinusoidal stellate cells (fat-storing cells, interstitial cells, lipocytes), their related structure in and around the liver sinusoids, and vitamin A-storing cells in extrahepatic organs. *Int. Rev. Cytol.* 66:303–353.

Wallace, R.A. (1978) Oocyte growth in non-mammalian vertebrates. *In:* Jones, R.E. (Ed.) *The vertebrate ovary.* Plenum, New York, pp 469–502.

Wester, P.W. and Canton, J.H. (1986) Histopathological study of *Oryzias latipes* (medaka) after long-term β-hexachlorocyclohexane exposure. *Aquat. Toxicol.* 9:21–45.

Wisse, E. (1974) Observations on the fine structure and peroxidase cytochemistry of normal rat liver Kupffer cells. *J. Ultrastruct. Res.* 38:528–562.

Wolke, R.E. (1992) Piscine macrophage aggregates: a review. *Ann. Rev. Fish Dis.* 2:91–108.

Yamamoto, M. and Enzan, H. (1975) Morphology and function of Ito cell (fat-storing cell) in the liver. *Rec. Adv. RES Res.* 15:54–75.

Youson, J.H. (1981) The liver. *In:* Hardisty, M.W. and Potter, I.C. (Eds) *The biology of lampreys.* Academic Press, London, pp 263–332.

Fish Ecotoxicology
ed. by T. Braunbeck, D. E. Hinton and B. Streit
© 1998 Birkhäuser Verlag Basel/Switzerland

Immunochemical approaches to studies of CYP1A localization and induction by xenobiotics in fish

Anders Goksøyr[1] and Astrid-Mette Husøy[2]

[1] *Department of Molecular Biology, University of Bergen, HIB, N-5020 Bergen, Norway*
[2] *Bergen College, N-5008 Bergen, Norway*

Summary. There is an increasing understanding that polynuclear aromatic hydrocarbons (PAHs) and organochlorine compounds (like polychlorinated biphenyls (PCBs), certain pesticides and dioxins) in the aquatic environment may lead to physiological and pathological effects such as immunological disturbances, effects on reproduction and development, and even neoplasms. Exposure to pollutants may have consequences at all levels in the biological organization, from the cellular level over effects on the individual organism, population, to the entire ecosystem. The cytochrome P450 system (CYP or P450) has an essential function in the biotransformation of endogenous and exogenous compounds. The fact that many different environmental pollutants induce *de novo* synthesis of cytochrome P450 1A (CYP1A) proteins in fish, gives these enzymes an interesting position in aquatic toxicology. Many investigations concerning the CYP1A system in fish have been performed over the last two decades, demonstrating its usefulness as a biomarker for aquatic pollution. A general overview of the biochemical and toxicological aspects concerning the cytochrome P450 system will be given here, followed by a more detailed description of CYP1A induction responses in fish. Ecotoxicological consequences of CYP1A induction and the use of immunochemical techniques for CYP1A detection as a biomarker in environmental monitoring will be discussed.

Introduction

Exposure to and toxic effects of contaminants in aquatic organisms can be detected by biochemical, physiological, pathological, or ecological responses. The need for early detection and assessment of the impacts of pollution in the aquatic environment has led to the development of biomarkers. Biomarkers have been defined as pollution-induced variations in a biological system (Peakall, 1992).

The cytochrome P450 system (CYP or P450) has an essential function in the biotransformation of endogenous and exogenous compounds. The fact that many different environmental pollutants induce *de novo* synthesis of cytochrome P450 1A (CYP1A) proteins in fish, gives these enzymes an interesting position in aquatic toxicology. A large number of investigations concerning the CYP system in fish have been published in the last 20 years, demonstrating its usefulness as a biomarker for aquatic pollution (for reviews, see Andersson and Förlin, 1992; Bucheli and Fent, 1995; Buhler and Williams, 1988; Goksøyr, 1995; Goksøyr and Förlin, 1992; Goksøyr et al., 1997; Lech and Bend, 1980; Livingstone, 1993; Payne et al., 1987; Stegeman et al., 1992; Stegeman and Hahn, 1994).

A more general overview of the biochemical and toxicological aspects concerning the cytochrome P450 system will be given in this introduction, followed by a more detailed description of CYP1A induction responses in fish. Consequences of CYP1A induction and the use of immunochemical techniques for CYP1A detection as a biomarker in environmental monitoring will be discussed.

The cytochrome P450 monooxygenase system

General aspects and molecular properties

The CYP monooxygenase system belongs to a family of structurally and functionally related heme proteins. Multiple families and subfamilies with a broad range of substrate specificities are recognized (Nelson et al., 1993). The biochemistry of cytochrome P450 isoenzymes including CYP1A has been extensively studied in mammals and several other organisms (for reviews, see Guengerich and Shimada, 1991; Nebert et al., 1991; Porter and Coon, 1991; Schenkman and Greim, 1993).

The CYP system is a central catalyst of oxidative reactions including hydroxylation, epoxidation, dealkylation and other oxidative reactions (reviewed by Guengerich and Shimada, 1991). Among the thousands of compounds metabolized by CYP isozymes are endogenous substances (i.e., fatty acids, steroids, prostaglandins, ketones) and exogenous compounds (i.e., food additives, drugs, polyaromatic hydrocarbons (PAHs), polychlorinated biphenyls (PCBs), dioxins). Although the majority of the CYP forms have a broad and overlapping substrate profile, the basic mechanism of substrate oxygenation is identical for all the isozymes. The overall monooxygenase reaction catalyzed by CYPs is:

$$RH + O_2 + NADPH + H^+ \rightarrow ROH + H_2O + NADP^+$$

The lipophilic substrate (RH) is converted to a more polar and water soluble product (ROH) after incorporation of one oxygen atom from molecular oxygen.

Nomenclature

According to the recommended nomenclature, the cytochrome P450 genes and cDNAs are abbreviated as italicized cytochrome P450 (CYP), followed by an Arabic number denoting the gene family, a letter indicating the subfamily and an Arabic numeral representing the individual gene, e.g., *CYP1A1* (Nebert et al., 1991). At the mRNA and protein level, non-italicized letters are recommended, i.e., CYP1A1.

CYP proteins with more than 40% identical amino acid sequences are defined as being within a single family. More than 55% similarity is defined as to belong to the same subfamily (Nelson et al., 1993). However, some problems may arise with this general rule in comparisons of orthologous genes from evolutionarily distant species.

Evolution and multiplicity

The cytochrome P450 system comprises an ever expanding superfamily of more than 400 individual forms, each belonging to one of at least 127 CYP gene subfamilies (Nebert et al., 1991; Nelson et al., 1993, 1996). (A continuously updated web-page with CYP genes is available at http://base.icgeb.trieste.it:80/p450/.)

Probably, the first CYP enzymes were involved in the synthesis of lipids and steroids essential for membrane integrity (Nebert and Gonzalez, 1985). CYPs responsible for metabolism of foreign compounds appear to have arisen about 400 to 500 million years ago (Nebert, 1991). A plant-animal "warfare" hypothesis has been suggested to explain the further evolution of CYPs (Gonzalez and Nebert, 1990; Nebert and Gonzalez, 1985). Plants may synthesize toxic chemicals (as a protective measure from being eaten by animals), and, in response, animals may have evolved CYPs to degrade such toxins. The most important inducible CYP subfamilies involved in xenobiotic transformations are presented in Table 1.

In teleost fish, P450 proteins have been isolated from liver, kidney and ovary. So far, twelve different P450s have been isolated, extensively purified and characterized from rainbow trout (*Oncorhynchus mykiss*, previously named *Salmo gairdneri*). Additionally, five different P450s have been isolated from scup (*Stenotomus chrysops*), four P450s from cod (*Gadus morhua* L.), and five P450 fractions were isolated from perch (*Perca fluviatilis*). In recent years, several additional fish CYP genes have been identified by cDNA cloning and sequencing.

Data on purified and cloned P450s from fish are summarized in Table 2 and discussed below.

Table 1. Nomenclature of selected P450 gene subfamilies, prominent inducers and common substrates

Subfamily	Prominent inducers	Common substrates
CYP1A	PAH, BNF, planar PCBs, dioxins, furans	PAH, ethoxyresorufin
CYP2B	Barbiturates, non-planar PCBs, DDT*	Barbiturates, steroids, ethyl-morphine
CYP2E	Ethanol, ketones, starvation, diabetes	Ethanol, alkylnitrosamines
CYP3A	PCN, glucocorticoids	Steroids (6β-hydroxylase)
CYP4A	Clofibrate, phthalates*	Lauric acid, arachidonic acid

Data based on Stegeman and Hahn (1994) and Goksøyr (1995); *, in mammals.

Table 2. Purified or cloned CYP family members from fish and their putative designation in the P450 superfamily

Subfamily	Species	Trivial name	Protein or cloned	References
CYP1A	Rainbow trout (*Oncorhynchus mykiss*)	LM4B	Protein + cloned	Williams and Buhler (1982), Williams and Buhler (1984), Heilman et al. (1988), Berndtson and Chen (1994)
	Scup (*Stenotomus chrysops*)	P450E	Protein + cloned	Klotz et al. (1983), Morrison et al. (1995)
	Cod (*Gadus morhua*)	P450c	Protein + cloned	Goksøyr, 1985); Richert, Goksøyr and Male, pers. comm.
	Perch (*Perca fluviatilis*)	P450V	Protein	Zhang et al. (1991)
	Rainbow trout (*Oncorhynchus mykiss*)	DS-3	Protein	Celander and Forlin (1991)
	Plaice (*Pleuronectes platessa*)		cloned	Leaver et al. (1993)
	Sea bass (*Dicentrarchus labrax*)		cloned	Stien et al. (1994)
	Toadfish (*Opsanus tau*)		cloned	Morrison et al. (1995)
	Tomcod (*Microgadus tomcod*)		cloned	Roy et al. (1995)
	Butterflyfish (*Chaetodon capistratus*)		cloned	Vrolijk and Chen (1994)
CYP2B	Rainbow trout (*Oncorhynchus mykiss*)	LMC1	Protein	Miranda et al. (1989)
	Scup (*Stenotomus chrysops*)	P450B	Protein	Klotz et al. (1986)
CYP2E	Topminnow (*Poeciliopsis spp.*)	P450pj	not isolated	Kaplan et al. (1991)
CYP2K	Rainbow trout (*Oncorhynchus mykiss*)	LMC2	Protein + cloned	Buhler et al. (1994), Miranda et al. (1989)
	Mummichog (*Fundulus heteroclitus*)		cloned	Oleksiak et al. (1997)
	Rainbow trout (*Oncorhynchus mykiss*)	KM2	Protein (from kidney)	Andersson (1992)
CYP2M	Rainbow trout (*Oncorhynchus mykiss*)	LMC1	Protein + cloned	Miranda et al. (1989), Yang et al. (1996)
CYP2N	Mummichog (*Fundulus heteroclitus*)		cloned	Oleksiak et al. (1997)
CYP2P	Mummichog (*Fundulus heteroclitus*)		cloned	Oleksiak et al. (1997)
CYP3A	Rainbow trout (*Oncorhynchus mykiss*)	LMC5, P450con	Protein + cloned	Miranda et al. (1989), Celander et al. (1989), Lee et al. (1998)
	Scup (*Stenotomus chrysops*)	P450A	Protein	Klotz et al. (1986)
	Cod (*Gadus morhua*)	P450b	Protein	Goksøyr (1985)
	Mummichog (*Fundulus heteroclitus*)	CYP3A30	cloned	Celander and Stegeman (1997)
CYP4T	Rainbow trout (*Oncorhynchus mykiss*)	CYP4T1	cloned	Falckh et al. (1997)
CYP7A	Atlantic salmon (*Salmo salar*)		cloned	Olsen and Walther, personal communication
CYP11	Rainbow trout (*Oncorhynchus mykiss*)	P450scc	cloned	Takahashi et al. (1993)
CYP17	Rainbow trout (*Oncorhynchus mykiss*)	P450c17	cloned	Sakai et al. (1992)
CYP19	Rainbow trout (*Oncorhynchus mykiss*)	P450arom	cloned	Tanaka et al. (1992)

Data based on Celander et al. (1996), Goksøyr et al. (1991a), as well as Stegeman and Hahn (1994).

The CYP1A gene subfamily

CYP1A is the most important subfamily in biotransformation and bio-activation of xenobiotics into reactive intermediates which may be more toxic than the parent compound. Furthermore, CYP1A may be an indicator of potential chemical carcinogenesis. Therefore, the CYP1A activity is also a factor that can place cells and host at risk (Gonzalez and Gelboin, 1994; Ioannides and Parke, 1993). Consequently, CYP1A has become the most studied and best characterized subfamily in the P450 system.

In mammals, the CYP1A subfamily consists of two isozymes: CYP1A1 and CYP1A2. Probably, CYP1A1 is the best conserved of the xenobiotic-metabolizing enzymes (Nebert et al., 1989). Nebert suggested that CYP1A1 may be involved in control of cell division and differentiation, metabolizing some unknown endogenous substrates. However, an endogenous substrate/ligand for the CYP1A protein has not yet been found (Beresford, 1993; Nebert, 1991).

CYP1A2 is constitutively expressed in liver (Gonzalez, 1990; Gonzalez and Gelboin, 1994). Induction of CYP1A2 by xenobiotics like arylamines and aflatoxin B1 has been demonstrated in rodents as well as in humans (Gonzalez, 1990; Gonzalez and Gelboin, 1994).

CYP1A1 and CYP1A2 induction is independently regulated in mammals, although they are both mediated by the Ah-receptor. The CYP1A2 gene lacks the xenobiotic regulating elements (XRE). CYP1A2-mRNA has been found to be stabilized by 3-methylcholanthrene and tetrachlorodibenzo-p-dioxin (TCDD; Gonzalez, 1990).

A new cytochrome P450 subfamily, CYP1B, has recently been identified in human fibroblasts and in mouse (Savas et al., 1994; Sutter et al., 1994). CYP1B1 is induced by β-naphthoflavone. CYP1B has been detected in mouse lung, uterus and kidney but not in brain (Savas et al., 1994).

CYP1A induction responses

The induction response of the CYP1A subfamily occurs *via* an intracellular cytosolic aryl hydrocarbon receptor (Ah-receptor; Poland and Knutson, 1982), predominantly found in liver, but also in extrahepatic tissues. The Ah-receptor has been identified in multiple species, organs and tissues (Hahn, 1996; Hahn et al., 1994; Okey and Harper, 1994; Safe, 1988; Stegeman and Hahn, 1994). In mammals, the Ah receptor complex consists of the TCDD-binding subunit and the heat shock protein Hsp90. The Ah-receptor has been identified as a member of the bHLH-PAS family of ligand-activated transcription factors. The PAS family includes the Ah-receptor nuclear translocator (arnt protein) and the *Drosophila* proteins Per and Sim (Burbach et al., 1992; Schmidt et al., 1993). The Ah-receptor shares a number of features with the steroid-thyroid hormone receptor

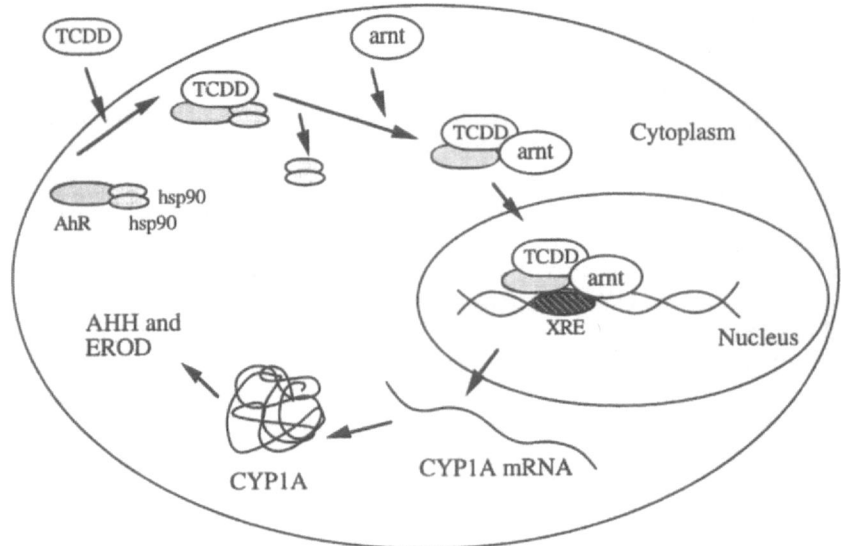

Figure 1. Molecular mechanism of induction of CYP1A gene expression by TCDD and related compounds via the Aryl hydrocarbon receptor (AhR).

family (Bresnick, 1993; Nebert, 1991). The molecular biology of the Ah-receptor and its functions have recently been reviewed (Okey et al., 1994).

CYP1A induction is initiated by binding of the inducer, e.g., TCDD, or other planar aromatic hydrocarbons, to the Ah-receptor. The association of the receptor with the ligand releases the heat shock protein (Hsp90). The arnt protein is a cytosolic protein responsible for subsequent translocation of the receptor complex to the nucleus (Hoffman et al., 1991). After translocation, the nuclear form of the Ah-receptor complex interacts with xenobiotic response elements (XREs), initiating the transcription of the CYP1A genes. CYP1A-mRNA is translated into protein, then the apoprotein binds heme and is inserted into the membrane of the endoplasmic reticulum, where it performs its catalytic activity. The induction response is illustrated in Figure 1.

In addition to induction of CYP1A gene expression, Ah-receptor ligands such as TCDD, 3-methylcholanthrene and related compounds enhance induction of CYP1A2 and some phase II enzymes (NAD(P)H:menadione oxidoreductase, aldehyde dehydrogenase, UDP glucuronosyltransferase (UDPGT), and glutathione S-transferase (GST) in mammals; Nebert and Gonzalez, 1987).

CYP1A in fish

The cytochrome P450 system in fish and aquatic animals has been thoroughly discussed in a number of reviews (Andersson and Förlin, 1992;

Buhler and Williams, 1989; Förlin et al., 1994; Goksøyr, 1995; Goksøyr and Förlin, 1992; Kleinow et al., 1987; Stegeman and Hahn, 1994; Stegeman and Kloepper-Sams, 1987).

The CYP1A1 and CYP1A2 genes probably separated 250 million years ago (Jaiswal et al., 1985; Nebert and Gonzalez, 1987). Fish and mammalian lines are thought to have diverged more than 400 million years ago, prior to the split in the CYP1A gene. The phylogenetic relationship has been established between mammalian CYP1A1 and CYP1A2 genes and fish CYP1A (Morrison et al., 1995), suggesting that the fish gene is ancestral both to mammalian CYP1A1 and CYP1A2.

Several research groups have attempted to find a second CYP1A form in fish, with varying success. In rainbow trout, two related CYP1A isozymes have been isolated, P450 LM4a and LM4b (Williams and Buhler, 1984). The first cDNA from fish was also sequenced from rainbow trout by Heilmann and co-workers (1988); reporting only a single gene, 57–59% of its amino acid residues were identical with mammalian 1A1 forms, and 51–53% were identical to the mammalian 1A2 forms. A single CYP1A gene was also reported from plaice (*Pleuronectes platessa*; Leaver et al., 1993), toadfish (*Opsanus tau*) and scup (Morrison et al., 1995), and butterflyfish (*Chaetodon capistratus*; Vrolijk and Chen, 1994). However, Berndtson and Chen (1994) recently found two CYP1A genes in rainbow trout. The CYP1A genes had 96% sequence similarity, indicating a relatively recent divergence (Morrison et al., 1995). These have been designated CYP1A1 and CYP1A3 to distinguish them from the mammalian CYP1A2 form (Berndtson and Chen, 1994).

Several CYP proteins, purified from liver of different fish species, can be classified to belong to the CYP1A subfamily. Trout P450LM4b (Williams and Buhler, 1982, 1984), scup P450E (Klotz et al., 1983), and cod P450c (Goksøyr, 1985) have similar physico-chemical, catalytic and regulatory properties. Specific antibodies to these fish CYP1A proteins cross-react with each other and with inducible forms in other fish species as well as with mammalian CYP1A forms (Goksøyr et al., 1991a; Stegeman, 1989).

Although mostly studied in liver, CYP1A induction has also been found in extrahepatic organs. Immunohistochemical analyses have uncovered cellular expression of CYP1A in fish. Induction in epithelial cells (liver, kidney, gill, and skin) and in vascular endothelial cells has frequently been seen (Husøy et al., 1994; Smolowitz et al., 1991; Stegeman and Hahn, 1994). Cellular localization of CYP1A will be further discussed below.

In fish, the CYP1A proteins are induced by similar compounds as in mammals, suggesting a similar mechanism. A TCDD-binding protein, corresponding to the Ah-receptor, was found in rainbow trout liver (Heilmann et al., 1988) as well as in trout hepatoma cells (Lorenzen and Okey, 1990). Hahn and colleagues (1994) have also detected Ah-receptor proteins in cartilaginous and bony fish. However, these authors could not detect Ah-receptor proteins in a jawless fish (hagfish) or in invertebrates (Stegeman and Hahn, 1994). Interestingly, two Ah-receptor genes were recently identified in fish (Hahn et al., 1998).

Regulation of the cytochrome P450 system in fish

The constitutive levels of CYP1A in fish are low. However, several factors, exogenous as well as endogenous, can affect the CYP1A expression in fish (recently reviewed by Andersson and Förlin, 1992; Goksøyr, 1995; Stegeman and Hahn, 1994).

Different fish species have a remarkable ability to adjust to a wide range of temperatures. The temperature influences the rate and magnitude of CYP1A induction. Increased CYP1A levels in fish acclimatized to lower temperatures have been reported (Andersson and Förlin, 1992), although the response may be less pronounced in some species (Skjegstad et al., 1994). A clear temperature-dependence of CYP1A levels was observed during summer months in dab (*Limanda limanda*) from the North Sea (Sleiderink et al., 1995).

Seasonal changes, photoperiod and sexual development may also affect the regulation of CYP1A in fish (Arukwe and Goksøyr, 1997; Edwards et al., 1988; Förlin and Haux, 1990; George et al., 1990; Larsen et al., 1992; Lindström-Seppä, 1985; Mathieu et al., 1991; Tarlebø et al., 1985). During sexual maturation, differences in P450 content between males and females may be found, being normally higher in mature males than in mature females (Arukwe and Goksøyr, 1997; Förlin, 1980; Förlin and Haux, 1990; Koivusaari et al., 1981; Larsen et al., 1992; Stegeman and Chevion, 1980). For a more detailed discussion of P450 regulation during sexual development, see Andersson and Förlin (1992).

In fish, several studies have indicated that biotransformation enzymes are active even in the early stages of development. Hendricks and co-workers (1980) reported that after treatment with aflatoxin B1 rainbow trout embryos metabolized reactive intermediates. Induction of CYP1A was reported in embryos (Binder and Stegeman, 1983; Laurén et al., 1990) and hatched embryos (Binder et al., 1985; Goksøyr et al., 1991b). Cellular localization in embryos has indicated CYP1A induction in several tissues and cells.

Adequate nutrition may be essential for biological defense against chemical and oxygen toxicity. Both diet and nutrition are known to influence CYP1A and CYP2E in mammals (Parke, 1991). Some studies have indicated that dietary factors may also influence xenobiotic metabolism in fish. The effects of dietary lipids (Ankley et al., 1989; George and Henderson, 1992), vitamins and antioxidants (Ankley et al., 1989; Williams et al., 1992), iron (Goksøyr et al., 1994a), terpenoids (Vrolijk and Chen, 1994), and carbohydrates (A. Goksøyr and O. Drægni, unpublished results) have been studied. Astaxanthin (a carotenoid) in the diet increased the CYP1A levels in salmon (*Salmo salar*) compared to control salmon (A.-M. Husøy, R. Christiansen, O. Torrisen and A. Goksøyr, unpublished results). Similar results have also been recently reported in rats (Astorg et al., 1994, 1995). The mechanisms involved in dietary regulations and their impact on toxicity remain to be established.

Endogenous substrates and physiological role

The physiological role of CYP1A is not known. In humans, CYP1A expression declines with increasing age, whereas higher levels are seen in neonates. Additionally, CYP1A1 is localized in epithelial cells and leukocytes, i.e., specialized cells with rapid turnover, suggesting that CYP1A may be involved in tissue proliferation (Beresford, 1993; Ioannides, 1990; Ioannides and Parke, 1993). .

Only scarce information is available concerning an endogenous CYP1A substrate in marine fish. A possible site of endogenous function of CYP1A is the mesenchymal cells or chondroid cells. Recently, high CYP1A expression was reported in these structures in untreated fathead minnow (*Pimephales promelas*), suggesting that CYP1A may be involved in growth and differentiation of bone or associated connective tissues (Stegeman et al., 1996).

In cultures of atrial endothelial cells from Atlantic salmon, high constitutive levels of CYP1A were also observed, suggesting an important physiological role in this cell type (Grøsvik et al., 1998).

Inhibition of CYP1A

Many dietary compounds or xenobiotics can bind to the active site of CYP enzymes, serving as substrates or competitive inhibitors. Flavonoids like α-naphthoflavone and β-naphthoflavone at high concentrations inhibit aryl hydrocarbon hydroxylase activity or 7-ethoxyresorufin-O-deethylase (EROD) activity both in mammals and fish (Celander et al., 1993; Haasch et al., 1993; Husøy et al., 1994). At high concentrations, environmental chemicals like benzo[a]pyrene and PCBs decreased CYP1A activity in liver microsomes (Boon et al., 1992; Goddard et al., 1987; Gooch et al., 1989; Hahn et al., 1993; Monosson and Stegeman, 1991; Zhu, 1995).

In addition to a reversible inhibition mechanism, many dietary chemicals can inactivate CYP enzymes by mechanism-based or suicide inhibition. The anaestethetic tricaine methyl-methane sulfonate (MS-222) has been suggested to have a similar effects on CYP1A in fish (Kleinow et al., 1986). Piperonyl butoxide, a CYP inhibitor used as a synergist in numerous insecticide formulations, is also able to induce CYP1A in fish liver (Erickson et al., 1988; Grøsvik et al., 1997).

Heavy metals may also affect the CYP system. Exposure to cadmium and other metals is followed by induction of heme oxygenase, resulting in an increased heme catabolism making it less available for incorporation in the CYP apoenzyme. In addition, heme oxygenase produces carbon monoxide, which is a direct inhibitor of CYP. In red carp (*Cyprinus carpio* L.), heme oxygenase activity was increased after injection of $CdCl_2$, whereas both total CYP activity and ethoxycoumarin-O-deethylase (ECOD) activity was decreased (Ariyoshi et al., 1990). Decreased CYP activities have also been observed in rainbow trout and plaice (Förlin et al., 1986; George, 1989) exposed to cadmium. Organometallics like tributyltin is another potent CYP inhibitor (Fent and Stegeman, 1991).

Biotransformation of xenobiotics in fish

Biotransformation

The biotransformation of xenobiotics includes a sequence of processes. Some exogenous chemicals may be excreted largely unchanged, whereas the majority are detoxified by increasing their polarity through phase I oxidation (mostly P450s), followed by phase II conjugation with either glutathione, sulfate, amino acids, or glucuronic acid (Goodman Gilman et al., 1990). In fish, the conjugated products are easily excreted in bile, urine or through the gills (Clarke et al., 1991; Nimmo, 1987).

In general, the highest levels of biotransformation enzymes have been found in liver. However, P450s and phase II enzymes are also present in extrahepatic organs (Andersson et al., 1993; Buchmann et al., 1993; George, 1994; Husøy et al., 1994; Smolowitz et al., 1991; vanVeld et al., 1988). For a discussion on localization of CYP1A, see below.

Detoxification and conjugation of xenobiotics in fish have recently been reviewed by George (1994). Generally, electrophilic compounds are conjugated with glutathione by glutathione S-transferases, while nucleophilic compounds are conjugated with glucuronic acid by UDP-glucuronyl transferases. Compared to the cytochrome P450 system, information about piscine phase II enzymes is limited (George, 1994).

Chemical activation

By metabolic activation and covalent binding to proteins, chemicals may alter cellular biochemistry; these are important steps in xenobiotic-induced cytotoxicity (Hinson et al., 1994). In humans, such chemical activation has been related to biotransformation by CYP1A1, CYP1A2, CYP2E1 and CYP3A.

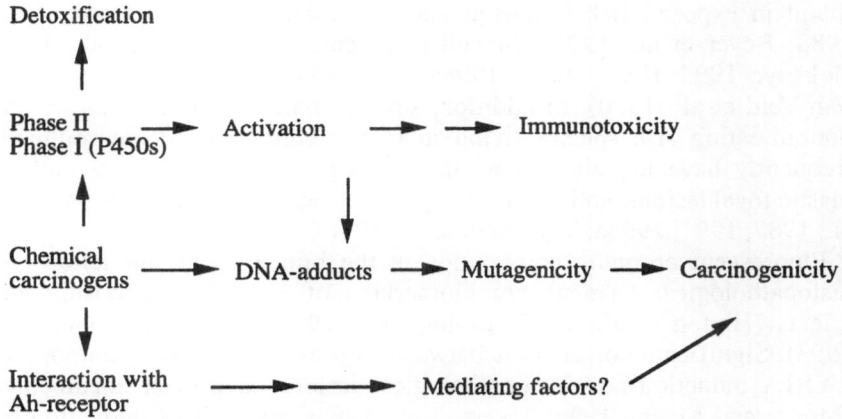

Figure 2. Pathways of chemical activation and detoxification.

In fish, CYP1A, UDP glucuronyl transferase and glutathione S-transferase are thought to be the main enzymes involved in biotransformation of xenobiotics. Activation of compounds like benzo[a]pyrene is similar to that described in rodent systems. Carcinogenic metabolites, such as benzo[a]pyrene-7,8-diol-9,10-epoxide with the sterically hindered "bay region," cannot be metabolized by epoxide hydroxylase or glutathione S-transferase. The epoxide forms adducts with DNA. The repair of DNA damage appears to be slow in fish (Maccubbin, 1994). In wild fish, two different forms of DNA adducts appear: direct binding of bulky aromatic compounds to DNA or oxidative damage related to chemical exposure (reviewed by Livingstone, 1993; Maccubbin, 1994). Pathways for metabolic activation and detoxification of xenobiotics are illustrated in Figure 2.

Biological effects and biomarkers

There is an increasing understanding that PAHs and organochlorine compounds (like PCBs, pesticides and dioxins) in the aquatic environment may lead to physiological and pathological effects such as immunological disturbances, effects on reproduction and development, and even neoplasms (reviewed in Baumann, 1989; Molven and Goksøyr, 1993; Moore and Myers, 1994). Exposure to pollutants may have consequences at all levels in the biological organization, from cellular level to effects on the individual organism, population, or the entire ecosystem. The actual consequences may be related both to agent, dose as well as time and duration of exposure.

In marine and freshwater systems contaminated with PAHs, PCBs and bleached kraft mill effluents, increased hepatic CYP1A levels have been found in exposed fish (Addison and Edwards, 1988; Andersson et al., 1988; Beyer et al., 1996; Bucheli and Fent, 1995; Förlin et al., 1991; Goksøyr, 1995; Husøy et al., 1996a; Lindström-Seppä and Oikari, 1990; Van Veld et al., 1990). In addition, several marine bottom-dwelling and bottom-eating fish species living in highly contaminated environments frequently have hepatic lesions including proliferative lesions, preneoplastic focal lesions, and neoplasms (Moore and Stegeman, 1994; Myers et al., 1987, 1991, 1994a; Vogelbein et al., 1990).

Fluorescent aromatic compounds in the bile, CYP1A induction and histopathological changes are biomarkers of xenobiotic exposure and effects (Hinton et al., 1992; Livingstone, 1993; Molven and Goksøyr, 1993). Significant correlations between fluorescent aromatic compounds, CYP1A induction and histopathological lesions have been documented (Moore and Myers, 1994; Myers et al., 1994a, b). CYP1A induction responses have been evaluated as an early warning signal of chemical exposure and of biological effect. Chemical carcinogenesis and toxicity associated with CYP1A induction make this biomarker especially interesting in the field of aquatic toxicology.

Methods for studying CYP1A expression in fish

Numerous laboratory and field studies have demonstrated induction of hepatic CYP1A in fish species exposed to planar organochlorine compounds and PAHs, and different approaches have been used to study CYP1A induction in fish (Förlin et al., 1994; Goksøyr, 1995; Goksøyr and Förlin, 1992; Stegeman and Hahn, 1994). Each step in the induction response, from gene transcription to mature enzyme, can be analyzed (Fig. 1, Tab. 3). A catalytic CYP1A-associated activity is the commonly used assay. Molecular tools such as immunochemical probes and nucleotide-based detection systems have also been developed.

The work discussed here is primarily based on recognition of CYP1A proteins by monoclonal and polyclonal antibodies produced in our laboratory. These antibodies, MAb-NP7 and PAb-Dy, were used to study cellular localization and regulation of CYP1A in cod and flounder.

Table 3. Detection methods for CYP1A in fish

Detection level	Probe	Assay
CYP1A enzyme	Catalytic assay	EROD, AHH
CYP1A protein	Antibody	Western blotting, ELISA, immunohistochemistry
CYP1A mRNA	cDNA, oligonucleotide	Northern blotting, RT-PCR, in situ hybridization

Catalytic activity

The catalytic activity of CYP1A can be measured by different substrates. 7-Ethoxyresorufin (EROD-assay) and benzo[a]pyrene (aryl hydrocarbon hydroxylase-assay) are well known and highly specific substrates for CYP1A (Stegeman and Kloepper-Sams, 1987). Pretreatment of fish with PAHs such as 3-methylcholanthrene and similar inducers, including TCDD, β-naphthoflavone and co-planar PCBs, significantly increase hepatic aryl hydrocarbon hydroxylase and EROD activities (Lech et al., 1982). Burke and Mayer (1974) described the first fluorimetric technique for EROD-activity. The CYP1A enzyme catalyzes the O-deethylation of 7-ethoxyresorufin, with the product resorufin being fluorometrically detected. Later, several modifications of the EROD assay have been described (Förlin et al., 1994; Hodson et al., 1991; Klotz et al., 1984; Stagg and Addison, 1995). In addition, a new and rapid plate-reader method has been developed (Eggens and Galgani, 1992).

The aryl hydrocarbon hydroxylase assay is an end-point method and uses benzo[a]pyrene or diphenyloxazole as the primary substrate (Hodson et al., 1991; Nebert and Gelboin, 1968). Methods that use radio-labeled benzo[a]pyrene have also been described (Binder and Stegeman, 1980). The major disadvantage of this method is that the substrate benzo[a]pyrene is a strong carcinogen.

However, catalytic activities are susceptible to denaturation or inactivation during sampling, storage, or during preparation of enzyme fractions. Furthermore, the catalytic activity may also be inhibited by exogenous (e.g., PAHs and PCBs) and endogenous (e.g., estradiol) compounds present in the samples, as discussed above.

CYP1A protein

CYP1A has now been purified from four fish species, as discussed above. For each of these, polyclonal or monoclonal antibodies have been prepared (Celander and Förlin, 1991; Goksøyr, 1985; Husøy et al., 1996b; Park et al., 1986; Williams and Buhler, 1984).

A new strategy to isolate CYP1A and prepare such antibodies was described in Husøy et al. (1996b). Hepatic cod CYP1A was purified by immunoadsorption chromatography. The purified CYP1A protein (m_r 58 kDa) gave a single band analyzed by Western blotting. Rabbits and mice were immunized with the immunoaffinity-purified CYP1A to generate polyclonal (denoted PAb-Dy) and monoclonal cod CYP1A antibodies (MAb-NP7), as discussed in Husøy et al. (1996b). The advantages of monoclonal antibodies are their high specificity, homogeneity, and their ability to be produced in unlimited quantities.

CYP1A levels are routinely detected and quantified by Western blotting (Goksøyr et al., 1991a; Kloepper-Sams et al., 1987; Varanasi et al., 1986) or semi-quantified by ELISA measurements (Celander and Förlin, 1991; Goksøyr, 1991). Cellular localization of CYP1A in fish has also aroused interest and was first reported in cod and rainbow trout (Goksøyr et al., 1987; Husøy et al., 1994, 1996a; Smolowitz et al., 1991). Immuno-chemical techniques are not dependent upon biologically active samples. Moreover, immunological techniques will be superior in situations, where the catalytic activities are inhibited or degraded (Goksøyr and Förlin, 1992; Goksøyr and Husøy, 1992). Immunological techniques may also be used as an additional quality control for enzyme activity (Collier et al., 1995).

Normally there is a close relationship between catalytic CYP1A activity (EROD and/or aryl hydrocarbon hydroxylase activity) and CYP1A levels measured as ELISA (Celander and Förlin, 1992; Collier et al., 1995; Husøy and Goksøyr, 1998; van der Weiden, 1993). In a β-naphthoflavone dose-response study, a statistically significant linear relationship between EROD and CYP1A ELISA was found with MAb-NP7 ($r^2 = 0.86$) and PAb-Dy ($r^2 = 0.83$) in cod liver (Husøy and Goksøyr, unpublished results; see Fig. 3). However, electron transport through NADPH-cytochrome P450 reductase appears to be rate-limiting for high EROD activities in β-naph-thoflavone-exposed salmon (Goksøyr et al., 1994a).

In the dose-response study with cod, immunochemical semi-quantifica-tion of CYP1A proteins seemed to have lower sensitivity compared to EROD measurements (Husøy and Goksøyr, 1997). Low EROD values could not be discriminated in ELISA probed with anti-cod CYP1A MAb-NP7. In our indirect ELISA system using MAb-NP7, the detection limit for EROD was at values around 25 pmol/min/mg protein. With the polyclonal antibodies, the detection limit was increased to 60 pmol/min/mg protein (Husøy and Goksøyr, 1997). An increased sensitivity in ELISA is attempted in further optimization of detection systems.

Monoclonal antibodies to scup CYP1A (MAb 1-12-3; Park et al., 1986) have a broad species cross-reactivity, which includes humans, rodents, marine mammals, reptiles, amphibians and fish (Stegeman and Hahn, 1994). The monoclonal anti-cod CYP1A MAb-NP7 described in Husøy et al. (1996b) was the first monoclonal antibody produced against fish spe-cies other than scup. The MAb-NP7 cross-reacts with CYP1A in several fish species. Apparently, MAb-NP7 cross-reacts better with CYP1A in gadoids, salmonids and cyprinids than with more distant species (Husøy et al., 1996b). Although this antibody does not have a cross-reactivity as broad as MAb 1-12-3, it works well in Western blotting, ELISA and immunohistochemistry with a number of fish species (Husøy and Goksøyr, 1998; Husøy et al., 1996a, b).

Recently, a monoclonal antibody raised against a synthetic peptide from a consensus region of the reported fish CYP1A genes was developed. Apparently, this antibody, similar to the scup MAb 1-12-3, has a broad cross-reactivity with a number of diverse fish species (Rice et al., 1998).

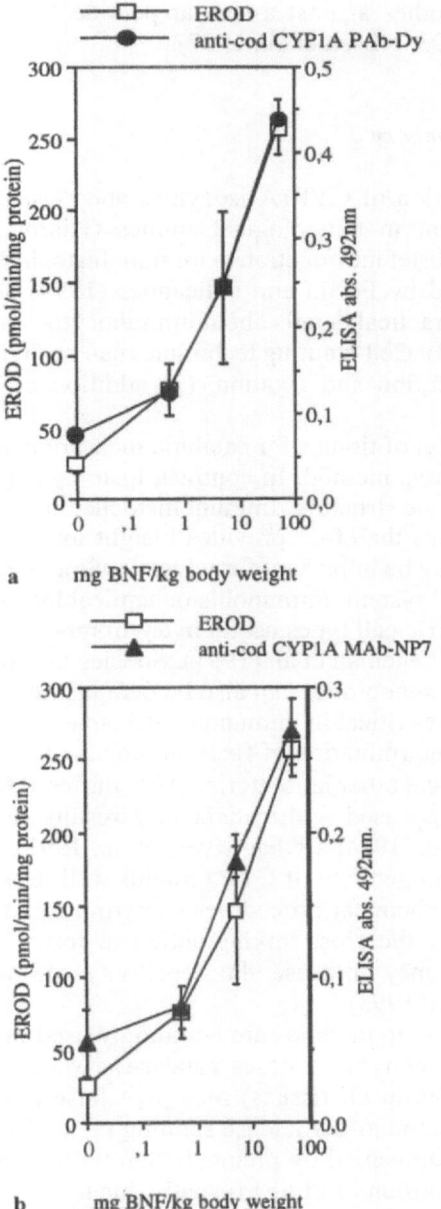

a   mg BNF/kg body weight

b   mg BNF/kg body weight

Figure 3. Dose-dependent CYP1A responses in liver microsomes from β-naphthoflavone treated Atlantic cod (*Gadus morhua*), detected by catalytic ethoxyresorufin-O-deethylase activity, and by ELISA using two different cod CYP1A antibodies: (a) PAb-Dy, (b) MAb-NP7.

Polyclonal antibodies against a similar peptide have also been reported (Myers et al., 1993; Nilsen et al., 1997).

Immunohistochemistry

Cellular localization of CYP1A isozymes and certain other proteins has become important in toxicological studies (Hinton, 1994; Moore and Myers, 1994). A brief introduction to immunohistochemical techniques has been summarized by Förlin and colleagues (1994) as well as by Husøy (1994); further practical details about immunocytochemistry are described by Larsson (1988). Cell-staining techniques can be divided into three steps: (1) tissue preparation and fixation, (2) addition of antibodies, and (3) detection.

Homogenization of tissues for catalytic measurements and ELISA analyses is a destructive method. In contrast, histological methods try to preserve cell and tissue structure. Immunohistochemical staining probed with CYP1A antibodies therefore provides insight into responses of different cell types that may be important for understanding the role and the mechanisms of the CYP system. Immunohistochemical localization may also call attention to specific cell types useful in environmental monitoring studies, as suggested by Stegeman et al. (1991). Species differences in distribution and responses to xenobiotics can also be demonstrated by this technique.

Fixation may be critical in immunohistochemical procedures. The choice of fixative and determination of fixation time is critical to prevent antigen masking. In our and other laboratories, 10% buffered formalin and Bouin's fixative have been used with satisfactory results (Husøy and Goksøyr, 1997; Husøy et al., 1994, 1996a; Myers et al., 1995). This fixative seems to preserve the antigenicity of CYP proteins well. Fixatives typically used for immunohistochemical procedures sometimes partially mask the antigenic epitopes by the cross-linking action of formaldehyde. Antigen retrieval methods may increase the sensitivity of immunohistochemical methods (Dapson, 1993).

Peroxidase staining methods are commonly used in immunohistochemistry. Both endogenous peroxidases, catalase as well as hemoglobin (present to various degrees in all tissues) may give false positive reactions with chromogens in immuno-peroxidase staining methods. However, this problem may be circumvented by preincubation with hydrogen peroxide, and nonspecific background staining (usually due to the highly charged collagen) can be eliminated after preincubation with non-immune serum.

CYP1A induction detected by immunohistochemistry may be quantified using computerized image analysis systems (Anulacion et al., 1994b; Van Veld et al., 1997). Immunohistochemistry analyses also provides opportunity for simultaneous histopathological evaluation, as demonstrated by Husøy et al. (1996a).

CYP1A mRNA

Analysis of mRNA can be used for measuring CYP1A expression. Although the handling of samples is critical in mRNA analyses, these techniques seem to be valuable when gene activation is suspected, but the enzyme is inhibited or degraded.

Several cDNA probes have been used to identify CYP1A mRNA in teleost fish (Heilmann et al., 1988; Leaver et al., 1993; Renton and Addison, 1992). CYP1A mRNA have been identified in hepatic and extrahepatic tissues in various fish species (Celander et al., 1993; Grøsvik et al., 1997; Haasch et al., 1989; Hahn et al., 1989; Kloepper-Sams and Stegeman, 1989; Leaver et al., 1993).

Pretreatment of fish with a single dose of β-naphthoflavone or a similar CYP1A-inducing agent involves enhancement of CYP1A mRNA, followed by increased protein levels and catalytic activities (Grøsvik et al., 1997; Haasch et al., 1989; Kloepper-Sams and Stegeman, 1989). These studies indicated that mRNA rapidly returned to control levels, whereas CYP1A protein level was maintained. However, prolonged exposure to β-naphthoflavone is followed by a prolonged elevation of CYP1A mRNA (Haasch et al., 1993).

In addition, different inducers result in a different duration of the CYP1A mRNA level. Halogenated and persistent inducers like TCDD maintain an elevated mRNA level, whereas β-naphthoflavone, which is rapidly metabolized, gives a more transient peak of CYP1A mRNA levels (Pesonen et al., 1992). In addition to increased hepatic CYP1A mRNA levels, Leaver and co-workers (1993) reported strong CYP1A mRNA induction in heart, kidney and the intestinal mucosa. CYP1A mRNA was also evident in fish blood (Leaver et al., 1993).

Tissue and cellular localization of CYP1A expression

Cellular localization of CYP1A induction has been described in multiple organs of several teleost species. The strong CYP1A induction in the liver is consistent with the role of hepatocytes in biotransformation of xenobiotics. However, strong CYP1A induction has also been reported in kidney (Lorenzana et al., 1988), heart (Stegeman et al., 1982), gut (Van Veld et al., 1990), and brain (Andersson and Goksøyr, 1994; Andersson et al., 1993). Following exposure to xenobiotics, CYP1A induction has also been found in other organs including gills, gonads, gall bladder, intestine, skin, spleen, and nasal tissues including the olfactory gland (Buchmann et al., 1993; Husøy et al., 1994; Lindström-Seppä et al., 1994; Smolowitz et al., 1991, 1992; Stegeman et al., 1991; Van Veld et al., 1997).

Liver

The teleost liver is the major site of biotransformation enzymes. In addition, the liver is also essential in synthesis of bile, storage of lipid and glycogen, as well as production of vitellogenin and eggshell proteins (Hinton et al., 1992; Oppen-Berntsen et al., 1992; Pelissero et al., 1993). Hepatocytes account for up to 80% of the total liver volume (Hampton et al., 1989). Biliary epithelial cells, endothelial cells, connective tissue elements and macrophages are also located in the liver. Furthermore, in some fish species, exocrine pancreatic tissues are located in the liver as well.

Hepatic microsomes are the most commonly used fraction to measure CYP1A induction responses in, and CYP1A has been purified from liver microsomes of several fish species. In teleosts, CYP1A is localized in hepatocytes, biliary epithelial cells and vascular endothelial cells (reviewed by Hinton, 1994, as well as by Stegeman and Hahn, 1994). By use of immuno-gold techniques, CYP1A has also been located on ultrastructural level in rainbow trout (Lester et al., 1993). In the liver, scup CYP1A was recognized in the granular endoplasmic reticulum, the nuclear envelope, and in the plasma membrane of the microvilli in the bile canaliculi. CYP1A was also expressed in plasma membranes of epithelial cells of the biliary duct system. In endothelial cells, CYP1A was present in the plasma membrane, the cytoplasm as well as in nuclei (Lester et al., 1993).

Cellular localization of CYP1A expression has been reported in several species, including rainbow trout (Miller et al., 1988), winter flounder (*Pseudopleuronectes americanus*; Smolowitz et al., 1989), scup (Smolowitz et al., 1991), topminnow (*Poeciliopsis* spp.; Smolowitz et al., 1992), cod (Husøy et al., 1994), flounder (Husøy et al., 1996a), English sole (*Pleuronectes vetulus*; Myers et al., 1995); toadfish (Collier et al., 1993), starry flounder (*Platichthys stellatus*; Anulacion et al., 1994a), and mummichog (*Fundulus heteroclitus*; Van Veld et al., 1997). Immunohistochemical studies have revealed species differences in CYP1A induction responses, where different cell types appear to respond differently in different species. In scup, rainbow trout and several flatfishes, CYP1A is homogeneously distributed throughout the liver (Lorenzana et al., 1989; Smolowitz et al., 1989, 1991).

In cod, CYP1A induction observed in hepatocytes, biliary epithelial cells and endothelial cells in liver after intraperitoneal β-naphthoflavone injections (Fig. 4). This species showed a strong multifocal CYP1A expression in the hepatocytes after β-naphthoflavone treatment. In addition a slight, diffuse induction was detected throughout the liver (Husøy et al., 1994). In a follow-up dose-response study, vascular endothelial cells were the first cells to demonstrate CYP1A induction in the cod liver after low-dose exposure. The strong multifocal induction response was not seen at this dose, but a weak staining intensity was observed in hepatocytes (Husøy and Goksøyr, 1998). In addition, cod caged in Sørfjorden demonstrated a similar induction pattern with strong induction in endothelial cells and biliary epithelial cells, but with a weak staining of hepatocytes (Husøy et al., 1996a).

The high lipid contents may contribute to this pattern of weak induction in cod hepatocytes (Husøy et al., 1994, 1996a). Exposure of cod to PCB and TCDD has revealed low CYP1A induction compared to rainbow trout, despite high levels of these compounds in the liver (Bernhoft et al., 1993; Hektoen et al., 1992, 1994; Ingebrigtsen et al., 1990). Moreover, low-dose

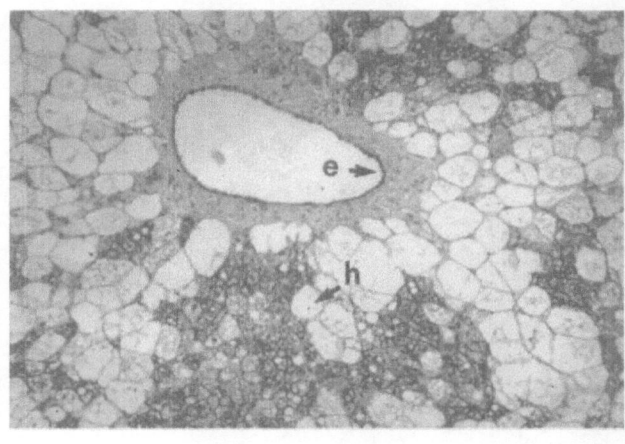

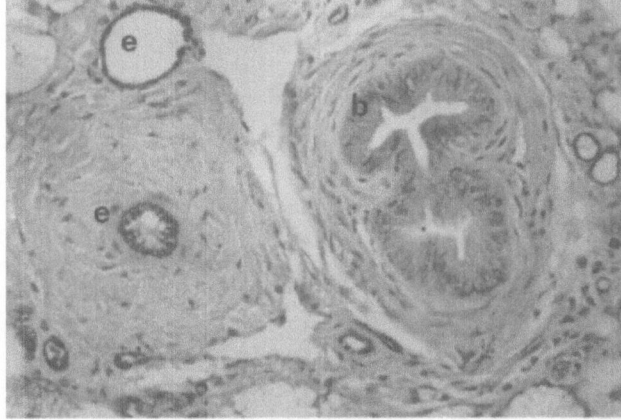

Figure 4. Immunohistochemical localization of CYP1A in Atlantic cod (*Gadus morhua*) and European flounder (*Platichthys flesus*).
(a) Distribution and localization of CYP1A in liver of β-naphthoflavone-exposed cod showing positive CYP1A staining in endothelium (e) and multifocal positive staining in hepatocytes (h). Hepatocytes shown contain abundant cytoplasmic lipid vacuoles (magnification × 250).
(b) Localization of CYP1A in vascular endothelium (e) and biliary epithelium (b) in liver from β-naphthoflavone-exposed cod (400 ×).

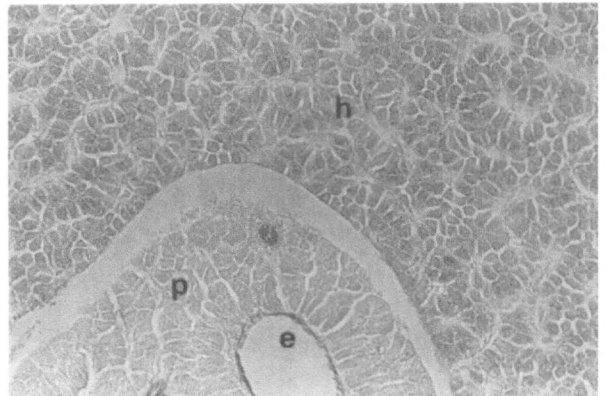

c

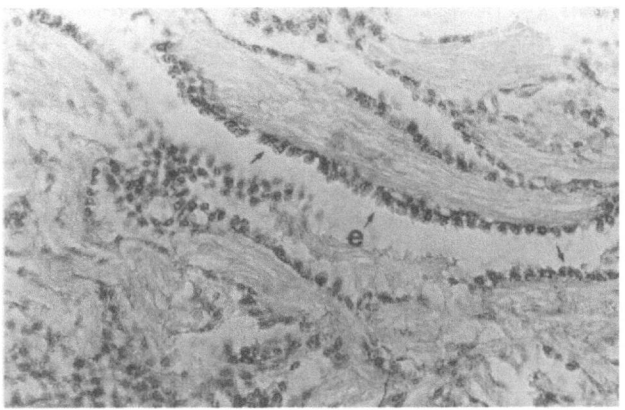

d

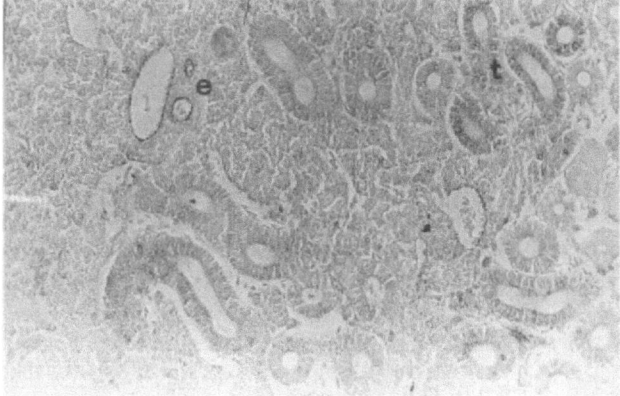

e

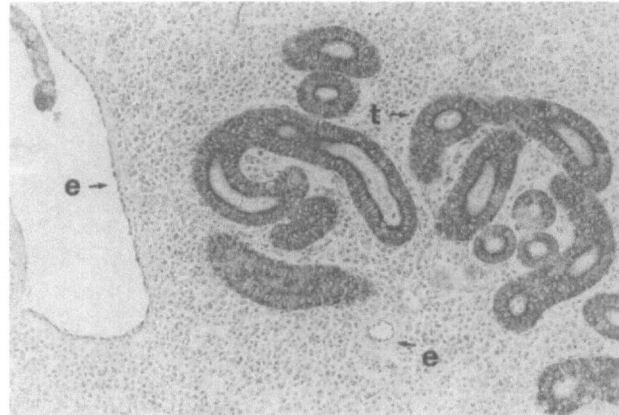

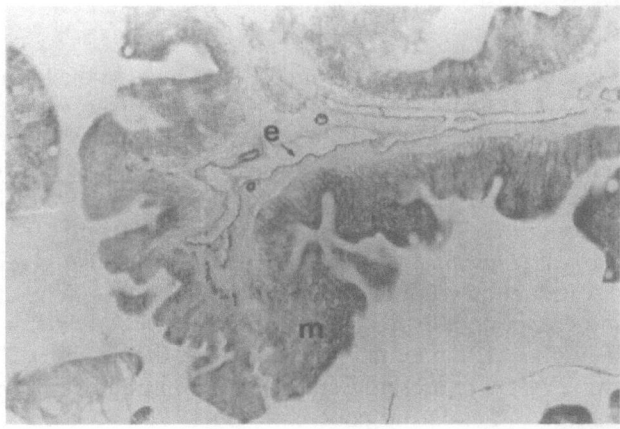

Figure 4. (c) Localization of CYP1A in flounder after caging in a contaminated Norwegian fjord. The section demonstrates strong CYP1A staining in endothelial cells (e) and hepatocytes (h). No CYP1A staining in exocrine pancreatic tissue (p; 250 ×).

(d) CYP1A localization in heart from cod after caging in a contaminated fjord. Section demonstrates strong CYP1A staining in endocardium (endothelial cells in the atrium (e); 400 ×).

(e) Distribution and localization of CYP1A in head kidney of cod exposed to a low dose of β-naphthoflavone (0.5 mg/kg). The section shows strong CYP1A staining in endothelial cells (e) and weak CYP1A staining in tubular epithelial cells (t; 250 ×).

(f) CYP1A staining in head kidney of cod exposed to 50 mg/kg β-naphthoflavone, showing strong staining in endothelial cells (e) and tubular epithelial cells (t; 250 ×).

(g) Localization of CYP1A in intestine of flounder exposed to contaminants by caging. Positive CYP1A staining in endothelial cells (e) and also in mucosa epithelial cells (m; 250 ×).

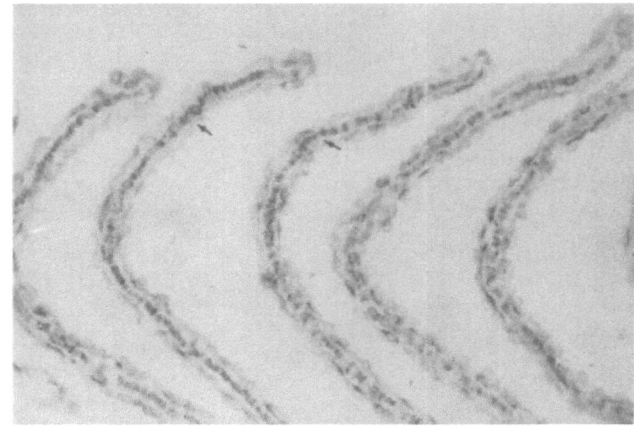

h

Figure 4. (h) Localization of CYP1A in gill tissue from β-naphthoflavone-exposed cod showing strong CYP1A staining in pillar cells (endothelial cells of the branchial capillaries; 250 ×).

exposure of cod to TCDD increased the CYP1A level in the kidney, but not in the liver (A.-M. Husøy, T. Andersson, K. Ingebrigtsen and A. Goksøyr, unpublished results). These results corroborate previous observations in cod (J. Beyer, J. Klungsøyr, and A. Goksøyr, unpublished data), and may suggest that the organ-specific distribution (and dose) of the inducer are important factors in the CYP1A induction response in cod.

Oyster toadfish (*Opsanus tau*), another gadoid, has also given similarly weak CYP1A induction responses (Collier et al., 1993). Whereas fluorescent aromatic compounds from bile and hepatic-DNA adducts correlated well with the high PAH contents in the sediment, the hepatic CYP1A level was still low. CYP1A expression studied by immunohistochemistry was only demonstrated in exocrine pancreatic ducts and vascular endothelial cells (Collier et al., 1993).

Interestingly, immunohistochemical analyses of English sole and starry flounder, two flatfish species from contaminated sites in Puget Sound, Washington (USA), demonstrated marked species differences in localization of CYP1A in hepatocytes and endothelial cells (Anulacion et al., 1994a, b). In English sole, CYP1A localization gave stronger signals in hepatocytes than in endothelial cells, while the converse was true for starry flounder. In these two species, tumor-incidence is much higher in English sole than in starry flounder. Similar species differences in CYP1A localization have also been demonstrated in Norwegian waters between lemon sole (*Microstomus kitt*) and flounder, as discussed in Husøy et al. (1996a).

The apparent species differences observed may be real, or they may be related to dose-response effects. Further studies are needed to clarify this issue. Species differences in CYP1A induction responses may be critical in selection of test organisms in biomonitoring programs.

Kidney

The renal function includes glomerular filtration to remove toxic waste products from the blood. In addition, renal cells host several biotransformation enzymes. In addition to the renal tubules, the teleost kidney contains hematopoietic, lymphoid and endocrine tissue, including interrenal cells, chromaffin cells and the corpuscles of Stannius.

Cellular localization of CYP1A in the kidney demonstrated CYP1A expression in tubular epithelial cells, glomerular endothelium, vascular and sinusoidal endothelium, as well as in nephronic and collecting ducts (Husøy et al., 1994; Smolowitz et al., 1991; Stegeman et al., 1991). Moreover, interrenal endocrine tissue showed positive staining with anti-cod CYP1A IgG. This was observed in unexposed fish and in β-naphthoflavone-stimulated fish (Husøy and Goksøyr, 1998; Husøy et al., 1994). Lorenzana and co-workers (1988) reported moderate immunostaining of interrenal cells, but interpreted this as immunological cross-reactivity to another constitutive P450. The CYP1A staining in interrenal cells in head kidney of cod may suggest an important endogenous role for this isoenzyme (Husøy et al., 1994).

The dose-response study visualized a strong CYP1A induction in the head kidney of cod. Increasing doses of β-naphthoflavone were followed by enhanced CYP1A staining in endothelial cells and tubular epithelial cells, indicating that the inducing parent compound was distributed to extrahepatic organs (Husøy and Goksøyr, 1998; cf. for Fig. 4).

Heart

The fish heart consists of four compartments: sinus venosus, atrium, ventricle, and bulbus arteriosus. The muscle walls of the atrium and ventricle are lined by endocardium (endothelial cells). In the teleost heart, the endothelial cells are specialized scavenger cells, capable of removing high molecular weight compounds such as connective tissue macromolecules, foreign pathogens (viruses), and toxic substances (Olsen et al., 1994; Smedsrød et al., 1995).

CYP1A induction in cardiac microsomes has been observed in teleosts (Husøy et al., 1994, 1996a; Miller et al., 1988; Smolowitz et al., 1992; Stegeman et al., 1982b, 1989). CYP1A was expressed in endothelial cells in the bulbous arteriosus, atrium and ventricle (Fig. 4). In addition, CYP1A induction was seen in the endothelium of coronary vessels and great vessels. In juvenile cod, CYP1A expression was demonstrated in endocardial cells in β-naphthoflavone-exposed as well as in unexposed fish (Husøy and Goksøyr, 1998).

Recently, primary cultures of these cells were established from Atlantic salmon and cod (Grøsvik et al., 1998). The endocardial cells expressed

CYP1A at high levels in both untreated and benzo[a]pyrene- or PCB-treated cells. Subcellular localization studies using immunohistochemistry and confocal microscopy demonstrated the CYP1A staining to be in the perinuclear region in association with the nuclear membrane, and in the cytoplasm in association with microtubuli- and microfilament-like structures. CYP1A seemed to be localized in a more sequestered pattern in cod than in salmon cells, due to a higher degree of vesicles in the cod endocard cells (ibid.).

Gastrointestinal tract

The intestine appears to be the major route of uptake of lipophilic compounds in fish (Van Veld, 1990). In general, the digestive tract of fish is not identical to that in mammals. Nonetheless, several biotransformation enzymes involved in detoxification and elimination are located in the fish intestine. CYP1A induction in the gut has been extensively reviewed (Van Veld, 1990).

Cellular localization of CYP1A in the intestinal tract demonstrate CYP1A expression in the vascular endothelium in the lamina propria (both in pyloric caeca and upper intestine) in addition to CYP1A staining in mucosal epithelial cells (Smolowitz et al., 1991, 1992; Stegeman et al., 1991). Moreover, CYP1A induction was also found in pancreatic acinar cells and pancreatic ductal epithelium.

Intraperitoneal injection of β-naphthoflavone in cod did not produce CYP1A expression in mucosal epithelial cells, probably due to the fact that β-naphthoflavone was rapidly metabolized (Husøy et al., 1994). In contrast, cod and flounder caged in Sørfjorden revealed CYP1A expression in vascular endothelium and mucosal epithelial cells, indicating that other routes of exposure and other inducers may produce CYP1A induction in the intestine (Husøy et al., 1996a; Fig. 4). For instance, mummichog exposed to benzo[a]pyrene *via* the diet showed strong CYP1A-staining in gut mucosal epithelium, whereas aqueous exposure gave low staining in this tissue (Van Veld et al., 1997).

Gill

The gill serves essential functions in respiration, excretion and osmoregulation. In addition to the respiratory epithelium, the secondary lamella consists of pillar cells (endothelial cells) and chloride cells.

CYP1A induction has been observed in gill endothelial (i.e., pillar cells) and/or (respiratory) epithelial cells (Husøy et al., 1994, 1996a; Miller et al., 1988; Smolowitz et al., 1992; Stegeman et al., 1991; Fig. 4).

Exposure to contaminants through the sediment or water may elicit a strong CYP1A induction in the endothelial and epithelial cells of the gill, whereas dietary exposure gives low induction (Van Veld et al., 1997). However, induction in these cells may also occur with inducer uptake *via* the circulation (Smolowitz et al., 1991, 1992).

Endothelial cells

The vascular endothelium in mammals is involved in a wide range of regulatory processes; including control of blood pressure, permeability, inflammation and metabolism of many vascular mediators, as reviewed by Thiemermann (1991). The endothelial cells produce and release important vasoactive products (e.g., nitrogen oxide (NO), prostaglandin (PGI2), endothelin-1 (ET-1), and metabolites from arachidonic acid (AA)). Cytochrome P450s involved in the metabolism of endogenous and exogenous substrates are also localized in the endothelial cells.

The endothelial cells provide a structural interface between blood-borne chemicals and the underlying target cells. Endothelial cells may detoxify or bioactivate foreign compounds (see above). Potentially, CYP1A induction and "first pass" metabolism in vascular endothelial cells may protect the underlying cells from toxic pollutants. Increased CYP1A levels in the endothelial cells could alter the hepatic pharmacokinetics and could also affect dose-response curves. Additionally, activated intermediates in the endothelial cells may affect both the cell itself and the underlying parenchymal cells. Activation of toxicants could contribute to endothelial damage and vascular diseases. Furthermore, CYP1A induction in endothelial cells may alter the metabolism of endogenous compounds like arachidonic acid and nitrogen oxide (Goksøyr, 1995; Stegeman and Hahn, 1994; Stegeman et al., 1994).

CYP1A has been located in the endothelium of several teleost species. In extrahepatic organs, and in situations with low dose of the inducer, the endothelial cells seem to be the target for CYP1A induction (Anulacion et al., 1994a; Collier et al., 1993; Husøy and Goksøyr, 1998; Husøy et al., 1994, 1996a; Lindström-Seppä et al., 1994; Miller et al., 1988; Myers et al., 1995; Smolowitz et al., 1991, 1992; Van Veld et al., 1997; see also Stegeman and Hahn, 1994). Still, the significance of strong CYP1A induction in endothelial cells is not yet quite clear.

Toxicity associated with cytochrome P450 function

Low-level and long-term exposure may produce sustained induction of CYP1A. The relationship between CYP1A induction and long-term effects still remains to be established. The consequences of CYP1A induction are related to the organ and the target cells involved. In addition to chemical carcinogenesis, CYP1A induction may also disturb hormone balance, signal transduction, and immune functions. Furthermore, many CYP1A inducers have high toxicity. Especially with dioxin-type inducers, this toxicity seems to be associated with Ah-receptor binding and CYP1A induction (Rifkind et al., 1990).

Hepatocytes and biliary epithelial cells are strongly responsive to CYP1A inducers. Toxicopathic liver lesions are observed in both cell types. Neoplastic and neoplasia-related lesions have been described in feral fish species (Myers et al., 1987). Furthermore, statistically significant relationships between CYP1A induction and toxicopathic lesions have been established (Myers et al., 1994b).

Potentially, increased CYP1A levels may reflect higher substrate turnover and possibly more reactive intermediates. Consequently, higher levels of DNA adducts may be formed and this may lead to chemical carcinogenesis. Interestingly, in abnormal and neoplastic cells the CYP1A expression is decreased compared to surrounding hepatocytes (Myers et al., 1995; Smolowitz et al., 1989). This observation has also been made in rat (Buchmann et al., 1985). This phenomenon is also accompanied with increased levels of multidrug resistance proteins (Köhler et al., 1998).

CYP1A induction in renal epithelial cells may affect the excretion of a xenobiotic. Toxicopathic lesions in the kidney, related to CYP1A induction, have not been reported. CYP1A induction in mucosa epithelial cells and endothelial cells may increase first pass metabolism of xenobiotics as well as drug metabolism. This may change the biological half-life ($T_{1/2}$) and the biological availability of a compound. Induction of biotransformation enzymes in the intestine may disrupt the effect of antibiotics and therapeutics used in aquaculture industries.

Increased CYP1A levels reported in the brain have been suggested to contribute to neurotoxicological or endocrinological effects (Andersson et al., 1993; Stegeman et al., 1991). CYP1A induction in the endocrine cells of kidney (Husøy et al., 1994) or in the pituitary (Andersson et al., 1993) may disturb hormonal control. Strong CYP1A induction in the olfactory epithelium in the nose of topminnows and dermal chemoreceptors (Smolowitz et al., 1992) have been suggested to be related to disturbances in perception and behaviour.

The dominant CYP1A induction in vascular endothelium indicate that CYP1A expression is essential and important; alterations may disturb metabolism of endogenous substrates. Observations of constitutive CYP1A expression in chondroid cells suggest that CYP1A may also be involved in important endogenous functions like regulation of growth and differentiation (Stegeman et al., 1996), as has previously been postulated by Nebert (1991) as well.

A direct linkage between CYP1A induction and toxic mechanisms remains to be established. Interactions between CYP and other biomarkers like heme oxygenase, stress proteins, phase-II enzymes, and metallothionein have recently been reviewed (Stegeman et al., 1992). The linkages between the CYP system and other biochemical systems are intricate, and this will complicate a clear interpretation of the significance of CYP1A induction.

CYP1A induction as a biomarker in the aquatic environment

It was early suggested that CYP1A may be a useful indicator of PAH exposure (Payne and Penrose, 1975). Oil compounds, PAHs, planar PCBs, dibenzo-*p*-dioxins and dibenzofurans (PCDD/PCDF), and other halogenated compounds such as pesticides and herbicides increase the CYP1A level in fish exposed to such organic pollutants. Several laboratory and field experiments have shown significant relationships between exposure to environmental pollution and CYP1A induction (see Bucheli and Fent, 1995; Collier et al., 1995; Förlin et al., 1994; Goksøyr, 1995; Kloepper-Sams and Benton, 1994).

CYP1A induction is now routinely analyzed in several laboratories around the world, and has been incorporated into national and international monitoring programmes (e.g., NOAA's National Status and Trends Program (USA) and the North Sea Task Force Monitoring Master Plan). In Norwegian field studies, CYP1A induction has been reported in cod, flounder and plaice. In Langesundfjord (southern Norway) and Sørfjorden, Hardanger (western Norway), a significant relationship was observed between CYP1A induction and chemical contamination (Addison and Edwards, 1988; Beyer et al., 1996; Goksøyr et al., 1994b; Husøy et al., 1996a; Stegeman et al., 1988). Increased EROD activity and CYP1A protein levels were also observed in fish downstream from the Glomma delta in the Hvaler Archipelago (Goksøyr et al., 1991c).

A combination of different CYP1A measurements may complement each other and give additional information on the CYP1A induction response (Collier et al., 1995). In combination with a battery of other biomarkers, such responses should provide us with a good system for detecting early warning signals of pollution by systematic use in surveys and monitoring programs, in accordance with the "precautionary principle".

Future studies

Increased use of biomarkers and improved monitoring tools will enhance the understanding of contamination and the consequences for the ecosystem. The relationship between short-term effects such as CYP1A induction and long-term effects like histopathological changes are not yet satisfactorily resolved. In addition, more research is needed to understand the connections between CYP1A and processes involved in signal transduction, growth, reproduction, and differentiation. Immunohistochemical analysis represents a convenient system for studies of CYP1A function and regulation, and for studies relating to carcinogenesis in liver and other organs. Immunohistochemistry can identify, localize and semi-quantify proteins within cells and organelles. With such techniques, it is possible to study the cellular and subcellular localization of CYP1A and other biomarkers, e.g., multi-drug resistant proteins, oncogenes, and phase II enzymes.

References

Addison, R.F. and Edwards, A.J. (1988) Hepatic microsomal mono-oxygenase activity in flounder *Platichthys flesus* from polluted sites in Langesundfjord and from mesocosms experimentally dosed with diesel oil and copper. *Mar. Ecol. Prog. Ser.* 46:51–54.

Andersson, T. (1992) Purification, characterization, and regulation of a male-specific cytochrome P450 in the rainbow trout kidney. *Mar. Environ. Res.* 34:109–112.

Andersson, T. and Förlin, L. (1992) Regulation of the cytochrome P450 enzyme system in fish. *Aquat. Toxicol.* 24:1–20.

Andersson, T. and Goksøyr, A. (1994) Distribution and induction of cytochrome P450 1A1 in the rainbow trout brain. *Fish Physiol. Biochem.* 13:335–342.

Andersson, T., Förlin, L., Härdig, J. and Larsson, Å. (1988) Physiological disturbances in fish living in costal water polluted with bleached kraft pulp mill effluents. *Can. J. Fish. Aquat. Sci.* 45:1525–1536.

Andersson, T., Förlin, L., Olsen, S., Fostier, A. and Breton, B. (1993) Pituitary as a target organ for toxic effects of P4501A1 inducing chemicals. *Mol. Cell. Endocrinol.* 91:99–105.

Ankley, G.T., Blazer, V.S., Plakas, S.M. and Reinert, R.E. (1989) Dietary lipid as a factor modulating xenobiotic metabolism in channel catfish (*Ictalurus punctatus*). *Can. J. Fish. Aquat. Sci.* 46:1141–1146.

Anulacion, B.F., Myers, M.S., Willis, M.L. and Collier, T.K. (1994a) *CYP1A expression in two flatfish species showing different prevalences of contaminant-induced hepatic disease.* Int. Symposium on Aquatic Animal Health, Seattle, WA, abstract.

Anulacion, B.F., Myers, M.S., Willis, M.L. and Collier, T.K. (1994b) *Quantification of immunohistochemically localized CYP1A in two flatfish species showing different prevalences of contaminant-induced neoplasia.* Pacific Northwest Association of Toxicologist (PANWAT) Annual Meeting, Newport, OR, abstract.

Ariyoshi, T., Shiiba, S., Hasegawa, H. and Arizono, K. (1990) Effects of the environmental pollutants on heme oxygenase activity and cytochrome P-450 content in fish. *Bull. Environ. Contam. Toxicol.* 44:189–196.

Arukwe, A. and Goksøyr, A. (1997) Changes in three hepatic cytochrome P450 subfamilies during a reproductive cycle in turbot (*Scophthalmus maximus* L). *J. Exp. Biol.* 277:313–325.

Astorg, P., Gradelet, S., Leclerc, J., Canivenc, M.-C. and Siess, M.-H. (1994) Effects of β-carotene and canthaxanthin on liver xenobiotic-metabolizing enzymes in rat. *Food Chem. Toxic.* 32:735–742.

Astorg, P., Gradelet, S., Leclerc, J. and Siess, M.-H. (1995) *Canthaxanthin, astaxanthin and β-apo-8'-carotenal are strong inducers of liver P450 1A1 and 1A2 in the rat.* The Gordon Conference on Carotenoids, Ventura, California, USA, abstract.

Baumann, P.C. (1989) PAH, metabolites, and neoplasis in feral fish populations. *In:* Varanasi, U. (Ed) *Metabolism of polycyclic aromatic hydrocarbons in the aquatic environment.* CRC Press, Boca Raton, Florida, pp 269–289.

Beresford, A.P. (1993) CYP1A1; friend of foe? *Drug Metab. Rev.* 25:503–517.

Berndtson, A.K. and Chen, T.T. (1994) Two unique CYP1 genes are expressed in response to 3-methylcholanthrene treatment in rainbow trout. *Arch. Biochem. Biophys.* 310:187–195.

Bernhoft, A., Hektoen, H., Skåre, J.U. and Ingebrigtsen, K. (1993) Tissue distribution and effects on hepatic xenobiotic metabolizing enzymes of 2,3,3',4,4'-pentachlorobiphenyl (PCB-105) in cod (*Gadus morhua*) and rainbow trout (*Oncorhynchus mykiss*). *Environ. Poll.* 85:351–359.

Beyer, J., Sandvik, M., Hylland, K., Fjeld, E., Egaas, E., Aas, E., Skåre, J.U. and Goksøyr, A. (1996) Biomarker Responses in Flounder (*Platichthys flesus*) and Atlantic Cod (*Gadus morhua*) Exposed by Caging to Contaminated Sediments in Sørfjorden, Norway. *Aquat. Toxicol.* 36:75–98.

Binder, R.L. and Stegeman, J.J. (1980) Induction of aryl hydrocarbon hydroxylase embryos of an estuarine fish. *Biochem. Pharmacol.* 29:949–951.

Binder, R.L. and Stegeman, J.J. (1983) Basal levels and induction of hepatic aryl hydrocarbon hydroxylase activity during the embryonic period of development in brook trout. *Biochem. Pharmacol.* 32:1324–1327.

Binder, R.L., Stegeman, J.J. and Lech, J.J. (1985) Induction of cytochrome P-450-dependent monooxygenase systems in embryos and eleutheroembryos of the killifish, *Fundulus heteroclitus*. *Chem. Biol. Interact.* 55:185–202.

Boon, J.P., Everaarts, J.M., Hillebrand, M.T.J., Eggens, M.L., Pijnenburg, J. and Goksøyr, A. (1992) Changes in levels of hepatic biotransformation enzymes and hemoglobin levels in female plaice (*Pleuronectes platessa*) after oral administration of a technical PCB mixture (Clophen A40). *Sci. Tot. Environ.* 114:113–133.

Bresnick, E. (1993) Induction of cytochromes P450 1 and P450 2 by xenobiotics. *In:* Schenkman, J. B. and Greim, H. (Eds) *Cytochrome P450. Handbook of experimental pharmacology*. Vol. 105. Springer, Heidelberg, Berlin, New York, pp 503–524.

Bucheli, T.D., and Fent, K. (1995) Induction of cytochrome P450 as a biomarker for environmental contamination in aquatic ecosystems. *Crit. Rev. Environ. Sci. Techn.* 25:201–268.

Buchmann, A., Kuhlmann, W., Schwarz, M., Kunz, W., Wolf, C., Moll, E., Freidberg, T. and Oesch, F. (1985) Regulation of expression of four cytochrome P-450 isoenzymes, NADPH-cytochrome P-450 reductase, the glutathione transferases B and C and microsomal epoxide hydrolase in preneoplastic and neoplastic lesions in rat liver. *Carcinogenesis* 6:513.

Buchmann, A., Wannemacher, R., Kulzer, E., Buhler, D.R. and Bock, K.W. (1993) Immunohistochemical localization of the cytochrome P450 isozymes LMC2 and LM4B (P4501A1) in 2,3,7,8-tetrachlorodibenzo-*p*-dioxin-treated zebrafish (*Brachydanio rerio*). *Toxicol. Appl. Pharmacol.* 123:160–69.

Buhler, D.R. and Williams, D.E. (1988) The role of biotransformation in the toxicity of chemicals. *Aquat. Toxicol.* 11:19–28.

Buhler, D.R. and Williams, D.E. (1989) Enzymes involved in metabolism of PAH by fishes and other aquatic animals: Oxidative enzymes (or phase I enzymes). *In:* Varanasi, U. (Ed) *Metabolism of polycyclic aromatic hydrocarbons in the aquatic environment*. CRC Press, Boca Raton, Florida, pp 151–184.

Buhler, D.R., Yang, Y.-H., Dreher, T.W., Miranda, C.L. and Wang, J.-L. (1994) Cloning and Sequencing of the major rainbow trout constitutive cytochrome P450 (CYP2K1): Identification of a new cytochrome P450 gen subfamily and its expression in mature rainbow trout liver and trunk kidney. *Arch. Biochem. Biophys.* 312:45–51.

Burbach, K.M., Poland, A.B. and Bradfield, C.A. (1992) Cloning of the Ah-receptor cDNA reveals a distinctive ligand-activated transcription factor. *Proc. Natl. Acad. Sci. USA* 89:8185–8189.

Burke, M.D. and Mayer, R.T. (1974) Ethoxyresorufin: Direct fluorometric assay of a microsomal O-dealkylation which is preferentially inducible by 3-methylcholanthrene. *Drug Metab. Disp.* 2:583–588.

Celander, M. and Förlin, L. (1991) Catalytic activity and immunochemical quantification of hepatic cytochrome P-450 in β-naphthoflavone and isosafrol treated rainbow trout (*Oncorhynchus mykiss*). *Fish Physiol. Biochem.* 9:189–197.

Celander, M. and Förlin, L. (1992) Quantification of cytochrome P450IA1 and catalytic activities in liver microsomes of isosafrol and β-naphthoflavone-treated rainbow trout (*Oncorhynchus mykiss*). *Mar. Environ, Res.* 34:123–126.

Celander, M. and Stegeman, J.J. (1997) Isolation of a cytochrome P450 3A cDNA sequence (CYP3A30) from the marine teleost *Fundulus heteroclitus* and phylogenetic analyses of CYP3A genes. *Biochem. Biophys. Res. Comm.* 236:306–312.

Celander, M., Ronis, M. and Förlin, L. (1989) Initial characterization of a constitutive cytochrome P-450 isoenzyme in rainbow trout liver. *Mar. Environ. Res.* 28:9–13.

Celander, M., Leaver, M.J., George, S.G. and Förlin, L. (1993) Induction of cytochrome P450 1A1 and conjugating enzymes in rainbow trout (*Oncorhynchus mykiss*) liver: A time course study. *Comp. Biochem. Physiol.* 106C:343–349.

Celander, M., Buhler, D.R., Förlin, L., Goksøyr, A., Miranda, C.L., Woodin, B.R. and Stegeman, J.J. (1996) Immunochemical relationships of cytochrome P450 3A-like proteins in teleost fish. *Fish Physiol. Biochem.*, 15:323–332.

Clarke, D.J., George, S.G. and Burchell, B. (1991) Glucuronidation in fish. *Aquat. Toxicol.* 20:35–56.

Collier, T.K., Stein, J.E., Goksøyr, A., Myers, M.S., Gooch, J.W., Huggett, R.J. and Varanasi, U. (1993) Biomarkers of PAH exposure in oyster toadfish (*Opsanus tau*) from the Elizabeth River, Virginia. *Environ. Sci.* 2:161–177.

Collier, T.K., Anulacion, B.F., Stein, J.E., Goksøyr, A. and Varanasi, U. (1995) A field evaluation of cytochrome P450 1A as a biomarker of contaminant exposure in three species of flatfish. *Environ. Toxicol. Chem.* 14:143–152.

Dapson, R.W. (1993) Fixation for the 1990s: a review of needs and accomplishments. *Biotechnic. Histochem.* 68:75–82.

Edwards, A.J., Addison, R.F., Willis, D.E. and Renton, K.W. (1988) Seasonal variation of hepatic mixed function oxidases in winter flounder (*Pseudopleuronectes americanus*). *Mar. Environ. Res.* 26:299–309.

Eggens, M.L. and Galgani, F. (1992) Ethoxyresorufin-O-deethylase (EROD) activity in flatfish: fast determination with a fluorescence in plate-reader. *Mar. Environ. Res.* 33:213–221.

Erickson, D.A., Goodrich, M.S. and Lech, J.J. (1988) The effect of piperonyl butoxide on hepatic cytochrome P-450-dependent monooxygenase activities in rainbow trout (*Salmo gairdneri*). *Toxicol. Appl. Pharmacol.* 94:1–10.

Falckh, P.H.J., Wu, Q.K. and Ahokas, J.T. (1997) CYP4T1 – a cytochrome P450 expressed in rainbow trout (*Oncorhynchus mykiss*) liver. *Biochem. Biophys. Res. Comm.* 236:302–305.

Fent, K. and Stegeman, J.J. (1991) Effects of tributyltin chloride *in vitro* on the hepatic microsomal monooxygenase system in the fish *Stenotomus chrysops*. *Aquat. Toxicol.* 20:159–168.

Förlin, L. (1980) Effects of Clophen A50, 3-methylcholanthrene, pregnenolone-16α-carbonitrile, and phenobarbital on the hepatic microsomal cytochrome P-450-dependent monooxygenase system in rainbow trout, *Salmo gairdneri*, of different age and sex. *Toxicol. Appl. Pharmacol.* 54:420–430.

Förlin, L. and Haux, C. (1990) Sex differences in hepatic cytochrome P-450 monooxygenase activities in rainbow trout during an annual reproductive cycle. *J. Endocrinol.* 124:207–213.

Förlin, L., Haux, C., Karlsson-Norrgren, L., Runn, P. and Larsson, Å. (1986) Biotransformation enzyme activities and histopathology in rainbow trout, *Salmo gairdneri*, treated with cadmium. *Aquat. Toxicol.* 8:51–64.

Förlin, L., Balk, L., Celander, M., Bergek, S., Hjelt, M., Rappe, C., de Witt, C. and Jansson, B. (1991) Biotransformation enzyme activities, and PCDD and PCDF levels in pike caught in a Swedish lake. *Mar. Environ. Res.* 34:169–173.

Förlin, L., Goksøyr, A. and Husøy, A.-M. (1994) Cytochrome P450 monooxygenase as indicator of PCB/dioxin like compounds in fish. *In:* Kramer, K. (Ed) *Biological monitoring of estuarine and coastal waters*. CRC Press, Boca Raton, Florida, pp 135–150.

George, S.G. (1989) Cadmium effects on plaice liver xenobiotic and metal detoxication systems: dose response. *Aquat. Toxicol.* 15:303–310.

George, S.G. (1994) Enzymology and molecular biology of phase II xenobiotic-conjugating enzymes in fish. *In:* Malins, D.C. and Ostrander, G.K. (Eds) *Aquatic toxicology: molecular, biochemical, and cellular perspectives*. Lewis Publisher, Boca Raton, pp 37–85.

George, S. and Henderson, J. (1992) Influence of dietary PUFA content on cytochrome P450 and transferase activities in Atlantic Salmon (*Salmo salar*). *Mar. Environ. Res.* 34:127–131.

George, S., Young, P., Leaver, M. and Clarke, D. (1990) Activities of pollutant metabolizing and detoxication systems in the liver of the plaice, *Pleuronectes platessa*: sex and seasonal variations in non-induced fish. *Comp. Biochem. Physiol.* 96C:185–192.

Goddard, K.A., Schultz, R.J. and Stegeman, J.J. (1987) Uptake, toxicity, and distribution of benzo[a]pyrene and monooxygenase induction in the topminnows *Poeciliopsis monacha* and *Poeciliopsis lucida*. *Drug Metab. Dispos.* 15:449–455.

Goksøyr, A. (1985) Purification of hepatic microsomal cytochromes P-450 from β-naphthoflavone-treated Atlantic cod (*Gadus morhua*), a marine teleost fish. *Biochim. Biophys. Acta* 840:409–417.

Goksøyr, A. (1991) A semi-quantitative cytochrome P450IA1 ELISA: A simple method for studying the monooxygenase induction response in environmental monitoring and ecotoxicological testing of fish. *Sci. Tot. Environ.* 101:255–262.

Goksøyr, A. (1995) Use of cytochrome P450 1A (CYP1A) in fish as a biomarker of aquatic pollution. *Arch. Toxicology Suppl.* 17:80–95.

Goksøyr, A. and Förlin, L. (1992) The cytochrome P450 system in fish, aquatic toxicology, and environmental monitoring. *Aquat. Toxicol.* 22:287–312.

Goksøyr, A. and Husøy, A.-M. (1992) The cytochrome P4501A1 response in fish: Application of immunodetection in environmental monitoring and toxicological testing. *Mar. Environ. Res.* 34:147–150.

Goksøyr, A., Andersson, T., Hansson, T., Klungsøyr, J., Zhang, Y. and Förlin, L. (1987) Species characteristics of the hepatic xenobiotic and steroid biotransformation systems of two teleost fish, Atlantic cod (*Gadus morhua*) and rainbow trout (*Salmo gairdneri*). *Toxicol. Appl. Pharmacol.* 89:347–360.

Goksøyr, A., Andersson, T., Buhler, D.R., Stegeman, J.J., Williams, D.E. and Förlin, L. (1991a) Immunochemical cross-reactivity of β-naphthoflavone-inducible cytochrome P-450 (P4501A) in liver microsomes from different fish species and rat. *Fish. Physiol. Biochem.* 9:1–13.

Goksøyr, A., Solberg, T.S. and Serigstad, B. (1991b) Immunochemical detection of cytochrome P4501A1 induction in cod (*Gadus morhua*) larvae and juveniles exposed to a water soluble fraction of North Sea crude oil. *Mar. Poll. Bull.* 22:122–127.

Goksøyr, A., Husøy, A.-M., Larsen, H. E., Klungsøyr, J., Wilhelmsen, S., Maage, A., Brevik, E.M., Andersson, T., Celander, M., Pesonen, M. and Förlin, L. (1991c) Environmental contaminants and biochemical responses in flatfish from the Hvaler archipelago in Norway. *Arch. Environ. Contam. Toxicol.* 21:486–496.

Goksøyr, A., Bjørnevik, M. and Maage, A. (1994a) Effects of dietary iron concentrations on the cytochrome P450 system of Atlantic salmon (*Salmo salar*). *Can. J. Fish. Aquat. Sci.* 51:315–319.

Goksøyr, A., Beyer, J., Husøy, A.-M., Larsen, H.E., Westrheim, K., Wilhelmsen, S. and Klungsøyr, J. (1994b) Accumulation and effects of aromatic and clorinated hydrocarbons in juvenile Atlantic cod (*Gadus morhua*) caged in a polluted fjord (Sørfjorden, Norway). *Aquat. Toxicol.* 29:21–35.

Goksøyr, A., Beyer, J., Egaas, E., Grøsvik, B. E., Hylland, K., Sandvik, M. and Skaare, J.U. (1997) Biomarker responses in flounder (*Platichthys flesus*) and their use in pollution monitoring. *Mar. Poll. Bull.* 33:36–45.

Gonzalez, F.J. (1990) Molecular genetics of the P-450 superfamily. *Pharmac. Ther.* 45:1–38.

Gonzalez, F.J. and Gelboin, H.V. (1994) Role of human cytochromes P450 in the metabolic activation of chemical carcinogens and toxins. *Drug Metab. Rev.* 26:165–183.

Gonzalez, F.J. and Nebert, D.W. (1990) Evolution of the P450 superfamily: animal-plant "warfare", molecular drive and human genetic differences in drug oxidation. *TIG* 6:182–186.

Gooch, J.W., Elskus, A.A., Kloepper-Sams, P.J., Hahn, M.E. and Stegeman, J.J. (1989) Effects of ortho- and non-ortho-substituted polychlorinated biphenyl congeners on the hepatic monooxygenase system in scup (*Stenotomus chrysops*). *Toxicol. Appl. Pharmacol.* 98:422–433.

Goodman Gilman, A., Rall, T.W., Nies, A.S. and Taylor, P. (1990) *Goodman and Gilman's The Pharmacological Basis of Therapeutics.* Pergamon Press, New York.

Grøsvik, B.E., Larsen, H.E. and Goksøyr, A. (1997a) Effects of piperonyl butoxide and β-naphthoflavone on cytochrome P450 1A expression and activity in Atlantic salmon (*Salmo salar*): a temperature and time study. *Environ. Toxicol. Chem.* 16:415–423.

Grøsvik, B.E., Vågnes, Ø., Holmqvist B. and Goksøyr A. (1998) Cytochrome P4501A protein expression and localization in primary cultures of atrial endothelial cells from fish. *Mar. Environ. Res.* 48:in press.

Guengerich, F.P. and Shimada, T. (1991) Oxidation of toxic and carcinogenic chemicals by human cytochrome p-450 enzymes. *Chem. Res. Toxicol.* 4:391–407.

Haasch, M.L., Wejksnora, P.J., Stegeman, J.J. and Lech, J.J. (1989) Cloned rainbow trout liver P_1450 complementary DNA as a potential environmental monitor. *Toxicol. Appl. Pharmacol.* 98:362–368.

Haasch, M.L., Quadokus, E.M., Sutherland, L.A., Goodrich, M.S. and Lech, J.J. (1993) Hepatic CYP1A1 Induction in rainbow trout by continuous flow-through exposure to β-naphthoflavone. *Fund. Appl. Toxicol.* 20:72–82.

Hahn, M.E. (1996) Ah receptors and the mechanism of dioxin toxicity: insights form homology and phylogeny. *In:* DiGiulio, R. T. and Monosson, E. (Eds) *Interconnections between human and ecosystem health.* Chapman and Hall, London, pp 9–27.

Hahn, M.E. and Stegeman, J.J. (1994) Regulation of Cytochrome P450IA1 in teleosts; sustained induction of CYP1A1 mRNA, protein, and catalytic activity by 2,3,7,8-tetrachlorodibenzofuran in the marine fish *Stenotomus chrysops*. *Toxicol. Appl. Pharmacol.* 127:187–198.

Hahn, M.E., Woodin, B.R. and Stegeman, J.J. (1989) Induction of Cytochrome P450E (P450IA1) by 2,3,7,8-tetrachlorodibenzofuran (2,3,7,8-TCDF) in the marine fish scup *(Stenotomus chrysops)*. *Mar. Environ. Res.* 28:61–65.

Hahn, M.E., Lamb, T., Schultz, M., Smolowitz, R. and Stegeman, J. (1993) Cytochrome P450IA induction and inhibition by 3,3′,4,4′-tetrachlorobiphenyl in an Ah receptor-containing fish hepatoma cell line (PLHC-1). *Aquat. Toxicol.* 26:185–208.

Hahn, M.E., Poland, A., Glover, E. and Stegeman, J.J. (1994) Photoaffinity labeling of the Ah receptor; phylogenetic survey of diverse vertebrate and invertebrate species. *Arch. Biochem. Biophys.* 310:218–228.

Hahn, M.E., Karchner, S.I., Perera, S.A. and Shapiro, M.A. (1998) Molecular evolution of the Aryl hydrocarbon receptor (AhR) and evidence for two AhR genes (AhR-1 and AhR-2) in early vertebrates. *Mar. Environ. Res.*, 48:in press.

Hampton, J.A., Lantz, C.R. and Hinton, D.E. (1989) Functional units in rainbow trout *(Salmo gairdneri*, Richardson) liver: III. Morphometric analysis of parenchyma, stroma, and component cell types. *Am. J. Anat.* 185:58–73.

Heilmann, L.J., Sheen, Y.-Y., Bigelow, S.W. and Nebert, D.W. (1988) Trout P450IA1: cDNA and deduced protein sequence, expression in liver, and evolutionary significance. *DNA* 7:379–387.

Hektoen, H., Ingebrigtsen, K., Brevik, E.M. and Oehme, M. (1992) Interspecies differences in tissue distribution 2,3,7,8-tetrachlorodibenzo-*p*-dioxin between cod *(Gadus morhua)* and rainbow trout *(Oncorhynchus mykiss)*. *Chemosphere* 24:581–587.

Hektoen, H., Bernhoft, A., Ingebrigtsen, K., Skaare, J.U. and Goksøyr, A. (1994) Response of hepatic xenobiotic metabolizing enzymes in rainbow trout *(Oncorhynchus mykiss)* and cod *(Gadus morhua)* to 2,3,7,8-tetrachlorodibenzo-*p*-dioxin (2,3,7,8-TCDD). *Aquat. Toxicol.* 28:97–106.

Hendricks, J.D., Wales, J.H., Sinnhuber, R.O., Nixon, J.E., Loveland, P.M. and Scanlan, RA. (1980) Rainbow trout *(Salmo gairdneri)* embryos: a sensitive animal model for experimental carcinogenesis. *Fed. Proc.* 39:3222–3229.

Hinson, J.A., Pumford, N.R. and Nelson, S.D. (1994) The role of metabolic activation in drug toxicity. *Drug Metab. Rev.* 26:395–412.

Hinton, D.E. (1994) Cells, cellular responses, and their markers in chronic toxicity of fishes. *In:* Malins, D.C. and Ostrander, G.K. (Eds) *Aquatic toxicology: molecular, biochemical, and cellular perspectives.* Lewis Publisher, Boca Raton, pp 207–239.

Hinton, D.E., Baumann, P.C., Gardner, G.R., Hawkins, W.E., Hendricks, J.D., Murchelano, R.A. and Okihiro, M.S. (1992) Histopathologic biomarkers. *In:* Huggett, R.J., Kimerle, R.A., Mehrle, Jr., P.M. and Bergman, H.L. (Eds) *Biomarkers: Biochemical, physiological and histological markers of anthropogenic stress.* Lewis Publisher, Boca Raton, pp 155–209.

Hodson, P.V., Kloepper-Sams, P. J., Munkittrick, K.R., Lockhart, W.L., Metner, D.A., Luxon, P.L., Smith, I.R., Gagnon, M.M., Servos, M. and Payne, J.F. (1991) Protocols for measuring mixed function oxygenases of fish liver. *Can. Tech. Rep. Fish. Aquat. Sci.* 1829.

Hoffman, E.C., Reyes, H., Chu, F.-F., Sander, F., Conley, L.H., Brooks, B.A. and Hankinson, O. (1991) Cloning of a factor required for activity of the ah (dioxin) receptor. *Science* 252:954–958.

Husøy, A.-M. (1994) Immunohistochemical localization of fish CYP1A, as a tool in fish histopathology. *In: Diseases and parasites of flounder in the Baltic sea. BMB Publ.* 15:109–115.

Husøy, A.-M. and Goksøyr, A. (1998) Cellular targets of CYP1A induction in juvenile cod *(Gadus morhua* L.) after administration with different doses of BNF. Submitted.

Husøy, A.-M., Myers, M.S., Willis, M.L., Collier, T.K., Celander, M. and Goksøyr, A. (1994) Immunohistochemical localization of CYP1A and CYP3A-like isozymes in hepatic and extrahepatic tissues of Atlantic cod *(Gadus morhua* L.), a marine fish. *Toxicol. Appl. Pharmacol.* 129:294–308.

Husøy, A.-M., Myers, M. and Goksøyr, A. (1996a) Cellular localization of cytochrome P450 (CYP1A) induction and histology in Atlantic cod (*Gadus morhua* L.) and European flounder (*Platichthys flesus*) after environmental exposure to contaminants by caging in Sørfjorden, Norway. *Aquat. Toxicol.* 36:53–74.

Husøy, A.-M., Strætkvern, K.O., Maaseide, N.P., Olsen, S.O. and Goksøyr, A. (1996b) Polyclonal and monoclonal antibodies against immunoaffinity purified Atlantic cod (*Gadus morhua* L.) CYP1A. *Mar. Mol. Biol. Biotech.* 5:84–92.

Ingebrigtsen, K., Hektoen, H., Andersson, T., Bergman, Å. and Brandt, I. (1990) Species-specific accumulation of the polychlorinated biphenyl (PCB) 2,3,3',4,4'-pentachlorobiphenyl in fish brain: A comparison between cod (*Gadus morhua*) and rainbow trout (*Oncorhynchus mykiss*). *Pharmacol. Toxicol.* 67:344–345.

Ioannides, C. (1990) Induction of cytochrome P-450 I and its influences in chemical carcinogenesis. *Biochem. Soc. Trans.* 18:32–34.

Ioannides, C. and Parke, D.V. (1993) Induction of cytochrome P450 I as an indicator of potential chemical carcinogenesis. *Drug Metab. Rev.* 25:485–501.

Jaiswal, A.K., Gonzalez, F.J. and Nebert, D.W. (1985) Human dioxin-inducible cytochrome P-450: Complementary DNA and amino sequence. *Science* 228:80–83.

Kaplan, L.A.E., Schultz, M.E., Schultz, R.J. and Crivello, J.F. (1991) Nitrosodiethylamine metabolism in the viviparous fish *Poeciliopsis*: evidence for the existence of liver P450pj activity and expression. *Carcinogenesis* 12:647–652.

Kleinow, K., Haasch, M.L. and Lech, J.J. (1986) The effect of tricaine anesthesia upon induction of select P-450 dependent monooxygenase activities in rainbow trout (*Salmo gairdneri*). *Aquat. Toxicol.* 8:231–241.

Kleinow, K.M., Melancon, M.J. and Lech, J.J. (1987) Biotransformation and induction: Implications for toxicity, bioaccumulation and monitoring of environmental xenobiotics in fish. *Environ. Health Persp.* 71:105–119.

Kloepper-Sams, P. J. and Benton, E. (1994) Exposure of fish to biologically treated bleach-kraft effluent. 2. Induction of hepatic cytochrome P4501A in mountain whitefish (*Prosopium williamsoni*) and other species. *Environ. Toxicol. Chem.* 13:1483–1496.

Kloepper-Sams, P.J. and Stegeman, J.J. (1989) The temporal relationships between P-450E protein content, catalytic activity, and mRNA levels in the teleost *Fundulus heteroclitus* following treatment with β-naphthoflavone. *Arch. Biochem. Biophys.* 268:525–535.

Kloepper-Sams, P.J., Park, S.S., Gelboin, H.V. and Stegeman, J.J. (1987) Specificity and cross-reactivity of monoclonal and polyclonal antibodies against cytochrome P-450E of the marine fish scup. *Arch. Biochem. Biophys.* 253:268–278.

Klotz, A.V., Stegeman, J.J. and Walsh, C. (1983) An aryl hydrocarbon hydroxylating hepatic cytochrome P-450 from the marine fish *Stenotomus chrysops*. *Arch. Biochem. Biophys.* 226:578–592.

Klotz, A.V., Stegeman, J.J. and Walsh, C. (1984) An alternative 7-ethoxyresorufin O-deethylase activity assay: a continuous visible spectrophotometric method for measurement of cytochrome P-450 monooxygenase activity. *Anal. Biochem.* 140:138–145.

Klotz, A.V., Stegeman, J.J., Woodin, B.R., Snowberger, E.A., Thomas, P.E. and Walsh, C. (1986) Cytochrome P-450 isozymes from the marine teleost *Stenotomus chrysops*: their roles in steroid hydroxylation and the influence of cytochrome b5. *Arch. Biochem. Biophys.* 249:326–338.

Koivusaari, U., Harri, M. and Hanninen, O. (1981) Seasonal variation of hepatic biotransformation in female and male rainbow trout (*Salmo gairdneri*). *Comp. Biochem. Physiol.* 70C:149–157.

Köhler, A., Lauritzen, B., Bahns, S., George, S.G., Förlin, L. and van Noorden, C.J.F. (1998) Clonal adaption of liver tumor cells in flatfish to environmental contamination by multidrug resistance and metabolic changes (G6PDH, CYP, GST). *Mar. Environ. Res.*, 48:in press.

Larsen, H.E., Celander, M. and Goksøyr, A. (1992) The cytochrome P450 system of Atlantic salmon (*Salmo salar*): II. Variations in hepatic catalytic activities and isozyme patterns during an annual reproductive cycle. *Fish Physiol. Biochem.* 10:291–301.

Larsson, L.-I. (1988) *Immunocytochemistry: Theory and Practice*. CRC Press, Boca Raton.

Laurén, D.J., Okihiro, M.S., Hinton, D.E. and Stegeman, J.J. (1990) Localization of cytochrome P450 IA1 induced by β-naphthoflavone (BNF) in rainbow trout (*Oncorhynchus mykiss*) embryos. *FEBS J.* 4:A739.

Leaver, M.J., Pirrit, L. and George, S.G. (1993) Cytochrome P450 1A1 cDNA from plaice (*Pleuronectes platessa*) and induction of P450 1A1 mRNA in various tissues by 3-methylcholanthrene and isosafrole. *Mol. Mar. Biol. Biotech.* 2:338–345.

Lech, J.J. and Bend, J.R. (1980) Relationship between biotransformation and the toxicity and fate of xenobiotic chemicals in fish. *Environ. Health. Persp.* 34:115–131.

Lech, J.J., Vodicnik, M.J. and Elcombe, C.R. (1982) Induction of monooxygenase activity in fish. *In*: Weber, L.J. (Ed) *Aquatic Toxicology*. Raven Press, New York, pp 107–148.

Lee, S.-J., Yang, Y.-H., Wang, J.-L., Miranda, C.L., Lech, J.J. and Buhler, D.C. (1998) Occurence of a CYP3A family member in fish. Cloning, sequencing, and expression of hepatic CYP3A27 in rainbow front (*Oncorhynchus mykiss*). In prep.

Lester, S.M., Braunbeck, T., Teh, S.J., Stegeman, J.J., Miller, M.R. and Hinton, D.E. (1993) Hepatic cellular distribution of cytochrome P450IA1 in rainbow trout (*Oncorhynchus mykiss*): An immunohisto- and -cytochemical study. *Cancer Res.* 53:3700–3706.

Lindström-Seppä, P. (1985) Seasonal variation of the xenobiotic metabolizing enzyme activities in the liver of male and female vendace (*Coregonus albula* L.). *Aquat. Toxicol.* 6:323–331.

Lindström-Seppä, P. and Oikari, A. (1990) Biotransformation activities of feral fish in waters receiving bleached pulp mill effluents. *Environ. Toxicol. Chem.* 9:1415–1424.

Lindström-Seppä, P., Korytko, P.J., Hahn, M. and Stegeman, J.J. (1994) Uptake of waterborne 3,3′,4,4′-tetrachlorobiphenyl and organ and cell-specific induction of cytochrome P4501A in adult and larval fathead minnow *Pimephales promelas. Aquat. Toxicol.* 28:147–167.

Livingstone, D.R. (1993) Biotechnology and pollution monitoring: use of molecular biomarkers in the aquatic environment. *J. Chem. Tech. Biotechnol.* 57:195–211.

Lorenzana, R.M., Hedström, O.R. and Buhler, D.R. (1988) Localization of cytochrome P-450 in head trunk kidney of rainbow trout (*Salmo gairdneri*). *Toxicol. Appl. Pharmacol.* 96:159–167.

Lorenzana, R.M., Hedström, O.R., Gallagher, J.A. and Buhler, D.R. (1989) Cytochrome P-450 isozyme distribution in normal and tumor-bearing hepatic tissue from rainbow trout (*Salmo gairdneri*). *Exp. Mol. Pathol.* 50:348–361.

Lorenzen, A. and Okey, A.B. (1990) Detection and characterization of (^{3}H)2,3,7,8-tetrachlorodibenzo-*p*-dioxin binding Ah-receptor in a rainbow trout hepatoma cell line. *Toxicol. Appl. Pharmacol.* 106:53.

Maccubbin, A.E. (1994) DNA adducts analysis in fish: laboratory and field studies. *In:* Malins, D.C. and Ostrander, G.K. (Eds) *Aquatic toxicology: molecular, biochemical, and cellular perspectives*. Lewis Publisher, Boca Raton, pp 267–294.

Mathieu, A., Lemaire, P., Carriere, S., Drai, P., Giudicelli, J. and Lafaurie, M. (1991) Seasonal and sex-linked variations in hepatic and extrahepatic biotransformation activities in striped mullet (*Mullus barbatus*). *Ecotox. Environ. Safety* 22:45–57.

Miller, M.R., Hinton, D.J., Blair, J.J. and Stegeman, J.J. (1988) Immunohistochemical localization of cytochrome P-450E in liver, gill and heart of scup (*Stenotomus chrysops*) and rainbow trout (*Salmo gairdneri*). *Mar. Environ. Res.* 24:37–39.

Miranda, C.L., Wang, J.L., Henderson, M.C. and Buhler, D.R. (1989) Purification and characterization of hepatic steroid hydroxylases from untreated rainbow trout. *Arch. Biochem. Biophys.* 268:227–238.

Molven, A. and Goksøyr, A. (1993) Biological effects and biomonitoring of organochlorides and polyaromatic hydrocarbons in the marine environment. *In*: Richardson, R.M. (Ed) *Ecotoxicology monitoring*. VCH, Weinheim, pp 137–162.

Monosson, E. and Stegeman, J.J. (1991) Cytochrome P450E (P450IA) induction and inhibition in winter flounder by 3,3′,4,4′-tetrachlorobiphenyl: comparison of response in fish from Georges Bank and Narragansett Bay. *Environ. Toxicol. Chem.* 10:765–774.

Moore, M.J. and Myers, M.S. (1994) Pathobiology of chemical-associated neoplasia in fish. *In:* Malins, D.C. and Ostrander, G.K. (Eds) *Aquatic toxicology: molecular, biochemical, and cellular perspectives*. Lewis Publisher, Boca Raton, pp 327–386.

Moore, M.J. and Stegeman, J.J. (1994) Hepatic neoplasms in winter flounder *Pleuronectes americanus* from Boston Harbor, Massachusetts, USA. *Dis. Aquat. Org.* 20:33–48.

Morrison, H., Oleksiak, M.F., Cornell, N.W., Sogin, M.L. and Stegeman, J.J. (1995) Identification of Cytochrome P450 1A genes from two teleost fish, toadfish (*Opsanus tau*) and scup (*Stenotomus chrysops*), and phylogenetic analysis of CYP1A genes. *Biochem. J.* 308:97–104.

Myers, M.S., Rhodes, L.D. and McCain, B.B. (1987) Pathologic anatomy and patterns of occurrence of hepatic neoplasms, putative preneoplastic lesions, and other idiopathic hepatic conditions in English sole (*Parophrys vetulus*) from Puget Sound, Washington. *J. Natl. Cancer Inst.* 78:33–363.

Myers, M.S., Landahl, J.T., Krahn, M.M. and McCain, B.B. (1991) Relationships between hepatic neoplasms and related lesions and exposure to toxic chemicals in marine fish from the U.S. West Coast. *Environ. Health Persp.* 90:7–15.

Myers, C.R., Sutherland, L.A., Haasch, M.L. and Lech, J.J. (1993) Antibodies to synthetic peptide that react specifically with rainbow trout hepatic cytochrome P450 1A. *Environ. Toxicol. Chem.* 12:1619–1616.

Myers, M.S., Stehr, C.M., Olson, O.P., Johnson, L.L., McCain, B.B., Chan, S.-L. and Varanasi, U. (1994a) Relationships between toxicopathic hepatic lesions and exposure to chemical contaminants in English sole (*Pleuronectes vetulus*), starry flounder (*Platichthys stellatus*), and white croaker (*Genyonemus lineatus*) from selected marine sites on the pacific coast, USA. *Environ Healt Persp.* 102:200–215.

Myers, M.S., Johnson, L.L., Hom, T., Collier, T.K., Stein, J.E. and Varanasi, U. (1994b) Toxicopathic hepatic lesions and other biomarkers of exposure to chemical contaminants in marine bottom fish species from the Northeast and Pacific Coasts, USA. *In: Diseases and parasites of flounder in the Baltic Sea, BMB Publ.* 15:81–98.

Myers, M.S., Willis, M.J., Husøy, A.-M., Goksøyr, A. and Collier, T.K. (1995) Immunohistochemical localization of cytochrome P4501A in multiple types of contaminant-associated hepatic lesions in English sole (*Pleuronectes vetulus*). *Mar. Environ. Res.* 39:283–288.

Nebert, D.W. (1991) Proposed role of drug-metabolizing enzymes: Regulation of steady state levels of the ligands that effect growth, homeostasis, differentiation and neuroendocrine functions. *Mol. Endocrinol.* 5:1203–1214.

Nebert, D.W. and Gelboin, H.V. (1968) Substrate-inducible microsomal aryl hydroxylase in mammalian cell culture. I. Assay and properties of induced enzyme. *J. Biol. Chem.* 243:6242–6249.

Nebert, D.W. and Gonzalez, F.J. (1985) Cytochrome P-450 gene expression and regulation. *Trends Pharmacol. Sci.* 6:160–164.

Nebert, D.W. and Gonzalez, F.J. (1987) P450 genes: Structure, evolution, and regulation. *Ann. Rev. Biochem.* 56:945–993.

Nebert, D.W., Nelson, D.R. and Feyereisen, R. (1989) Evolution of the cytochrome P-450 genes. *Xenobiotica.* 19:1149–1160.

Nebert, D.W., Nelson, D.R., Coon, M J., Estabrook, R.W., Feyereisen, R., Fujii-Kuriyama, Y., Gonzalez, F., Guengerich, F.P., Gunsalus, I.C., Johnson, E. F., Loper, J.C., Sato, R., Waterman, M.R. and Waxman, D.J. (1991) The P450 superfamily: Update on new sequences, gene mapping, and recommended nomenclature. *DNA Cell Biol.* 10:1–14.

Nelson, D.R., Kamataki, T., Waxman, D.J., Guengerich, F.P., Estabrook, R.W., Feyereisen, R., Gonzalez, F.J., Coon, M.J., Gunsalus, I.C., Gotoh, O., Okuda, K. and Nebert, D.W. (1993) The P450 superfamily – update on new sequences, gene mapping, accession numbers, early trivial names of enzymes, and nomenclature. *DNA Cell Biol.* 12:1–51.

Nelson, D.R., Kamataki, T., Waxman, D.J., Guengerich, F.P., Estabrook, R.W., Feyereisen, R., Gonzalez, F.J., Coon, M.J., Gunsalus, I.C., Gotoh, O., Okuda, K. and Nebert, D.W. (1996) The P450 superfamily – update on new sequences, gene mapping, accession numbers, early trivial names of enzymes, and nomenclature. *Pharmacogenetics* 6:1–42.

Nimmo, I.A. (1987) The glutathione S-transferases of fish. Fish Physiol. *Biochem.* 3:163–172.

Nilsen, B.M., Berg K., Husøy A.-M. and Goksøyr A. *The monoclonal antibody NP-7 against cod cytochrome P450 1A cross-reacts with CYP1A from several other fish species.* SECOTOX 96, Metz, France, August 1996, abstract.

Okey, A.B. and Harper, P.A. (1994) The Ah receptor: mediator of toxicity of 2,3,7,8-tetrachlorodibenzo-*p*-dioxin (TCDD) and related compounds. *Toxicol. Lett.* 70:1–22.

Okey, A.B., Riddick, D.S. and Harper, P.A. (1994) Molecular biology of aromatic hydrocarbon (dioxin) receptor. *TIPS* 15:227–232.

Oleksiak, M.F., Wu, S., Zeldin, D.C. and Stegeman, J.J. (1997) *Characterization of members of the novel cytochrome P450 subfamilies CYP2N and CYP2P from the fish Fundulus heteroclitus.* 9th Int. Symposium on Responses of Marine Org. to Pollutants, PRIMO 9, Bergen, Norway, abstract.

Olsen, R., Sveinbjørnssen, B. and Smedsrød, B. (1994) *Endocytotic function of endothelial cells of cod (Gadus morhua). Histological and ultrastructural studies.* ICEM 13, Paris, France, abstract.

Oppen-Berntsen, D.O., Hyllner, S.J., Haux, C., Helvik, J.V. and Walther, B.T. (1992) Eggshell zona radiata-proteins from cod (*Gadus morhua*) extra-ovarian origin and induction by estradiol-17β. *Int. J. Dev. Biol.* 36:247–254.

Park, S.S., Miller, H., Klotz, A.V., Kloepper-Sams, P.J., Stegeman, J.J. and Gelboin, H.V. (1986) Monoclonal antibodies to liver microsomal cytochrome P-450E of the marine fish *Stenotomus chrysops* (scup): Cross reactivity with 3-methylcholanthrene induced rat cytochrome P-450. *Arch. Biochem. Biophys.* 249:339–350.

Parke, D. (1991) Nutritional requirements for detoxication of environmental chemicals. *Food Add. Contam.* 8:381–396.

Payne, J.F. and Penrose, W.R. (1975) Induction of aryl hydrocarbon (benzo[a]pyrene) hydroxylase in fish by petroleum. *Bull. Environ. Cont. Toxicol* 14:112–116.

Payne, J.F., Fancey, L.L., Rahimtula, A.D., and Porter, E.L. (1987) Review and perspective on the use of mixed-function oxygenase enzymes in biological monitoring. *Comp. Biochem. Physiol.* 86C:233–245.

Peakall, D. (1992) *Animal biomarkers as pollution indicators.* Chapman and Hall, London, New York, Tokyo, Melbourne, Madras, 291 pp.

Pelissero, C., Flouriot, G., Foucher, J.L., Bennetau, B., Dunogues, J., Gac, F.L. and Sumpter, J.P. (1993) Vitellogenin synthesis in cultured hepatocytes; an *in vitro* test for the estrogenic potency of chemicals. *J. Steroid Biochem. Molec. Biol.* 44:263–272.

Pesonen, M., Goksøyr, A. and Andersson, T. (1992) Expression of P4501A1 in a primary culture of rainbow trout hepatocytes exposed to β-naphthoflavone or 2,3,7,8-tetrachloro-*p*-dioxin. *Arch. Biochem. Biophys.* 292:228–233.

Poland, A. and Knutson, J.C. (1982) 2,3,7,8-tetrachlorodibenzo-*p*-dioxin and related halogenated aromatic hydrocarbons: examination of the mechanism of toxicity. *Ann. Rev. Pharmacol. Toxicol.* 22:517–554.

Porter, T.D. and Coon, M.J. (1991) Cytochrome P-450. Multiplicity of isoforms, substrates, and catalytic and regulatory mechanisms. *J. Biol. Chem.* 266:13469–13472.

Renton, K.W. and Addison, R.F. (1992) Hepatic microsomal mono-oxygenase activity and P4501A mRNA in North Sea dab *Limanda limanda* from contaminated sites. *Mar. Ecol. Prog. Ser.* 91:65–69.

Rice, C.D., Schlenk, D. and Goksøyr, A. (1998) Cross reactivity of monoclonal antibodies against peptide 277–294 of rainbow trout P4501A1 with microsomal P450 1A among fish. *Mar. Environ. Res.*, 48:in press.

Rifkind, A.B., Gannon, M. and Gross, S.S. (1990) Arachidonic acid metabolism by dioxin-induced cytochrome P-450: a new hypothesis on the role of P-450 in dioxin toxicity. *Biochem. Biophys. Res. Commun.* 172:1180–1188.

Roy, N.K., Konkle, B.A., Kreamer, G.-L., Grunwald, C. and Wirgin, I. (1995) Characterization and prevalence of a polymorphism in the 3'-untranslated region of cytochrome P4501A1 in cancer-prone Atlantic tomcod. *Arch. Biochem. Biophys.* 322:204–213.

Safe, S.H. (1988) The aryl hydrocarbon (Ah) receptor. ISI Atlas of Science. *Pharmacology* 2:78–83.

Sakai, N., Tanaka, M., Adachi, S., Miller, W.L. and Nagahama, Y. (1992) Rainbow trout cytochrome P-450c17 (17a-hydroxylase/17,20-lyase) cDNA cloning, enzymatic properties and temporal pattern of ovarian P-450c17 mRNA expression during oogenesis. *FEBS Lett.* 301: 60–64.

Savas, Ü., Bhattacharyya, K.K., Christou, M., Alexander, D.L. and Jefcoater, C.R. (1994) Mouse cytochrome P-450EF, representative for a new 1B subfamily of cytochrome P-450. *J. Biol. Chem.* 269:14905–14911.

Schenkman, J.B. and Greim, H. (Eds) (1993) Cytochrome P450. *Handbook of experimental pharmacology.* Vol. 105. Springer, Berlin, Heidelberg, New York, pp 739 ff.

Schmidt, J.V., Carver, L.A. and Bradfield, C.A. (1993) Molecular characterization of the murine Ahr gene: organization, promoter analysis, and chromosomal assignment. *J. Biol. Chem.* 268:22203–22209.

Skjegstad, N.L., Garatun-Tjeldstø, O., Grahl-Nielsen, O. and Goksøyr, A. (1994) *Temperature adaptation of turbot (Scophthalmus maximus L): biochemical studies of cytochrome P450 and fatty acid profiles in liver membranes.* 15th Annual Conf. of ESCPB, Genova, Italy, abstract.

Sleiderink, H.M., Beyer, J., Scholtens, E., Goksøyr, A., Nieuwenhuize, J., Van Liere, J.M., Ever-aarts, J.M. and Boon, J.P. (1995) Influence of temperature and polyaromatic contaminants on CYP1A levels in North Sea dab (*Limanda limanda*). *Aquat. Toxicol*. 32:189–209.

Smedsrød, B., Olsen, R. and Sveinbjørnsson, B. (1995) Circulating collagen is catabolized by endocytosis mainly in endothelial cells of endocardium of cod (*Gadus morhua*) *Cell Tiss. Res*. 280:39–48.

Smolowitz, R.M., Moore, M.J. and Stegeman, J.J. (1989) Cellular distribution of cytochrome P-450E in Winter flounder liver with degenerative and neoplastic disease. *Mar. Environ. Res*. 28:441–446.

Smolowitz, R.M., Hahn, M.E. and Stegeman, J.J. (1991) Immunohistochemical localization of cytochrome P-450IA1 induced by 3,3′,4,4′-tetrachlorobiphenyl and by 2,3,7,8-tetrachlorodi-benzo-furan in liver and extrahepatic tissues of the teleost *Stenotomus chrysops* (scup). *Drug Met. Disp*. 19:113–123.

Smolowitz, R.M., Schultz, M.E. and Stegeman, J.J. (1992) Cytochrome P4501A induction in tissues, including olfactory epithelium, of topminnows (*Poeciliopsis* spp.) by waterborne benzo[a]pyrene. *Carcinogenesis* 13:2395–2402.

Stagg, R.M. and Addison, R.F. (1995) An inter-laboratory comparison of measurements of ethoxyresorufin o-deethylase activity in dab (*Limanda limanda*) liver. *Mar. Environ. Res*. 40: 93–108.

Stegeman, J.J. (1989) Cytochrome P450 forms in fish: catalytic, immunological and sequence similarities. *Xenobiotica*. 19:1093–1110.

Stegeman, J.J. and Chevion, M. (1980) Sex differences in cytochrome P-450 and mixed-function oxygenase activity in gonadally mature trout. *Biochem. Pharmacol*. 29:553–558.

Stegeman, J.J. and Kloepper-Sams, P.J. (1987) Cytochrome P-450 isoenzymes and monooxy-genase activity in aquatic animals. *Environ. Health Persp*. 71:87–95.

Stegeman, J.J. and Hahn, M.E. (1994) Biochemistry and molecular biology of monooxygen-ases: Current perspectives on forms, functions and regulation of Cytochrome P450 in aquatic species. *In:* Malins, D.C. and Ostrander, G.K. (Eds) *Aquatic toxicology: molecular, biochemical, and cellular perspectives*. Lewis Publisher, Boca Raton, pp 87–204.

Stegeman, J.J., Woodin, B.R., Klotz, A.V., Wolke, R.E. and Orme-Johnson, N.R. (1982) Cyto-chrome P-450 and monooxygenase activity in cardiac microsomes from the fish *Stenotomus chrysops*. *Mol. Pharmacol*. 21:517–526.

Stegeman, J.J., Woodin, B.R. and Goksøyr, A. (1988) Apparent cytochrome P-450 induction as an indication of exposure to environmental chemicals in the flatfish *Platichthys flesus*. *Mar. Ecol. Prog. Ser*. 46:55–60.

Stegeman, J.J., Miller, M.R. and Hinton, D.E. (1989) Cytochrome P450IA1 induction and loca-lization in endothelium of vertebrate (teleost) heart. *Mol. Pharmacol*. 36:723–729.

Stegeman, J.J., Smolowitz, R.M. and Hahn, M.E. (1991) Immunohistochemical localization of environmentally induced cytochrome P450IA1 in multiple organs of the marine teleost *Stenotomus chrysops* (scup). *Toxicol. Appl. Pharmacol*. 110:1–17.

Stegeman, J.J., Brouwer, M., Richard, T.D.G., Förlin, L., Fowler, B.A., Sanders, B.M. and Van Veld, P.A. (1992) Molecular responses to environmental contamination: Enzyme and protein systems as indicators of chemical exposure and effect. *In:* Huggett, R.J., Kimerle, R.A., Mehrle, Jr., P.M. and Bergman, H.L. (Eds) *Biomarkers: Biochemical, physiological and histological markers of anthropogenic stress*. Lewis Publisher, Boca Raton, pp 235–335.

Stegeman, J.J., Hahn, M.E., Weisbrod, R., Woodin, B.R., Joy, J.J., Najibi, S. and Cohen, R.A. (1994) Induction of cytochrome P4501A by aryl hydrocarbon receptor agonists in porcine aorta endothelial cells in culture and cytochrome P4501A1 activity in intact cells. *Mol. Phar-macol*. 47:296–306.

Stegeman, J.J., Lindström-Seppa, P., Knipe, T., Suter, S., Smolowitz, R.M. and Hestermann, E. (1996) Cytochrome P4501A expression in teleost chondroid cell: a possible site of endo-genous function of the Ah-receptor-CYP1A loop. *Mar. Environ. Res*. 42:306.

Sutter, T.R., Tang, Y.M., Hayes, C.L., Wo, Y.-Y. P., Wang Jabs, E., Li, X., Yin, H., Cody, C.W. and Greenlee, W.F. (1994) Complete cDNA sequence of a human dioxin-inducible mRNA identifies a new gene subfamily of cytochrome P450 that maps to cytochrome 2. *J. Biol. Chem*. 269:13092–13099.

Tarlebø, J., Solbakken, J.E. and Palmork, K.H. (1985) Variation in hepatic aryl hydrocarbon hydroxylase activity in flounder, *Platichthys flesus*: A baseline study. *Helgol. Wiss. Meer. Unters*. 39:187–199.

Thiemermann, C. (1991) Biosynthesis and interaction of endothelium-derived vasoactive mediators. *Eicosanoids* 4:187–202.

van der Weiden, M.E.J. (1993) *Cytochrome P450 1A induction in carp as a biological indicator for the aquatic contamination of chlorinated polyaromatics,* Utrecht University, Ph.D.-Thesis, The Netherlands.

Van Veld, P.A. (1990) Absorption and metabolism of dietary xenobiotics by the intestine of fish. *Aquat. Sci.* 2:185–203.

Van Veld, P.A., Stegeman, J.J., Woodin, B.R., Patton, J.S. and Lee, R.F. (1988) Induction of monooxygenase activity in the intestine of spot (*Leiostomus xanthurus*), a marine teleost, by dietary polycyclic aromatic hydrocarbons. *Drug. Metab. Disp.* 16:659–665.

Van Veld, P.A., Westbrook, D.J., Woodin, B.R., Hale, R.C., Smith, C.L., Huggett, R.J. and Stegeman, J.J. (1990) Induced cytochrome P-450 in intestine and liver of spot (*Leiostomus xanthurus*) from a polycyclic aromatic hydrocarbon contaminated environment. *Aquat. Toxicol.* 17:119–132.

Van Veld, P.A., Vogelbein, W.K., Cochran, M.K., Goksøyr, A. and Stegeman, J.J. (1997) Route-specific cellular expression of cytochrome P450 1A (CYP1A) in fish (*Fundulus heteroclitus*) following exposure to aqueous and dietary benzo[a]pyrene. *Toxicol. Appl. Pharmacol.* 142:348–359.

Varanasi, U., Collier, T. K., Williams, D. E. and Buhler, D. R. (1986) Hepatic cytochrome P-450 isozymes and aryl hydrocarbon hydroxylase in English sole (*Parophrys vetulus*). *Biochem. Pharmacol.* 35:2967–2971.

Vogelbein, W.K., Fournie, J.W., Van Veld, P.A. and Huggett, R.J. (1990) Hepatic neoplasms in the mummichog *Fundulus heteroclitus* from a creosote-contaminated site. *Cancer Res.* 50:5978–5986.

Vrolijk, N. and Chen, T.T. (1994) *Characterization and expression of CYP1A in the butterflyfish, Chaetodon capistratus, and its response to β-naphthoflavone and terpenoid treatment.* 3rd International Marine Biotechnology Conference, Tromsø, Norway, Abstract.

Williams, D.E. and Buhler, D.R. (1982) Purification of cytochromes P-448 from β-naphthoflavone-treated rainbow trout. *Biochim. Biophys. Acta* 717:398–404.

Williams, D.E. and Buhler, D.R. (1984) Benzo[a]pyrene-hydroxylase catalyzed by purified isozymes of cytochrome P-450 from β-naphthoflavone-fed rainbow trout. *Biochem. Pharmacol.* 33:3743–3753.

Williams, D.E., Carpenter, H.M., Buhler, D.R., Kelly, J.D. and Dutchuk, M. (1992) Alternations in lipid peroxidation, antioxidant enzymes, and carcinogen metabolism in liver microsomes of vitamin E-deficient trout and rat. *Toxicol. Appl. Pharmacol.* 116:78–84.

Yang, Y.-H., Wang, J.-L. and Buhler, D.R. (1996) Cloning and characterization of a new cytochrome P450 form from rainbow trout. *Mar. Environ. Res.* 42:27 (abstract).

Zhang, Y.S., Goksøyr, A., Andersson, T. and Förlin, L. (1991) Initial purification and characterization of hepatic microsomal cytochrome P-450 from BNF-treated perch (*Perca fluviatilis*). *Comp. Biochem. Physiol.* 98B, 97–103.

Zhu L. (1995) *Organochlorines in aquatic organisms: Laboratory kinetics of contamination and biochemical response.* Ph. D. Thesis, Vrije Universiteit Brussel.

Fish Ecotoxicology
ed. by T. Braunbeck, D. E. Hinton and B. Streit
© 1998 Birkhäuser Verlag Basel/Switzerland

"Oxidative stress" in fish by environmental pollutants

Reinhard Lackner

Department of Zoology and Limnology, University of Innsbruck, Technikerstr. 25, A-6020 Innsbruck, Austria

Summary. Evolution of aerobic life on earth faced the problem of having to minimize the constant threat of damage through reactive oxygen species. As a response, mechanisms for safe handling of oxygen and its metabolites have evolved. There is a constant production of reactive oxygen species in all living cells in which roughly up to 1% of the total oxygen consumption of an animal may be attributed to reactive oxygen species generation and detoxification. Mitochondrial electron transport chain is a major source for reactive oxygen species. Interferences with this pathway even increase the cells' reactive oxygen species burden, inducing oxidative stress. Metabolism of xenobiotic pollutants may result in the additional formation of reactive oxygen species. During oxidation of xenobiotics ambient oxygen is activated which may be released from the enzymes, damaging cellular structures and functions. Furthermore, reduced moieties of pollutants or their metabolites react with oxygen, producing superoxide (redox cycling). Reactive oxygen species thus represent a common side product of xenobiotic metabolism and many pollutants actually exert part of their toxicity through the formation of reactive oxygen species.

Introduction

Aerobic life is firmly linked with oxygen-dependent oxidation processes with the concurrent danger of cellular damage by reactive oxygen species. The term "oxidative stress" was introduced in biological sciences to denote a disturbance in the prooxidant-antioxidant balance in favour of the former (Sies, 1985).

Biological utilisation of oxygen involves its activation, which proceeds *via* consecutive reductions. The complete reduction of oxygen with four electrons yields water, but, depending on the reaction, various intermediates may occur: the superoxide radical anion, hydrogen peroxide and the hydroxyl radical. The reaction potentials for these step-by-step reductions are given by Imlay and Linn (1988). These very reactive intermediates of oxygen reduction eventually undergo protonation or lead directly to the formation of organic and inorganic peroxides and radicals. Organic radicals are characterized by the occurrence of single electrons. A chemical bond consists of two electrons with opposite spin. The single electron of a radical thus represents half a chemical bond with a strong tendency to complete the configuration by forming a complete chemical bond. If chemical bonds are formed by joining two radicals together the electron spin is undefined and excited states of molecules may be formed (triplet state, e.g.,

Table 1. Reactive oxygen species and oxygen centred radicals of biological interest

Compound	Source
Ozone (O_3)	Electric discharges and radiation, disinfection
Singlet oxygen ($^1O_2$; also $^1\Delta_g O_2$, $O_2{}^*$)	Excited state of oxygen, 22 kcal/mol above ground state, radiation, nonenzymatic dismutation of $O_2{}^{\cdot-}$ and H_2O_2
Superoxide anion radical ($O_2{}^{\cdot-}$; also protonated as $HO_2{}^{\cdot}$)	One-electron reduction product of oxygen, O_2
Hydrogen peroxide (H_2O_2)	Dismutation of $O_2{}^{\cdot-}$
Hydroxyl radical ($OH^{\cdot}$; also $HO^{\cdot}$)	Fenton reaction, metal (iron)-catalysed Haber-Weiss reaction out of $O_2{}^{\cdot-}$ and H_2O_2
Alkoxy radical ($RO^{\cdot}$)	Lipid peroxidation
Peroxy radical ($ROO^{\cdot}$)	Lipid peroxidation
Organic hydroperoxide ($ROOH$)	Lipid peroxidation
Excited carbonyl (^3RO; also RO^*)	Excited state of organic compound, may also be produced by $^1O_2$

^3RO). In contrast to most other molecules, the excited state of oxygen is the singlet state ($^1O_2$). The chemistry of radicals and excited states is characterized by the exchange of single electrons or energy with other molecules. The toxicity of the parent or the produced compound correlates with its chemical energy, and the formation of products with low free energy detoxifies radicals.

Most of the oxygen-centred radicals and excited states produce similar symptoms *in vivo*. They are summarised as reactive oxygen species (Tab. 1) including also excited states of oxygen, organic compounds (triplet state) and ozone. In fact, the term "reactive oxygen species" has become almost synonymous to "oxidative stress".

Sources of reactive oxygen species

Depending on the specific reactions leading to the activation of molecular oxygen, different concentrations ranging from hypoxic to hyperoxic favour reactive oxygen species production (Jones, 1985). Hypoxia may induce a very reduced state of the cells ("reductive stress") resulting in reactive oxygen species formation when O_2-concentrations increase. Redox cycles are one of the most important sources for superoxide radicals in tissues and many pollutants exert part of their toxicity *via* this mechanism (Kappus,

1987). In a redox cycle, a xenobiotic (e.g., quinone, aromatic nitro or amine) compound is reduced by a reductase or reductant such as NADPH in a one-electron step to a xenobiotic free radical which, in turn, is able to reduce O_2 to $O_2^{\cdot-}$. These reactions form cycles which produce superoxide at the expense of intracellular redox equivalents (notably NADPH is depleted), but depending on the system, NADH, glutathione and ascorbic acid may also be the reducing equivalents. Important reductase systems involved in redox cycling are, for example, flavoprotein reductases including microsomal cytochrome P450 reductase (Livingstone et al., 1990), but even cytochrome P450 (and cytochrome b_5) itself may be a source for superoxide and hydrogen peroxide (Bainy et al., 1996; Ortiz de Montellano, 1986; Premereur et al., 1986). The cytochrome P450-mediated reactive oxygen species generation may be considered an undesirable "side-reaction" of the enzymes, but it frequently occurs with other oxidases as well. Redox cycles are frequently used to introduce reactive oxygen species under experimental conditions using model compounds like paraquat (methyl viologen; gramoxone; 1,1'-dimethyl-4,4'dipyridylium dichloride) which produces superoxide radical anions ($O_2^{\cdot-}$).

Polycyclic aromatic hydrocarbons are priority pollutants in aquatic ecosystems. They are readily taken up from the water column, sediments and food sources into the liver and other tissues of fish (Walker and Livingstone, 1992). During metabolism, quinones may be formed which have the potential to produce reactive oxygen species by redox cycling (Lemaire and Livingstone, 1997; Lemaire et al., 1994; Washburn and DiGiulio, 1989). Reducing equivalents for this cycles are readily available from the flavoprotein NAD(P)H-dependent cytochrome P450 reductase, NADH being the preferred substrate (Lemaire and Livingstone, 1994, 1997).

Hydrogen peroxide is the first detoxification product of $O_2^{\cdot-}$ by superoxide dismutase. However, it serves as precursor of the highly toxic hydroxyl radical, $OH^\cdot$, through the Haber-Weiss reaction (1).

$$O_2^{\cdot-} + H_2O_2 \rightarrow O_2 + OH^\cdot + OH^- \qquad (1)$$

This reaction proceeds very slowly unless catalytically active metal ions, particularly iron, are present. Usually iron is required for a significant production of hydroxyl radicals *in vivo* through a Fenton-type reaction (2, 3).

$$Fe(III) + O_2^{\cdot-} \rightarrow Fe(II) + O_2 \qquad (2)$$

$$Fe(II) + H_2O_2 \rightarrow Fe(III) + OH^\cdot + OH^- \qquad (3)$$

The net result of the redox cycling of the iron catalyst (2, 3) is identical to (1). If Fe(II) is present, $OH^\cdot$ may be formed directly by (3). Redox cycling of transition metals is not limited to the production of $OH^\cdot$, but may also be an independent source for reactive oxygen species involving similar reactions as mentioned above for organic compounds. Thus reactive oxygen species are an important facet of metal toxicity.

Mitochondria contribute significantly to cellular O_2^{-} production. Ubisemiquinone serves as the primary direct electron donor (80% of mitochondrial O_2^{-} production), the remaining 20% being contributed by the NADH dehydrogenase flavoprotein. The production of H_2O_2 by mitochondria results from the dismutation of O_2^{-} catalyzed by mitochondrial superoxide dismutase (Boveris and Cadenas, 1982). Reactive oxygen species production is highest in state 4 respiration, where the absence of the phosphate acceptor ADP sets a slow respiratory rate with the components of the respiratory chain being highly reduced. Toxicants interfering with membrane integrity increase reactive oxygen species production and uncoupling of the mitochondria (e.g., with 2,4-dinitrophenol) may increase H_2O_2 production up to 50-fold (Boveris and Cadenas, 1982).

Damage by reactive oxygen species

Even without pollution and xenobiotic metabolism there is a constant production of reactive oxygen species in all living cells. Due to the high reactivity of reactive oxygen species most components of cellular structure and function are likely to be targets of oxidative damage (Kappus, 1987; Fig. 1).

Damage to the DNA

One of the most important reactions is damage to nucleic acids. The characterization of these alterations has been complicated by the variety of direct and indirect effects observed and by the diversity of reactive oxygen species generating systems that have been used (Imlay and Linn, 1988). Pollution-mediated damage includes oxidation and modification of the bases, with both pyrimidines and purines being affected. Damage to the

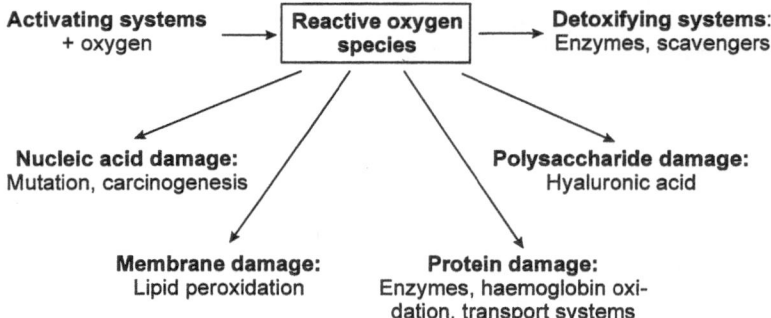

Figure 1. Generalized scheme for oxidative injury to macromolecules (after Jones, 1985).

deoxyribose part usually results in strand breaks. Additionally, the DNA is modified by adduct formation with the xenobiotic, its metabolites or with products of lipid peroxidation. Guanosine appears to be especially sensitive against oxidation leading to the formation of 8-hydroxydeoxyguanosine. This may result in misreading during replication at the residue itself and at adjacent positions (Kuchino et al., 1987) causing G:T and A:C substitutions (Cheng et al., 1992). Pollution-induced mutagenesis may also occur by adduct formation with products of xenobiotic metabolism or aldehydic compounds resulting from lipid peroxidation. Among the products, malondialdehyde is a less effective toxicant than the mono- and dihydroperoxides and epoxides of unsaturated fatty acids (Fujimoto et al., 1984). The possibility that reactive oxygen species are mutagenic has inspired much medical research, and reactive oxygen species-induced mutations are considered to induce and promote carcinogenesis. However, the relationship between oxidative stress and carcinogenesis has not been documented in fish. Nevertheless, pollution-related damage to the DNA and an increased incidence for cancer are known (Sindermann, 1979). Pollution causes oxidative damage to the DNA (DiGiulio et al., 1993; Nishimoto et al., 1991), and the frequency of strand breaks increases in bluegill sunfish (*Lepomis macrochirus*) and fathead minnow (*Pimephales promelas*), when the fish are exposed to 1 µg/L benzo[a]pyrene in the water (Shugart, 1988).

Lipid peroxidation and malondialdehyde

The most typical reaction during reactive oxygen species-induced damage involves the peroxidation of unsaturated fatty acids (Kappus, 1987). The reaction sequence starts with a radical (e.g., OH·) which removes one proton from the hydrocarbon tail of the fatty acid leaving the radical of the acid. This radical undergoes isomerization and oxidation with molecular oxygen yielding a peroxy radical of the fatty acid. Peroxy radicals remove protons from other molecules and become hydroperoxides. Since this proton may originate from another fatty acid, a new cycle is started and lipid peroxidation proceeds *via* a chain reaction, until the chain is interrupted by either the dimerisation of two radicals or until the proton is removed from a substance which forms relatively stable radicals with low reactivity (radical scavengers). Through this chain reaction, one initiating radical may lead to the peroxidation of hundreds of fatty acids. The resulting hydroperoxides are unstable and decompose by chain cleavage to a very complex mixture of aldehydes, ketones, alkanes, carboxylic acids and polymerisation products (Esterbauer et al., 1982). Hydroperoxides and decomposition products are toxic and may form fluorescent adducts with DNA (Fujimoto et al., 1984).

The only mechanism which produces malondialdehyde in biological systems is lipid peroxidation. Malondialdehyde is not the major product of lipid peroxidation, but a typical degradation product. This and the high sensitivity of the thiobarbituric acid test have greatly inspired reactive oxygen species research. Many variations of this test have been published, but in all modifications samples are heated in an acid environment together with thiobarbituric acid. During this procedure, a pink to red colour is formed. However, part or probably all of the malondialdehyde is formed during the acid heating step and variations of the procedure differ in the extent of this additionally formed malondialdehyde. Because the pink reaction product is not necessarily based on malondialdehyde in the samples, the term "thiobarbituric acid-reactive substances" has been introduced, lipid peroxides being the major source for thiobarbituric acid-reactive substances. Thus the amount of thiobarbituric acid-reactive substances in a sample may be used as a measure for reactive oxygen species-induced lipid peroxidation, although direct and more specific assays for malondialdehyde give consistently lower readings than the thiobarbituric acid assay (Lopez-Torres et al., 1993).

Increased amounts of thiobarbituric acid-reactive substances have been found in fish under experimental conditions using paraquat to induce reactive oxygen species in silver carp (*Hypophthalmichthys molitrix*; Matkovics et al., 1984), but carp (*Cyprinus carpio*) erythrocytes had unchanged levels (Matkovics et al., 1987). Thiobarbituric acid-reactive substances in different tissues of carp correlated with ambient oxygen concentration (Radi et al., 1988) and increased when the animals were exposed to copper sulfate; only muscle thiobarbituric acid-reactive substances were elevated in the presence of zinc sulfate (Radi and Matkovics, 1988). Experimental exposure of channel catfish to different concentrations of bleached kraft mill effluents did not increase lipid peroxidation in liver (Mather-Mihaich and DiGiulio, 1991). Likewise, the organophosphorous herbicide S,S,S-tri-*n*-butyl phosphorotrithioate (DEF, a defoliant) and its metabolite n-butyl mercaptan did not increase lipid peroxidation in liver and gills of channel catfish (Mather-Mihaich and DiGiulio, 1986). Thiobarbituric acid-reactive substances were unchanged in *Mugil* spec. from a polluted site together with low lipohydroxide levels compared to specimens from an unpolluted site (Rodriguez-Ariza et al., 1993). In contrast to these findings, sole from a polluted harbour showed high lipid peroxidation (DiGiulio et al., 1993). Elasmobranchs have higher thiobarbituric acid-reactive substances than marine teleosts and much higher values than freshwater teleosts (Filho, 1996). Lipid peroxidation tends to be lower in herbivorous fish than in omnivorous species, correlating with lower glutathione peroxidase and catalase activities, although the herbivorous species have higher superoxide dismutase activity (Radi et al., 1985). When compared with other vertebrates, fish (*Salmo trutta*) have similar lipid peroxidation levels (Lopez-Torres et al., 1993).

Protein damage and methemoglobin formation

Erythrocytes are prone to oxidative damage due to their high content of polyunsaturated fatty acids in the membrane in combination with an active metalloprotein, hemoglobin, which can function as an oxidase and a peroxidase. High oxygen tension in many areas of the circulation favours reactive oxygen species formation and membrane proteins and other functional proteins are cross-linked (Bainy et al., 1996; Falcioni et al., 1987; Stern, 1985). The oxidative denaturation of the erythrocyte membrane is considered a major cause of hemolytic processes involving both lipid peroxidation of the membrane and degradation of the cytoskeleton (Caprari et al., 1995; Falcioni et al., 1987). Ozone is a possible agent which damages the erythrocyte membrane (Fukunaga et al., 1992). In the cytoplasm hemoglobin is oxidized to methemoglobin which can be further oxidised to hemichromes leading to the formation of Heinz bodies. Red cells contain methemoglobin reductase, which regenerates hemoglobin from methemoglobin, if oxidant stress is not too high.

Methemoglobin concentrations in fish blood vary with ambient temperature, with higher concentrations observed at elevated temperatures, independent of the age of the fish (Härdig and Höglund, 1983). In general, erythrocytes are well protected against oxidative damage, which is observed only if all intracellular thiols like glutathione are depleted (Bainy et al., 1996; Stern, 1985). Reactive oxygen species are not the only cause for methemoglobin formation, but elevated levels may also be caused by xenobiotics such as nitrite, phenols, chlorine and hydrazine, which oxidise hemoglobin more specifically (Härdig and Höglund, 1983). Perch and four-horn sculpin (*Myxocephalus quadricornis*) exposed to bleached kraft mill effluents showed increased methemoglobin levels (Andersson et al., 1988; Härdig et al., 1988). It is concluded that bleached kraft mill effluents contain substances which increase the oxidation of hemoglobin to methemoglobin and/or inhibit the reducing system responsible for the reverse reaction. The organophosphorous herbicide DEF is metabolized in the stomach, n-butylmercaptan being the major product. The parent compound did not increase methemoglobin levels in channel catfish, but its metabolite n-butylmercaptan, which is known for its hemolytic toxicity, induced higher levels (Mather-Mihaich and DiGiulio, 1986). However, all other indicators for oxidative stress (malondialdehyde, glutathione, superoxide dismutase, glutathione peroxidase) remained unchanged.

Enzymatic defense systems

The key enzymes for the detoxification of reactive oxygen species in all organisms are superoxide dismutase (EC 1.15.1.1), glutathione peroxidase (EC 1.11.1.9), peroxidase (EC 1.11.1.7), and catalase (EC 1.11.1.6), all of them being abundant in fish tissues (DiGiulio et al., 1989). This battery of antioxidant enzymes is supplemented by ancillary systems providing reducing equivalents needed for detoxifying activity (e.g., glucose 6-phosphate dehydrogenase, EC 1.1.1.49; glutathione reductase, EC 1.6.4.2). Since most of these enzyme activities are induced by reactive oxygen species, they may be useful indicators of oxidative stress. However, their suitability as biomarkers still requires interspecies comparisons (Petrivalsky et al., 1997).

Catalase

Catalase detoxifies hydrogen peroxide to O_2 and water (4).

$$2 H_2O_2 \rightarrow 2 H_2O + O_2 \tag{4}$$

This enzyme primarily occurs in peroxisomes. Its activity increases together with other peroxisomal enzymes in channel catfish (*Ictalurus punctatus*) liver upon exposure of the animals to bleached kraft mill effluents for 14 days, suggesting that a general peroxisomal proliferation is induced (Mather-Mihaich and DiGiulio, 1991). Since other indicators of peroxisomal function (lauroyl-CoA oxidase and palmitoyl-CoA oxidase) increase 3- to 7-fold, while catalase activity increases only about 2-fold, it has been suggested that increased peroxisomal β-oxidation of fatty acids leads to higher concentrations of H_2O_2 in bleached kraft mill effluents-exposed channel catfish than in controls (Mather-Mihaich and DiGiulio, 1991). Increased activities were also observed in *Mugil* sp. from a polluted site together with elevated glucose 6-phosphate dehydrogenase and glutathione-reductase activities (Rodriguez-Ariza et al., 1993).

Peroxisomal proliferation is the major cause for elevated catalase activities in tissues, and reactive oxygen species probably do not induce higher activities *per se*, although peroxisomal proliferation will result in increased reactive oxygen species. Bucher and co-workers (1993) found no increase in catalase activity in bullheads (*Cottus gobio*) caught in a river downstream of a paper mill, indicating that stimulators for peroxisomal proliferation were removed by the waste water treatment. Catalase activity also remained unchanged in carp erythrocytes, when the animals were exposed to paraquat (Matkovics et al., 1987). Probably catalase (as an indicator of peroxisomal proliferation) is always elevated, when other indicators of oxidative stress and xenobiotic metabolism are increased as in rainbow trout injected tetrachlorobiphenyl (Otto and Moon, 1995). Reactive oxygen species do not induce catalase activity (Aksnes and Njaa, 1981).

Some pollutants may inhibit catalase. High concentrations of copper (25–50 mg/L $CuSO_4$ in the water) inhibited catalase in liver, gill and muscle, and 100 mg/L $ZnSO_4$ in gill and muscle, but not in liver of carp after 24 h of exposure (Radi and Matkovics, 1988). Even nitrite may inhibit catalase (Arrillo and Melodia, 1991). Ozone exposed Japanese char (*Salvelinus leucomaenis*) were found to have lower catalase and glutathione peroxidase activities in the erythrocytes (Fukunaga et al., 1992). Enzyme activity also correlates with feeding habits and tends to be lower in herbivorous fish than in omnivorous fish (Radi et al., 1985).

Superoxide Dismutase

Superoxide dismutase catalyzes the dismutation of the superoxide anion radical to hydrogen peroxide plus water (5).

$$2 \, O_2^{\cdot-} + 2 \, H^+ \rightarrow O_2 + H_2O_2 \tag{5}$$

The copper-containing superoxide dismutase protein was first isolated by McCord and Fridovich (1969) from bovine blood. In most animal tissues, two forms of superoxide dismutase occur, a Cu/Zn-containing superoxide dismutase in the cytoplasm and a Mn-containing enzyme in the mitochondria. Mitochondrial superoxide dismutase contributes up to 60% to total tissue activity (Radi et al., 1985), but usually much less. Enzyme concentrations tend to be very high because of the high reactivity of superoxide radicals.

In fish liver, at least five different isoforms of Cu/Zn-superoxide dismutase have been found (*Mugil* sp.) which change in frequency in fish from polluted areas (Pedrajas et al., 1993). The different isoforms may be partly due to oxidative modification of the proteins. Superoxide dismutase activity rises very fast in carp erythrocytes, when the fish are exposed to 10 mg/L paraquat in the water reaching maximum induction after 12 h (Matkovics et al., 1987), followed by a decrease below control levels after 48−96 h. Likewise, activities in the gills are induced within 8 h by 10 mg/L paraquat (Vig and Nemcsok, 1989). The lower activities after prolonged exposure may be explained by the inhibition of superoxide dismutase by its product H_2O_2 or by $OH^{\cdot}$ (Bray et al., 1974). The oxidation of superoxide dismutase may be of biological and ecological significance, since the resulting superoxide dismutase isoforms with lower pI resemble those found in fish from polluted areas (Pedrajas et al., 1993). Oxidation of superoxide dismutase may also explain the observed lower activities of this enzyme after exposure of carp to copper and zinc sulfate, because these metal ions generate reactive oxygen species as visualized by the increased lipid peroxidation in these experiments (Radi and Matkovics, 1988). Similarly in silver carp, superoxide dismutase can be induced by low concentrations of paraquat (0.1 mg/L in the water), but increasing concentrations reverse this effect, whereas lipid peroxidation remains high (Matkovics et al.,1984). Hypoxic exposure of carp induced superoxide dismutase in gills, liver, and brain, but a combination of 10 mg/L paraquat together and hypoxia showed additive effects in the gills only. Injection of tetrachlorobiphenyl, a well-known inducer of cytochrome P450-dependent enzyme activities, stimulates superoxide dismutase activity in rainbow trout (Otto and Moon, 1995).

Fish (*Mugil* sp.) and molluscs from an iron/copper-polluted estuary also containing organic xenobiotics like industrial pollutants and pesticides showed increased superoxide dismutase activities in the tissues (Rodriguez-Ariza et al., 1991, 1992, 1993). The same is true for spot (*Leiostomus xanthurus*) from a polluted site, when compared with fish from a clean reference site (Roberts et al., 1987) and Nile tilapia (*Oreochromis niloticus*; Bainy et al., 1996). However, bullheads caught in a river 300 m below the discharge point of a paper mill had lower superoxide dismutase activities in the liver than those caught 4 km upstream (Bucher et al., 1993), and activities were unchanged in channel catfish liver, when fish were exposed to bleached kraft mill effluents (Mather-Mihaich and DiGiulio, 1991). Nevertheless, superoxide dismutase tends to be higher in animals from polluted areas, suggesting that its activity can be used as a measure for the severity of environmental impact. It is induced very quickly within a few hours especially if production of reactive oxygen species is not too high. Superoxide dismutase shows great differences in activity between tissues and species (Aceto et al., 1994) but tends to be higher in herbivore fish (Radi et al., 1985) and to be lower in freshwater fish compared to marine species (Filho, 1996).

Glutathione peroxidase and peroxidase

Glutathione peroxidase is the most important peroxidase for the detoxification of hydroperoxides. It catalyzes the glutathione-dependent reduction of hydroperoxides (6) and of hydrogen peroxide (7).

$$2\ GSH + ROOH \rightarrow GSSG + ROH + H_2O \qquad (6)$$

$$2\ GSH + H_2O_2 \rightarrow GSSG + 2\ H_2O \qquad (7)$$

with GSH – glutathione, GSSG – glutathione disulfide.

It is characterized by its contents of functional selenium and by its ability to reduce hydrogen peroxide and a large number of organic hydroperoxides. The Se contents make this enzyme strongly dependent on nutritional uptake of this trace element, its activity varying as a logarithmic function of Se in the diet (Tappel et al., 1982). Rainbow trout fed a diet depleted in Se have reduced glutathione peroxidase activities (Bell et al., 1986). However, signs of Se-deficiency in fish may remain absent, as long as the diet contains adequate levels of vitamin E (Hilton, 1989). The Se-containing glutathione peroxidase may be inhibited by $O_2{}^-$, but full activity is regained after reduction with glutathione (Blum and Fridovich, 1985), while the inhibition of glutathione peroxidase in erythrocytes during oxidative stress is probably due to hemichromes resulting from the oxidation of hemoglobin (Grelloni et al., 1991). Also Se-independent glutathione peroxidase has also been described, but this activity is probably a side reactivity of glutathione-S-transferase (Tappel et al., 1982). By the reduction of peroxides, glutathione peroxidase provides protection against oxidative damage and accumulation of free radical products. The system is located in the cytoplasm as well as in the mitochondrial matrix (Boveris and Cadenas, 1982) and shares the ability to detoxify hydrogen peroxide with catalase. However, catalase is largely restricted to peroxisomes, most of the cytoplasmic and mitochondrial peroxides thus being detoxified by glutathione peroxidase, although tissue glutathione peroxidase activity is one thousandth of that of catalase (Aksnes and Njaa, 1981).

Glutathione peroxidase and other glutathione metabolizing enzyme activities are strongly dependent on tissue, species and developmental stage (Aceto et al., 1994; Hasspieler et al., 1994). Rainbow trout and brook trout (*Salvelinus fontinalis*) have lower glutathione peroxidase activities than bluegill sunfish (*Lepomis macrochirus*) and much lower activities than carp (Tappel et al., 1982). Among cyprinids highest activity is found in carp, with lower values for grass carp (*Ctenopharyngodon idella*), silver carp (*Hypophthalmichthys molitrix*), barbel (*Barbus barbus*) and crucian carp (*Carassius carassius*; Radi et al., 1985). It has been suggested that glutathione peroxidase activities reflect feeding behaviour of the fish, being highest in the omnivorous carp. However, Tappel et al. (1982) did not find such a relationship.

Glutathione peroxidase activity may be induced by environmental pollutants. Activity increases together with glutathione reductase in rainbow trout after injection of tetrachlorobiphenyl (Otto and Moon, 1995) and in carp after exposure to copper, but not to zinc sulfate (Radi et al., 1988). However, in carp erythrocytes, glutathione peroxidase and peroxidase were unaffected by exposure of the animals to paraquat (Matkovics et al., 1987).

Increased activities were found in *Mugil* sp. from a polluted site together with elevated glucose 6-phosphate dehydrogenase and glutathione reductase activities (Rodriguez-Ariza et al., 1993). On the other hand, activities remained unchanged in the liver of channel catfish exposed to bleached kraft mill effluents (Mather-Mihaich and DiGiulio, 1991) and were depressed in the liver of annular sea bream (*Diplodus annularis*) from a polluted harbor (Bagnasco et al., 1991) as well as in Nile tilapia from a polluted lake (Bainy et al., 1996), although other detoxifying systems were found to have increased. Protection by glutathione peroxidase against reactive oxygen species may be questionable (Hasspieler et al., 1994), but fish being more susceptible to oxidative damage have generally higher glutathione peroxidase activities (Aksnes and Njaa, 1981; Hasspieler et al., 1994).

Low molecular weight scavengers

A clue to the understanding of reactive oxygen species detoxification is that the high chemical energy of these compounds needs to be converted into heat. This is not necessarily based on single reactions, but energy may drive complex pathways, overall detoxification being a step-by-step process. At this level, low molecular weight antioxidants and radical scavengers contribute to tissue protection against reactive oxygen species damage. Typical antioxidants are compounds which easily take up or release single electrons. In this context, even molecular oxygen may function as a scavenger for very reactive radicals, the resulting superoxide ($O_2^{\cdot-}$) being efficiently detoxified by superoxide dismutase, followed by catalase and peroxidases (Imlay and Linn, 1988). Additional electrons of organic radicals may be dislocated between several chemical bonds. Highly conjugated systems of double bonds (e.g., in phenolic compounds) sometimes containing hydroxy groups capable of keto-enol tautomerism (e.g., ascorbic and uric acid) are typical scavengers. If these compounds take up one additional electron (e.g., from $OH^{\cdot}$ or $O_2^{\cdot-}$), an antioxidant radical is formed with lower reactivity, lower chemical energy and lower toxicity than the parent radical. The lower reactivity does not imply that these radicals are harmless. The ascorbate radical, for example, is known to promote lipid peroxidation (Bachowski et al., 1988). Antioxidants may also store vibrational energy within extensively conjugated double bonds (e.g., the reaction of β-carotene with $^1O_2$), thereby reducing toxicity of excited states.

The transfer of electrons from one radical to an antioxidant and from an antioxidant radical to another antioxidant possibly involves several steps, and complex cascades are likely. An example is given in Figure 2. These cascades are the only pathway allowing the scavenging and detoxification of radicals inside a lipid bilayer or lipid droplet. Not all steps are necessary, and reactive oxygen species may enter at any level. However, the terminal importance of glutathione and its regeneration at the expense of NADPH are typical. In all other steps, different antioxidants may contribute and increase the detoxification capacity.

As a response to environmental pollution, all low molecular weight antioxidants appear to respond similarly by enhanced concentrations in the liver (Fig. 3). The increased environmental impact is well documented by the increase of mixed function oxygenase (aryl hydrocarbon hydroxylase) activity. Thus, protection against reactive oxygen species seems to require all components of the detoxification systems.

Ascorbic acid (vitamin C)

Although the ability to synthesize ascorbic acid has been documented for some fish species, most of them depend on nutritional uptake of this vitamin (Thomas et al., 1985). It is not only a dietary requirement for fish ensuring optimal growth rates as well as collagen and hormone synthesis (Halver et al., 1969), but it is also a powerful antioxidant in aqueous media and the major protective agent against oxidative damage in blood plasma

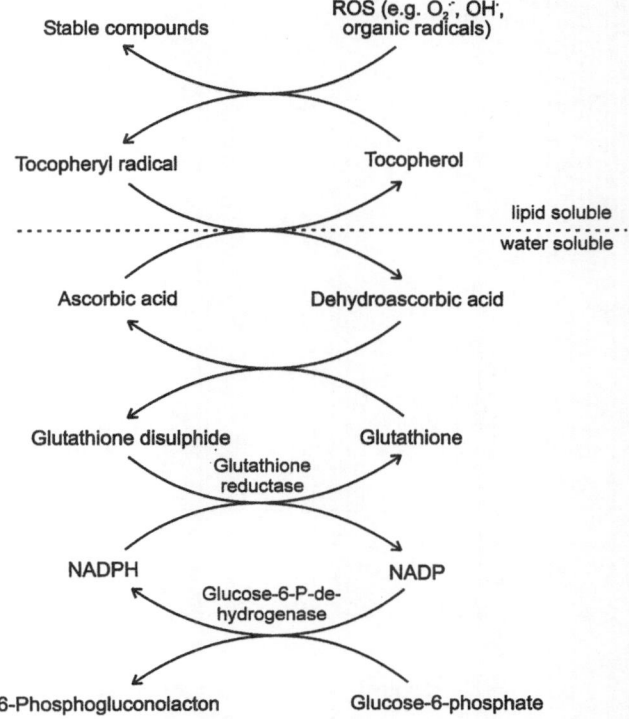

Figure 2. Schematic drawing of the detoxification of reactive oxygen species (ROS) and radicals by scavengers. Detoxification proceeds *via* a cascade of reactions and lipophilic antioxidants exchange electrons with water soluble scavengers. Moreover, other antioxidants not included in this scheme may exchange electrons with the listed compounds and replace them in some cases. Thus reactive oxygen species may enter the cascade at any level.

(Frei et al., 1989). Ascorbic acid deficiency may reduce the activity of xenobiotic metabolizing enzymes (Andersson et al., 1988; Zannoni and Sato, 1975). Ascorbic acid ameliorates copper and cadmium toxicity and is required for trace element homeostasis (Hilton, 1989; Thomas et al., 1982). Apart from dietary intake (Dabrowski et al., 1990), tissue ascorbic acid levels are influenced by developmental stage (Dabrowski, 1992) and display distinct seasonal patterns (Thomas et al., 1985). It is released into the digestive tract, but reabsorbed almost quantitatively, thus conserving the animals' ascorbic acid pools (Dabrowski, 1990). Ascorbic acid half-life in the tissues depends on dietary intake and tissue concentration and may be as long as 30 days in rainbow trout (Tucker and Halver, 1986).

During the acute response to different stressors, ascorbic acid is depleted (Tucker and Halver, 1986; Wedemeyer, 1969). The increased incidence of disease and skeletal deformations observed in fish exposed to bleached kraft mill effluents have been attributed to depletion caused by the increas-

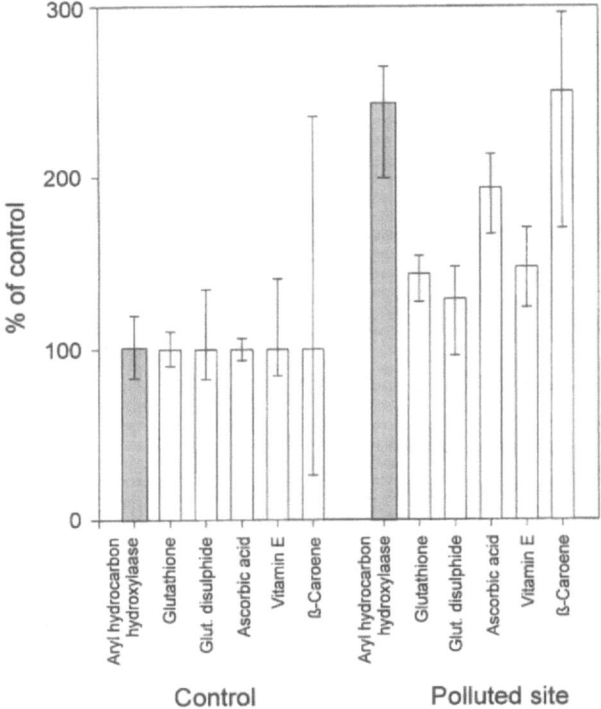

Figure 3. Antioxidant concentrations in the liver of graylings (*Thyllamus thyllamus*) from a polluted site downstream a bleached kraft mill effluent location compared with fish from an upstream control site. Each value represents median ±25%, 75% quartiles, n = 15.

ed requirement for detoxification processes (Lehtinen et al., 1990). Lower concentrations were found in bullhead liver from a polluted site (Bucher et al., 1993) and in *Mugil cephalus* liver and gill, but not in kidney, if exposed to cadmium for up to 6 weeks (Thomas et al., 1982) or exposed to oil (Thomas, 1987).

Adaptative responses are also known, when ascorbic acid concentrations exceed control values (Fig. 3). Perch (*Perca fluviatilis*) living in a bleached kraft mill effluents-polluted area of the Baltic sea showed accumulation of ascorbic acid in the liver in a dose-dependent manner (Andersson et al., 1988).

Glutathione

Glutathione is the most abundant thiol in most tissues. Its function is two-fold: it is (1) an antioxidant, which scavenges e.g., O_2^-, and (2) a cofactor for enzymatic reactions like the conjugation of xenobiotics by glutathione-

S-transferase. When glutathione scavenges a radical, a thiol radical is formed, which undergoes dimerisation to oxidized glutathione disulfide, or mixed disulfides with proteins are formed. Both glutathione disulfide and mixed disulfides with proteins are indicators of increased reactive oxygen species, and both are reduced by glutathione reductase, which restores original glutathione concentrations at the expense of cellular NADPH. Glutathione is necessary for glutathione peroxidase, which detoxifies organic and inorganic peroxides, and for glutathione-S-transferase, which are involved in xenobiotic conjugation and excretion. The latter pathway depletes the cellular glutathione pool, because glutathione-conjugates of xenobiotics are usually degraded to mercapturic acids and excreted *via* the bile. However, all animals are able to synthesize glutathione in sufficient amounts depending on nutritional intake of sulfur containing amino acids and selenium. Under normal circumstances, glutathione turnover is low and glutathione levels remain unchanged in channel catfish liver for 3 days if synthesis is inhibited (Gallagher et al., 1992). The importance of glutathione for the protection of tissues is documented by the fact that toxicant induced damages occur only if intracellular glutathione is depleted. This is true for oxidative damage to erythrocytes (Stern, 1985) and for the hepatotoxicity of allyl formate to trout liver (Droy et al., 1989). When glutathione is oxidized, glutathione disulfide is formed, which is released into blood plasma. Most of this oxidation is due to glutathione peroxidase, since in glutathione peroxidase-depleted rainbow trout liver no peroxide-stimulated glutathione disulfide release has been observed (Bell et al., 1986).

In fish exposed to pollutants, both types of response may occur: first, a decrease in glutathione is an acute reaction; secondly, an increase in glutathione concentration is an adaptation to increased detoxification activity. A decrease in glutathione was observed in climbing perch (*Anabas testudineus*) exposed to industrial pollutants after short- and long-term exposure (Chatterjee and Bhattacharya, 1984), in bullhead liver from a polluted site (Bucher et al., 1993) and in channel catfish exposed to bleached kraft mill effluents (Mather-Mihaich and DiGiulio, 1991). However, in the latter study, concentrations recovered after 7 to 14 days of exposure.

Adaptive responses were observed in rainbow trout after injection of tetrachlorobiphenyl inducing higher glutathione and glutathione disulfide concentrations in liver and kidney, while ratios between glutathione disulfide and glutathione remained unchanged (Otto and Moon, 1995). A similar response was observed in graylings (*Thymallus thymallus*) from a polluted site compared to a clean reference site (Fig. 3). The glutathione disulfide/glutathione ratio is a measure of the intracellular redox state, higher values indicating oxidative stress. An oxidized glutathione redox status was found in *Mugil* sp. from a polluted area (Rodriguez-Ariza et al., 1993). Fish with an increased glutathione disulfide level appear prone to pollution-mediated damages and carcinogenesis (Hasspieler et al., 1994).

Lipid-soluble antioxidants

The concentration of lipid-soluble antioxidants is much more dependent on nutritional intake than those of the water soluble. The most important are tocopherol (vitamin E) and β-carotene (provitamin A). Actually, eight different compounds have vitamin E activity, four tocopherols and four tocotrienols, but α-tocopherol is by far the most abundant in most animal tissues. The major function of vitamin E is its ability to function as an antioxidant in lipid membranes, lipoproteins and lipid droplets. Its ability to protect polyunsaturated fatty acids (PUFA) from peroxidation explains the increased requirement of fish for this vitamin when fed a diet enriched with fish oil (Hilton, 1989). Vitamin E is absorbed in the intestine together with lipids and transported to the tissues in lipoproteins. In the liver, a specific protein retains α-tocopherol (Wolf, 1994), while other tocopherols and tocotrienols are excreted *via* the bile. Due to the unspecific uptake, tissue concentrations correlate very much with nutritional intake (Frigg et al., 1990). Oxidized vitamin E is regenerated *in vivo* by ascorbic acid (Fig. 2; Hilton, 1989).

β-Carotene is one of about 650 different compounds differing in their antioxidant activity, some of which have almost no antioxidant function like astaxanthin, canthaxanthin and zeaxanthin, but some appear superior, like lycopene. As with vitamin E, absorption in the digestive tract occurs together with lipids, but there does not appear to exist a specific retention mechanism. Nevertheless, specific uptake mechanisms have been suggested (Guillou et al., 1992; Schiedt et al., 1985). Both vitamin E and carotenoids detoxify peroxy-, fatty-acid- and superoxideradicals. Especially carotenoids are necessary to detoxify excited states of molecules ($^1O_2$, ^3RO) and are usually more effective in this reaction than vitamin E.

The question arises as to by which mechanism fish from polluted sites gain higher tissue concentrations of lipid soluble antioxidants. Algae increase carotenoid synthesis during oxidative stress (Kobayashi et al., 1993), and crustaceans (copepods exposed to UV-radiation) increase carotenoid pigmentation (Byron, 1982). Thus, adaptation could occur at lower levels of the food chain, and fish might profit from the antioxidant defense of their prey.

Concluding remarks

Fish appear to possess the same biochemical pathways to deal with the toxic effects of endogenous and exogenous agents as do mammalian species. Similarly to mammals, fish tissues consistently produce reactive oxygen species, which may be useful agents for certain biochemical pathways or may damage cells. One of the most important sources for native reactive oxygen species production is the mitochondrial respiratory chain, especially at the ubiquinone or NADH site, where about 1% of the oxygen is reduced to O_2^- and H_2O_2 (Barja, 1993). It is often assumed that this represents some kind of "inefficiency" of conversion by the respiratory chain. However, this is not a strong argument, because enzymes are known that approach 100% conversion efficiency. The best example is mitochondrial cytochrome oxidase, for which a sequential reduction of oxygen to water has been suggested with the concomitant production of reactive oxygen species. However, no reactive oxygen species are liberated at this site. Thus, the consistent production of reactive oxygen species in the cells may have some beneficial functions. Furthermore, it has to be assumed that the production of reactive oxygen species is regulated so as to proceed at the required rate for ensuring the "beneficial" effects of these radicals (Barja, 1993).

Environmental pollution may increase the cellular reactive oxygen species load, because many xenobiotics exert part of their toxicity *via* the formation of reactive oxygen species. Especially redox cycling gains significance, if compared to mitochondrial radical formation (Kappus, 1987; Lehtinen et al., 1990; Lemaire and Livingstone, 1994, 1997; Lemaire et al., 1994; Livingstone et al., 1990). Pollution-induced oxidative stress has clearly been shown for bleached kraft mill effluents and other industrial pollutants. Depending on the pulping process, discharge water may be contaminated with resin acids, fatty acids, and chlorinated phenols, guaiacols, and catechols (Oikari et al., 1985). Bleached kraft mill effluents contain a complex mixture of chemicals and not all of the components have been identified. Mixed function oxygenase (i.e., the cytochrome P450 system) is induced in a dose-dependent manner, and with time peroxisomes proliferate in the liver (Mather-Mihaich and DiGiulio, 1991). Synergistic, additive or antagonistic effects of the chemicals present in the effluent are possible. Bleached kraft mill effluents contain compounds which could potentially undergo redox cycling and form reactive oxygen species. Even if the parent compounds do not promote redox cycling, their metabolites (often products of mixed function oxygenase oxidation) may have this ability.

The additional load of reactive oxygen species requires that the animals adapt to the specific situation and increase detoxification capacity including both detoxifying enzymes and low molecular scavengers. Adaptation is not restricted to reactive oxygen species, but all systems dealing with xenobiotics are induced. The cytochrome P450-dependent enzymes (phase I metabolism) increase in activity. Other xenobiotic metabolizing enzymes and those which further process metabolites of phase I (phase II metabolism) such as glutathione-S-transferase and transport systems are also induced. In general, fish have lower activities of reactive oxygen species detoxifying enzymes (Perez-Campo et al., 1993; Tappel et al., 1982) and lower concentrations of ascorbic acid, glutathione, and uric acid than other vertebrates, but lipid peroxidation is similar (Lopez-Torres et al., 1993). This may be explained by the higher oxidative capacity of other vertebrates (Lopez-Torres et al., 1993). Especially warm blooded animals have a higher metabolic flux and presumably a higher mitochondrial reactive oxygen species generation requiring higher detoxification capacities.

The same is true for more active fish species, which possess an enhanced antioxidant protection, if compared to sluggish species (Filho, 1996). However, species differences exist, and different fish species may differ in their sensitivity to environmental pollution. An example is given by Hasspieler et al. (1994), who compared channel catfish with brown bullhead. In the latter species, neoplasia is common in contaminated systems, whereas channel catfish seems to be protected by substantially higher hepatic glutathione concentrations together with higher enzyme activities for glutathione synthesis, glutathione reductase and glutathione-S-transferase. In brown bullhead, higher activities of glutathione peroxidase have been found than in channel catfish. However, this protection appeared less effective than in channel catfish, because the level of glutathione disulfide remains high.

It might be desirable to measure pollution-mediated reactive oxygen species production directly in the tissues of exposed animals. The sophisticated equipment needed for this purpose (electron spin resonance spectroscopy or bioluminescence measurement) is usually not available, when fish are sampled in the field. Moreover, reactive oxygen species are very short-lived compounds (10^{-12} to 10 s). Thus, scientists usually resort in measuring specific damage caused by reactive oxygen species (malondialdehyde, thiobarbituric acid-reactive substances, modified fatty acids, methemoglobin, mixed protein-glutathione disulfides, accumulation of fluorescent material). It appears that these measures are valuable laboratory methods applicable to isolated systems only. In field studies, they frequently fail to correlate with the degree of pollution.

The evaluation of adaptive responses to pollution, i.e., the increase of detoxification capacity, proved to be a much more valuable indicator than the measurement of individual xenobiotics or metabolites. Cytochrome P450-related enzyme activities and their induction are well accepted indicators for pollution by organic compounds. Metallothionein concentration increases upon exposure to metals like cadmium, and reactive oxygen species induce increased antioxidant capacity. Thus, the complexity of environmental pollution is reflected in the activity of different detoxifying systems. It is reasonable to assume that fish living in a polluted environment for prolonged periods are adapted to the specific situation, and it has been suggested that antioxidant enzymes together with P450-dependent enzyme activities and other xenobiotic metabolizing enzymes such as glutathione-S-transferase are suitable for biomonitoring (Rodriguez-Ariza et al., 1993) as well as the measurement of low molecular antioxidants probably depict more accurately the degree of reactive oxygen species-formation than enzyme activities alone.

Acknowledgements
The author wishes to express his thanks to Profs. Dr. W. Wieser and Dr. B. Pelster for critically reading the manuscript and for helpful discussions. This work was supported by the Austrian Science Foundation, project P10594-BIO.

References

Aceto, A., Amicarelli, F., Saccheta, P., Dragani, B., Bucciarelli, T., Masciocco, L., Miranda, M. and Di Ilio, C. (1994) Developmental aspects of detoxifying enzymes in fish (*Salmo irideus*). *Free Radic. Res.* 21:285–294.

Aksnes, A. and Njaa, L.R. (1981) Catalase, glutathione peroxidase and superoxide dismutase in different fish species. *Comp. Biochem. Physiol.* 69B:893–896.

Andersson, T., Förlin, L., Hardig, J. and Larsson A. (1988) Physiological disturbances in fish living in costal water polluted with bleached kraft pulp mill effluents. *Can. J. Fish. Aquat. Sci.* 45:1525–1536.

Arrillo, A. and Melodia, F. (1991) Nitrite oxidation in *Eisenia foetida* (Savigny): Ecological implications. *Funct. Ecol.* 5:629–634.

Bachowski, G.J., Thomas, J.P. and Girotti, A.W. (1988) Ascorbate-enhanced lipid peroxidation in photooxidized cell membranes: Cholesterol product analysis as a probe of reaction mechanism. *Lipids* 23:580–586.

Bainy, A.C.D., Saito, E., Carvalho, P.S.M. and Junqueira, V.B.C. (1996) Oxidative stress in gill, erythrocytes, liver and kidney of Nile tilapia (*Oreochromis niloticus*) from a polluted site. *Aquat. Toxicol.* 34:151–162.

Bagnasco, M., Camoirano, A., De Flora, S., Melodia, F. and Arillo, A. (1991) Enhanced liver metabolism of mutagens and carcinogens in fish living in polluted seawater. *Mutat. Res.* 262:129–137.

Barja, G. (1993) Commentary: Oxygen radicals, a failure or a success of evolution? *Free Rad. Res. Comms.* 18:63–70.

Bell, J.G., Adron, J.W. and Cowey, C.B. (1986) Effect of selenium deficiency on hydroperoxide-stimulated release of glutathione from isolated perfused liver of rainbow trout (*Salmo gairdneri*). *Br. J. Nutr.* 56:421–428.

Blum, J. and Fridovich, I. (1985) Inactivation of glutathione peroxidase by superoxide radical. *Arch. Biochem. Biophys.* 240:500–508.

Boveris, A. and Cadenas, E. (1982) Production of superoxide radicals and hydrogen peroxide in mitochondria. *In:* Oberley, L.W. (Ed) *Superoxide Dismutase*. Vol. 2. CRC Press, Boca Raton, pp 15–30.

Bray, R.C., Cockle, S.A., Fielden, E.M., Roberts, P.B., Rotilio, G. and Calabresel, L. (1974) Reduction and inactivation of superoxide dismutase by hydrogen peroxide. *Biochem. J.* 139:43–48.

Bucher, F., Hofer, R., Krumschnabel, G. and Doblander, C. (1993) Disturbances in the pro-oxidant – antioxidant balance in the liver of bullhead (*Cottus gobio* L.) exposed to treated paper mill effluents. *Chemosphere* 27:1329–1338.

Byron, E.R. (1982) The adaptive significance of calanoid copepod pigmentation: A comparative and experimental analysis. *Ecology* 63:1871–1886.

Caprari, P., Bozzi, A., Malorni, W., Bottini, A., Iosi, F., Santini, M.T. and Salvati, A.M. (1995) Junctional sites of erythrocyte skeletal proteins are specific targets of *tert*-butylhydroperoxide oxidative damage. *Chem.-Biol. Inter.* 94:243–258.

Chatterjee, S. and Bhattacharya, S. (1984) Detoxication of industrial pollutants by the glutathione-S-transferase system in the liver of *Anabas testudineus* (Bloch). *Tox. Lett.* 22:187–198.

Cheng, K.C., Cahill, D.S., Kasai, H., Nishimura, S. and Loch, L. (1992) 8-Hydroxyguanine, an abundant form of oxidative DNA damage, causes G T and A C substitutions. *J. Biol. Chem.* 267:166–172.

Dabrowski, K. (1990) Gastro-intestinal circulation of ascorbic acid. *Comp. Biochem. Physiol.* 95A:481–486.

Dabrowski, K. (1992) Ascorbate concentration in fish ontogeny. *J. Fish Biol.* 40:273–279.

Dabrowski, K., Lackner, R. and Doblander, C. (1990) Effect of dietary ascorbate on the concentration of tissue ascorbic acid, dehydroascorbic acid, ascorbic sulfate, and activity of ascorbic acid sulfate sulfohydrolase in rainbow trout (*Oncorhynchus mykiss*). *Can. J. Fish. Aquat. Sci.* 47:1518–1525.

DiGiulio, R.T., Washburn, P.C., Wenning, R.J., Winston, G.W. and Jewell, C.S. (1989) Biochemical responses in aquatic animals: A review of determinants of oxidative stress. *Environ. Toxicol. Chem.* 8:1103–1123.

DiGiulio, R.T., Habig, C. and Gallagher, E.P. (1993) Effects of Black Rock Harbour sediments on indices of biotransformation, oxidative stress, and DNA integrity in channel catfish. *Aquat. Toxicol.* 26:1–22.

Droy, B.F., Davis, M.E. and Hinton, D.E. (1989) Mechanism of allyl formate-induced hepato-toxicity in rainbow trout. *Toxicol. Appl. Pharmacol.* 98:313–324.

Esterbauer, H., Cheeseman, K.H., Dianzani, M.U., Poli, G. and Slater, T.F. (1982) Separation and characterisation of the aldehydic products of lipid peroxidation stimulated by ADP-Fe^{2+} in rat liver microsomes. *Biochem. J.* 208:129–140.

Falcioni, G., Cincola, G. and Brunori, M. (1987) Glutathione peroxidase and oxidative hemo-lysis in trout red blood cells. *FEBS Lett.* 221:355–358.

Filho, D.W. (1996) Fish antioxidant defences – A comparative approach. *Braz. J. Med. Biol. Res.* 29:1735–1742.

Frei, B., England, L. and Ames, B.N. (1989) Ascorbate is an outstanding antioxidant in human blood plasma. *Proc. Natl. Acad. Sci.* 86:6377–6381.

Frigg, M., Prabuchi, A.L. and Ruhdel, E.U. (1990) Effect of dietary vitamin E levels on oxida-tive stability of trout fillets. *Aquaculture* 84:145–158.

Fujimoto, K., Neff, W.E. and Frakel, E.N. (1984) The reaction of DNA with lipid oxidation pro-ducts, metals and reducing agents. *Biochim. Biophys. Acta* 795:100–107.

Fukunaga, K., Suzuki, T., Hara, A. and Takama, K. (1992) Effect of ozone on the activities of reactive oxygen scavenging enzymes in RBC of ozone exposed Japanese charr (*Salvelinus leucomaenis*). *Free Radic. Res. Com.* 17:327–333.

Gallagher, E.P., Hasspieler, B.M. and DiGiulio, R.T. (1992) Effects of buthionine sulfoximine and diethyl maleate on glutathione turnover in the channel catfish. *Biochem. Pharmacol.* 43:2209–2215.

Grelloni, F., Gabbianelli, R. and Falcioni, G. (1991) Inactivation of glutathione peroxidase fol-lowing hemoglobin oxidation. *Biochem. Int.* 25:789–795.

Guillou, A., Choubert, G. and Noüe, B. (1992) Absorption and blood clearance of labeled caro-tenoids ([^{14}C] astaxanthin, [^{3}H] canthaxanthin and [^{3}H] zeaxanthin) in mature female rainbow trout (*Oncorhynchus mykiss*). *Comp. Biochem. Physiol.* 103A:301–306.

Halver, J.E., Ashley, L.M. and Smith, R.E. (1969) Ascorbic acid requirements of coho salmon and rainbow trout. *Trans. Amer. Fish. Soc.* 98:762–771.

Härdig, J. and Höglund L.B. (1983) Seasonal and ontogenetic effects on methemoglobin and reduced glutathione contents in the blood of reared Baltic salmon. *Comp. Biochem. Physiol.* 75A:27–34.

Härdig, J., Andersson, T., Bengtsson, B.E., Forlin, L. and Larsson, A. (1988) Long-term effects of bleached kraft mill effluents on red and white blood cell status, ion balance, and vertebral structure in fish. *Ecotoxicol Environ. Safety* 15:96–106.

Hasspieler, B.M., Behar, J.V. and DiGiulio R.T. (1994) Glutathione-dependent defense in chan-nel catfish (*Ictalurus punctatus*) and brown bullhead (*Ameiurus nebulosus*). *Ecotoxicol. Environ. Saf.* 28:82–90.

Hilton, J.W. (1989) The interaction of vitamins, minerals and diet composition in the diet of fish. *Aquaculture* 79:223–214.

Imlay, J.A. and Linn, S. (1988) DNA damage and oxygen radical toxicity. *Science* 240:1302–1309.

Jones, D.P. (1985) The role of oxygen concentration in oxidative stress: hypoxic and hyperoxic models. *In:* Sies, H. (Ed) *Oxidative stress*. Academic Press, London, Orlando, pp 151–195.

Kappus, H. (1987) Oxidative stress in chemical toxicity. *Arch. Toxicol.* 60:144–149.

Kobayashi, M., Kakizono, T. and Nagai, S. (1993) Enhanced carotenoid biosynthesis by oxida-tive stress in acetate-induced cyst cells of a green unicellular alga, *Haematococcus pluvialis*. *Appl. Environm. Microbiol.* 59:867–873.

Kuchino, Y., Mori, F., Kasai, H., Inoue, H., Iwai, S., Miura, K., Ohtsuka, E. and Nishimura, S. (1987) Misreading of DNA templates containing 8-hydroxydeoxyguanosine at the modified base and adjacent residues. *Nature* 327:77–79.

Lehtinen, K.J., Kierkegaard, A., Jakobson, E. and Wändell, A. (1990) Physiological effects in fish exposed to effluents from mills with six different bleaching processes. *Ecotox. Environ. Safety* 19:33–46.

Lemaire, P. and Livingstone, D.R. (1994) Inhibition studies on the involvement of flavoprotein reductases in menadione- and nitrofurantoin-stimulated oxyradical production by hepatic microsomes of flounder (*Platichthys flesus*). *J. Biochem. Toxicol.* 9:87–95.

Lemaire, P. and Livingstone, D.R. (1997) Aromatic hydrocarbon quinone-mediated reactive oxygen species production in hepatic microsomes of the flounder (*Platichthys flesus* L.). *Comp. Biochem. Physiol.* 117C:131–139.

Lemaire, P., Matthews, A., Förlin, L. and Livingstone, D.R. (1997) Stimulation of oxyradical production of hepatic microsomes of flounder (*Platichthys flesus*) and perch (*Perca fluviatilis*) by model and pollutant xenobiotics. *Arch Environ. Contam. Toxicol.* 26:(1994) 191–201.

Livingstone, D.R., Garcia Martinez, P., Michel, X., Narbonne, J.F., O'Hara, S., Ribera, D. and Winston, G.W. (1990) Oxyradical production as a pollution-mediated mechanism of toxicity in the common mussel, *Mytilus edulis* L., and other molluscs. *Functional Ecology* 4: 415–424.

Lopez-Torres, M., Perez-Campo, R., Cadenas, S., Rojas, C. and Barja, G. (1993) A comparative study of free radicals in vertebrates – II. Non-enzymatic antioxidants and oxidative stress. *Comp. Biochem. Physiol.* 105B:757–763.

McCord, J.M. and Fridovich, I. (1969) Superoxide dismutase: an enzymic function for erythrocuprein (hemocuprein). *J. Biol. Chem.* 244:6049–6055.

Mather-Mihaich, E. and DiGiulio, R.T. (1986) Antioxidant enzyme activities and malondialdehyde, glutathione and methemoglobin concentrations in channel catfish exposed to DEF and n-butyl mercaptan. *Comp. Biochem. Physiol.* 85C:427–432.

Mather-Mihaich, E. and DiGiulio, R.T. (1991) Oxidant, mixed-function oxidase and peroxisomal responses in channel catfish exposed to a bleached kraft mill effluent. *Arch. Environ. Contam. Toxicol.* 20:391–397.

Matkovics, B., Szabo, L., Varga, S.I., Barabas, K., Berencsi, G. and Nemcsok, J. (1984) Effects of a herbicide on the peroxide metabolism enzymes and lipid peroxidation in carp fish (*Hypophthalmichthys molitrix*). *Acta Biol. Hung.* 35:91–96.

Matkovics, B., Witas, H., Gabrielak, T. and Szabo, L. (1987) Paraquat as an agent affecting antioxidant enzymes of common carp erythrocytes. *Comp. Biochem. Physiol.* 87C:217–219.

Nishimoto, M. Roubal, W.T., Stein, J.E. and Varanasi, U. (1991) Oxidative DNA damage in tissues of English Sole (*Parophrys vetulus*) exposed to nitrofurantoin. *Chem. Biol. Interact.* 80:317–326.

Oikari, A., Holmbom, B., Anas, E., Miilunpalo, M., Kruzynski, G. and Castren, M. (1985) Ecotoxicological aspects of pulp and paper mill effluents discharged to an inland water system: Distribution in water and toxicant residues and physiological effects in caged fish (*Salmo gairdneri*). *Aquat. Toxicol.* 6:219–239.

Ortiz de Montellano, P.R. (1986) Oxygen activation and transfer. In: Ortiz de Montellanao, P.R. (Ed) *Cytochrome P-450. Structure, mechanism and biochemistry*. Plenum Press, New York, pp 217–271.

Otto, D.M. and Moon, T.W. (1995) 3,3′,4,4′-Tetrachlorobiphenyl effects on antioxidant enzymes and glutathione status in different tissues of rainbow trout. *Pharmacol. Toxicol.* 77:281–287.

Pedrajas, J.R., Peinado, J. and Lopez-Barea, J. (1993) Purification of Cu, Zn-superoxide dismutase isoenzymes from fish liver: Appearance of new isoforms as a consequence of pollution. *Free Rad. Res. Com.* 19:29–41.

Perez-Campo, R., Lopez-Torres, M., Rojas, C., Cadenas, S. and Barja, G. (1993) A comparative study of free radicals in vertebrates – I. Antioxidant enzymes. *Comp. Biochem. Physiol.* 105B:749–755.

Petrivalsky, M., Machala, M., Nezveda, K., Piacka, V., Svobodova, Z. and Drabek, P. (1997) Glutathione-dependent detoxifying enzymes in rainbow trout liver: search for specific biochemical markers of chemical stress. *Environ. Toxicol. Chem.* 16:1417–1421.

Premereur, N., van den Branden, C. and Roels, F. (1986) Cytochrome P-450-dependent H_2O_2 production demonstrated *in vivo*. Influence of phenobarbital and allylisoprpylacetamide. *FEBS Lett.* 199:19–22.

Radi, A.A.R. and Matkovics, B. (1988) Effects of metal ions on the antioxidant enzyme activities, protein content and lipid peroxidation of carp tissues. *Comp. Biochem. Physiol.* 90C: 69–72.

Radi, A.A.R., Hai, D.Q., Matkovics, B. and Gabrielak, T. (1985) Comparative antioxidant enzyme study in freshwater fish with different types of feeding behaviour. *Comp. Biochem. Physiol.* 81C:395–399.

Radi, A.A.R., Matkovics, B. and Csengeri, I. (1988) Effects of various oxygen concentrations on antioxidant enzymes and the quantity of tissue phospholipid fatty acids in the carp. *Acta. Biol. Hung.* 39:109–119.

Roberts, M.H., Jr., Sved, D.W. and Felston, S.P. (1987) Temporal changes in AHH and SOD activities in feral spot from the Elizabeth River, a polluted subestuary. *Mar. Environ. Res.* 23:89–101.

Rodriguez-Ariza, A., Dorado, G., Peinado, J., Pueyo, C. and Lopez-Barea, J. (1991) Biochemical effects of environmental pollution in fishes from Spanish south-Atlantic littoral. *Biochem. Soc. Trans.* 19:301S.

Rodriguez-Ariza, A., Abril, N., Navas, J.I., Dorado, G., Lopez-Barea, J. and Pueyo, C. (1992) Metal, mutagenicity, and biochemical studies on bivalve molluscs from Spanish coasts. *Environm. Molec. Mutagen.* 19:112–124.

Rodriguez-Ariza, A., Peinado, J., Pueyo, C. and Lopez-Barea, J. (1993) Biochemical indicators of oxidative stress in fish from polluted littoral areas. *Can. J. Fish. Aquat. Sci.* 50:2568–2573.

Schiedt, K., Levenberger, F.J., Vecchi, M. and Glinz, E. (1985) Absorbance retention and metabolic transformation of carotenoids in rainbow trout, salmon, and chicken. *Pure and Appl. Chem.* 57:685–692.

Shugart, L.R. (1988) Quantitation of chemically induced damage to DNA of aquatic organisms by alkaline unwinding assay. *Aquatic Toxicol.* 13:43–52.

Sies, H. (1985) Oxidative stress: Introductory Remarks. *In:* Sies, H. (Ed) *Oxidative stress.* Academic Press, London, Orlando, pp 1–8.

Sindermann, C.J. (1979) Pollution-associated diseases and abnormalities of fish and shellfish: A review. *Fishery Bull.* 76:717–749.

Stern, A. (1985) Red cell oxidative damage. *In:* Sies, H. (Ed) *Oxidative stress.* Academic Press, London, Orlando, pp 331–349.

Tappel, M.E., Chaudiere, J. and Tappel A.L. (1982) Glutathione peroxidase activities of animal tissues. *Comp. Biochem. Physiol.* 73B:945–949.

Thomas, P. (1987) Influence of some environmental variables on the ascorbic acid status of striped mullet, *Mugil cephalus* Linn., tissues. III. Effects of exposure to oil. *J. Fish Biol.* 30:485–494.

Thomas, P., Bally, M. and Neff, J.M. (1982) Ascorbic acid status of mullet, *Mugil cephalus* Linn., exposed to cadmium. *J. Fish Biol.* 20:183–196.

Thomas, P., Bally, M.B. and Neff, J.M. (1985) Influence of some environmental variables on the ascorbic acid status of mullet, *Mugil cephalus* L., tissues. II. Seasonal fluctuations and biosynthetic ability. *J. Fish Biol.* 27:47–57.

Tucker, B.W. and Halver, J.E. (1986) Vitamin C metabolism in trout. *Comp. Pathol. Bull.* 18:1–6.

Vig, E. and Nemcsok, J. (1989) The effects of hypoxia and paraquat on the superoxide dismutase activity in different organs of carp, *Cyprinus carpio* L. *J. Fish Biol.* 35:23–25.

Walker, C.H. and Livingstone D.R. (1992) *Persistent pollutants in marine ecosystems.* SETAC Special Publication. Pergamon Press, Oxford.

Washburn, P.C., DiGiulio, R.T. (1989) Stimulation of superoxide production by nitrofurantoin, *p*-nitrobenzoic acid and *m*-dinitrobenzoic acid in hepatic microsomes of three species freshwater fish. *Environ. Toxicol. Chem.* 8:171–180.

Wedemeyer, G. (1969) Stress-induced ascorbic acid depletion and cortisol production in two salmonid fishes. *Comp. Biochem Physiol.* 29:1247–1251.

Wolf, G. (1994) Structure and function of an α-tocopherol binding protein. *Nutr. Rev.* 52:97–98.

Zannoni, V. G. and Sato, P.H. (1975) Effects of ascorbic acid on microsomal drug metabolism. *Ann. N.Y. Acad. Sci.* 258:119–131.

Fish Ecotoxicology
ed. by T. Braunbeck, D. E. Hinton and B. Streit
© 1998 Birkhäuser Verlag Basel/Switzerland

Origin of cadmium and lead in clear softwater lakes of high-altitude and high-latitude, and their bioavailability and toxicity to fish

Günter Köck and Rudolf Hofer

Department of Zoology and Limnology, University of Innsbruck, Technikerstr. 25, A-6020 Innsbruck, Austria

Summary. As a consequence of atmospheric deposition, effects of pollutants such as acidification and metal contamination are evident even in remote aquatic ecosystems of mountain and polar regions. Due to similar environmental characteristics (e.g., oligotrophy, low buffering capacity, long ice-cover, high precipitation rates), clearwater high altitude and high latitude lakes represent very sensitive ecosystems, which are extremely susceptible to even slight changes of the environment. Thus, the environmental relevance of Cd and Pb for both types of lakes is discussed in relation to their extraordinary sensitivity to environmental changes. The impact of Cd and Pb on fish from high altitude and high latitude lakes is reviewed and biotic and abiotic factors controlling bioavailability and toxicity of metals to fish are summarized. Apart from direct toxic effects of low pH, acidification increases the bioavailability of metals for fish. Furthermore, low concentrations of dissolved organic carbon and suspended particles take influence on the uptake and toxicity of metals in fish from clear high altitude and high latitude lakes. Since even very low concentrations of Cd and Pb may result in high metal concentrations in fish, evaluation of critical metal loads for clear high altitude and high latitude lakes is of major importance.

Introduction

Since the 1970s, one of the major interests in limnology has focused on the acidification of surface waters (Baker et al., 1991; Hendriksen et al., 1992; Marchetto et al., 1994; Psenner, 1994; Psenner and Catalan, 1994; Schindler, 1988). Areas sensitive to acidification usually have crystalline bedrock with small amounts of easily soluble minerals and, thus, slow weathering rates and low buffering capacities of soils and sediments (Psenner and Catalan, 1994). The low permeability of crystalline rocks favors the formation of surface waters such as brooks, ponds and lakes. As a consequence of the limited or nearly lacking acid neutralizing capacity, the low natural water pH of these ecosystems often is additionally decreased by acid atmospheric deposition and CO_2 accumulation during periods of ice-cover.

Anthropogenic acidification is mainly restricted to moderate or Arctic climate zones of the northern hemisphere. Due to the rapid industrialization during this century, with escalating emissions of SO_x and NO_x, large areas of Europe and North America suffer from acid deposition by direct precipitation and leaching processes of previously dry-deposited sulfates from the catchment.

Acidic precipitation has also become a complex key factor in both the input of metals into surface waters and their toxicity: (1) Elements attached to aerosols are more soluble at low pH (Colin et al., 1990; Losno et al., 1988), (2) acidic rain events mobilize trace metals previously deposited from the atmosphere and temporarily fixed in soils (Steinnes, 1990), and (3) acidic precipitation favors weathering and leaching processes, which lead to an increased input of abundant crustal elements, such as Al, Fe and Mn, into surface waters (Howells et al., 1990; Psenner and Catalan, 1994). Finally, (4) acidification of less buffered water systems in crystalline bedrock impacts metal speciation and thus their availability and toxicity for limnic organisms.

Among the most striking biological consequences of acidification are effects on fish populations (Rosseland and Hedriksen, 1990) including sublethal effects, failure of reproduction and acute mortality (Campbell and Stokes, 1985; Haines, 1981; Nelson and Mitchell, 1992; Sayer et al., 1993; Spry and Wiener, 1991; Wren and Stephenson, 1991). Since acidification and metal peaks typically occur in irregular and often short pulses (heavy rain fall and snow melt), the vulnerability of surface waters is inversely related to their volume and their water exchange rate. In consequence, direct and indirect effects of acidification are frequently seen in brooks and small precipitation-determined lakes which lack permanent inflows. Since they receive most of their water from direct precipitation and run-off, the biota of these limnic ecosystems are particularly endangered by acid and metal exposure. Concentrations of Cd and Pb in circumneutral and acidified lakes are shown in Table 1.

Since the chemical buffering capacity in clear softwater lakes is extremely low, metals largely occur in their highly available forms and can easily

Table 1. Water concentrations of cadmium and lead in circumneutral and acidified lakes

Location, number and characteristics of waters	pH	Cadmium (µg/L)	Lead (µg/L)	Reference
USA, 6 New Jersey lakes	3.7–5.8	0.09–0.48	<0.5–3.6	Sprenger et al. (1988)
USA, 3 Adirondack lakes	4.9–6.8	0.6–0.7	2.0–3.0	Stripp et al. (1990)
Northern Sweden, brooks	5.5–7.0	0.007		Johansson et al. (1995)
Southern Sweden, brooks	4–5	0.046		Johansson et al. (1995)
Southern Sweden, rivers	6.2–7.5	0.009		Johansson et al. (1995)
Sweden, 17 lakes	4.7–7.4	0.004–0.12	0.2–1.1	Borg (1983)
Norway, 9 lakes	4.3–7.2	0.3–1.0	1.0–2.2	Fjerdingstad and Nilssen (1983)
Finland, 14 lakes	4.8–7.0	<0.01–0.06	<0.01–0.18	Iivonen et al. (1992)
Finland, 256 lakes	6.0	0.02	0.08	Mannio et al. (1993)
Finland, 36 Lapland lakes	4.9–6.4	0.02	0.25	Mannio et al. (1995)
Germany, 12 brooks	4.8–6.8	0.05–0.39		Miller (1987)
Austrian Alps, 3 lakes	5.4–7.1	0.04–0.12	0.2–0.5	Köck et al. (1995)
Austrian Alps, 3 lakes	6.1–6.45		1.1–3.0	Honsig-Erlenburg and Psenner (1986)

penetrate biological membranes (Campbell and Stokes, 1985; Spry and Wiener, 1991). Thus, metals may accumulate in specific organs to extremely high concentrations, although water levels are low (Köck et al., 1995). Apart from stress caused by low ionic water, acidification or metals, biota are additionally exposed to an extreme seasonality of other environmental factors. The winter with low temperature and substantially reduced light intensities is extended (more than 6 months) during which precipitation-determined lakes are almost isolated from external impacts and significant shifts from O_2 to CO_2 may occur. The period of snowmelt and ice break (lasting up to 3 months) follows with still winter-like conditions. The first pulses of snowmelt usually contain high concentrations of acids and metals accumulated during the winter (Jefferies et al., 1979; Köck et al., 1996a). Only in the second half of this period, the ice-cover becomes gradually leaky, favoring exchange processes. The relatively short ice-free season (2–4 months) is distinguished by a rapid increase of water temperature and pH as well as by a high radiation including UV-A and UV-B, which easily penetrates the clear water column (Schindler et al., 1996). Particularly in the mountains, the high UV intensities may lead to several chemical and biological consequences which are still less investigated (Hoigne, 1990; Sturzenegger, 1989).

Among acid-sensitive lakes, three main types can be distinguished: (1) clearwater lakes of high transparency, (2) brown water lakes with high levels of organic matter, and (3) glacier-fed lakes with high contents of suspended solids. The aim of this study was to review the impact of Cd and Pb on fish from clear softwater lakes of high-altitude and high-latitude. Polar lakes exhibit environmental characteristics similar to those of many mountain lakes: long ice-cover, oligotrophy, low buffering capacity, poorly vegetated catchments and high precipitation rates. Both high-altitude and high-latitude lakes represent very sensitive ecosystems. Thus, the environmental relevance of Cd and Pb will be discussed for both types of lakes in relation to their high sensitivity to environmental changes.

Origin of trace metals in remote softwater lakes

Atmospheric deposition of trace metals

Increased levels of toxicants such as mercury, DDT or PCBs have been found even in food chains of Arctic and Antarctic regions as well as other remote sites (Lockhart et al., 1992). Without significant pollution sources of its own, these contaminants are released elsewhere on the globe and deposited after long-range transport *via* the atmosphere (Ottar, 1989; Pain, 1995; Scorer, 1992). While the existence of organic pesticides in the atmosphere is exclusively attributed to human activities, both anthropogenic and natural sources are responsible for metal emissions. The atmosphere of remote areas contains Cd and Pb in the range of 0.002–0.1 and 0.01–5.0 ng/m³ air, respectively; in industrial regions, however, their concentration may rise to up to 10 ng/m³ Cd and 1, 400 ng/m³ Pb (Nriagu, 1990). Major natural sources of atmospheric Cd and Pb are wind-borne soil particles, volcanoes and biogenic sources, whereas sea salt spray and forest fires are less important (Nriagu, 1979, 1990). Since the beginning of the industrial revolution between 1850 and 1900, however, anthropogenic emissions of Cd, Pb and other "atmosphilic" trace metals such as As, Cu, Hg, Ni, V, Zn into the atmosphere have gradually increased (Hong et al., 1996; Nriagu, 1979, 1996) and exceed now the natural emission (Fig. 1; Migon and Caccia, 1990). This development is reflected by the increasing metal contamination of ice cores at pole sites and in sediment cores of remote lakes (Boutron et al., 1991; Norton et al., 1990). Lead is by far the most dominant metal released into the atmosphere by human activities, and it is assumed that more than 95% of the lead within the biosphere is of anthropogenic origin (Smith and Flegal, 1995). Reductions of lead in petrol (banned in the USA in 1975) resulted in a marked decline of lead concentrations in local aerosols (Smith and Flegal, 1995). In fact, lead contamination of Greenland snow pack has shown a 7.5-fold decrease over the past 20 years (Boutron et al., 1991). In many other countries including European ones, however, "unleaded" petrol (<26 mg/L Pb) was introduced much later (1987–1989); nevertheless, within a few years (until 1992), significant lead reductions by about 25% could be measured in the Mediterranean atmosphere, while Cd, Zn and Cu remained constant (Migon and

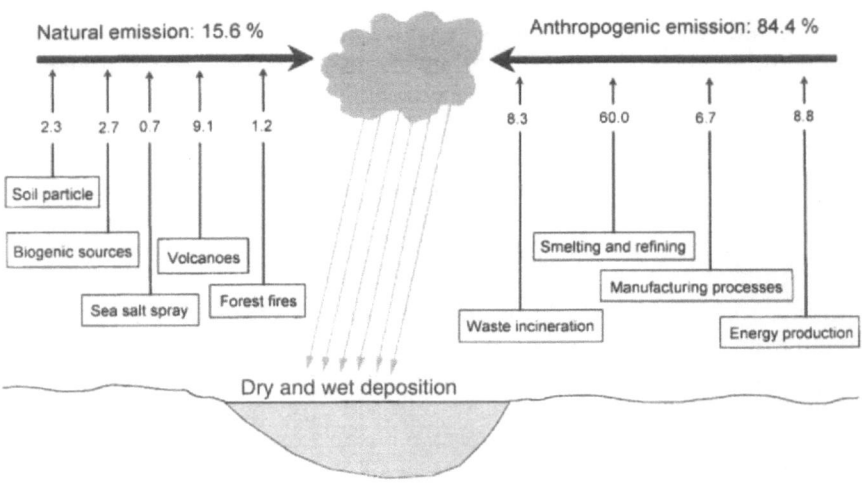

Figure 1. Worldwide atmospheric emissions of Cd from natural and anthropogenic sources (data from Nriagu (1990) were transformed to % of total atmospheric emissions).

Caccia, 1990). Despite the significant reduction in the atmosphere, contamination of biota decreases only slowly, since lead is highly persistent in the environment, and world-wide lead production is still increasing (Smith and Flegal, 1995).

Unlike mercury, which occurs mostly in the gas phase (Berrie et al., 1992), Cd and Pb are attached to small aerosol particles (Boutron and Patterson, 1987; Wagenbach et al., 1988), which results in a long-range transport and low deposition velocity (Arimoto and Duce, 1986). Approximately 40% of Cd and 60% of Pb are associated with particles less than 1 μm (Injuk et al., 1992). On the other hand, the small size of anthropogenic particles favors their solubility in rainwater. Since metal pollution is usually associated with gaseous acidic emissions (SO_x and NO_x), low pH values additionally support the dissolution of atmospheric matter in rainwater. In fact, acidic precipitation is generally enriched in dissolved trace metals (Chester et al., 1990; Nordberg, 1985). The contamination of aerosol particles in the atmosphere is highly variable as a function of both time and location (Injuk et al., 1992). Thus, the concentration of elements in precipitation varies extremely from one event to another, depending on the origin of air masses (Atteia, 1994; Boutron et al., 1993; Ross, 1990). The concentration of cadmium in rainwater may differ in the range of 6, 300, 900 and 10 000 ng/L at remote, rural, urban and industrial (metallurgic) sites, respectively (Mart, 1983; Nguyen et al., 1990). Injuk and co-workers (1992) found highest metal concentrations in the atmospheric layer between 100 and 200 m above the Northern Seas and a significantly decreasing contamination at higher altitudes. In mountains, however, where air masses are forced to rise along the slopes, the situation is more complex. In fact, Atteia (1994) reported negligible effects of altitudes (900–2100 m) in element concentrations of precipitation, and heavy metal concentrations (Pb, Cd) accumulated in mosses even increased with altitude (Zechmeister, 1995). This may be due to higher precipitation rates with rising altitudes (Barros and Lettenmaier, 1993) and, to a lesser extent, due to higher dust deposition with increasing wind speeds and higher portions of indigenous soils above the timber line (Clough, 1975).

Weathering and leaching processes in the catchment area

One major consequence of soil acidification is the dissolution of aluminum (the most abundant element of the earth crust) from its mineral sources and its transfer to surface waters (Hendershot et al., 1986). In the presence of soluble organic acids, i.e., humic acids, leaching processes are even enhanced, forming organometal complexes (Pohlmann and McColl, 1986). However, soils and sediments usually do not significantly contribute to the lead and cadmium load of clear high-altitude and high-latitude lakes.

Behavior of metals in low-alkalinity lakes

Bioavailability and toxicity of metals are largely controlled by a variety of physical and chemical water parameters such as pH, alkalinity, hardness, dissolved organic carbon (DOC), particulate matter, temperature or oxygen concentrations, which, directly or indirectly, separately or in combination with others, may impact metal fluxes between water, sediment and biota. Metals may occur in particulate (adsorbed onto or contained within particles) or dissolved form such as simple hydrated metal ions, as inorganic and organic complexes, in differing valency states as well as absorbed to colloidal particles. The effects of metal pollution on fish strongly depend on the physicochemical properties of the chemical form in which the element occurs (chemical speciation; Florence et al., 1992; Förstner and Wittman, 1983). Although the diet can be a significant source of metal accumulation (particularly for essential trace elements), it is broadly accepted that gills are the major site of uptake for many metals, since they account for up to approximately 90% of the total surface and are permeable for many substances. Generally, bioavailability of free and/or dissolved trace metals is higher than of those bound to either organic or inorganic ligands. In consequence, both chemical speciation and concentration of metals in the water directly affect their toxicity to fish. However, due to the high complexity of the aquatic environment, biological consequences of metal contamination in different water bodies are highly variable. In eutrophic hardwater systems, most of the metals are not directly available, and the metal uptake by fish is hindered by the protective action of calcium ions. On the other hand, oligotrophic acidic softwater lakes such as high-altitude and high-latitude lakes are most vulnerable systems, in which a large portion of metals is present in a soluble and highly toxic ionic form, and biological membranes are more permeable.

Variables controlling solubility and chemical speciation of cadmium and lead

To a large extent, the distribution of trace metals in an aqueous system is influenced by adsorption and desorption reactions (e.g., physical adsorption, chemisorption, electrostatic interaction), complexation as well as precipitation and coprecipitation processes (Förstner and Wittman, 1983; Merian, 1991; Stumm and Morgan, 1981). Various hydrochemical parameters such as pH, redox conditions, alkalinity, the presence of natural and synthetic chelators, the presence of suspended particles, and concentrations of DOC affect these processes and thereby take influence on speciation, mobility and solubility of trace metals in a very complex manner (Lehman and Mills, 1994; Stumm and Schnoor, 1995). Among these variables, soft water and low concentrations of particulates and dissolved matter were revealed as major factors for metal toxicity in low alkalinity clearwater lakes. Hence, the presented paper focuses on these environmental variables.

The effect of pH and alkalinity on solubility and chemical speciation of cadmium and lead

For the vast majority of freshwater lakes, a positive correlation exists between pH and alkalinity or Ca concentration. The effects of these closely linked variables on the metal speciation are very complex and still not clearly understood. For many metals, bioavailability and, in consequence, toxicity are determined by the concentrations of the free metal ion rather than by its total concentrations (Campbell and Stokes, 1985; Vymazal, 1989). Complexation of metals by inorganic ligands is mainly controlled by pH as well as bicarbonate and carbonate systems of natural waters (Davies et al., 1993). At circumneutral or alkaline water conditions, many metals form soluble and insoluble complexes. In hardwater lakes, Ca will be the major inorganic ligand. Thus, for many metals the proportion of free metal ions to the total metal concentration is low in neutral or alkaline waters (Fig. 2). Since metal complexes with inorganic and organic ligands tend to dissociate with decreasing pH due to competition by H^+, acidification leads to an increase in the proportion of free metal ions (Campbell and Stokes, 1985; Ledin et al., 1989; Waiwood and Beamish, 1978; Waller and Pickering, 1993). Trace metals have different sensitivities to pH-mediated changes of speciation (Campbell and Evans, 1987; Campbell and Stokes,

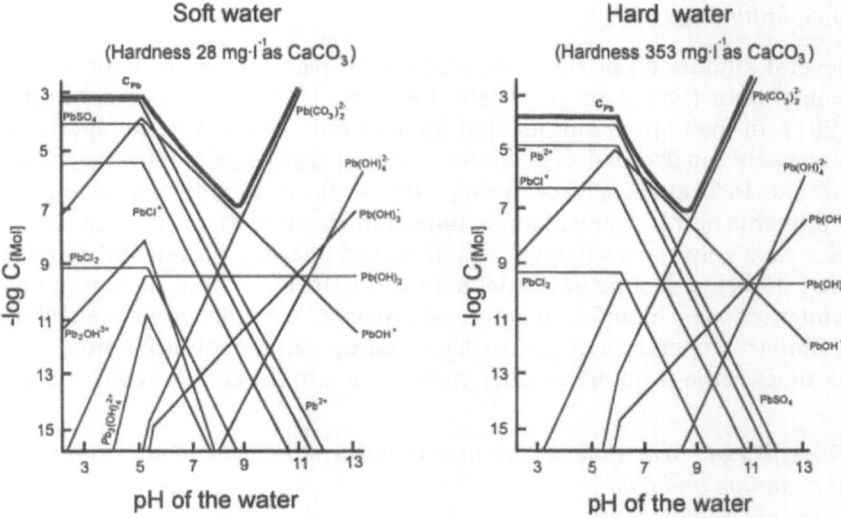

Figure 2. Solubility (C_{Pb}) and species distribution of Pb (II) in soft and hard water at different pH values (redrawn from Davies et al., 1976).

1985). Over the environmentally significant range of pH 4–7, the speciation of Pb, Al, Cu and Hg is highly affected by pH changes, whereas Cd and Zn are only slightly sensitive to pH alterations (Campbell and Stokes, 1985; Chakoumakos et al., 1979; Nelson and Campbell, 1991).

In natural waters, both Cd and Pb form complexes with a variety of substances. Major inorganic ligands are carbonates, hydroxides, chloride, sulfates and, particularly for Pb, phosphates and sulfides (Block, 1991; Campbell et al., 1988; Förstner and Wittman, 1983; Jaworski, 1978). Across the range of pH 7–4, Cd mainly exists as free divalent ion (Campbell and Stokes, 1985; Cusimano et al., 1986; Gardiner, 1974). At pH > 7, the proportion of the free Cd ion decreases with increasing pH, and formation of $CdCO_3$ gradually becomes important (Salomons and Kerdijk, 1986; Turner et al., 1981). At pH <5.5, the free Pb-aquo ion is the dominant Pb species in soft waters. With increasing pH, however, other Pb forms (e.g., $Pb(CO_3)_2^{2-}$ and $Pb(OH)_3^-$) become increasingly important (Campbell and Stokes, 1985; Davies et al., 1976; Jaworski, 1978).

In hard waters, bioavailability of Cd and Pb is lowered by complexation (e.g., with bicarbonate and carbonate ligands), adsorption and precipitation (Carrier and Beitinger, 1988; McCarty et al., 1978; O'Shea and Mancy, 1978; Pärt et al., 1985). The importance of coprecipitation of metals with Fe/Mn-oxides, sulfides and carbonates, an important sink at neutral and alkaline pH, decreases with progressing acidification, since solubility of these complexes increases. Thus, in acidic lakes the rate of metals deposited in the sediment can be reduced.

The effect of temperature on solubility and chemical speciation of cadmium and lead

Several authors reported metal sorption to particulates to increase with temperature (Byrne et al., 1988; Johnson, 1990; Stumm and Morgan, 1981). In many high-altitude and high-latitude lakes, water temperatures range between 0° and 4°C in winter, and temperature maxima vary between 10° and 14°C at the surface during a few weeks in summer. However, even within this narrow range, temperature significantly affects the distribution of metals between particulate and dissolved phases (Warren and Zimmerman, 1994): a decline in temperature from 10° to 1°C during autumn and winter led to an eightfold increase of dissolved Cd. The authors postulated a similar temperature effect for lakes during summer stratification, when particles settle from the warmer epilimnion into the colder hypolimnion.

The effect of particulate matter on solubility and chemical speciation of cadmium and lead

In many natural waters of neutral or alkaline pH, only a small percentage of the dissolved trace metals such as Cu, Pb, Cd or Zn is present as free (aquo) metal ion. Most of the metal is adsorbed to colloidal and/or particulate matter or combined into complexes, thereby suspended in the water column or deposited in the sediment (Florence et al., 1992). Settling particles are of major concern for binding heavy metals and "immobilizing" them in the sediment. However, many high-altitude and high-latitude lakes (except glacier-fed lakes with high turbidity) are clearwater lakes, having low concentrations of particulate matter. Thus, scavenging of metals by sinking particles and hence clearance of metals from the water column is of minor importance. Solid phases such as Fe-, Mn-, Al-oxides and -carbonates represent important ligands for trace metals. In particular, hydrous Fe/Mn-oxides, having high adsorption capacity for many metals, play a dominant role in the control of the distribution of metals in the aquatic environment (Balistrieri et al., 1992; Bradley and Cox, 1988; Förstner and Wittman, 1983; Fu et al., 1991; Mantei and Foster, 1991; Vymazal, 1989). Affinity of Pb^{2+} to Mn-oxides is higher than compared to other trace metals (Pb > Cu > Cd > Zn) and increases with pH (Balikungeri and Haerdi, 1988). Furthermore, Pb has a high affinity for clay minerals and biogenic particles (Campbell et al., 1988; Förstner and Wittman, 1983; Jaworski, 1978). Thus, in many natural waters Pb is bound to a large extent to organic and inorganic particles. However, in a clear acidic high mountain lake partitioning of dissolved Pb to total Pb amounted to 42–71% (G. Köck, unpublished data).

In contrast to other metals, binding to particles is of less importance for Cd (Breder, 1988; Malm et al., 1988). More than 80% of Cd have been reported to occur in dissolved forms in Lakes Constance, Zürich and Michigan (Breder, 1988; Sephard et al., 1980). In a clear acidic high mountain lake, 80–100% of total Cd were dissolved (G. Köck, unpublished data).

Biological surfaces, having high affinity for divalent ions, are particularly efficient scavengers – probably better ones than mineral surfaces for heavy metals. Hence the productivity of a lake is a major factor controlling the sedimentation rate of biogenic particles and thus removing "biophilic" heavy metals. In eutrophic lakes adsorption of trace metals onto phytoplankton and other biogenic particles will be considerably higher than in ultra-oligotrophic mountain lakes.

The effect of dissolved organic carbon (DOC) and dissolved organic matter (DOM) on solubility and chemical speciation of cadmium and lead

Bioavailability of dissolved trace elements is usually reduced by complexation with dissolved organic material (Hutchinson and Sprague, 1987; Landers et al., 1994; Spry and Wiener, 1991; Vymazal, 1989; Welsh et al., 1993). A negative correlation between organic contents of lakes and Cd and Pb concentrations in fish has been reported (Iivonen et al., 1992). Organic loading of a water body can strongly affect the environmental conditions which control metal speciation, especially at the sediment-water interface. To a large amount, acid and metal loads of surface waters are controlled by both the atmospheric deposition and the catchment. The amount of organic acids received by surface waters largely depends on soil thickness and terrestrial vegetation (Howells et al., 1990; Psenner and Catalan, 1994; Sparling, 1995). While wooded catchments of unbuffered lakes of Scandinavia and North America are rich in organic acids, the vegetation of high-altitude and high-latitude regions is poor or even absent, resulting in DOC concentrations of surface waters below 2 mg/L (Bernes, 1991; Schindler et al., 1996; Spry and Wiener, 1991). Dissolved organic matter (DOM) is important for chelation, flocculation and changes in mobility of trace metals (Schindler et al., 1996). DOC in waters is mainly composed of humic substances (humic and fulvic acids) (Lehman and Mills, 1994; Reuter and Perdue, 1977). Since humic substances are weak buffers, lakes that are high in DOC are typically less sensitive to acid deposition than are clearwater lakes (Sparling, 1995). In natural waters, the majority of metal complexation can be attributed to interactions with humic substances (Hering and Morel, 1990). However, the impact of interactions between metals and humic substances on availability and toxicity of metals to aquatic life is not yet fully understood. Various trace metals such as Cu, Pb, Hg and Al readily form complexes with dissolved organic matter (Campbell and Evans, 1987; Senesi, 1992). Compared to other metals, the tendency of Cd to form organic complexes is low (Breder, 1988). In surface waters, the extent of organic complexation varies from nearly 100% for Cu to negligible values for Mn and Cd (Campbell and Stokes, 1985; Hering and Morel, 1990). In lakes, the proportion of organic Cd complexes to total dissolved Cd has been reported to account for less than 4% (Moore and Ramamoorthy, 1984; Sephard et al., 1980). Even in waters with high concentration of humic substances, Cd is predominantly present as dissolved cationic forms, whereas other metals such as Al and Fe precipitate as solids (Pettersson et al., 1993; Sephard et al., 1980). Complexation of Pb with humic acids is higher than for Cd (Campbell and Evans, 1987; LaZerte, 1989; Spry and Wiener, 1991).

Since lake clarity often increases with declining pH, many low-alkalinity lakes are clearwater lakes with low concentrations (<2 mg/L) of dissolved organic carbon (Schindler et al., 1996; Sparling, 1995; Spry and Wiener, 1991). Furthermore, photodegradation may lead to declines in DOC contents of lake waters (Bernes, 1991; Schindler, 1996). Since UV-B radiation increases with altitude, mountain lakes receive a much higher UV-B dose than lowland lakes (Blumenthaler et al., 1992, 1993; Cabrera et al., 1995; Schindler et al., 1996). Thus, susceptibility of high mountain lakes to metal input may be higher than compared to that of other lakes. Due to the high water transparency of high-altitude lakes, potentially harmful intensities of UV-B radiation can reach depths of several meters

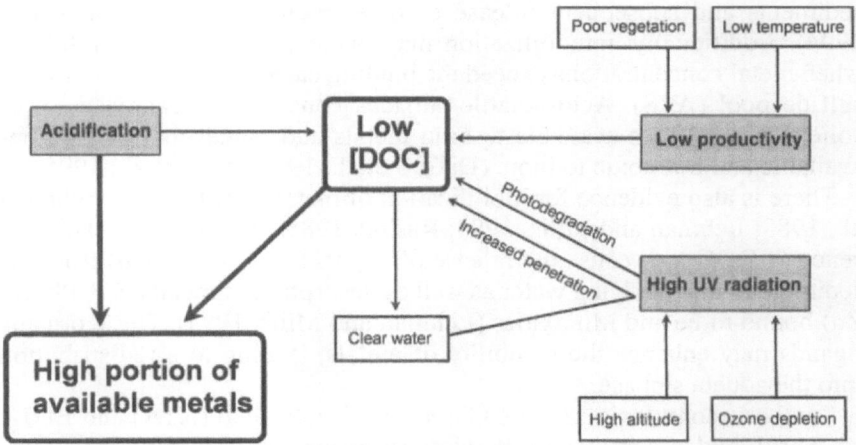

Figure 3. Factors controlling dissolved organic carbon (DOC) in clearwater lakes.

in these lakes (Schindler et al., 1996; Sommaruga and Psenner, 1997). Sommaruga and Psenner (1996) reported 10% of the subsurface UV-B radiation to reach 10 m depth in a high mountain lake. Furthermore, evidence for an increasing trend in UV-B radiation has been found (Blumenthaler and Ambach, 1990; Kerr and McElroy, 1993). However, increasing degradation of DOC would also enlarge environmental sensitivity of high-altitude and high-latitude lakes. Factors controlling DOC in clearwater lakes are shown in Figure 3.

The effect of remobilization on solubility and chemical speciation of cadmium and lead

Due to precipitation, sorption and particulate-settling processes, the major part of trace metals is deposited and "immobilized" in the sediments of the aquatic environment. The interactions of metals with sediments, interstitial and overlying waters are very complex and not yet clearly understood. However, pH reduction, low redox potential and the presence of complexing agents facilitates the release of metals into the overlying water. The special climatic situation of high-altitude and high-latitude lakes causes distinct seasonal variations in water chemistry, making these lakes particularly sensitive. The long ice cover leads to CO_2 oversaturation and a concomitant decrease of pH. Since carbonates tend to be more soluble in the presence of CO_2, the CO_2 oversaturation influences the remobilization of metals from the sediment. Full water circulation occurring after the ice-break can distribute remobilized metals into the water body. Furthermore, bottom waters of many of these lakes exhibit anoxic conditions during winter, which can contribute to dissolution of Fe hydroxides from surface sediments and subsequent release of other metals (Lehman and Mills, 1994). Additionally, remobilization may occur under anoxic conditions, when metal concentrations exceed the binding capacity of the acid-volatile sulfide pool (AVS). Acid-volatile sulfides constitute a reactive pool of solid-phase sulfides available to bind metals and render that portion unavailable and non-toxic to biota (DiToro et al., 1992; Rand et al., 1995).

There is also evidence for mobilization of metals by DOC (Campbell et al., 1988; Lehman and Mills, 1994; Rashid, 1985; Weber, 1988). The presence of DOC may cause the release of metals (Fe, Mn, As, Cu) from the sediment to the overlying water as well as desorption of metals (Cd, Pb, Cr, Zn) bound to Fe and Mn oxides (Lehman and Mills, 1994). These organic ligands may enhance the solubility of metals, leading to a redistribution into the aqueous phase.

Synthetic complexing agents like nitrilotriacetic acid (NTA) and EDTA have profound effects on sorption/desorption processes, leading to increased desorption of Cd and Pb from suspended solids and sediments and, hence, to increased metal concentrations in the water (Förstner and Wittman, 1983; Lietz and Galling, 1989; Salomons and Kerdijk, 1986). The presence of synthetic organic substances such as dithiocarbamates or xanthates, forming lipophilic complexes with trace metals, may enhance metal bioavailability due to facilitated penetration of these complexes through the gill membrane (Block, 1991; Block et al., 1991, 1992; Tjälve and Gottofrey, 1991). However, environmental significance of these chemicals might be negligible in remote areas.

Macrobenthic (bioturbation) and microbenthic (microbial processes) activities in the sediment may alter the physicochemical conditions at the sediment/water interface, leading to enhanced remobilization of metals (Francis and Dodge, 1990; Rand et al., 1995). Furthermore, biomethylation of Pb may occur under anoxic conditions (Jaworski, 1978).

Apart from direct toxic action of water-borne metals, fish can also be affected by metal contamination *via* the food chain (Dallinger and Kautzky, 1985; Dallinger et al., 1987; Köck et al., 1991; Patrick and Loutit, 1978). In acidic waters, benthic invertebrates may contain elevated levels of metals. Since they are important food sources for many fish species, they can contribute to increased metal concentrations in fish (Czarnezki, 1985; Köck et al., 1991; Wren and Stephenson, 1991). Furthermore, fish might be affected by metal-related changes of their food spectrum. Negative effects of both metal pollution and acidification on species diversity, taxonomic distribution, biomass and abundances of planktonic and benthic macroinvertebrate communities are well documented (Baker and Christensen, 1991; Clements et al., 1988; Deniseger et al., 1990; Haines, 1981; Wren and Stephenson, 1991). Since many fish species are highly specific in their food selection (prey size, food spectrum), the disappearance of a certain prey size or species could have profound effects on fish populations (Munkittrick and Dixon, 1988). Furthermore, metal pollution may lead to a shift in the community structure to metal-tolerant organisms, which are often capable of accumulating large amounts of metals (Dallinger et al., 1987). Thus, fish are forced to feed mainly on highly contaminated food organisms. For Arctic char from an acidic high mountain lake, pronounced seasonal changes in both main components and metal load of the diet has been reported (Köck et al., 1996a, 1998; Triendl, 1994). During the period of ice-cover, fish fed exclusively on highly contaminated autochthonous prey (chironomid larvae, oligochaetes). During summer, however, they switched to low-contaminated allochthonous prey ingested from the water surface ("Anflug").

Parameters affecting the uptake of metals by biological membranes

It is widely accepted that toxic action of a trace metals to fish is mainly a function of the free metal ion concentration in the water (Campbell and Stokes, 1985; Pagenkopf, 1983; Roy and Campbell, 1995). However, a biological response from the organism requires interaction between metals and ligands located at the cell surface (e.g., gill membranes), leading to direct toxic action at the gill surface or passive or to active uptake over the cell membrane. Any mechanism that prevents metal adsorption on the gills either by reduction of the ambient concentration of free metal ions or the number of binding sites at the gill surface will reduce bioavailability and toxicity of metals (Playle et al., 1993a). Hydrochemical variables such as pH, alkalinity and temperature are relevant for both processes.

The influence of alkalinity and hardness on metal uptake across membranes

Major effects of alkalinity or Ca concentration on uptake and toxicity of metals are a consequence of alterations of both binding of metals on and transfer through biological membranes. Competition between divalent metal ions and Ca^{2+} or Mg^{2+} for binding sites on the gill surface and the passage through ion-sensitive channels may decrease uptake and toxicity of metals in hardwater (McDonald et al., 1989; Pärt, 1990; Pagenkopf, 1983; Verbost et al., 1987; Wicklund and Runn, 1988). For uptake, passive diffusion of Cd through Ca^{2+} channels in the apical membrane of branchial chloride cells followed by competition with Ca^{2+} for active transport sites at the basolateral membrane has been proposed (Lockhart et al., 1992; Losno et al., 1988; Verbost et al., 1987, 1989; Wicklund Glynn et al., 1994). Ca^{2+}-mediated changes in the permeability of gill membranes have also been suggested to influence both metal uptake and toxicity (Calamari et al., 1980; Pärt et al., 1985). In a set of oligotrophic high mountain lakes which differed from each other in several physicochemical water characteristics, Cd and Pb accumulations in fish organs have been observed to increase

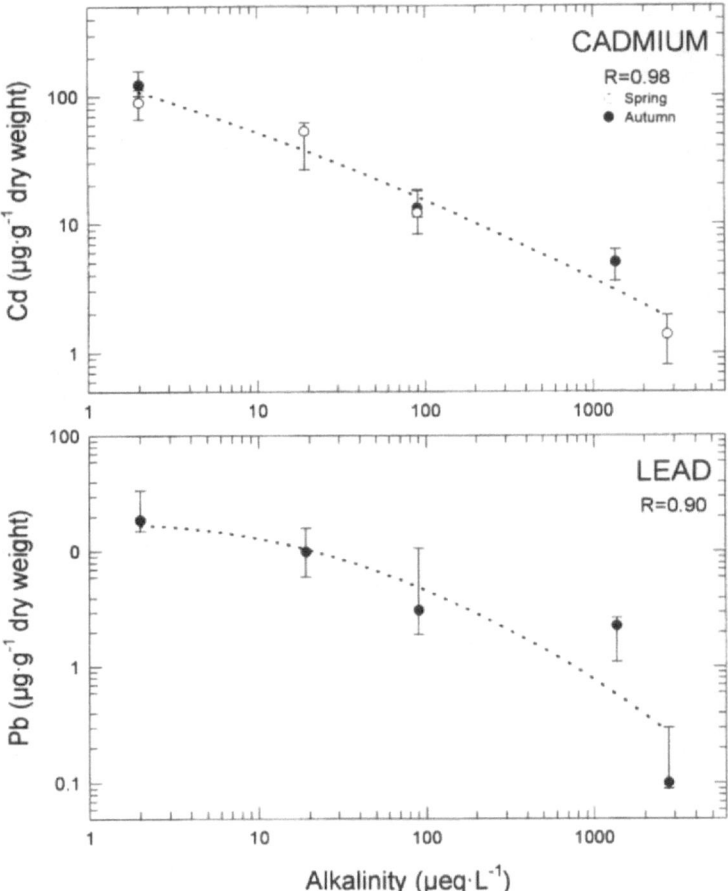

Figure 4. Accumulation of Cd and Pb in the kidney of Arctic char (*Salvelinus alpinus*) from 5 oligotrophic alpine lakes in relation to lake alkalinity. Median and percentiles (25 and 75%) are given (modified from Köck et al., 1995).

with a decrease in alkalinity, Ca^{2+} concentrations and pH, with the highest significance for alkalinity (Fig. 4; Köck et al., 1995). Metal concentrations in fish from the lake with the lowest alkalinity were, despite relatively low metal concentrations in the water, comparable with those of fish in highly polluted environments.

The influence of pH on metal uptake across membranes

The increased solubility and, for some metals (e.g., Pb), also the shift in chemical speciation towards free ions is only one aspect of water acidification which may lead to an increased uptake by fish. However, also reduction of metal uptake and toxicity as a consequence of competition between metals ions and H^+ for binding sites at the gill surface has been observed (Campbell and Stokes, 1985; Cusimano, 1986). Although it has been shown that uptake and toxicity of Pb increase with decreasing pH (Hamilton and Haines, 1989; Köck and Wögrath, 1995; Merlini and Pozzi, 1977; Stripp et al., 1990), studies on the interaction between Cd accumulation in fish and pH have yielded somewhat contradictory results. Some authors reported Cd accumulation and toxicity to decrease with decreasing pH (Campbell and Stokes, 1985; Cusimano et al., 1986; Pärt et al., 1985; Roy and Campbell, 1995). In other studies, Cd accumulation correlated negatively with pH and alkalinity (Haines and Brumbaugh, 1994; Hamilton and Haines, 1989; Iivonen et al., 1992; Köck et al., 1995; Sprenger et al., 1988; Stripp et al., 1990). Since laboratory studies have shown that Cd uptake by the gills is to a large extent influenced by aqueous Ca concentrations (a strong correlate of alkalinity) and is not dependent on pH over a range from 5 to 7 (Pärt et al., 1985; Wicklund and Runn, 1988), it may be assumed that alkalinity is a key factor for Cd accumulation (Köck and Wögrath, 1995).

The influence of temperature on metal uptake across membranes

Although metal sorption to particulate matter increases with rising temperature and, thus, bioavailability for fish is reduced (see above), an increase in temperature is generally associated with enhanced metal toxicity (Alabaster and Lloyd, 1980; Cairns et al., 1975; Hodson and Sprague, 1975; McGeachy and Dixon, 1989; Roch and Maly, 1979). Since temperature is an important controlling factor for poikilothermic animals (Fry, 1971; Hochachka and Somero, 1971; Schmidt-Nielsen, 1979), rising water temperature results in increased gill ventilation rates due to the higher oxygen demand for metabolic requirement (Douben, 1990; Heath, 1987; Niimi, 1987). Significantly improved feeding conditions during the ice-free season, gonadal development during summer and spawning activities in late autumn may additionally stimulate metabolic activity (Hofer and Medgyesy, 1997; Köck et al., 1996a; Olsson et al., 1987; Povlsen et al., 1990). This results in a higher volume of water passing the gill surface and, in consequence, to an increased uptake of metals from ambient water. Increased metal uptake due to increased ventilation as a consequence of hypoxic conditions during winter may occur only in shallow lakes.

Apart from data on Hg (Bodaly et al., 1993; Hoffmann, 1987; MacLeo and Pessah, 1973; Reinert et al., 1974), only few data are available on temperature-dependent accumulation of trace metals in fish. Some studies revealed Cd uptake rates to increase with temperature (Douben, 1989; Phillips and Russo, 1978). Cd and Pb accumulation in liver and kidney of Arctic char from an acidic high mountain lake has been shown to increase with

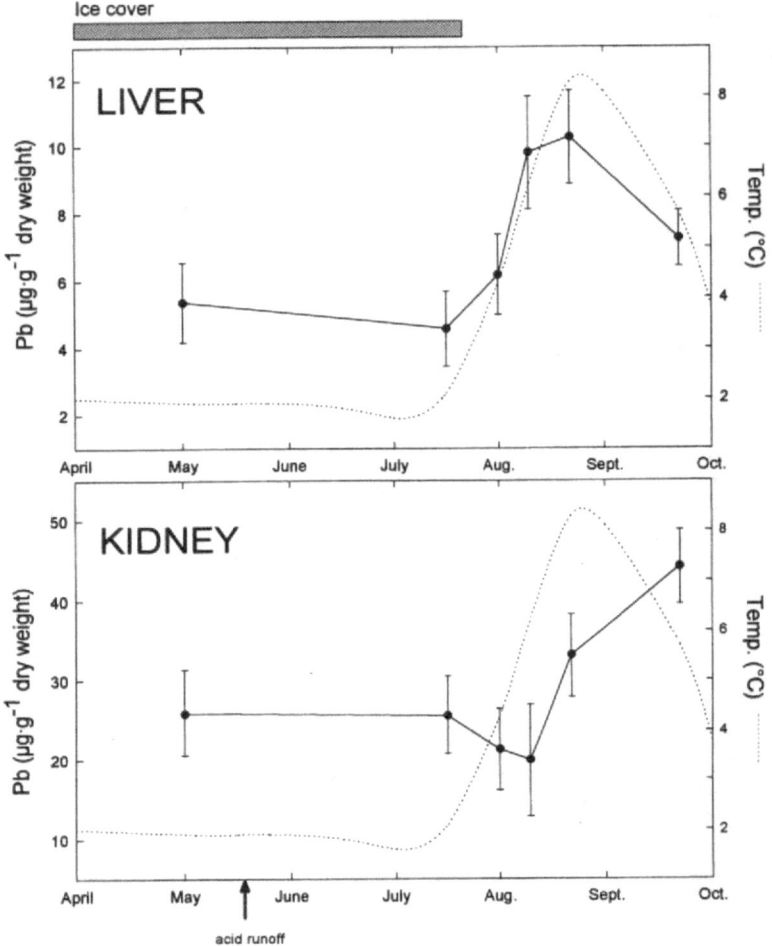

Figure 5. Seasonal variation of Pb concentrations in liver and kidney of Arctic char (*Salvelinus alpinus*) from an acidic oligotrophic alpine lake. Means ± SE are given. Dotted line: seasonal variation of water temperature (modified from Köck et al., 1996a).

rising temperature during the short ice-free period in summer and autumn (Fig. 5; Köck et al., 1996a). During winter, however, metal excretion rates apparently exceeds uptake, thus leading to a gradual decrease of tissue metal burden to a minimum at the time of ice-break. However, since Cd and Pb are not completely eliminated after accumulation as a consequence of increased metabolic rates during summer, tissue concentrations of both metals increase with the age of fish. In high-altitude and high-latitude lakes, temperature appears to be the driving force of metal accumulation in fish, whereas seasonal variations of metal concentrations in lake water and the diet are of minor importance (Köck et al., 1996a). In lakes with distinct algal blooms during summer or seasonally varying DOC concentrations, however, temperature effects on metal availability might be partly compensated by increased sorption and complexation.

Accumulation and toxicity of heavy metals

Cadmium accumulation and toxicity

Cadmium is accumulated predominantly in the kidney, liver, gills and digestive tract, with highest concentrations usually being found in the kidney. Numerous studies reported a positive correlation between tissue Cd and length, weight and age of fish (Douben, 1989; Douben and Koeman, 1989; Köck et al., 1995, 1996a; Mehrle et al., 1982; Ney and vanHassel, 1983; Sprenger et al., 1988). In many high-altitude and high-latitude lakes fish populations are aged with individuals of 15 to 25 years as a consequence of lacking predation, leading to high metal concentrations in tissues and severe histological alterations in liver and kidney (Hofer et al., 1994; Köck et al., 1995; Reimers, 1979).

Cadmium exposure causes a broad spectrum of toxic effects in fish, including pathological changes of blood and gills, enzymatic inhibition, reproductive failure, skeletal deformities and behavioral changes. The main toxic action of Cd is the impairment of ionic regulation rather than respiratory or nervous functions (Larsson et al., 1981; Reid and McDonald, 1988; Roch and Maly, 1979). Toxic effects of Cd in low alkalinity waters are summarized in Table 2.

Fish species and their developmental stages differ considerably in their sensitivity to Cd. Salmonids are considered to be more sensitive to Cd and other metals than cyprinids (Alabaster and Lloyd, 1980; Köck, 1996). Among salmonids, brown trout appears to be more Cd-tolerant than rainbow trout (Alabaster and Lloyd, 1980; Brown et al., 1994). Eggs and yolk-sac stages are considered to be less sensitive than swim-up fry or adult fish (Alabaster and Lloyd, 1980; Buhl and Hamilton, 1991; Chester et al., 1990; Eaton et al., 1978; Hwang et al., 1995; Kuroshima et al., 1993). Eggs have been shown to be more resistant to Cd, since the chorion is an effective barrier to protect the embryo (Beattie and Pascoe, 1978; Nakagawa and Ishio, 1988a, b, 1989a, b, c, d).

Metallothionein plays a major role in homeostatic regulation and detoxification of metals in fish (Dallinger et al., 1997; George and Langston, 1994; Roesijadi, 1992, 1994). Induction of metallothionein as a consequence of exposure to metals in the water has been observed in many studies (Hogstrand et al., 1989, 1990; Klaverkamp et al., 1984; Roch et al., 1982). Furthermore, increased tolerance to toxic metals has often been attributed to induction of metallothioneins. In the liver of Arctic char from a set of oligotrophic Alpine lakes, concentrations of both Cd and metallo-

Table 2. Effects of cadmium exposure (in mg/L Cd) in soft water to selected fish species

Species	pH	Hardness	Concentration	Effect	Reference
Oncorhynchus mykiss		4	0.011	75% Mortality within 4 days	Bilinski and Jonas (1973)
		4	1.12 (24 h)	Gill damage	Bilinski and Jonas (1973)
	4.8	3	0.73 (24 h)	Inhibition of Ca^{2+} influx across the gills	Reid and McDonald (1988)
Salvelinus fontinalis		low	0.0007	Loss of weight, increase of AChE activity	Christensen (1975)
		low	0.034	Increase of ALP and PGOT activities	Christensen (1975)
		low	0.010 (21 d)	Gonadal damage	Sangalang and O'Halloran (1972)
		low	0.025 (24 h)	Gonadal damage, impairment of androgen synthesis	Sangalang and O'Halloran (1972)
	6.2	6.6	0.005 (30 d)	Increased mortality (age of fish: 3–4 months)	Hamilton et al. (1987b)
	6.1	14.5	0.0036 (7 d)	Increased mortality (age of fish: 3–4 months)	Hamilton et al. (1987a)
	7–8	44	0.0034	3 Generations were Cd-exposed: increased mortality 2nd and 3rd generation during spawning; hyperactivity	Benoit et al. (1976)
	6.3–7.6	20	0.008	NOEC for test endpoint "survival"	Jop et al. (1995)
			0.018	NOEC for test endpoint "growth"	
	6.3–7.6	20	0.018	LOEC for test endpoint "survival"	Jop et al. (1995)
	6.3–7.6	20	0.012	MATC for test endpoint "survival"	Jop et al. (1995)
	7–8	44	0.0017–0.0034	MATC for test with three generations	Benoit et al. (1976)
	7.6	45	0.0011–0.0038	NOEC for embryonal and larval stages	Eaton et al. (1978)
	6.5–7.2	37	0.001–0.003	MATC (60 d)	204
Carassius auratus	7.4–7.8	21	0.090 (21 d)	Decreases in lymphocyte and thrombocyte numbers; increases in the numbers of neutrophils, eosinophils and basophils	Murad and Houston (1988)
Carassius auratus	7.4–7.8	21	0.085 (21 d)	Karyorrhexis of erythrocytes	Houston et al. (1983)

Hardness is given as mg/L $CaCO_3$. Threshold concentrations (No Observed Effect Concentration – NOEC, Lowest Observed Effect Concentration – LOEC and Maximum Acceptable Tolerance concentration – MATC) for soft water are also presented.
Abbreviations: ALP – Alkaline Phosphatase, Pgot – Plasma Glutamic Oxalacetic Transaminase. AChE – acetylcholine esterase.

thioneins were negatively correlated to alkalinity (Dallinger et al., 1997; Köck et al., 1995).

Lead accumulation and toxicity

Lead is accumulated in bony structures, kidney, liver and gills. Numerous studies reported Pb to be preferentially accumulated in calcareous structures (bones, scales) of fish (Dwyer et al., 1988; Hodson et al., 1978; Köck et al., 1996b; Stripp et al., 1990). This "bone-seeking" phenomenon was interpreted as a detoxification mechanism for Pb (Miyahara et al., 1983; Ney and vanHassel, 1983; Settle and Patterson, 1980). In the opercula of Arctic char from oligotrophic lakes, Pb concentrations correlated negatively with alkalinity, Ca concentration and pH of the lakes, thereby reflecting increased uptake of the metal in low-alkalinity lakes (Köck et al., 1995). However, Pb concentrations in the otoliths did not correlate with any of these water characteristics. The tissue-specific differences in the accumulation of Pb indicate different metabolic pathways for Pb incorporation into otoliths and opercula. Studies on the correlation between tissue Pb and biometric parameters of fish yielded somewhat contradictory results. Some authors observed Pb concentrations in liver, kidney and calcified tissues to correlate positively to age (Köck et al., 1995, 1996b; Ney and vanHassel, 1983; Pagenkopf and Neumann, 1974). Others reported a negative correlation between tissue Pb and body weight (Bohn and Fallis, 1978; Wiener and Giesy, 1979). However, in most studies no correlation between tissue Pb and length or weight of fish could be detected (Czarnezki, 1985; Hamilton and Haines, 1989; Heit et al., 1989; Ney and vanHassel, 1983; Vinikour et al., 1980).

Although acute toxicity of Pb is relatively low, chronic Pb exposure may affect activities of different enzymes and causes disturbances of ionic regulation and carbohydrate metabolism (Bengtsson and Larsson, 1986; Haux and Larsson, 1982, 1986; Jackim et al., 1970, 1973). However, symptoms of chronic Pb toxicity are mainly expressed as neurophysiological and hematological responses. Neurophysiological symptoms include blackening of the skin in the caudal region ("black tails"), muscle atrophy of the tail region and skeletal abnormalities, e.g., spinal and vertebral deformities particularly in the tail region (Bengtsson and Larsson, 1986; Hodson et al., 1978, 1984; Holcombe et al., 1976; Newsome and Piron, 1982). In soft water, long-term exposure to very low concentrations of 7.6 µg/L Pb caused occurrence of blacktails (Davies et al., 1976). For neurotoxic action, a blood Pb concentration of 300 µg/L Pb has been reported (Hodson et al., 1984). In fish from Pb-contaminated waters, negative effects on density, strength, Ca contents and collagen composition of bones have been observed (Dwyer et al., 1988; Hamilton et al., 1988, 1989). Toxic effects of Pb in low alkalinity waters are summarized in Table 3.

Table 3. Effects of lead exposure (in mg/L Pb) in soft water to selected fish species

Species	pH	Hardness	Concentration	Effect	Reference
Oncorhynchus mykiss	6.85	32	1.77	LC_{50} 96 h	Davies et al. (1976)
		50	1.1	LC_{50} 96 h	Brown (1968)
	6.9–7.4	35	0.672	Reduced hatching success	Sauter et al. (1976)
	6.85	32	1.17	LC_{50} 96 h	Davies et al. (1976)
	6.7–7.3	28	0.0041	NOEC (no fish with "black tails")	Davies et al. (1976)
			0.0076 (19 m)	4.7% of fish exhibited "black tails" MATC (test starting with eyed eggs) lies between these concentrations	Davies et al. (1976)
	6.6–7.3	28	0.0072	NOEC (no fish with "black tails"); 41.3 % of fish exhibited "black tails"; MATC (test starting with swim-up fry) between these concentrations	Davies et al. (1976)
			0.0146 (19 m)		
	6.9–7.4	35	0.071–0.146 (60 d)	MATC (test starting with yolk-sac larvae)	Sauter et al. (1976)
Salvelinus fontinalis		low	0.132 (16 d)	Reduction of larval growth	Christensen (1975)
Salvelinus fontinalis		low	0.53 (21 d)	Increases of AChE and alkaline phosphatases activities	Christensen (1975)
Salvelinus fontinalis	7.5	45	0.08	LOEC (for "coughing")	Drummond and Carlson (1977)
	7.0–7.5	45	3.4	LC_{50} 96 h	Holcombe et al. (1976)
Salvelinus fontinalis	7.0–7.5	45	0.039–0.084 (266 d)	MATC	Holcombe et al. (1976)
Carassius auratus		20	20.6	LC_{50} 96 h	Pickering and Henderson (1966)
Pimephales promelas	7.5	20	5.6 (96 h)	LC_{50} 96 h	Pickering and Henderson (1966)
Lepomis macrochirus	7.5	20	24 (96 h)	LC_{50} 96 h	Pickering and Henderson (1966)

Hardness is given as mg/L $CaCO_3$. Threshold concentrations (NOEC, LOEC and MATC) for soft water are also presented.

Organic Pb compounds have much higher toxicities than inorganic Pb (Babich and Borenfreund, 1990; Hodson et al., 1984; Wong et al., 1981). By fish, organic Pb is absorbed more readily than inorganic Pb, and it is preferentially accumulated in body lipids. Hodson et al. (1984) reported LC50 values of 50–230 µg/L Pb for organic Pb, and Wong et al. (1978) observed tetramethyl-Pb concentrations of 3.5 µg/L to cause death of rainbow trouts.

The major toxic action of inorganic Pb in the blood is the inhibition of erythrocyte δ-aminolevulinic acid dehydratase (ALA-D) activity (Hodson et al., 1984). The enzyme catalyzes the formation of porphobilinogen (a precursor of hemoglobin) from aminolevulinic acid. The enzyme's activity is inhibited *in vitro* by many metals, but only Pb is inhibitory *in vivo*. Due to its high sensitivity and selectivity to Pb, the ALA-D activity has frequently been used as an indicator of sublethal effects of Pb on fish (Hodson et al., 1976, 1977; Johansson-Sjöbeck and Larsson, 1979; Köck et al., 1991; Schmitt et al., 1984, 1993). Inhibition of ALA-D has been proposed to occur at blood threshold concentrations of approximately 300 µg/L Pb (Hodson et al., 1984). ALA-D activity is negatively correlated to Pb concentrations in the blood. Likewise, a negative correlation between ALA-D activity in the blood and Pb concentrations in the kidney has been observed (Köck et al., 1991). Köck et al. (1991) demonstrated a depression of ALA-D activity by 36% to coincide with a significant reduction of hemoglobin concentrations in the blood. However, despite profound inhibition of ALA-D activity, negative effects on hemoglobin concentration, hematocrit or erythrocyte counts could not be found by other authors (Haux et al., 1986; Johansson-Sjöbeck and Larsson, 1979; Larsson et al., 1985; Schmitt et al., 1984). Since ALA-D is apparently in excess in the blood, inhibition of the enzyme might not necessarily lead to anemic response (Dixon et al., 1985; Posner et al., 1980; Schmitt et al., 1984). Reduction of hemoglobin contents has been shown to occur only at high water-borne Pb concentrations (≥ 300 µg/L Pb^{2+}; Johansson-Sjöbeck and Larsson, 1979). However, inhibition of ALA-D activity is observed at very low concentrations of Pb in the water: in low-alkalinity water (10–20 mg/L $CaCO_3$), concentrations of 0.5–4.5 µg/L Pb led to enzyme activity depression by 88% (Haux et al., 1986). In hard waters, ALA-D activity was inhibited by about 65% at concentrations of 5–20 µg/L Pb (Schmitt et al., 1984). However, ALA-D activity seems to be relatively insensitive to organic Pb components (Dixon et al., 1985).

Summary and conclusions

As a consequence of atmospheric deposition, effects of pollutants such as acidification and metal contamination are evident even in remote aquatic ecosystems of mountain and polar regions.

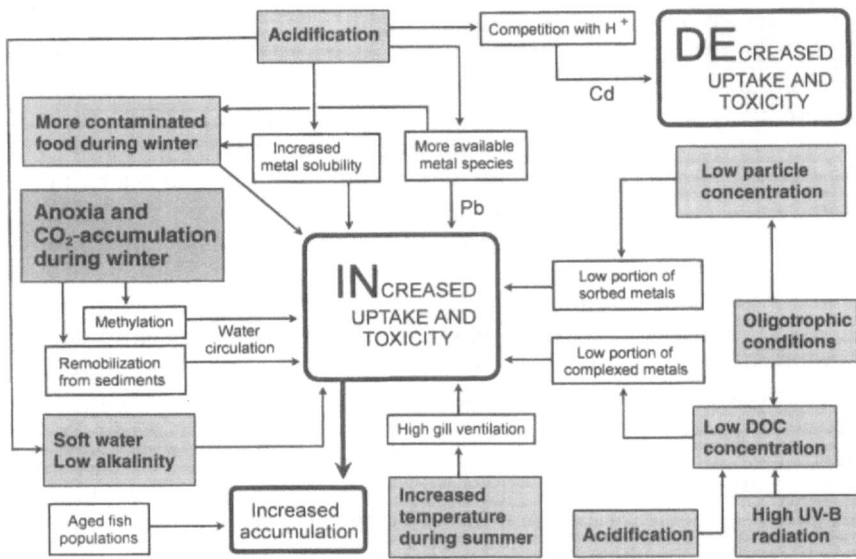

Figure 6. Major biotic and abiotic factors influencing uptake and toxicity of metals in high-altitude and high-latitude lakes.

Due to similar environmental characteristics (e.g., oligotrophy, low buffering capacity, long ice-cover, high precipitation rates) clearwater high altitude and high latitude lakes are very sensitive ecosystems, being extremely susceptible to even slight changes of the environment. Apart from direct toxic effects of low pH, acidification increases the bioavailability of metals for fish. Furthermore, low concentrations of DOC and suspended particles influence uptake and toxicity of metals in fish from clear high-altitude and high-latitude lakes. Biotic and abiotic factors controlling bioavailability of metals to fish are summarized in Figure 6. Since even very low concentrations of Cd and Pb may lead to high metal concentrations in fish, evaluation of critical metals loads for clear high-altitude and high-latitude lakes is of major importance. Comparison between lake types (clearwater lakes, brown water lakes, glacier-fed lakes) would increase our knowledge of the significance of DOC and particulate matter as controlling factors of metal toxicity to fish from softwater lakes.

Acknowledgments
This work was funded by the "Fonds zur Förderung der wissenschaftlichen Forschung in Österreich", project no. P08970-BIO. We thank Drs. W. Wieser, R. Psenner, and R. Dallinger for a critical reading of the manuscript.

References

Alabaster, J.S. and Lloyd, R. (1980) Water quality criteria for freshwater fish. *Butterworths*, London.

Arimoto, R. and Duce, R.A. (1986) Dry deposition models and the air-sea exchange of trace elements. *J. Geophys. Res.* 88:2787–2792.

Atteia, O. (1994) Major and trace elements in precipitation on western Switzerland. *Atmosph. Environ.* 28:3617–3624.

Babich, H. and Borenfreund, E. (1990) *In vitro* cytotoxicities of inorganic lead and dialkyl and trialkyl lead compounds to fish cells. *Bull. Environ. Contam. Toxicol.* 44:456–460.

Balikungeri, A. and Haerdi, W. (1988) Complexing abilities of hydrous manganese oxide surfaces and their role in the speciation of heavy metals. *Intern. J. Anal. Chem.* 34:215–225.

Balistrieri, L.S., Murray, J.W. and Paul, B. (1992) The biochemical cycling of trace metals in the water column of Lake Sammamish, Washington: Response to seasonally anoxic conditions. *Limnol. Oceanogr.* 37:529–548.

Baker, J.P. and Christensen, S.W. (1991) Effects of acidification on biological communities in aquatic ecosystems. *In:* Charles, D.F. (Ed) *Acidic deposition and aquatic ecosystems.* Springer-Verlag, New York, Berlin, Heidelberg, pp 83–106.

Baker, L.A., Herlihy, A.T., Kaufmann, P.R. and Eilers, J.M. (1991) Acidic lakes and streams in the United States: the role of acidic deposition. *Science* 252:1151–1154.

Barros, A.P. and Lettenmaier, D.P. (1993) Dynamic modeling of the spatial distribution of precipitation in remote mountainous areas. *Monthly Weather Rev.* 21:1195–1214.

Beattie, J.H. and Pascoe, D. (1978) Cadmium uptake by rainbow trout, *Salmo gairdneri*, eggs and alevins. *J. Fish Biol.* 13:631–637.

Bengtsson, B.E. and Larsson, A. (1986) Vertebral deformities and physiological effects in four-horn sculpin (*Myoxocephalus quadricornis*) after long-term exposure to a simulated heavy metal-containing effluent. *Aquat. Toxicol.* 9:215–229.

Benoit, D.A., Leonard, E.N., Christensen, G.M. and Fiandt, J.T. (1976) Toxic effects of cadmium on three generations of brook trout (*Salvelinus fontinalis*). *Trans. Amer. Fish. Soc.* 4:550–560.

Bernes, C. (1991) Acidification and liming of Swedish freshwaters. *Swedish Environmental Protection Agency*, Monitor 12, Solna, Sweden.

Berrie, L.A., Gregor, D., Hargrave, B., Lake, R., Muir, D., Shearer, R., Tracey, B. and Bidleman, T. (1992) Arctic contaminants: sources, occurrence and pathways. *Sci. Tot. Environ.* 122:1–74.

Bilinski, E. and Jonas, R.E.E. (1973) Effects of cadmium and copper on the oxidation of lactate by rainbow trout (*Salmo gairdneri*) gills. *J. Fish. Res. Board Can.* 30:1553–1558.

Block, M. (1991) Uptake of cadmium in fish – Effects of xanthates and diethyldithiocarbamate. Comprehensive summaries of Uppsala dissertations from the Faculty of Science, Uppsala, Sweden. *Acta Universitatis Uppsaliensis* 326:37 pp.

Block, M. and Wicklund Glynn, A. (1992) Influence of xanthates on the uptake of Cd-109 by Eurasian dace (*Phoxinus phoxinus*) and rainbow trout (*Oncorhynchus mykiss*). *Environ. Toxicol. Chem.* 11:873–879.

Block, M., Wicklund Glynn, A. and Pärt, A. (1991) Xanthate effects on cadmium uptake and intracellular distribution in rainbow trout (*Oncorhynchus mykiss*) gills. *Aquat. Toxicol.* 20:267–284.

Blumenthaler, M. and Ambach, W. (1990) Indication of increasing solar ultraviolet-B radiation flux in alpine regions. *Science* 248:206–208.

Blumenthaler, M., Ambach, W. and Rehwald, W. (1992) Solar UV-A and UV-B radiation fluxes at tow alpine stations at different altitudes. *Theo. Appl. Climatol.* 46:39–44.

Blumenthaler, M., Ambach, W. and Huber, M. (1993) Altitude effect of solar UV radiation dependent on albedo, turbidity and solar elevation. *Meteorol. Z.* 2:116–120.

Bodaly, R.A., Rudd, J.W.M., Fudge, R.J.P. and Kelly, C.A. (1993) Mercury concentrations in fish related to size of remote Canadian shield lakes. *Can. J. Fish. Aquat. Sci.* 50:980–987.

Bohn, A. and Fallis, B.W. (1978) Metal concentrations (As, Cd, Cu, Pb and Zn) in shorthorn sculpins, *Myoxocephalus scorpius* (L.) and Arctic char, *Salvelinus alpinus* (L.), from the vicinity of Strathacona Sound, Northwest Territories. *Wat. Res.* 12:659–663.

Borg, H. (1983) Trace metals in Swedish natural freshwaters. *Hydrobiologia* 101:27–34.

Boutron, C.F. and Patterson, C.C. (1987) Relative levels of natural and anthropogenic lead in recent Antarctic snow. *J. Geophys. Res.* 92:8454–8464.

Boutron, C.F., Görlach, U., Candelone, J.P., Bolshov, M. and Delmas, R. (1991) Concentrations of lead, copper, cadmium and zinc in Greenland snow since the late 1960's. *Nature* 353:153–156.

Boutron, C.F., Ducroz, M., Görlach, U., Jaffrezo, J.L., Davidson, C.I. and Bolshov, A. (1993) Variation in heavy metal concentrations in fresh Greenland snow from January to August 1989. *Atmosph. Environ.* 27A:2773–2779.

Bradley, S.B. and Cox, J.J. (1988) The potential availability of cadmium, copper, iron, lead, manganese, nickel and zinc in standard river sediment (NBS 1645). *Environ. Technol. Let.* 9:733–739.

Breder, R. (1988) Cadmium in European Inland waters. *In:* Stoeppler, M. and Piscator, M. (Eds) *Environmental Toxins Series 2.* Cadmium – 3rd IUPAC Cadmium Workshop, Jülich, FRG, August 1985. Springer, Berlin, pp 159–169.

Brown, V.M. (1968) The calculation of the acute toxicity of mixtures of poisons to rainbow trout. *Wat. Res.* 2:723–733.

Brown, V., Shurben, D., Miller, W. and Crane, M. (1994) Cadmium toxicity to rainbow trout *Oncorhynchus mykiss* (Walbaum) and brown trout *Salmo trutta* (L.) over extended exposure periods. *Ecotox. Environ. Safety* 29:38–46.

Buhl, K.J. and Hamilton, S.J. (1991) Relative sensitivity of early life stages of Arctic grayling, coho salmon and rainbow trout to 9 inorganics. *Ecotox. Environ. Safety* 22:184–197.

Byrne, R.H., Kump, L.R. and Cantrell, K.J. (1988) The influence of temperature and pH on trace metal speciation in seawater. *Mar. Chem.* 25:163–181.

Cabrera, S., Bozzo, S. and Fuenzalida, H. (1995) Variations in UV radiation in Chile. *J. Photochem. Photobiol. B: Biol.* 28:137–142.

Cairns, J., Heath, A.G. and Parker, B.C. (1975) Temperature influence on chemical toxicity to aquatic organisms. *J. Water Pollut. Control Fed.* 47:267–280.

Calamari, D., Marchetti, R. and Vailati, G. (1980) Influence of water hardness on cadmium toxicity to *Salmo gairdneri* Rich. *Water Res.* 14:1421–1426.

Campbell, J.H. and Evans, R.D. (1987) Inorganic and organic ligand binding of lead and cadmium and resultant implications for bioavailability. *Sci. Tot. Environ.* 62:219–217.

Campbell, P.G.C. and Stokes, P.M. (1985) Acidification and toxicity of metals of aquatic biota. *Can. J. Fish. Aquat. Sci.* 42:2034–2049.

Campbell, P.G.C., Lewis, A.G., Chapman, P.M., Crowder, A.A., Fletcher, W.K., Imber, B., Luoma, S.N., Stokes, P.M. and Winfrey, M. (1988) Biologically available metals in sediments. National Research Council of Canada – *Scientific Committee on Scientific Criteria for Environmental Quality*, NRCC No. 27694, Ottawa, Canada, 298 pp.

Carrier, R. and Beitinger, T.L. (1988) Resistance of temperature tolerance ability of green sunfish to cadmium exposure. *Bull. Environ. Contam. Toxicol.* 40:475–480.

Chakoumakos, C., Russo, R.C. and Thurston, R.V. (1979) Toxicity of copper to cutthroat trout (*Salmo clarkii*) under different conditions of alkalinity, pH and hardness. *Environ. Sci. Technol.* 13:213–219.

Chapman, G.A. (1978) Toxicities of cadmium, copper and zinc to four juvenile stages of chinook salmon and steelhead. *Trans. Amer. Fish. Soc.* 107:841–847.

Chester, R., Nimmo, M., Murphy, K.J.T. and Nicolas, E. (1990) Atmospheric trace metals transported to the western Mediterranean: data from a station on Cap Ferrat. *Water Pollution Research Reports* 20:597–612.

Christensen, G.M. (1975) Biochemical effects of methylmercuric chloride, cadmium chloride and lead nitrate on embryos and alevins of the brook trout (*Salvelinus fontinalis*). *Toxic. Appl. Pharmacol.* 32:191–197.

Clements, W.H., Cherry, D.S. and Cairns Jr., J. (1988) Impact of heavy metals on insect communities in streams: a comparison of observational and experimental results. *Can. J. Fish. Aquat. Sci.* 45:2017–2025.

Clough, W.S. (1975) The deposition of particles on moss and grass surfaces. *Atmosph. Environ.* 9:1113–1119.

Colin, J.L., Jaffrezo, J.L. and Gros, J.M. (1990) Solubility of major species in precipitation: factors of variation. *Atmosph. Environ.* 24A:537–544.

Cusimano, R.F., Brakke, D.F. and Chapman, G.A. (1986) Effects of pH on the toxicities of cadmium, copper and zinc to steelhead trout (*Salmo gairdneri*). *Can. J. Fish. Aquat. Sci.* 43:1497–1503.

Czarnezki, J.M. (1985) Accumulation of lead in fish from Missouri streams impacted by lead mining. *Bull. Environ. Contam. Toxicol.* 34:736–745.

Dallinger, R. and Kautzky, H. (1985) The importance of contaminated food for the uptake of heavy metals by rainbow trout (*Salmo gairdneri*): a field study. *Oecologia* 67:82–89.

Dallinger, R., Prosi, F., Segner, H. and Back, H. (1987) Contaminated food uptake of heavy metals by fish: a review and a proposal for further research. *Oecologia* 73:91–98.

Dallinger, R., Egg, M., Köck, G. and Hofer, R. (1997) The role of metallothionein in cadmium accumulation of Arctic char (*Salvelinus alpinus*) from high alpine lakes. *Aquat. Toxicol.* 38:47–66.

Davies, P.H., Goettl, J.P., Sinley, J.R. and Smith, N.F. (1976) Acute and chronic toxicity of lead to rainbow trout, *Salmo gairdneri*, in hard and soft water. *Wat. Res.* 10:199–206.

Davies, P.H., Gorman, W.C., Carlson, C.A. and Brinkman, S.F. (1993) Effect of hardness on bioavailability and toxicity of cadmium to rainbow trout. *Chem. Spec. Bioavail.* 5:67–77.

Deniseger, J., Erickson, L.J., Austin, A., Roch, M. and Clark, M.J.R. (1990) The effects of decreasing heavy metal concentration on the biota of Butle Lake, Vancouver Island, British Columbia. *Wat. Res.* 24:403–416.

DiToro, D.M., Mahony, J.D., Hansen, D.J., Scott, K.J., Carlson, A.R. and Ankley, G.T. (1992) Acid volatile sulfide predicts the acute toxicity of cadmium and nickel in sediments. *Environ. Sci. Technol.* 26:96–101.

Dixon, D.G., Hodson, P.V., Klaverkamp, J.F., Lloyd, K.M. and Roberts, J.R. (1985) The role of biochemical indicators in the assessment of ecosystem health – their development and validation. *National Research Council of Canada*, No. 24371, Ottawa, Canada.

Douben, P.E.T. (1989) Uptake and elimination of waterborne cadmium by the fish *Noemacheilus barbatulus* L. (Stone loach). *Arch. Environ. Contam. Toxicol.* 18:576–586.

Douben, P.E.T. (1990) A mathematical model for cadmium in the stone loach (*Noemacheilus barbatulus* L.) from the River Ecclesbourne, Derbyshire. *Ecotox. Environ. Safety* 19:160–183.

Douben, P.E.T. and Koeman, J.H. (1989) Effect of sediment on cadmium and lead in the stone loach (*Noemacheilus barbatulus* L.). *Aquat. Toxicol.* 15:253–268.

Drummond, R.A. and Carlson, R.W. (1977) Procedures for measuring cough (gill purge) rates of fish. *US-EPA, Ecological Research Series*, EPA 600/3-77-133, Duluth.

Dwyer, F.J., Schmitt, C.J., Finger, S.E. and Mehrle, P.M. (1988) Biochemical changes in long ear sunfish, *Lepomis megalotis*, associated with lead, cadmium and zinc from mining tailings. *J. Fish Biol.* 33:307–317.

Eaton, G.J., McKim, J.M. and Holcombe, G.W. (1978) Metal toxicity to embryos and larvae of seven freshwater fish species – I. Cadmium. *Bull. Environ. Contam. Toxicol.* 19:95–103.

Fjerdingstad, E. and Nilssen, J.P. (1983) Heavy metals distribution in Norwegian acidic lakes: a preliminary record. *Arch. Hydrobiol.* 96:190–204.

Florence, T.M., Morrison, G.M. and Stauber, J.L. (1992) Determination of trace element speciation and the role of speciation in aquatic toxicity. *Sci. Tot. Environ.* 125:1–13.

Förstner, U. and Wittman, G. (1983) *Metall pollution in the aquatic environment*. Springer Verlag, Berlin, Heidelberg, New York, 485 pp.

Francis, A.F. and Dodge, C. (1990) Anaerobic microbial remobilization of toxic metals coprecipitated with iron oxide. *Environ. Sci. Technol.* 24:373–378.

Fry, F.E.J. (1971) The effect of environmental factors on the physiology of fish. *In:* Hoar, W.S. and Randall, D.J. (Eds) *Fish physiology*. Vol. VI. Academic Press, New York, London, pp 1–99.

Fu, G., Allen, H.E. and Cowan, C.E. (1991) Adsorption of cadmium and copper by manganese oxide. *Soil Science* 152:72–81.

Gardiner, J. (1974) The chemistry of cadmium in natural waters. I. A study of cadmium complex formation using the cadmium specific-ion electrode. *Wat. Res.* 8:23–30.

George, S.G. and Langston, W.J. (1994) Metallothionein as an indicator of water quality – assessment of the bioavailability of cadmium, copper, mercury and zinc in aquatic animals at the cellular level. *In:* Sutcliffe, D.W. (Ed) *Water quality and stress indicators in marine and freshwater ecosystems: linking levels of organization (individuals, populations, communities).* Freshwater Biological Association Special Publications. Freshwater Biological Association, Ambleside, pp 138–153.

Haines, T.A. (1981) Acidic precipitation and its consequence for aquatic ecosystems: a review. *Trans. Amer. Fish. Soc.* 110:669–707.

Haines, T.A. and Brumbaugh, W.G. (1994) Metal concentration in the gill, gastrointestinal tract and carcass of white suckers (*Catostomus commersoni*) in relation to lake acidity. *Water Air Soil Pollut.* 73:265–274.

Hamilton, S.J. and Haines, T.A. (1989) Bone characteristics and metal concentrations in white suckers (*Catostomus commersoni*) from one neutral and three acidified lakes in Maine. *Can. J. Fish. Aquat. Sci.* 46:440–446.

Hamilton, S.J. and Reash, R.J. (1988) Bone development in creek chub from a stream chronically polluted with heavy metals. *Trans. Amer. Fish. Soc.* 117:48–54.

Hamilton, S.J., Mehrle, P.M. and Jones, J.R. (1987a) Evaluation of metallothionein measurement as a biological indicator of stress from cadmium in brook trout. *Trans. Amer. Fish. Soc.* 116:551–560.

Hamilton, S.J., Mehrle, P.M. and Jones, J.R. (1987b) Cadmium-saturation technique for measuring metallothionein in brook trout. *Trans. Amer. Fish. Soc.* 116:541–550.

Haux, C. and Larsson, A. (1982) Influence of inorganic lead on the biochemical blood composition in the rainbow trout, *Salmo gairdneri*. *Ecotox. Environ. Safety* 6:28–34.

Haux, C., Larsson, A., Lithner, G. and Sjöbeck, M.L. (1986) A field study of physiological effects on fish in lead-contaminated lakes. *Environ. Toxicol. Chem.* 5:283–288.

Heath, A.G. (1987) *Water pollution and fish physiology*. CRC Press, Boca Raton, 245 pp.

Heit, M., Schofield, C., Driscoll, C.T. and Hodgkiss, S.S. (1989) Trace element concentrations in fish from three Adirondack lakes with different pH values. *Water Air Soil Pollut.* 44:9–30.

Hendershot, W.H., Dufresne, A., Lalande, H. and Courchesne, F. (1986) Temporal variation in aluminium speciation and concentration during snowmelt. *Water Air Soil Pollut.* 31:231–237.

Hendriksen, A., Kämäri, J., Posch, M. and Wilander, A. (1992) Critical loads of acidity: nordic surface waters. *Ambio* 21:356–363.

Hering, J.G. and Morel, F.M.M. (1990) The kinetics of trace metal complexation: implications for metal reactivity in natural waters. *In*: Stumm, W. (Ed) *Aquatic Chemical Kinetics*. Wiley, New York, pp 145–171.

Hochachka, P.W. and Somero, G.N. (1971) Biochemical adaptation to the environment. *In:* Hoar, W.S. and Randall, D.J. (Eds) *Fish physiology*. Vol. VI. Academic Press, New York, London, pp 100–156.

Hodson, P.V. (1976) δ-Aminolevulinic acid dehydratase activity of fish blood as an indicator of a harmful exposure to lead. *J. Fish. Res. Board Can.* 33:268–271.

Hodson, P.V. and Sprague, J.B. (1975) Temperature induced changes in acute toxicity of zinc to Atlantic salmon (*Salmo salar*). *J. Fish. Res. Board Can.* 32:1–10.

Hodson, P.V., Blunt, B.R., Spry, D.J. and Austen, K. (1977) Evaluation of erythrocyte δ-aminolevulinic acid dehydratase activity as a short-term indicator in fish af an harmful exposure to lead. *J. Fish. Res. Board Can.* 34:501–508.

Hodson, P.V., Blunt, P.R. and Spry, D.J. (1978) pH-Induced changes in blood lead of lead-exposed rainbow trout (*Salmo gairdneri*). *J. Fish. Res. Board Can.* 35:437–445.

Hodson, P.V., Blunt, B.R. and Whittle, D.M. (1984a) Monitoring lead exposure of fish. *In:* Cairns, V.W., Hodson, P.V. and Nriagu, J.O. (Eds) *Contaminant effects on fisheries*. Wiley, New York, Toronto, pp 87–98.

Hodson, P.V., Whittle, D.M., Wong, P.T.S., Borgmann, U., Thomas, R.L., Chau, Y.K., Nriagu, J.O. and Hallet, D.J. (1984b) Lead contamination of the Great Lakes and its potential effects on aquatic biota. *In:* Nriagu, J.O. and Simmons, M.S. (Eds) *Toxic contaminants in the great lakes*. Adv. Environ. Sci. Technol., Vol. 14. Wiley, New York, Toronto, 527 pp.

Hofer, R. and Medgyesy, N. (1997) Growth, reproduction and feeding of dwarf Arctic char, *Salvelinus alpinus*, from an Alpine high mountain lake. *Arch. Hydrobiol.* 138:509–524.

Hofer, R., Pittracher, H., Köck, G. and Weyrer, S. (1994) Metal accumulation by Arctic char (*Salvelinus alpinus*) in a remote acid alpine lake. *In:* Müller, R. and Lloyd, R. (Eds) *Sublethal and chronic effects of pollutants on freshwater fish*. Fishing New Books, Blackwell Science, Oxford, pp 294–300.

Hoffmann, H.J. (1987) Untersuchungen der Einflüsse von Temperatur, Strömung und Sauerstoffgehalt auf das Akkumulationsverhalten von Quecksilber in Karpfen (*Cyprinus carpio* L.). *In:* Lille, K., de Haar, U., Elster, H.J., Karbe, L., Schwoerbel, I. and Simonis, W. (Eds) *Bioakkumulation in Nahrungsketten*. VCH, Weinheim, New York, pp 211–218.

Hogstrand, C. and Haux, C. (1990) Metallothionein as an indicator of heavy–metal exposure in two subtropical fish species. *J. Exp. Mar. Biol. Ecol.* 138:69–84.

Hogstrand, C., Lithner, G. and Haux, C. (1989) Relationship between metallothionein, copper and zinc in perch (*Perca fluviatilis*) environmentally exposed to heavy metals. *Mar. Environ. Res.* 28:179–182.

Hoigne, J. (1990) Formulation and calibration of environmental reaction kinetics: oxidation by aqueous photooxidants as an example. *In:* W. Stumm (Ed) *Aquatic chemical kinetics.* Wiley, New York, pp 43–70.

Holcombe, W., Benoit, D.A., Leonard, D.N., McKim, J.M. (1976) Long-term effects of lead exposure on three generations of brook trout (*Salvelinus fontinalis*). *J. Fish. Res. Board Can.* 33:1731–1741.

Hong, S., Candelone, J.P., Patterson, C.C. and Boutron, C.F. (1996) History of ancient copper smelting pollution during roman and medieval times recorded in Greenland ice. *Science* 272:246–249.

Honsig-Erlenburg, W. and Psenner, R. (1986) Zur Frage der Versauerung von Hochgebirgsseen in Kärnten. *Carinthia* II 176:443–461.

Howells, G., Dalziel, T.R.K., Reader, J.P. and Solbe, J.F. (1990) EIFAC water quality criteria for European freshwater fish: report on aluminum. *Chem. Ecol.* 4:117–173.

Hutchinson, N.J. and Sprague, J.B. (1987) Reduced lethality of Al, Zn and Cu mixtures to American flagfish by complexation with humic substances in acidified soft waters. *Environ. Toxicol. Chem.* 6:755–765.

Hwang, P. P., Lin, S. W. and Lin, H. C. (1995) Different sensitivities to cadmium in tilapia larvae (*Oreochromis mossambicus*; Teleostei). *Arch. Environ. Contam. Toxicol.* 29:1–7.

Iivonen, P., Piepponen, S. and Verta, M. (1992) Factors affecting trace-metal bioaccumulation in Finnish headwater lakes. *Environ. Pollut.* 78:87–95.

Injuk, J., Otten, P., Laane, R., Maenhaut, W. and Van Grieken, R. (1992) Atmospheric concentrations and size distributions of aircraft-sampled Cd, Cu, Pb and Zn over the Southern Bight of the North Sea. *Atmosph. Environ.* 26A:2499–2508.

Jackim, E. (1973) Influence of lead and other metals on fish δ-aminolevulinate dehydrase activity. *J. Fish. Res. Board Can.* 30:560–562.

Jackim, E., Hamlin, J.M. and Sonis, S. (1970) Effects of metal poisoning on five liver enzymes in the killifish (*Fundulus heteroclitus*). *J. Fish. Res. Board Can.* 27:383–390.

Jaworski, J.F. (1978) *Effects of lead in the environment – quantitative aspects.* National Research Council Canada, No. 16736, Ottawa, Canada.

Jefferies, D.S., Cox, C.M. and Dillon, P.J. (1979) Depression of pH in lakes and streams in central Ontario during snowmelt. *J. Fish. Res. Board Can.* 36:640–646.

Johansson, K., Bringmark, E., Lindevall, L. and Wilander, A. (1995) Effects of acidification on the concentration of heavy metals in running waters in Sweden. *Water Air Soil Pollut.* 85:779–784.

Johansson-Sjöbeck, M.L. and Larsson, A. (1979) Effects of inorganic lead on δ-aminolevulinic acid dehydratase activity and hematological variables in the rainbow trout, *Salmo gairdneri.* *Arch. Environ. Contam. Toxicol.* 8:419–431.

Johnson, B.B. (1990) Effect of pH, temperature and concentration on the adsorption of cadmium on goethite. *Environ. Sci. Technol.* 24:112–118.

Jop, K.M., Askew, A.M. and Foster, R.B. (1995) Development of a water-effect ratio for copper, cadmium and lead for the Great Works River in Maine using *Ceriodaphnia dubia* and *Salvelinus fontinalis.* *Bull. Environ. Contam. Toxicol.* 54:29–35.

Kerr, J. B. and McElroy, C. T. (1993) Evidence for large upward trends of ultraviolet-B radiation linked to ozone depletion. *Science* 262:1032–1034.

Klaverkamp, J.F., MacDonald, W.A., Duncan, D.A. and Wageman, R. (1984) Metallothionein and acclimation to heavy metals in fish: a review. *In:* Cairns, V.W., Hodson, P.V. and Nriagu, J.O. (Eds) *Contaminant effects on fisheries.* Wiley, New York, Toronto, pp 100–113.

Köck, G. (1996) Die toxische Wirkung von Kupfer, Cadmium, Quecksilber, Chrom, Nickel, Blei und Zink auf Fische – Beiträge zur Festlegung vom Immissionsbereichen für Metalle für österreichische Gewässer. *In:* Steinberg, C., Calmano, W., Klapper, H. and Wilken, R.D. (Eds) *Handbuch Angewandte Limnologie.* Ecomed, Landsberg, pp 1–167.

Köck, G., Bucher, F. and Hofer, R. (1991) *Schwermetalle und Fische – Anforderungen an die Wassergüte.* Report of the Austrian Ministry of Agriculture and Forestry, Vienna, Austria, 324 pp.

Köck, G., Hofer, R. and Wögrath S. (1995) Accumulation of trace metals (Pb, Cd, Cu, Zn) in Arctic char (*Salvelinus alpinus*) from oligotrophic Alpine lakes: relation to lake alkalinity. *Can. J. Fish. Aquat. Sci.* 52:2367–2376.

Köck, G., Triendl, M. and Hofer, R. (1996a) Seasonal patterns of metal accumulation in Arctic char (*Salvelinus alpinus*) from an oligotrophic Alpine lake related to temperature. *Can. J. Fish. Aquat. Sci.* 53:780–786.

Köck, G., Noggler, M. and Hofer, R. (1996b) Pb in otoliths and opercula of Arctic char (*Salvelinus alpinus*) from oligotrophic lakes. *Wat. Res.* 30:1919–1923.

Köck, G., Triendl, M. and Hofer, R. (1998) Lead (Pb) uptake in Arctic char (*Salvelinus alpinus*) from oligotrophic alpine lakes: gills *versus* digestive tract. *Water Air Soil Pollut.*: 102:303–312.

Kuroshima, R., Kimura, S., Date, K. and Yamamoto, Y. (1993) Kinetic analysis of cadmium toxicity to Red Sea bream, *Pagrus major. Ecotox. Environ. Safety* 25:300–314.

Landers, D.H., Bayley, S.E., Ford, J., Gunn, J.M., Lükewille, A., Norton, S.A., Steinberg, C.E.W., Vesely, J. and Zahn, M.T. (1994) Group report: interactions among acidification, phosphorus, contaminants and biota in freshwater ecosystems. *In:* Steinberg, C. and Wright, R.F. (Eds) *Acidification of freshwater ecosystems – implications for the future.* Dahlem Workshop Reports, Environmental Sciences Series Report 14. Wiley, New York, pp 185–200.

Larsson, A., Bengtsson, B.E. and Haux, C. (1981) Disturbed ion balance in flounder, *Platichthys flesus* L. exposed to sublethal levels of cadmium. *Aquat. Toxicol.* 1:19–35.

Larsson, A., Haux, C. and Sjöbeck, M. L. (1985) Fish physiology and metal pollution: results and experiences from laboratory and field studies. *Ecotox. Environ. Safety* 9:250–281.

LaZerte, B., Evans, D. and Grauds, P. (1989) Deposition and transport of trace metals in an acidified catchment of central Ontario. *Sci. Tot. Environ.* 87/88:209–221.

Ledin, A., Petterson, C., Allard, B. and Aastrup, M. (1989) Background concentration ranges of heavy metals in Swedish groundwaters from crystalline rocks: a review. *Water Air Soil Pollut.* 47:419–426.

Lehman, R.M. and Mills, A.L. (1994) Field evidence for copper mobilization by dissolved organic matter. *Wat. Res.* 28:2487–2497.

Lietz, W. and Galling, G. (1989) Metals from sediments. *Wat. Res.* 23:247–252.

Lockhart, W.L., Wagemann, R., Tracey, B., Sutherland, D. and Thomas, D.J. (1992) Presence and implications of chemical contaminants in the freshwaters of Canadian Arctic. *Sci. Tot. Environ.* 122:165–243.

Losno, R., Bergametti, G. and Buat-Menard, P. (1988) The partitioning of zinc in Mediterranean rainwater. *Geophys. Res. Lett.* 15:1389–1392.

MacLeod, J.C. and Pessah, E. (1973) Temperature effects on mercury accumulation, toxicity and metabolic rate in rainbow trout (*Salmo gairdneri*). *J. Fish. Res. Board Can.* 30:485–492.

Malm, O., Pfeiffer, W.C., Fiszman, M. and Azcue, J.M. (1988) Transport and availability of heavy metals in the Paraiba do Sul-Guandu River system, Rio de Janeiro State, Brazil. *Sci. Tot. Environ.* 75:201–209.

Mannio, J., Verta, M. and Järvinen, O. (1993) Trace metal concentrations in the water of small lakes, Finland. *Appl. Geochem.* 2:57–59.

Mannio, J., Järvinen, O., Tuominen, R. and Verta, M. (1995) Survey of trace elements in lake waters of Finnish Lapland using the ICP-MS technique. *Sci. Tot. Environ.* 160/161:433–439.

Mantei, E.J. and Foster, M.V. (1991) Heavy metals in stream sediments: effects of human activities. *Environ. Geol. Water Sci.* 18:95–104.

Marchetto, A., Mosello, R., Psenner, R., Barbieri, A., Bendetta, G., Tait, D., Tartari, G.A. (1994) Evaluation of the level of acidification and the critical load for Alpine lakes. *Ambio* 23:150–154.

Mart, L. (1983) Seasonal variations of Cd, Pb, Cu, Ni in snow from the eastern Arctic Ocean. *Tellus* 35 B: 131–134.

McCarty, L.S., Henry, J.A.C. and Houston, A.H. (1978) Toxicity of cadmium to goldfish, *Carassius auratus* in hard and soft water. *J. Fish. Res. Board Can.* 35:35–42.

McDonald, D.G., Reader, J.P. and Dalziel, T.K.R. (1989) The combined effects of pH and trace metals on fish ion regulation. *In:* Morris, R., Brown, D.J.A., Taylor, E.W. and Brown, J.A. (Eds) *Acid Toxicity and Aquatic Animals.* Society for Experimental Biology Seminar Series. Cambridge University Press, Cambridge, pp 221–242.

McGeachy, S. and Dixon, D.G. (1989) The impact of temperature on the acute toxicity of arsenate and arsenite to rainbow trout (*Salmo gairdneri*). *Ecotox. Environ. Safety* 17:86–93.

Mehrle, P.M., Haines, T.A., Hamilton, S., Ludke, J.L., Mayer, F.L. and Ribickm, M.A. (1982) Relationship between body contaminants and bone development in east-coast striped bass. *Trans. Amer. Fish. Soc.* 111:231–241.

Merian, E. (1991) *Metals and their compounds in the environment – occurrence, analysis and biological relevance.* VCH, Weinheim, New York, 1438 pp.

Merlini, M. and Pozzi, G. (1977) Lead and freshwater fishes: Part 1 – Lead accumulation and water pH. *Environ. Pollut.* 12:167–172.

Migon, C. and Caccia, J.-L. (1990) Separation of anthropogenic and natural emissions of particulate heavy metals in the Western Mediterranean atmosphere. *Atmosph. Environ.* 24A: 399–405.

Miller, H. (1987) *Metal contents in organs of salmonids from acidified surface waters.* Ph.D. thesis, Department of Zoology and Hydrobiology, University of Munich, Germany.

Miyahara, T., Oh-E, Y., Takaine, E. and Kozuka, H. (1983) Interaction between cadmium and zinc, copper, or lead in relation to the collagen and mineral content of embryonic chick bone in tissue culture. *Toxicol. Appl. Pharmacol.* 67:41–48.

Moore, J.W. and Ramamoorthy, S. (1984) *Heavy metals in natural waters – applied monitoring and impact assessment.* Springer, New York, Berlin, Heidelberg, Tokyo, 268 pp.

Munkittrick, K.R. and Dixon, D.G. (1988) Growth, fecundity and energy stores of white sucker (*Catostomus commersoni*) from lakes containing elevated levels of copper and zinc. *Can. J. Fish. Aquat. Sci.* 45:1355–1365.

Murad, A. and Houston, A.H. (1988) Leucocytes and leucopoietic capacity in goldfish, *Carassius auratus*, exposed to sublethal levels of cadmium. *Aquat. Toxicol.* 13:141–154.

Nakagawa, H. and Ishio, S. (1988a) Toxicity of cadmium and its accumulation on the egg and larva of medaka *Oryzias latipes*. *Nippon Suisan Gakkaishi* 54:2153–2158.

Nakagawa, H. and Ishio, S. (1988b) Aspects of accumulation of cadmium ion in the egg of medaka *Oryzias latipes*. *Nippon Suisan Gakkaishi* 54:2159–2164.

Nakagawa, H. and Ishio, S. (1989a) Aspects of accumulation of copper, manganese and zinc ions in the egg of medaka *Oryzias latipes*. *Nippon Suisan Gakkaishi* 55:117–121.

Nakagawa, H. and Ishio, S. (1989b) Properties as a cation exchanger of chorions of medaka *Oryzias latipes* and rainbow trout *Salmo gairdneri*. *Nippon Suisan Gakkaishi* 55:123–129.

Nakagawa, H. and Ishio, S. (1989c) Effects of water pH on the toxicity and accumulation of cadmium in eggs and larvae of medaka *Oryzias latipes*. *Nippon Suisan Gakkaishi* 55:327–331.

Nakagawa, H. and Ishio, S. (1989d) Effects of water hardness on the toxicity and accumulation of cadmium in eggs and larvae of medaka *Oryzias latipes*. *Nippon Suisan Gakkaishi* 55:321–326.

Nelson, J.A. and Mitchell, G.S. (1992) Blood chemistry response to acid exposure in yellow perch (*Perca flavescens*) – comparison of populations from naturally acidic and neutral environments. *Physiol. Zool.* 65:493–514.

Nelson, W.O. and Campbell, P.G.C. (1991) The effects of acidification on the geochemistry of Al, Cd, Pb and Hg in freshwater environments: a literature review. *Environ. Pollut.* 71:91–130.

Newsome, C.S. and Piron, R.D. (1982) Aetiology of skeletal deformities in the zebra danio fish (*Brachydanio rerio* Hamilton Buchanan). *J. Fish Biol.* 21:231–237.

Ney, J.J. and vanHassel, J.H. (1983) Sources of variability in accumulation of heavy metals by fishes in a roadside stream. *Arch. Environ. Contam. Toxicol.* 12:701–706.

Nguyen, V.D., Merks, A.G.A. and Valent, P. (1990) Atmospheric deposition of acid, heavy metals, dissolved organic carbon and nutrient in the Dutch delta area in 1980–1986. *Sci. Tot. Environ.* 99:77–91.

Niimi, A.J. (1987) Biological half-lives of chemicals in fishes. *Rev. Environ. Contam. Toxicol.* 99:1–46.

Nordberg, G.F., Goyer, R.A. and Clarkson, T.W. (1985) Impact of effects of acidic precipitation on toxicity of metals. *Environ. Health Perspec.* 63:169–180.

Norton, S.A., Dillon, P.J., Evans, R.D., Mierle, G. and Kahl, J.S. (1990) The history of atmospheric deposition of Cd, Hg and Pb in North America: evidence from lake and peat bog sediment. *In:* Lindberg, S.E. and Page, A.L. (Eds) *Sources, deposition and canopy interactions. Vol. 3: Acidic precipitation.* Springer, New York, Berlin, Heidelberg, pp 73–102.

Nriagu, J.O. (1979) Global inventory of natural and anthropogenic emissions of trace metals to the atmosphere. *Nature* 279:409–411.

Nriagu, J.O. (1990) Global metal pollution: Poisoning the biosphere? *Environment* 32:7–33.

Nriagu, J. O. (1996) A History of Global Metal Pollution. *Science* 272:223–224.

Olsson, P.E., Haux, C. and Förlin, L. (1987) Variations in hepatic metallothionein, zinc and copper levels during an annual reproductive cycle in rainbow trout, *Salmo gairdneri*. *Fish Physiol. Biochem*. 3:39–47.

O'Shea, T.A. and Mancy, K.H. (1978) The effect of pH and hardness metal ions on the competitive interactions between trace metal ions and inorganic complexing agents found in natural waters. *Wat. Res*. 12:703–711.

Ottar, B. (1989) Arctic air pollution: a Norwegian perspective. *Atmosph. Environ*. 223:2349–2356.

Pagenkopf, G.K. and Neuman, D.R. (1974) Lead concentration in native trout. *Bull. Environ. Contam. Toxicol*. 12:70.

Pagenkopf, G.K. (1983) Gill surface interaction model for trace metal toxicity to fish: role of complexation, pH and water hardness. *Environ. Sci. Toxicol*. 17:347–352.

Pain, D.J. (1995) Lead in the Environment. *In:* Hoffman, D.J., Rattner, B.A., Allen Burton, Jr., G. and Cairns, Jr., J. (Eds) *Handbook of ecotoxicology*. Lewis Publishers, Boca Raton, Ann Arbor, London, Tokyo, pp 356–391.

Pärt, P. (1990) The perfused fish gill preparation in studies of the bioavailability of chemicals. *Ecotox. Environ. Safety* 19:106–115.

Pärt, P., Svanberg, O. and Kiessling, A. (1985) The availability of cadmium to perfused rainbow trout gills in different water qualities. *Wat. Res*. 19:427–434.

Patrick, F.M. and Loutit, M.W. (1978) Passage of metals to freshwater fish from their food. *Wat. Res*. 12:395–398.

Pettersson, C., Hakanson, K., Karlsson, S. and Allard, B. (1993) Metal speciation in a humic surface water system polluted by acidic leachates from a mine deposit in Sweden. *Wat. Res*. 27:863–871.

Phillips, G.R. and Russo, R.C. (1978) *Metal bioaccumulation in fishes and aquatic invertebrates: a literature review*. EPA-600/3-78–103, United States Environmental Protection Agency, Duluth, 115 pp.

Pickering, Q.H. and Henderson, C. (1966) The acute toxicity of some heavy metals to different species of warm water fishes. *Air Water Pollut*. 10:453–463.

Playle, R.C., Dixon, D.G., Burnison, K. (1993a) Copper and cadmium binding to fish gills: modificaion by dissolved organic carbon and synthetic ligands. *Can. J. Fish. Aquat. Sci*. 50:2667–2677.

Playle, R.C., Dixon, D.G., Burnison, K. (1993b) Copper and cadmium binding to fish gills: estimates of metal-gill stability constants and modeling of metal accumulation. *Can. J. Fish. Aquat. Sci*. 50:2678–2687.

Pohlman, A.A. and McColl, J.G. (1986) Kinetics of metal dissolution from forest soils by soluble organic acids. *J. Environ. Qual*. 15:86–92.

Posner, H.S., Damstra, T. and Nriagu, J.O. (1980) Human health effects of lead. *In:* Nriagu, J.O. (Ed) *The biochemistry of lead in the environment. Part b: biological effects*. Elsevier, Amsterdam, pp 173–221.

Povlsen, A.F., Korsgaard, B. and Bjerregaard, P. (1990) The effect of cadmium on vitellogenin metabolism in estradiol-induced flounder (*Platichthys flesus*, L.) males and females. *Aquat. Toxicol*. 17:253–262.

Psenner, R. (1989) Chemistry of high mountain lakes in the Central Eastern Alps. *Aquat. Sci*. 51:108–128.

Psenner, R. (1994) Environmental impacts on freshwaters: acidification as a global problem. *Sci. Tot. Environ*. 143:53–61.

Psenner, R. and Catalan, J. (1994) Chemical composition of lakes in crystalline basins: a combination of atmospheric deposition, geologic background, biological activity and human action. *In:* Margalef, R. (Ed) *Limnology now: a paradigm of planetary problems*. Elsevier, Amsterdam, pp 255–314.

Rand, G.M., Wells, P.G. and McCarty, L.S. (1995) Introduction to Aquatic Toxicology. *In:* Rand, G.M. (Ed) *Fundamentals of aquatic toxicology – effects, environmental fate and risk assessment*. 2nd ed., Taylor and Francis, pp 3–67.

Rashid, M.A. (1985) *Geochemistry of Marine Humic Substances*. Chpts. 4 und 7. Springer-Verlag, New York, 300 pp.

Reid, S.D., McDonald, D.G. (1988) Effects of cadmium, copper and low pH on ion fluxes in the rainbow trout, *Salmo gairdneri*. *Can. J. Fish. Aquat. Sci*. 45:244–253.

Reimers, N. (1979) A history of a stunted brook trout population in an alpine lake: life span of 24 years. *Calif. Fish Game* 65:196–215.

Reinert R.E., Stone, L.J. and Wilford, W.A. (1974) Effect of temperature on accumulation of methyl mercuric chloride and p, p-DDT by rainbow trout (*Salmo gairdneri*). *J. Fish. Res. Board Can.* 31:1649–1652.

Reuter, J.A. and Perdue, E.M. (1977) Importance of heavy metal-organic matter interactions in natural waters. *Geochom. Cosmochim. Acta* 41:325–334.

Roch, M. and Maly, E. (1979) Relationship of cadmium-induced hypocalcemia with mortality in *Board Can.* 36:1297–1303.

Roch, M., McCarter, J.A., Matheson, A.T., Clark, M.J.R. and Olafson, R.W. (1982) Hepatic metallothionein in rainbow trout (*Salmo gairdneri*) as an indicator of metal pollution in the Campbell River system. *Can. J. Fish. Aquat. Sci.* 39:1596–1601.

Roesijadi, G. (1992) Metallothioneins in metal regulation and toxicity in aquatic animals. *Aquat. Toxicol.* 22:81–114.

Roesijadi, G. (1994) Metallothionein induction as a measure of response to metal exposure in aquatic animals. *Environmental Health Perspectives* 102:91–100.

Ross, H. (1990) Trace metal wet deposition in Sweden: insight gained from daily wet only collection. *Atmosph. Environ.* 24 A:1929–1938.

Rosseland, B.O. and Hedriksen, A. (1990) Acidification in Norway – Loss of fish populations and the 1000–lake survey 1986. *Sci. Tot. Environ.* 96:45–56.

Roy, R.R. and Campbell, P.G.C. (1995) Survival time of exposure of juvenile Atlantic salmon (*Salmo salar*) to mixtures of aluminum and zinc in soft water at low pH. *Aquat. Toxicol.* 33:155–176.

Salomons, W. and Kerdijk, H.N. (1986) Cadmium in fresh and estuarine waters. *Experientia Suppl. Ser.* 50:24–28.

Sangalang, G.B. and O'Halloran, M.J. (1972) Cadmium induced testicular injury and alterations of androgen synthesis in brook trout. *Nature* 240:470–471.

Sauter, S., Buxton, K.S., Macek, K.J. and Petrocelli, S.R. (1976) Effects of exposure to heavy metals in selected fresh water fish: toxicity of copper, cadmium, chromium and lead of eggs and fry of seven fish species. *US-EPA-600/3*, 76–105, Duluth.

Sayer, M.D.J., Reader, J.P. and Dalziel, T.R.K. (1993) Freshwater acidification: effects on early life stages of fish. *Rev. Fish Biol. Fish.* 3:95–132.

Schindler, D.W. (1988) Effects of acid rain on freshwater ecosystems. *Science* 239:149–157.

Schindler, D.W., Jefferson Curtis, P., Parker, B.R. and Stainton, M.P. (1996) Consequences of climate warming and lake acidification for UV-B penetration in North American boreal lakes. *Nature* 379:705–708.

Schmidt-Nielsen, K. (1979) *Animal physiology: adaptation and environment*. Cambridge University Press, Cambridge, 699 pp.

Schmitt, C.J., Dwyer, F.J. and Finger, S.E. (1984) Bioavailability of Pb and Zn from mine tailings as indicated by erythrocyte δ-aminolevulinic acid dehydratase (ALA-D) activity in suckers (Pisces: Catostomidae). *Can. J. Fish. Aquat. Sci.* 41:1030–1040.

Schmitt, C.J., Wildhaber, M.L., Hunn, J.B., Nash, T., Tieger, M.N. and Steadman, B.L. (1993) Biomonitoring of lead-contaminated Missouri streams with an assay for erythrocyte δ-aminolevulinic acid dehydratase activity in fish blood. *Arch. Environ. Contam. Toxicol.* 25:464–475.

Scorer, R.S. (1992) Deposition of concentrated pollution at large distance. *Atmosph. Environ.* 26A:793–805.

Senesi, N. (1992) Metal-humic substance complexes in the environment. Molecular and mechanistic aspects by multiple spectroscopic approach. *In:* Adriano, D.C. (Ed) *Biogeochemistry of trace metals*. Lewis Publ., Boca Raton, Ann Arbor, London, Tokyo, pp 429–496.

Sephard, B.K., McIntosh, A.W., Atchinson, G.J. and Nelson, D.W. (1980) Aspects of the aquatic chemistry of cadmium and zinc in a heavy metal contaminated lake. *Wat. Res.* 14:1061–1066.

Settle, D.M. and Patterson, C.C. (1980) Lead in Albacore: Guide to lead pollution in americans. *Science* 207:1167–1176.

Smith, D.R. and Flegal, A.R. (1995) Lead in the biosphere: recent trends. *Ambio* 24:21–23.

Sommaruga, R. and Psenner, R. (1997) Ultraviolet radiation in a high mountain lake of the Austrian Alps: air and underwater measurements. *Photochem. Photobiol.* 65:957–963.

Sparling, D.W. (1995) Acidic deposition: a review of biological effects. *In:* Hoffman, D.J., Rattner, B.A., Allen Burton, Jr., G. and Cairns, Jr., J. (Eds) *Handbook of ecotoxicology.* Lewis Publishers, Boca Raton, Ann Arbor, London, Tokyo, pp 301–329.

Sprenger, M.D., McIntosh, A.W. and Hoenig, S. (1988) Concentrations of trace elements on yellow perch (*Perca flavescens*) from six acidic lakes. *Water Air Soil Pollut.* 37:375–388.

Spry, D.J. and Wiener, J.G. (1991) Metal bioavailability and toxicity to fish in low-alkalinity lakes: a critical review. *Environ. Pollut.* 71:243–304.

Steinnes, E. (1990) Lead, cadmium and other metals in Scandinavian surface waters, with emphasis on acidification and atmospheric deposition. *Environ. Toxicol. Chem.* 9:825–831.

Stripp, R.A., Heit, M., Bogen, D.C., Bidanset, J. and Trombetta, L. (1990) Trace element accumulation in the tissues of fish from lakes with different pH values. *Water Air Soil Pollut.* 51:75–87.

Stumm, W. and Morgan, J.J. (1981) *Aquatic chemistry: an introduction emphasizing chemical equilibria in natural waters.* 2nd ed. Wiley, New York, 780 pp.

Stumm, W. and Schnoor, J. (1995) Atmospheric depositions: impact of acids on lakes. *In:* Lerman, A., Imboden, D.M. and Gat, J.R. (Eds) *Physics and chemistry of lakes.* Springer, Berlin, Heidelberg, New York, pp 185–215.

Sturzenegger, V.T. (1989) *Wasserstoffperoxid in Oberflächengewässern: Photochemische Produktion und Abbau.* Ph.D. Thesis, ETH Zürich.

Tjälve, H. and Gottofrey, J. (1991) Effects of lipophilic complex formation on the uptake and distribution of some metals in fish. *Pharmacology and Toxicology* 69:430–439.

Triendl, M. (1994) *Metal uptake in fish from acidic high mountain lakes.* M.Sc. thesis, Dept. of Zoology and Limnology, University of Innsbruck, Innsbruck, Austria, 40 pp.

Turner, D. R., Whitfield, M. and Dickson, A. G. (1981) The equilibrium speciation of dissolved components in fresh water and seawater at 25°C and 1 atm. pressure. *Geochim. Cosmochim. Acta* 45:855–881.

Verbost, P.M., Flik, G., Lock, R.A.C. and Wendelaar Bonga, S.E. (1987) Cadmium inhibition of Ca^{2+} uptake in rainbow trout gills. *Am. J. Physiol.* 253:216–221.

Verbost, P.M., van Rooil, J., Flik, G., Lock, R.A.C. and Wendelaar Bonga, S.E. (1989) The movement of cadmium through freshwater trout branchial epithelium and its interference with calcium transport. *J. Exp. Biol.* 145:185–197.

Vinikour, W.S., Goldstein, R.M. and Anderson, R.V. (1980) Bioconcentration pattern of zinc, copper, cadmium and lead in selected fish species from the Fox River, Illinois. *Bull. Environ. Contam. Toxicol.* 24:727–734.

Vymazal, J. (1989) Size fractions of heavy metals in waters. *Acta hydrochim. hydrobiol.* 17:309–313.

Wagenbach, D., Görlach, U. and Münnich, K.O. (1988) Coastal Antarctic aerosol: the seasonal pattern of its chemical composition and radionucleid content. *Tellus* 40B:426–436.

Waiwood, K.G. and Beamish, F.W.H. (1978) Effects of copper, pH and hardness on the critical swimming performance of rainbow trout (*Salmo gairdneri* Richardson). *Wat. Res.* 12:611–619.

Waller, P.A. and Pickering, W.F. (1993) The effect of pH on the lability of lead and cadmium on humic particles. *Chem. Spec. Bioavail.* 5:11–22.

Warren, L.A. and Zimmerman, A.P. (1994) The influence of temperature and NaCl on cadmium, copper and zinc partitioning among suspended particulate and dissolved phases in an urban river. *Wat. Res.* 28:1921–1931.

Weber, J.H. (1988) Binding and transport of metals by humic materials. *In:* Frimmel, F.H. and Christman, R.F. (Eds) *Humic substances and their role in the environment.* Wiley, New York, Toronto, pp 165–178.

Welsh, P.G., Skidmore, J.F., Spry, D.J., Dixon, D.G., Hodson, P.V., Hutchinson, N.J. and Hickie, B.E. (1993) Effect of pH and dissolved organic carbon on the toxicity of copper to larval fathead minnow (*Pimephales promelas*) in natural lake waters of low alkalinity. *Can. J. Fish. Aquat. Sci.* 50:1356–1362.

Wicklund, A. and Runn, P. (1988) Calcium effects on cadmium uptake, redistribution and elimination in minnows, *Phoxinus phoxinus*, acclimated to different calcium concentrations. *Aquat. Toxicol.* 13:109–122.

Wicklund Glynn, A., Norrgren, L., Müssener, A. (1994) Differences in uptake of inorganic mercury and cadmium in the gills of the zebrafish, *Brachydanio rerio. Aquat. Toxicol.* 30:13–26.

Wiener, J.G. and Giesy, J.P. (1979) Concentrations of Cd, Cu, Mn, Pb and Zn in fishes in a highly organic softwater pond. *J. Fish. Res. Board Can.* 36:270–279.

Wong, P.T.S., Silverberg, B.A., Chau, Y.K. and Hodson, P.V. (1978) Lead and the aquatic biota. *In:* Nriagu, J.J. (Ed) *Biogeochemistry of lead.* Elsevier, New York, pp 279–342.

Wong, P.T.S., Chau, Y.K., Kramar, O. and Bengert, G.A. (1981) Accumulation and depuration of tetramethyl lead by rainbow trout. *Wat. Res.* 15:621–625.

Woodward, D.F., Brumbaugh, W.G., Delonay, A.J., Little, E.E. and Smith, C.E. (1994) Effects on rainbow trout fry of a metals-contaminated diet of benthic invertebrates from Clark Fork River, Montana. *Trans. Am. Fish. Soc.* 123:51–62.

Wren, C.D. and Stephenson, G.L. (1991) The effect of acidification on the accumulation and toxicity of metals to freshwater invertebrates. *Environ. Pollut.* 71:205–241.

Zechmeister, H.G. (1995) Correlation between altitudes and heavy metal deposition in the Alps. *Environ. Pollut.* 89:73–80.

Fish Ecotoxicology
ed. by T. Braunbeck, D. E. Hinton and B. Streit
© 1998 Birkhäuser Verlag Basel/Switzerland

Effects of organotin compounds in fish: from the molecular to the population level

Karl Fent

Swiss Federal Institute for Environmental Science and Technology (EAWAG) and Swiss Federal Institute of Technology (ETH), Überlandstrasse 133, CH-8600 Dübendorf, Switzerland

Summary. Organotin compounds are ubiquitous contaminants in the environment. The high toxicity of trisubstituted derivatives towards aquatic organisms has resulted in deleterious impacts on aquatic ecosystems. Although regulations were effective in some respects, tributyltin concentrations remain high enough to cause toxicity to aquatic and benthic organisms. In this communication, the ecotoxicology of organotins is critically reviewed with emphasis on fish as key organisms in aquatic ecosystems. Emphasis is put on effects at different biological levels and the links between levels. The influence of chemical speciation on bioavailability and basic modes of toxic action are discussed. A more complete understanding of the ecotoxicity of organotins can be achieved both by linking environmental chemical and toxicological aspects and by interrelating effects at various biological levels. Molecular and biochemical processes, as well as the modes of toxicant action are of pivotal importance; however, the correlation with higher levels of the biological hierarchy, including organism and population levels, remain to be elucidated. This review gives insights into the potential hazard associated with organotin pollution in aquatic ecosystems, discusses interrelations between the effects at different biological levels, and allows, to a certain extent, some generalizations of the ecotoxicological effects of pollutants in fish.

Introduction

General ecotoxicology

Ecotoxicology investigates the effects of environmental chemicals (xenobiotics) in ecosystems. Fish are key species in aquatic systems and their protection is important for both ecological and economic reasons, as they represent a major protein source for the nutrition of mankind. Ecotoxicological investigations cannot be restricted to pure effect studies, but must also include investigations on environmental fate. The fate of chemicals must be related to their interactions with biota. The core of ecotoxicology, however, is the study of toxicological effects, which should essentially be integrated at various biological levels from the gene, cell and organism up to populations and entire ecosystems (Fig. 1). Understanding of the ecotoxicological effects of selected pollutants requires an interdisciplinary effort, considering chemical, biological and molecular processes, as well as mechanisms of toxicity and ecological processes. Environmental chemistry, toxicology and ecology are linked with each other; therefore, it is of crucial importance that ecotoxicology is aimed at an understanding of phenomena and processes.

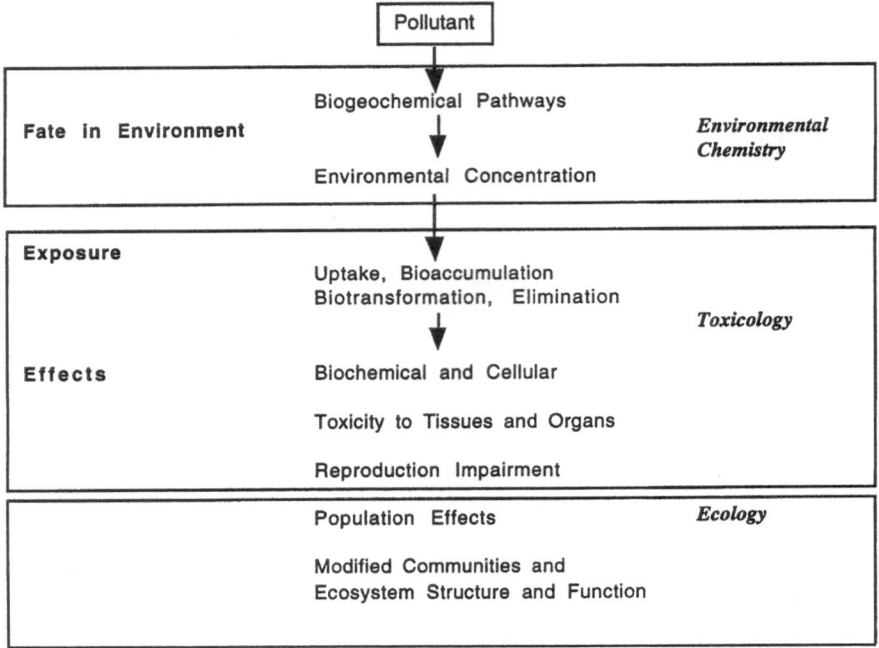

Figure 1. Ecotoxicology is an integration of environmental chemical, toxicological and ecological concepts.

The first effects of pollutants take place at the molecular and cellular levels of target organs and tissues, before effects become visible at a higher level of biological organization. Insights into the modes of action are important, as they allow a mechanistic understanding of ecotoxicological processes. However, effects at the various biological levels are linked with each other (Fig. 2). This is a keystone of ecotoxicology, however, it is often very difficult to interrelate the effects at the various levels. Mechanistic studies can rarely be extrapolated to the organism or population level. In this review, an attempt is being made to bridge some of the gaps by discussing hypotheses on correlations between molecular and histological effects in fish.

The connection between different levels of the biological hierarchy is a challenge not only in toxicology, but even more so in ecotoxicology. Most studies address toxic effects at one level, such as at the molecular or organismic level, but interrelations are rarely made. Straightforward connections between mechanistic processes at the molecular level and the organ or organism level are difficult and often associated with hypotheses and speculations. However, the ultimate goal of ecotoxicology is to reach a more in-depth understanding of the toxic effects of environmental chemicals. Here, some hypotheses are formulated on the connection of molecular effects such as the inhibition of the energy production and alteration in

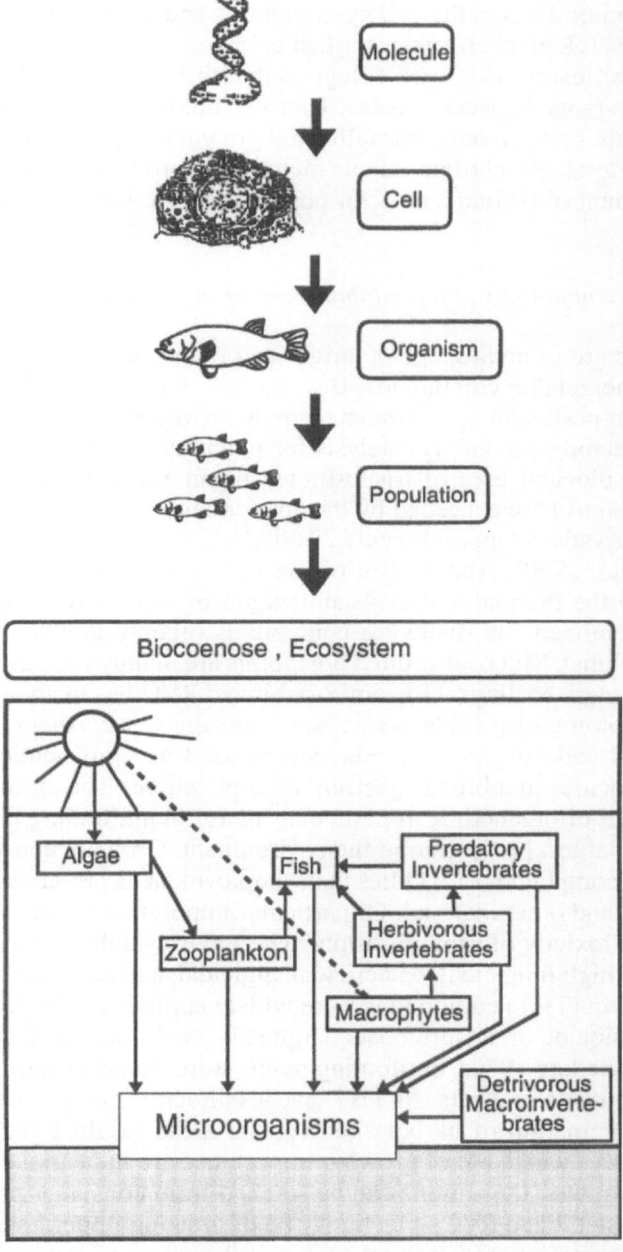

Figure 2. Ecotoxicology encompasses all biological levels and major emphasis should be placed on interrelationships between them for understanding effects of pollutants such as organotins. From Fent (1998).

calcium homeostasis in the cell by organotins and the resulting morphological and histological effects described below.

A general feature in ecotoxicology is that the effects of pollutants vary over a wide range between species. This also holds true for organotin compounds (Tab. 4). Generally, mortality and growth inhibition are determined in toxicity tests, but chronic effects including reproductive failures, which have pronounced consequences for populations, are less often considered.

Organotin compounds and contamination of the aquatic environment

Organotins are of anthropogenic origin and range among the most widely used organometallic compounds. Besides use as pesticides, they have important non-pesticidal applications, namely inclusion in polyvinyl chloride (PVC) as stabilizers and as catalysts for polyurethane and silicone elastomers. The biocidal uses of trisubstituted organotin compounds (approximately 8000 t/y) are exceeded by the applications of the di- and monosubstituted derivatives (approximately 27 000 t/y), used as stabilizers and catalysts (WHO, 1990). About 70% of the total annual world production is devoted to the thermal and UV-stabilization of rigid and semirigid PVC. Typical stabilizers are dialkyltin compounds (mainly dimethyl-, dibutyl-, and dioctyltins). Monosubstituted organotins are mainly used as PVC stabilizers and glass coatings. Organotin-stabilized PVC has many applications including piping of potable, waste, and drainage water. About 23% of the total world-wide organotin production is used as agrochemicals and as general biocides in a broad spectrum of applications. The agricultural and biocidal use of organotins, in particular its use in antifouling paints, gives rise to the largest proportion in the environment. Growing consumption of tributyltin compounds (TBT) lies in its employment as preservative for timber, wood, and other material. Of particular importance to the environment is the high toxicity of tributyl-, triphenyl-, and tricyclohexyltin derivatives. They have high fungicidal, bactericidal, algicidal, and acaricidal properties.

Tributyltin (TBT) compounds were widely applied as effective antifouling paint biocide on leisure boats, large ships and docks in the 1970s and 1980s. In the late 1970s, antifouling paints were found to cause detrimental environmental impacts. As TBT leaches directly from paints into water, high contamination of harbors and coastal areas resulted (Fent, 1996a). Other sources which are of growing importance are industrial and municipal waste waters. Organotins can be leached from consumer products and organotin-stabilized PVC, which may be of growing importance, since the variety of material protected by TBT and the range of industrial applications is increasing (Fent and Müller, 1991; Maguire, 1991). This leads to the contamination of waste waters and sewage sludges by butyltins (Becker-van Slooten et al., 1994; Chau et al., 1992; Fent, 1996b; Fent and Müller, 1991; Müller, 1987; Schebek et al., 1991).

Triphenyltin (TPT) compounds have also been employed as a co-toxicant with TBT in some antifouling paints. However, the major employment of TPT compounds lies in agriculture, where they are used as fungicides in crop protection (potato, celery, sugar beet, coffee, and rice; Stäb et al., 1992). Triphenyltin compounds enter the aquatic environment *via* leaching and runoff from agricultural fields.

The use of TBT containing antifouling paints resulted in widespread and significant TBT-pollution of marine (Alzieu et al., 1986; Seligman et al., 1989; Wade et al., 1988; Waldock et al., 1988) and freshwater eco-systems (Becker-van Slooten and Studer, 1993; Fent and Hunn, 1991, 1995; Maguire, 1987). Dibutyltin (DBT) and monobutyltin (MBT) were also generally found in water, sediment and biota, and TPT was also present in marine and freshwaters. The fate of organotins in harbors is the result of a combination of removal processes (dilution and scavenging to sediments), microbial and algal biodegradation, and bioaccumulation in biota (Fent, 1996a). In freshwaters, TBT occurs predominantly in the dis-solved phase (95 to 99%), but sorption to particulates is important for the removal into sediments. Sedimentation and persistence in anoxic lake sediments play a pivotal role in the fate of these compounds in lakes. Biodegradation in the water column is also important, in particular during summer (Lee et al., 1989; Seligman et al., 1986).

Figure 3 shows typical ranges of TBT contamination in different compart-ments of freshwater harbors. TBT-containing antifouling paints were con-

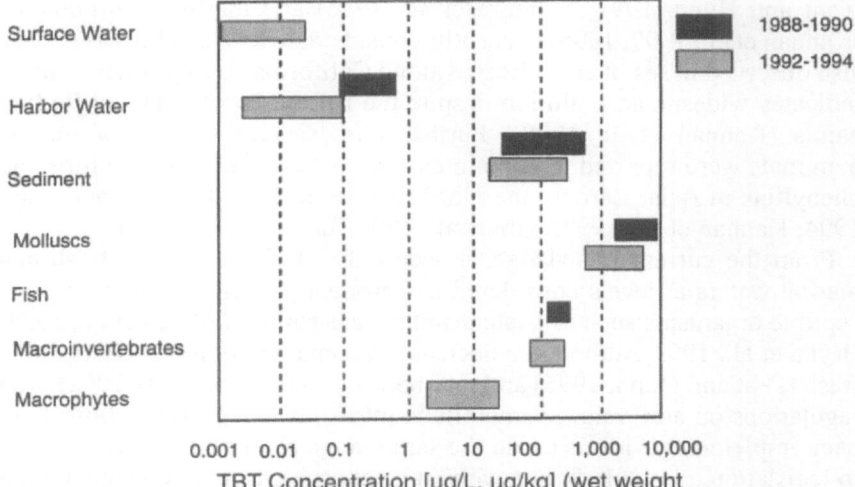

Figure 3. Concentrations of tributyltin in different compartments and biota of freshwater systems found in Lake Lucerne, prior to 1988–1990 and after 1992–1994 the sales ban of organotin containing antifouling paints. Data from Fent and Hunn (1991) as well as Fent and Hunn (1995). Tributyltin contamination of freshwaters is still widespread, although decreasing. Of particular importance is the considerable contamination of both sediments which act as reservoirs, and biota, of which molluscs are most heavily contaminated.

trolled or banned in many countries, which resulted in a decrease of the TBT contamination at many locations. However, it becomes increasingly evident that the regulations have only been partially effective (Chau et al., 1997; Fent, 1996a). In many sites, particularly in marine environments, the TBT concentrations in water and sediment remain high enough to cause toxicity to aquatic and benthic organisms. Therefore, TBT-containing antifouling paints remain important sources to the aquatic environment due to their continued use in large vessels and applications in countries without regulations. It appears that large harbors with heavy shipping traffic continue to have a TBT-contamination. Recently, considerable contamination was also reported in several Arabic (Hasan and Juma, 1992), Asian and Oceanian countries (Kannan et al., 1995; Tanabe et al. 1998).

Considerable levels of butyltin, phenyltin and methyltin compounds were also found in marine and freshwater organisms including fish (Fent, 1996a). This holds in particular for TBT and TPT, which have a considerable potential for bioaccumulation. Residues in fresh and seawater species show a great variability; TBT concentrations typically fall in a range between 5–5000 ng/g wet weight. In general, molluscs have high residues of TBT and TPT. In freshwater, zebra mussels (*Dreissena polymorpha*) showed highest values of up to 9.4 µg/g (Fent and Hunn, 1995); but a variety of organisms (algae, invertebrates, benthic invertebrates, zebra mussels, benthivorous, omnivorous and predatory fish, birds) covering different levels of the lake food chain contained butyl- and phenyltins (Stäb et al., 1994, 1996).

TBT residues in fish are in the range of 0.06–0.3 µg/g in both freshwater (Fent and Hunn, 1991; Maguire et al., 1986) and marine environments (Kannan et al., 1997, 1995). Recently, considerable levels of butyltins were also observed in sea otters collected along California coastal waters, which indicates widespread pollution despite the regulation of TBT antifouling paints (Kannan et al. 1998). Furthermore, several species of marine mammals were reported to contain considerable residues of butyltins and phenyltins in Asia, Europe, the North Pacific and the USA (Iwata et al., 1994; Kannan et al., 1997; Kim et al., 1996; Tanabe et al. 1998).

From the current knowledge, it seems that TBT persists in fresh and marine waters at levels considered as chronically toxic for the most susceptible organisms such as oysters and neogastropods (Alzieu et al., 1991; Bryan et al., 1993), although a decrease in contamination was observed in fresh (Fent and Hunn, 1995) and marine waters (Huggett et al., 1992) after regulations on antifouling paints. In countries, where no regulations have been implemented, levels are in the same range as in other countries prior to legislation, particularly in developing countries. Hence, TBT contamination of harbors and environments with heavy shipping activity, but also offshore locations (Law et al., 1994) remains a global problem. TBT contamination of harbor sediments, typically at 0.1–1 mg/kg (dry weight), persists. As the degradation half-life of TBT in sediments is on the order of years, sediments remain contaminated for longer periods of time (Fent and

Hunn, 1991, 1995; Fent et al., 1991a). Anoxic harbor sediments represent long-term reservoirs of TBT and TPT, and remobilisation may occur. Hence, the current ecotoxicological hazards of organotins are related to contaminated sediments. An in-depth discussion on the contamination and fate of TBT and TPT in aquatic systems can be found in Fent (1996a).

Toxicokinetics of organotins in fish

Bioavailability and bioconcentration

The physicochemical properties of environmental chemicals, such as hydrophobicity and speciation, are among the general factors that determine their reactivity and fate in the aquatic environment. They are also important for bioconcentration and effects in biota. The availability of pollutants to organisms is a key determinant for uptake, bioaccumulation and ecotoxicity. For neutral organic compounds, the octanol-water partition coefficient (log K_{OW}) serves as a good estimate of the bioconcentration in aquatic biota. The speciation of TBT and TPT in water shows a strong pH-dependence (Arnold et al., 1997). Based on the pKa of 6.25 of TBT, the dominant tributyltin species at pH < pKa is the cation, whereas at pH > 6.25, tributyltin dominates as neutral TBTOH. For TPT, a pKa of 5.20 was determined. The reported log K_{OW} for the predominant hydroxide species in water are 4.10 (TBTOH) and 3.53 (TPTOH), respectively (Arnold et al., 1997). The two species (cations or hydroxides) show a different partitioning into octanol. Whether or not this is reflected in the uptake and bioconcentration behavior of these species in biota has been investigated with different aquatic organisms.

Studies in *Daphnia magna, Chironomus riparius* and the yolk sac larvae of fish (*Thymallus thymallus*) indicate that the chemical speciation of TBT influences the bioavailability, and thus the bioaccumulation and toxicity of this compound (Fent and Looser, 1995; Looser et al., 1998). Uptake and bioconcentration of TBT in *Daphnia* and *Chironomus* were higher at pH 8.0 than at pH 6.0 and 5.0, respectively. At pH 8.0, TBT predominates as the neutral TBTOH species, whereas it predominates as the cation at pH 6.0. Similarly, a higher bioaccumulation was found at pH 7.8 than 6.0 in fish for both TBT and TPT compounds (Tsuda et al., 1990b).

The dissolved organotins in water seem to be available for uptake, while the presence of dissolved organic matter ameliorates bioaccumulation and toxicity. The presence of relatively high concentrations of dissolved organic carbon (humic acids) in exposure waters led to a related reduction in the TBT bioaccumulation in *Daphnia* and fish larvae *T. thymallus* (Fent and Looser, 1995; Looser et al., 1998). These studies show that both pH and the presence of humic substances influence the bioavailability of TBT and TPT and thus the bioconcentration and toxicity in surface waters.

Bioaccumulation

TBT and TPT are rapidly accumulated in aquatic organisms, as shown in fish yolk sac larvae of European minnows (*Phoxinus phoxinus*; Fig. 4) and graylings (*Thymallus thymallus*; Fent, 1991). In various aquatic organisms, bioconcentration of these compounds can be approximately described by a first-order uptake process.

Bioconcentration factors (BCF) vary between organisms and life stages. Typical BCFs are in the range of 500–1000 in fish and most aquatic invertebrates, but 10 000 or higher in molluscs (Fent, 1991; Fent et al., 1991b; Meador, 1997; Semlitsch et al., 1995). Bioconcentration factors (BCF) were species specific, as shown for fish (Tab. 1). Tissue concentrations in many species were not correlated with the lipid content (Looser et al., 1998; Meador, 1997), which is in contrast to other persistent lipophilic pollutants such as organochlorine pesticides (Iwata et al., 1995; Kannan et al., 1997; Kim et al., 1996; Takahashi et al., 1998). Interspecific differences can partly be explained by differences in uptake mechanisms, physiology, metabolism (uptake rate and elimination rate constants), and differences in lipid contents (*Daphnia* up to 0.3%, and fish larvae up to 3%). Metabolism of TBT can result in lower BCFs than expected on the basis of the K_{OW}, which has been observed in *Chironomus* (Looser et al., 1998). Additionally, a correlation between the acute toxicity and BCF was found in five marine species (Meador, 1997).

Bioaccumulation is important with respect to biomagnification within the food web and contamination of the human diet. Our studies in Lake Lucerne, Switzerland, indicated that benthic organisms, which are prey for other organisms including fish, have considerable TBT residues, but accu-

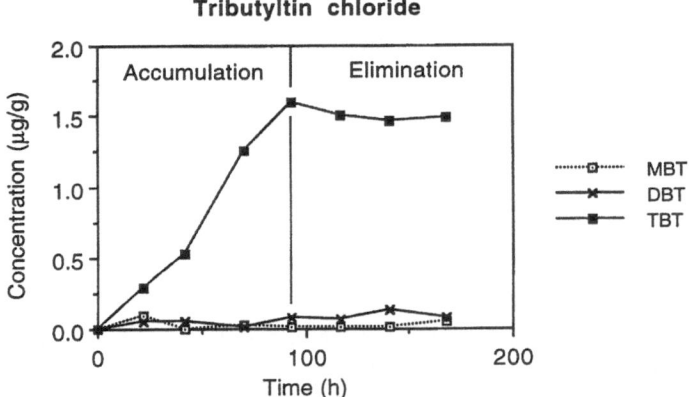

Figure 4. Uptake and elimination of tributyltin by yolk sac larvae of minnows *P. phoxinus* at average aqueous concentrations of 2.13 µg/L. Metabolites di- and monobutyltin (DBT, MBT) occur at only low concentrations. A similar toxicokinetic behavior occurs with triphenyltin. From Fent (1991) as well as Fent et al. (1991b).

Table 1. Bioconcentration factors (BCF) of organotins in fish

Organotin	Species	Exposure concentration (µg/L)	Exposure (days)	BCF	Reference
TBTCl	*Phoxinus phoxinus* larvae	2.13	4	410	Fent (1991)
		6.60	4	540	
	Thymallus thymallus larvae	4.50	7	2020	Fent and Looser (1995)
	Cyprinus carpio	1.8	14	2400	Tsuda et al. (1990b)
	Poecilia reticulata	0.54 (freshwater)	14	460	Tsuda et al. (1990a)
		0.28 (seawater)	14	240	
TBTO	*Cyprinodon variegatus*	1.61	58	2600	Ward et al. (1981)
	Oncorhynchus mykiss	0.51	64	410	Martin et al. (1989)
	P. major	0.04	56	2400–9400	Yamada and Takayanagi (1992)
	R. ercodes	0.12	56	3200	Yamada and Takayanagi (1992)
TPTCl	*Phoxinus phoxinus* larvae	4.97	4	460	Fent et al. (1991b)
		4.61	6	930	
	Thymallus thymallus larvae	3.2	7	2200	Looser et al. (1998)
	P. major	0.06	56	3100	Yamada and Takayanagi (1992)
	R. erodes	0.15	56	4100	Yamada and Takayanagi (1992)
	Cyprinus carpio	1.1	14	600	Tsuda et al. (1990b)
TPTH	*Oncorhynchus mykiss* larvae	2.8	4	80	Tas (1989)
	Poecilia reticulata	6.1	8	630	Tas (1989)

TBTCl: tributyltin chloride.
TPTCl: triphenyltin chloride.
TBTO: bis(tributyltin)oxide.
TPTOH: triphenyltin hydroxide.

mulation in the food web was not examined (Fent, 1996a). A recent study in a lake food web in the Netherlands comes to the conclusion that some minor biomagnification for TPT may take place, but it is rather unlikely for TBT (Stäb et al. 1996). Biomagnification of TBT and TPT in aquatic system seems to be of only minor importance. However, there is a need to further investigate the biomagnification in more detail.

Distribution

In fish, TBT and TPT partition into different organs and the highest concentrations are found in lipid-rich tissues including the liver (Martin et al., 1989; Tsuda et al., 1986, 1988). High TBT residues are found in the peritoneal fat (9.2 μg Sn/g), liver, gall bladder, and kidneys (3.1–3.7 μg Sn/g), whereas lower levels occur in all other tissues (0.5–1.5 μg Sn/g; Martin et al., 1989). About 3.5–52 times higher TBT levels have also been observed in the blood of fish than in the muscle tissue (Oshima et al., 1997). The high residues in these compartments can be explained by the lipophilic property of trisubstituted organotins, but it may also relate to the sites where metabolism and clearance of these compounds occur. The latter holds particularly true for liver, gall bladder and kidney. This may be the reason for the low hepatic butyltin concentrations in sharks compared to other tissues, particularly the kidney.

A similar distribution and accumulation was found for TPT in fish (Schwaiger et al., 1996; Tsuda et al., 1987). Metabolites (DPT and MPT) were generally low in all tissues, but increased slightly with time in liver, gall bladder and kidneys. Figure 5 shows that the highest TPT concentra-

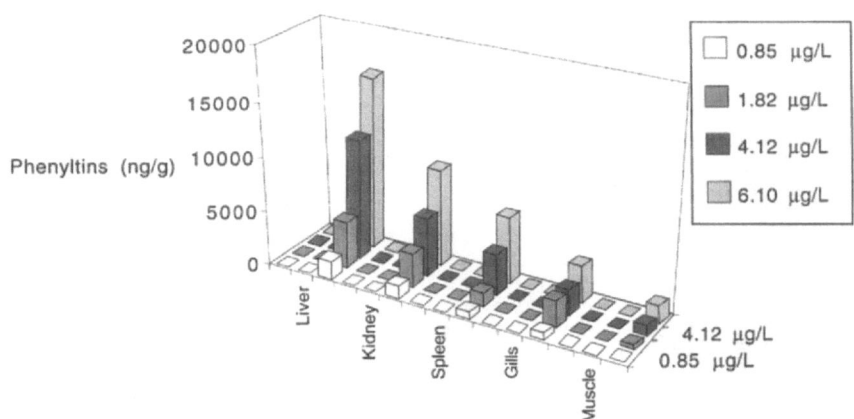

Figure 5. Distribution of phenyltins in rainbow trout after exposure to four aqueous concentrations of 0.85 to 6.10 μg/L TPT for up to 28 days. Triphenyltin (highest residues), diphenyl- and monophenyltin concentrations (ng/g wet weight) are given from right to left for each organ. After Fent (1996a) and Schwaiger et al. (1996).

tions of up to 16.1 µg/g were found in liver, 9.0 µg/g in kidney, and 1.5 µg/g in muscle tissue. Bioconcentration factors at the different exposure concentrations were 2120–2690 in liver, 1310–1470 in kidneys, 630–1000 in spleen, and 140–240 in muscles. The occurrence of only low levels of transformation products (MPT and DPT) may indicate a slow metabolism, or alternatively, that the metabolites were rapidly cleared. In experimentally exposed carp, the concentration of TPT in muscle tissue was higher than that of TBT, being in the range of 0.005–0.148 µg/g TPT and 0.011–0.046 µg/g TBT (Tsuda et al., 1991).

Metabolism

Metabolism of TBT in juvenile and adult fish proceeds mainly in the liver, where this compound is transformed. Metabolites are transferred to the bile, where they are eliminated. Butyl(3-hydroxybutyltin) was detected in the liver of fish reared in TBT painted nets (Ishizaka et al., 1989), and increases of DBT and MBT with exposure time were recorded in spot (Lee, 1986). Furthermore, increased proportions of DBT or MBT relative to TBT were recorded in liver, gall bladder and kidneys of different fish (Martin et al., 1989; Tsuda et al., 1988; Ward et al., 1981).

No notable biotransformation of TBT and TPT compounds was indicated in early life stages of *Phoxinus phoxinus* (Fent, 1991; Fent et al., 1991b) and *Thymallus thymallus* (Fent and Hunn, 1993; Fent and Looser, 1995). Metabolism was absent in embryos and extremely slow in fish yolk sac larvae. This was concluded from the general low occurrence of the respective transformation products (Fig. 4). The low rate of TBT and TPT metabolism in embryos and yolk sac larvae is assumed to be caused either by inhibitory effects of these organotins upon hepatic cytochrome P450 dependent monooxygenases responsible for metabolism of these compounds or its low occurrence in this life stage. The latter is indicated in tadpoles of frogs (*Rana esculenta*), where also a rapid uptake of triphenyltin chloride after short-term exposure occurs (Semlitsch et al., 1995).

Biotransformation of TBT and TPT was also investigated in marine invertebrates, where these compounds get metabolized to a certain extent (Lee, 1991, 1993). Significant metabolism of TBT, but not of TPT, was observed in *Chironomus* (Looser et al., 1998). Due to this process, the BCF for TBT (310) is more than a factor of two lower than the BCF for TPT (680). Based on the log K_{OW}, a lower BCF would be expected for TPT than for TBT. This demonstrates that metabolism occurs even in a lower animal, and that estimates based on K_{OW} must be regarded with caution. Whether *Chironomus* is an atypical invertebrate species, and whether generally different metabolism rates occur in invertebrates than in vertebrates should further be investigated.

In summary, TBT and TPT have a considerable potential for bioaccumulation in fish and other aquatic organisms, as uptake is rapid and clearance

and biotransformation slow. The high variability observed in the bioaccumulation of these compounds is only partly based on differences in the lipid
contents of organisms, but probably more so on differences in the toxicokinetics. Therefore, the log K_{OW} can only be a rough estimate of the bioaccumulation potential in organisms. In particular, slow metabolism may
be one of the factors for the high susceptibility of early life stages of fish
and amphibians. The high susceptibility of certain marine molluscs such as
oysters and dogwhelks may also be related to the low activity of their
metabolizing enzymes (CYP). This may be the reason for the high bioaccumulation found in these organisms.

Ecotoxicity of organotins

Molecular and cellular effects and modes of action

Organotin compounds exert a number of important cellular, biochemical
and molecular effects (Tab. 2). The basic modes of action are similar in
aquatic organisms and mammals, but specific actions are also evident, as
indicated by the high susceptibility of certain aquatic organisms. The following principal modes of action of organotins are known: (1) perturbation
of cellular calcium homeostasis; (2) inhibition of energy production in
cells, e.g., the inhibition of mitochondrial oxidative phosphorylation; (3)
damage of plasma membranes and inhibition of ion pumps and effects on
intracellular sulfhydryl-containing proteins. The alteration of calcium
homeostasis seems to be a possible ultimate basis for the cytotoxic effects
of organotins. This mechanism is also responsible for the apoptosis of
thymocytes, which results in immunotoxicity (Pieters et al., 1989). In addition, intracellular proteins, such as the cytochrome P450-dependent monooxygenases are negatively affected due to the coordination of the trisubstituted organotin molecule with the amino acids cysteine and histidine.

Perturbation of calcium homeostasis
Alteration of calcium homeostasis represents a basic mode of action leading to various molecular, biochemical and cellular effects observed from
organotin compounds. These include apoptosis of thymocytes, neurotoxicity, cytotoxicity, and effects on mitochondria. For instance, TBT was
found to induce a sustained elevation of cytosolic free calcium in isolated
rainbow trout hepatocytes, before the loss of cell viability was detectable
(Reader et al., 1993).

Elevation of calcium concentration appears to be responsible for the
thymocyte killing and stimulation of apoptosis by TBT. An increase in the
cytosolic free Ca^{2+} concentration in rat thymocytes resulted in the direct
opening of plasma membrane Ca^{2+} channels, and was associated with
nuclear chromatin condensation and extensive DNA fragmentation, mediat-

Table 2. Molecular and biochemical effects of organotins

Effect	Compound	Conc. (μM)	(μg/L)	Reference
1. Cytotoxicity				
Fish hepatoma cells (PLHC-1), IC_{50}	TBT, TPT	0.1, 0.7	32; 270	Brüschweiler et al. (1995)
2. Disturbance of Ca^{2+} homeostasis and induction of apoptosis in thymocytes				
Increase in cytosolic Ca^{2+}	TBT	1–10	326–3255	Aw et al. (1990)
Induction of apoptosis in rat thymocytes	TBTO	0.1–1	3.1–31	Raffray and Cohen (1991)
DNA fragmentation, loss of viability, cytotoxicity	TBTO	5	1530	
Alteration in Ca^{2+} homeostasis in human neutrophils	TPT	1–10	385–3855	Miura and Matsui (1991)
3. Inhibition of mitochondrial oxidative phosphorylation and ATP-synthesis				
Binding to ATP-synthase complex				
First detectable Inhibition of ATP production IC_{50}	TET	0.1–0.2	24–48	Aldridge and Rose (1969)
	TBT	1.1	358	
Uncoupling (disturbance of existing proton gradient)	Trialkyltins	0.8–630	–	Aldridge (1976)
	Trialkyltins	0.2–2.5	–	Selwyn et al. (1970)
Swelling of mitochondria (structural damage)	TBT, TPT	0.2, 0.5	65, 193	Wulf and Byington (1975)
	DBT, DPT	2 > 100	608 > 34 381	
Impairment of energy metabolism in vitro (thymocytes)	Dialkyltins	5–120	1200–38 000	Penninks and Seinen (1980)
4. Inhibition and uncoupling of photophosphorylation in chloroplasts				
Inhibition of O_2-production	Trialkyltins	0.1–100	–	Watling-Payne and Selwyn (1974)
Inhibition of ATP-production IC_{50}		20		
Inhibition of electron transport (20%)		100–2000		

Table 2 (continued)

Effect	Compound	Conc. (μM)	(μg/L)	Reference
5. Inhibition of ion pumps and cell membrane damage				
Inhibition of Na$^+$K$^+$-ATPase in heart (IC$_{50}$)	TBT	0.6	195	Cameron et al. (1991)
Inhibition of cardiac Ca^{2+}-dependent ATPase	TBT	0.4	130	Pinkney et al. (1989)
Inhibition of Na$^+$K$^+$-ATPase (fish gills)	TBT	0.08	26	Selwyn (1978)
Inhibition of Ca^{2+}-, Na$^+$K$^+$-pumps	TBT	≥10	≥3255	Raffray and Cohen (1991)
Disturbance of membrane integrity (thymocytes)	TBTO	1–10	306–3060	Gray et al. (1987)
Hemolysis (erythrocytes)	TBT	≥5	≥1627	Zucker et al. (1988)
Fixation (protein denaturation, decreasing mobility)	TBT	2.5	814	
6. Inhibition of cytochrome P450 system				
Cytochrome P450 monooxygenase in fish	TBT, TPT	50	16274; 19273	Fent and Stegeman (1991)
Enzyme content and EROD activity,		10	3255	
Detectable inhibition				
IC$_{50}$	TBT, TPT	180–830	58600–350000	Fent and Bucheli (1994)
Inhibition of P450 protein and activity in rats	TBTO	200		Rosenberg et al. (1981)
7. Inhibition of intracellular enzymes				
Glutathione S-transferases, IC$_{50}$	TBT	0.02–30	8–11564	George and Buchanan (1990)
8. Heme metabolism				
Induction of heme oxygenase (in vivo)	TBTO	3.8 mg/kg		Rosenberg et al. (1981)
Inhibition of ALA-synthetase	TBTO	200		

DBT: dibutyltin.   TPT: triphenyltin.
DPT: diphenyltin.   TET: triethyltin.
TBT: tributyltin.   IC$_{50}$: concentration that induces 50% inhibition.
TBTO: bis(tributyltin)oxide.

ed by the activation of Ca^{2+}-dependent endonucleases (Aw et al., 1990). Tributyltin at concentrations of $1-10$ µM caused a rapid and sustained increase in the cytosolic free Ca^{2+} concentration by mobilizing intracellular Ca^{2+}, activating Ca^{2+} entry, and inhibiting its efflux in thymocytes prior to apoptosis (Chow et al., 1992; Jantzen and Wilken, 1991). Other organotins including trimethyltin, TPT and DBT were less potent in the alteration of calcium homeostasis in thymocytes, which seems to indicate a more selective immunotoxic activity of TBT (Aw et al., 1990).

Inhibition of oxidative phosphorylation
The inhibition of oxidative phosphorylation in mitochondria represents a key process and basic mode of action of di- and trisubstituted organotins and was identified in different groups of organisms and tissues. Both di- and trialkyltin compounds are inhibitors of oxygen uptake into tissues and mitochondria of cells and potent inhibitors of ATP synthesis. Tetrasubstituted organotins have no direct selective action on mitochondria, but they are converted *in vivo* to trisubstituted moieties, which may cause the effects. Thus, trialkyltins are potent inhibitors of mitochondrial oxidative phosphorylation.

Inhibition of mitochondrial oxidative phosphorylation, or of ATP synthesis, by trisubstituted and disubstituted organotins was first described in rats (Aldridge, 1958, 1976; Aldridge and Street, 1964), and subsequently in other organisms. For a series of organotins the order of effectiveness in causing inhibition of ATP production was triethyltin > tripropyltin > tributyltin > tricyclohexyltin > trimethyltin (Aldridge, 1976). These compounds were also found to disrupt ion (proton) gradients and produce swelling of mitochondria (Wulf and Byington, 1975). The inhibition of ATP-synthesis, which is a basic cellular process, has many consequences that lead to various toxic effects.

The effects of trisubstituted organotins to uncouple or directly inhibit oxidative phosphorylation in mitochondria have been attributed to three basic modes of action. The first is a direct inhibition of the basic energy conservation system involved in the synthesis of ATP, e.g., inhibition of the ATPase complex which is manifested in the loss of ATPase and ATP synthase activities (Aldridge and Street, 1964). This appears to be a result of the interaction with the F-component of the F_1F_0-ATPase, causing the inhibition of proton transfer. The second mode is to induce a chloride/hydroxide exchange activity across the inner mitochondrial membrane which causes an equilibration of respiration-generated pH differences (uncoupling; Selwyn, 1978). Thus, organotins act as ionophores able to catalyze Cl^-/OH^- exchange, and inhibit the anion uniport pathway as well as anion channels of the inner membranes of mitochondria, and hence to inhibit the proton flow through these components. Additional studies indicated that trisubstituted organotins possess an uncoupling effect, which is independent of Cl^-/OH^- exchange; hence they act directly by inhibition of the ATPase complex (Connerton and Griffith, 1989). The third mode of action is to cause gross swelling of mitochondria accompanied by the loss of most energetic functions (Aldridge, 1976).

Inhibition of ion transport and cell membrane damage
Organotins are also potent inhibitors of ion transport proteins in the cell membrane, such as ATPases (Na^+K^+-ATPase, Ca^{2+}-ATPase) (Selwyn et al., 1970). At ≥ 10 µM TBT shuts down calcium and sodium-potassium pumps, prior to inducing cell damage (Selwyn, 1978). However, trisubstituted organotin compounds also act as potent cell membrane toxicants leading to perturbations of plasma membranes and membrane bound enzymes involved in the cellular transport of nutrients. These compounds affect ion and pH gradients across membranes.

In rainbow trout erythrocytes, TBT was found to inhibit adrenergically activated sodium/proton exchange and produce cell swelling (Virkki and Nikinmaa, 1993). Organotins may also have negative consequences for the exchange of ions through the gills of fish. Inhibition of the Na^+/K^+-ATPase activity was observed in *Fundulus heteroclitus* gill homogenates at concentrations of 25 µg/L (78 nM) and at higher levels in striped bass (Pinkney et al., 1989). This ion translocation enzyme is concentrated in mitochondria-rich gill chloride cells, and the activity of this enzyme is correlated with osmoregulatory activity. In addition, Mg^{2+} ATPase which is important in oxidative phosphorylation and anion transport was inhibited at similar concentrations *in vitro*. This indicates that TBT places stress on the ionic and osmotic regulatory system of fish. Negative effects on osmoregulatory functions of cell membranes may also account for the pronounced and characteristic vacuolization of cells observed in fish early life stages after exposure to TBT and TPT (Fent and Meier, 1992, 1994).

Inhibition of the cytochrome P450 dependent monooxygenase system
Organotins interact with various intracellular proteins by coordination with amino acids such as histidine and cysteine. Histidine coordinates to the tributyl- and triethyltin ions as a monodentate ligand through the amino group, cysteine and glutathione coordinate mainly *via* the S-group (complex formation; Shoukry, 1993).

One important enzyme family with which organotins interact, are the microsomal cytochrome P450 dependent monooxygenases (CYP). They are mainly localized in the endoplasmic reticulum or mitochondria of liver and several other organs of fish (Stegeman and Hahn, 1994). This enzyme system is of crucial importance for the metabolism of xenobiotic organic substances and plays a major role in the biotransformation of various endogenous substances. Basically, it consists of two linked enzymes, NADPH-cytochrome P450 reductase and cytochrome P450 (CYP). Cytochrome b5 and its reductase, NADH-cytochrome b5 reductase, are related to CYP by transfer of electrons to this enzyme.

Trisubstituted organotins lead to the inhibition of hepatic microsomal cytochrome P450 (CYP) system in fish (Fent, 1996a). The P450 protein and enzyme activity and the reductases are all affected. Organotins act by direct mechanisms and not by a suicide mechanism, in which a metabolite is the toxic moiety. Moreover, the isoform cytochrome P450 1A1 (CYP1A1), which is induced by important environmental organic chemicals, seems to be selectively affected in fish by TBT and TPT *in vitro* and *in vivo*, but other P450 forms, including those with testosterone hydroxylase activity, are also affected at high concentrations. It should be noted that inhibition of CYP by organotins occurs at high concentrations, which are close to acute levels. Hence, it remains open, as no long-term experiments with low concentrations have been performed, whether these inhibitory effects causatively contribute to the organotin toxicity or whether they are a consequence of a more general toxicity, affecting a variety of sulfhydryl-containing proteins.

In vivo studies on the inhibition of the cytochrome P450 dependent monooxygenase system

Tributyltin and triphenyltin chloride were shown to have a strong inhibitory effect *in vitro* and *in vivo* on hepatic microsomal P450 in marine scup *Stenotomus chrysops* (Fent and Stegeman, 1991,1993; Fent et al., 1998). In fish injected with single doses of 1.9–19.3 mg/kg tributyltin or triphenyltin chloride, the EROD activity, catalyzed by CYP1A1, was decreased (Fig. 6). In case of TBT, P450 was converted to its degraded form, cyto-

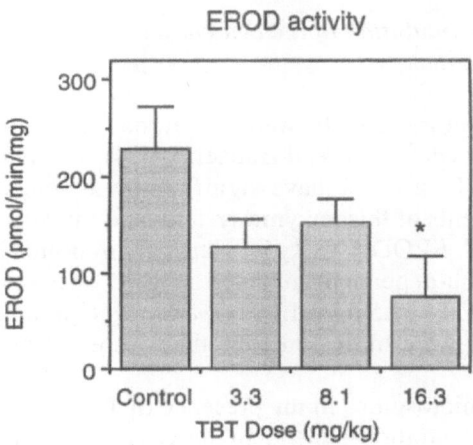

Figure 6. Inhibition of ethoxyresorufin-O-deethylase (EROD) activity in liver microsomes *in vivo* after treatment with three doses of tributyltin for 24 h. Values ± standard error of 4–5 fish per dose group. From Fent and Stegeman (1993). * p < 0.05.

chrome P420 occurred as well. No effect was noted after exposure to TBT and TPT on cytochrome b_5. TPT was also found to inhibit both NAD(P)H cytochrome c reductases.

As the inhibition of EROD activity and the reduction of spectral total P450 did not occur in parallel, this could represent a selective action on the CYP1A1 enzyme. This was demonstrated by the analysis of the interactions of TBT with three CYP forms by immunoblotting with specific mono- and polyclonal antibodies: CYP1A1, the aryl hydrocarbon-inducible form; CYP3A-like protein, the major contributor to microsomal testosterone 6β-hydroxylase activity; and CYP2B-like protein, which oxidizes testosterone at several different sites including the 15α-position. A decrease of CYP1A1 protein content occurred at all TBT doses, with a significant loss at 16.3 mg/kg, which was almost identical to the EROD activity pattern. At this high dose, the CYP3A-like and CYP2B-like proteins were decreased as well. A similar specificity towards CYP1A1 was found in scup with TPT, but the other CYP-forms were not altered (Fent et al., 1998).

The mechanism responsible for the loss and inactivation of cytochrome P450 forms *in vivo* is assumed to be based on the direct destruction of P450 and formation of P420 with subsequent rapid degradation and breakdown of the apoprotein by proteases. *In vitro* data with microsomes showed that binding of TBT or TPT resulted in the destruction of native P450 and loss of EROD activity. These compounds produced a denaturation of P450 directly, and not indirectly through heme oxygenase. The inhibition of P450 synthesis is not primarily responsible for the loss of P450 forms, but it might be a secondary effect in addition to the degradation of these enzymes.

In vitro studies on inhibition of the cytochrome P450 dependent monooxygenase system

Pronounced inhibitory effects were observed in ecologically different freshwater fish *in vitro* (Fent and Bucheli, 1994). The data provide strong evidence that TBT and TPT have significant and selective effects upon different components of this enzyme system, and marked species differences were observed. EROD activity was strongly inhibited by TBT and TPT in a concentration-dependent manner (Fig. 7). Significant inhibitions occurred at 0.1 mM (32.5 µg/mL) TBT and (38.6 µg/mL) TPT. Rainbow trout microsomes were more sensitive than were eel or bullhead microsomes.

Incubation of microsomes in the presence of TBT and TPT also showed a time- and concentration-dependent decrease in total spectrally-determined microsomal CYP content, indicating that the protein was destroyed directly (Fig. 7). TPT led to a greater inactivation of CYP than TBT and induced a 50% loss in all fish at 0.08 mM.

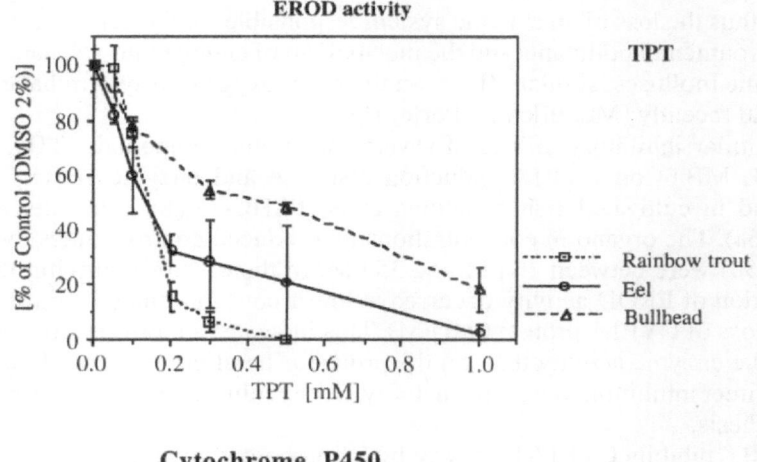

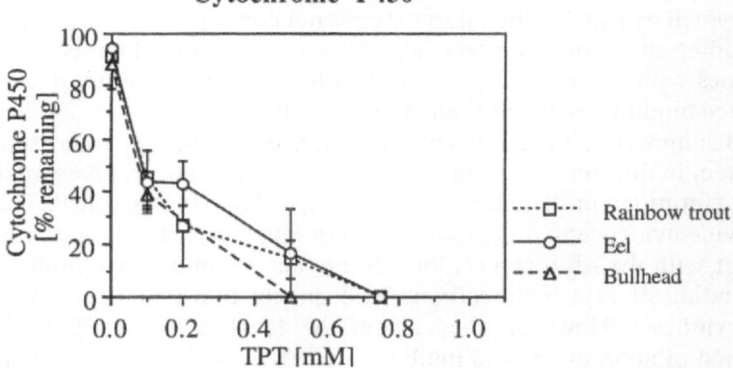

Figure 7. Concentration-dependent inhibition of ethoxyresorufin-O-deethylase (EROD) activity by triphenyltin chloride (TPT) in rainbow trout (*Oncorhynchus mykiss*), eel (*Anguilla anguilla*), and bullhead (*Cottus gobio*) liver microsomes (top). Concentration-dependent decrease of spectrally determined total cytochrome P450 and formation of P420 after 5 min incubation at 30°C in liver microsomes with triphenyltin. Averages ± standard error of at least three separate determinations are given for rainbow trout, eel and bullhead (below). From Fent and Bucheli (1994).

Organotins inhibit other components of the microsomal monooxygenase system as well. TBT and TPT acted *in vitro* on the flavoproteins NADH cytochrome b_5 reductase and NADPH cytochrome P450 reductase. There was a selectivity towards the different reductases in rainbow trout and eel, and to a lesser extent in bullhead. TBT selectively inhibited NADH cytochrome b_5 activity, whereas TPT selectively inhibited the NADPH cytochrome P450 reductase.

These *in vitro* studies show that in fish TBT and TPT strongly interact with microsomal monooxygenase systems, resulting in the inhibition of CYP1A1 activity, inhibition of NAD(P)H cytochrome c reductase activity, and thus the loss of an enzyme system responsible for the detoxification of environmental pollutants and the metabolism of endogenous substances. In marine molluscs, similar effects on the monooxygenase system have been found recently (Morcillo and Porte, 1997).

Similar inhibitory effects of several organotin compounds (TBT, TPT, DBT, MBT) on CYP1A induction response and enzyme activity were found in cultivated fish hepatoma cells (PLHC-1) (Brüschweiler et al., 1996a). The organotin concentrations that reduced control values by 50% (EC50) were between 16 μM and 35 μM. In the case of dibutyltin the inhibition of EROD activity occurred at lower concentrations (1.2 μM) than the loss of CYP1A protein (9.0 μM). This indicates a more selective action on the enzyme activity than on the protein. The effects are mainly caused by direct inhibition of enzyme activity, not by inhibition of CYP1A protein synthesis.

TBT inhibits CYP1A1 activity by a non-competitive mechanism (K_i at 12 μM), which means that the inhibitor does not compete with the substrate on the binding site (Brüschweiler et al., 1996a). Incubation of rainbow trout microsomes with 100 μM TBT *in vitro* resulted in the formation of a type I difference binding spectrum (Fent and Bucheli, 1994).

Table 3 summarizes the effects of organotins on the hepatic microsomal CYP system in different fish and fish cell lines, and Figure 8 gives inhibitory concentrations in different *in vitro* assays. The findings in PLHC-1 cells provide evidence that organotins do not interfere with the binding of an inducer with the Ah receptor; they do not act as competitive inhibitors to the binding of aryl hydrocarbons, and do not interfere with CYP1A protein synthesis. However, they act at the level of the CYP protein (destruction of apoprotein) and inhibit the catalytic activity. Heavy metals including Cd(II), Co(II), Cu(II), Ni(II), Pb(II) and Zn(II) have been shown to act similarly (Brüschweiler et al., 1996b).

The similar effects of TBT and TPT in different fish and PLHC-1 cells indicate a general mode of action towards CYP. Organotins seem to bind to amino acids such as cysteine at the active site, or on other sites (histidine and cysteine amino acids) of the enzyme. The lipophilic TBT and TPT likely penetrate the hydrophobic membrane environment, in which cytochromes P450 are embedded, and thereby gain access to these enzymes. Interestingly, the action is directed to CYP, as cytochrome b_5 is unaffected and the reductases are affected differently by TBT and TPT. Hence, an unspecific action on the membrane of the endoplasmic reticulum can be ruled out. The specific mechanism of action of TBT and TPT on CYP, however, should be investigated further.

Figure 9 summarizes the effects on the hepatic microsomal monooxygenase system in fish *in vitro*. In several fish, organotins act on different components of the microsomal electron transport system, the hemoprotein CYP and the flavoproteins (reductases). Effects on CYP1A also occur *in vivo*, and TPT also acts on both reductases.

Table 3. Effects on the fish hepatic microsomal CYP system and fish cell line PLHC-1

Species	Effect	Concentration	Reference
Fish			
Stenotomus chrysops[1]	Total CYP content	50 µM TBT[1]	Fent and Stegeman (1993)
	EROD	50 µM TBT[1]	
	CYP1A1	50 µM TBT[1]	
	NADH cytochrome c reductase	500 µM TBT[2]	Fent and Stegeman (1991)
	NADPH cytochrome c reductase	no effect with TBT[2]	
	Total CYP contents	500 µM TPT[2]	Fent et al. (1998)
	EROD	250 µM TPT[2]	
	CYP1A1	25 µM TPT[1]	
	NADH cytochrome c reductase	5 µM TPT[1]; no effect[2]	
	NADPH cytochrome c reductase	5 µM TPT[1]; 100 µM TPT[2]	
Oncorhynchus mykiss[2]	Total CYP content (50% loss)	180 µM TBT; 80 µM TPT	Fent and Bucheli (1994)
	EROD	100 µM TBT or TPT	
	NADH cytochrome c reductase	200 µM TBT; minor effect TPT	
	NADPH cytochrome c reductase	minor effect TBT; 200 µM TPT	
Anguilla anguilla[2]	Total CYP contents (50% loss)	830 µM TBT; 80 µM TPT	Fent and Bucheli (1994)
	EROD	100 µM TBT or TPT	
	NADH cytochrome c reductase	200 µM TBT; no effect TPT	
	NADPH cytochrome c reductase	no effect TBT; 200 µM TPT	
Cottus gobio[2]	Total CYP contents (50% loss)	300 µM TBT; 80 µM TPT	Fent and Bucheli (1994)
	EROD	100 µM TBT or TPT	
	NAD(P)H cytochrome c reductase	200 µM TBT or TPT	
PLHC-1 cells[2]	EROD (50% loss)	0.2 µM TBT or TPT	Brüschweiler et al. (1996a)
		1 µM DBT	
		600 µM MBT	
	CYP1A1 (50% loss)	0.4 µM TBT	
		0.2 µM TPT	
		9 µM DBT	
		no effect MBT	

DBT: dibutyltin.
MBT: monobutyltin.
TBT: tributyltin.
TPT: triphenyltin.

[1] in vivo.
[2] in vitro.

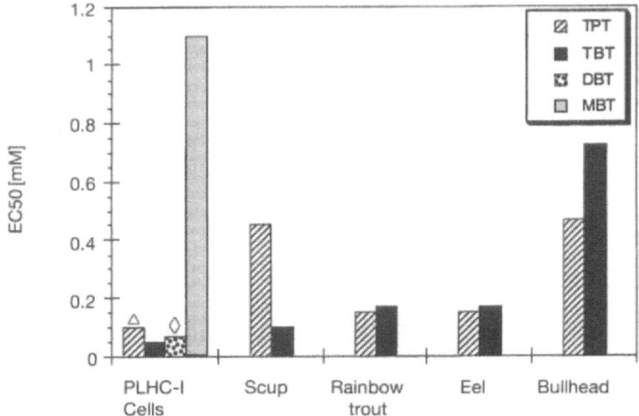

Figure 8. Comparison of the inhibition of EROD activity (50% inhibition, EC_{50} values) in PLHC-1 cells and in different fish *in vitro*. In PLHC-1 cells, values for dibutyltin and triphenyltin could not be determined exactly, because the compound could not be further dissolved in the medium ($\triangle$ TPT, $\lozenge$ DBT).

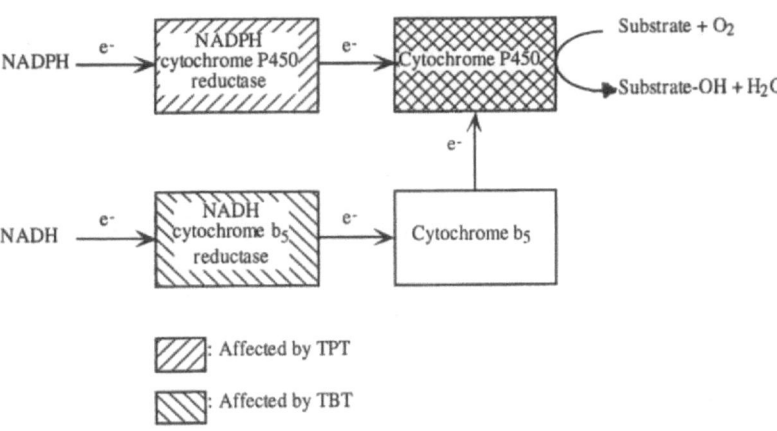

Figure 9. Effects of tributyltin and triphenyltin *in vitro* on components of hepatic microsomal monooxygenase system in freshwater and marine fish. From Fent and Bucheli (1994).

Cytotoxicity in fish hepatoma cell line (PLHC-1)

Cytotoxicity is among the first signs related to acute toxicity and takes place after toxicants have been bound to cell receptors or destroyed plasma membranes. Investigations on the cytotoxicity of a series of 21 organotins in cultured fish hepatoma cells (PLHC-1) indicated that permanent fish cell lines may serve as useful tools for the assessment of the acute toxicity

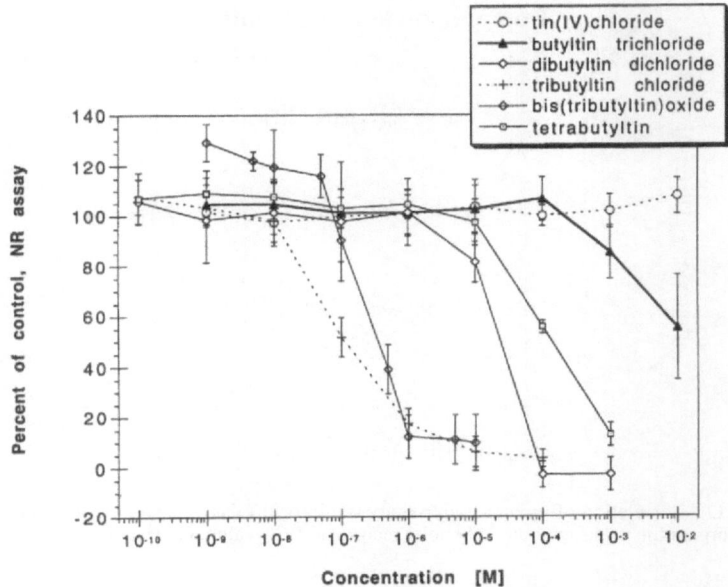

Figure 10. Concentration-response curves of a series of butyltins in PLHC-1 cells. Cells were incubated at 30°C for 24 h, and cytotoxicity was determined by neutral red uptake inhibition. Treatments causing membrane damage inhibit the accumulation of this dye. Data as mean percentage of control ± SD. From Brüschweiler et al. (1995).

of organotins to fish (Brüschweiler et al., 1995; Fent and Hunn, 1996). The PLHC-1 hepatoma cell line, derived from topminnow (*Poeciliopsis lucida*; Hightower and Renfro, 1988), contains an aryl hydrocarbon (Ah) receptor (Hahn et al., 1993). Cytotoxic effects of organotins were found at concentrations between 10 µM and 1 mM. Trisubstituted compounds were most cytotoxic, followed by disubstituted and tetrasubstituted organotins. The lowest cytotoxicity was detected for monosubstituted moieties and inorganic tin (Fig. 10).

Whether this *in vitro* system is a useful tool in the determination of acute toxicities of environmental chemicals to fish depends on the correlation between *in vivo* and *in vitro* data. A correlation between the cytotoxicity of organotins and the *in vivo* toxicity (48 h LC_{50}) for the red killifish (medaka; *Oryzias latipes*; Nagase et al., 1991) was found (Fig. 11), which indicates that the cytotoxicity is a good estimate for the acute toxicity to fish (Fent and Hunn, 1996). This *in vitro* technique may thus be useful in the prescreening of new chemicals and environmental samples, and may replace some fish test in the regulatory context.

Tri- and disubstituted organotin compounds also exhibited a significant correlation between octanol-water partition coefficients (log K_{OW}) and effect concentrations that inhibit the uptake of neutral red by 50% (EC_{50} values). K_{OW} values of disubstituted and trisubstituted organotin com-

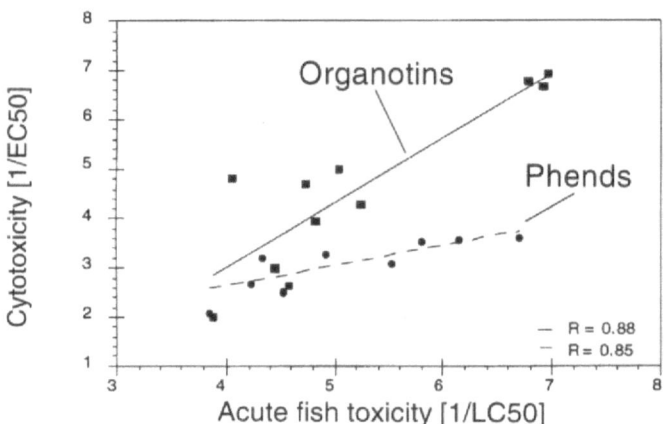

Figure 11. Correlation of *in vitro* cytotoxicity (reciprocal EC_{50} values) and *in vivo* acute toxicity of organotins and substituted phenols (reciprocal LC_{50} values) in fish. From Fent and Hunn (1996).

pounds with up to six carbon atoms per substituent serve as suitable measures for cytotoxicity. Therefore, hydrophobicity seems to be a significant parameter for the acute toxicity of organotins.

Effects on fish

Organotins induce adverse histological effects on various organs and tissues of fish, several of which are shared among other species including mammals. The trisubstituted organotins, which exert highest toxicity are trimethyl- and triethyltin that are primarily neurotoxic, and tripropyl-, tributyl-, triphenyl-, and tricyclohexyltin that are essentially immunotoxic (Snoeij et al., 1985). As a general feature, considerable species differences between aquatic biota occur (Table 4).

Effects on early life stages of fish

Survival and growth

Early life stages of fish are typically more susceptible to environmental contamination than adults. Skeletal abnormalities observed in embryonic and larval fish have often been associated with polluted waters. As these life stages are prone to be affected most by organotin compounds, we studied the influence and effects of TBT and TPT in representative European freshwater fish that spawn and recruit at the time of the year when organotin pollution is highest. Toxic effects of tributyltin and triphenyltin

Table 4. Toxicity of tributyltin compounds to sensitive organisms

Organism	Parameter	Compound	Conc. (mg/L)	Reference
Algae				
Marine diatoms	acute (EC_{50}, 72 h)	TBTO	0.3–0.4	Walsh et al. (1985)
	chronic (growth)	TBTO	0.1	Beaumont and Newman (1986)
Zooplankton				
Acartia tonsa	acute (LC_{50}, 96 h)	TBTO	1.0	U'ren (1983)
	chronic (death, 144 h)	TBTO	0.3	U'ren (1983)
(Nauplii)	chronic (survival, 6 d)	TBTCl	0.023	Bushong et al. (1990)
Eurytemora affinis	acute (LC_{50}, 72 h)	TBT	0.6	Hall Jr. et al. (1988a)
(Neonates)	chronic (21% survival, 6 d)	TBT	0.1	
Daphnia magna	acute (LC_{50}, 96 h)	TBTO	1.7	Cardwell (1986)
Amphipods				
Gammarus spec. (young)	acute (LC_{50} 96 h)	TBT	1.3	Bushong et al. (1988)
	chronic, 24 d, no lethality	TBT	0.53	
Bivalves				
Crassostrea gigas	chronic (LC_{100}, 12 d)	TBT acetate	0.18	His and Robert (1980, 1985)
Crassostrea gigas spat	chronic (growth)	TBTO	0.01	Lawler and Aldrich (1987)
C. gigas, M. edulis, V. decussata	chronic (growth)	TBT	0.25	Thain (1986)
Mytilus edulis, spat	chronic (7 d growth)	TBTO	0.4	Stromgren and Bongard (1987)
Mytilus edulis, larvae	chronic (LC_{50} 15 d)	TBTO	0.04	Beaumont and Budd (1984)
Mytilus edulis, larvae	chronic (25 d growth)	TBT	0.05	Lapota et al. (1993)
Mercenaria mercenaria, embryos	acute (LC_{50} 48 h)	TBTCl	1.1	Roberts (1987)
Gastropods				
Nucella lapillus	chronic (sterilization)	TBTO	0.002	Bryan et al. (1986), Gibbs et al. (1988)
Oligochaetes				
Tubifex tubifex	acute (LC_{50} 24 h)	TBTO	5.5	Polster and Halacka (1971)
Crustaceans				
Mysid shrimp, juvenile	acute (LC_{50} 96 h)	TBTO	0.3	Valkirs et al. (1987)

DBTCl: dibutyltin chloride.   TBTO: bis(tributyltin)oxide.
DPTCl: diphenyltin chloride.   TPTCl: triphenyltin chloride.
TBTCl: tributyltin chloride.   TPTOH: triphenyltin hydroxide.

chloride on hatching, survival, behavior, morphology, and histology were investigated in minnows (*Phoxinus phoxinus*; Fent, 1992). Effects on growth and survival are less sensitive toxicity parameters than histopathology (Fent and Meier, 1992, 1994). Effects of both compounds were found to be similar, which indicates a more general effect pattern of trisubstituted organotins on early life stages of fish.

During embryogenesis, the fish embryo may be affected by environmental pollutants, resulting in altered embryonic development, teratogenesis, and delayed or inhibited hatching. Survival during the embryonic phase and hatching time of minnows were not negatively affected up to 9.33 µg/L TBT, but 10–30% of the larvae at ≥4.1 µg/L TBT, and ≥8.00 µg/L TPT were unable to perform complete hatching, because of paralysis. Significantly reduced hatching rate, delayed hatching, and overall retardation of development were found at 15.9 µg/L TPT.

Short-term exposure of minnow eggs and larvae to 0.7–18.2 µg/L tributyltin and triphenyltin chloride induced dose-related morphological and behavioral effects on larvae. Striking morphological effects were noted at ≥4.0 µg/L TBT and TPT. Severe malformations of the body axis were observed in hatched larvae of minnows. Many of the yolk sac larvae were unable to uncurl, and most of them had a crippled shape due to an abnormal vertebral axis (bent tails). Severely deformed body axes were often associated with retarded yolk sac resorption. Eyes of larvae exposed to ≥3.5 µg/L TBT and ≥ 8.0 µg/L TPT became opaque. Whereas the skeletal malformations are non-specific, effects on the eyes are due to the irritating properties of these organotins.

Behavioral toxicity

Furthermore, considerable dose-related behavioral effects were observed with TBT and TPT. After hatching, reduced motility, erratic swimming behavior, uncoordinated muscle contractions and immobilization occurred in one-third of the larvae at doses of 4 µg/L and higher. Reduced swimming and feeding activity have also been observed in tadpoles of *Rana esculenta* after short term exposure to 20 µg/L triphenyltin chloride (Semlitsch et al., 1995). As swimming is generally related to foraging, predator avoidance, and thermoregulation, the reduced activity likely results in lower food intake, which has negative effects on growth and development of amphibian larvae.

Mortality increased in a dose-related manner after TBT and TPT exposure (Fig. 12). Toxicity was higher when larvae were exposed to TBT and TPT (larval exposure) than when eggs were exposed (embryonic-larval exposure); hence, the chorion reduces toxicity. Survival of *Phoxinus* was significantly reduced at ≥ 4.1 µg/L TBT and TPT in the embryonic-larval, and 3.5 µg/L TBT and 1.8 µg/L TPT, respectively, in larval exposures. Even though survival was not reduced at 0.7 mg/L TBT, the incidence of up to 50% of abnormal larvae at water temperatures of 21°C (deformations, erratic swimming, motionlessness) indicate embryo toxicity at this concentration.

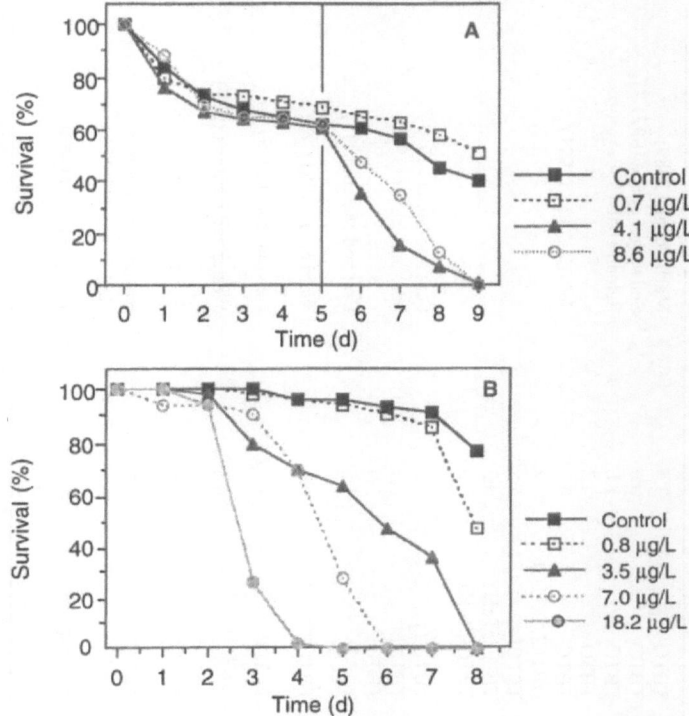

Figure 12. Survival of *Phoxinus phoxinus* after embryonic-larval exposure (A) and larval exposure (B) to various concentrations of tributyltin chloride. In A, hatching is marked by a vertical line. From Fent and Meier (1992).

As shown in Table 5, minnows are as sensitive to TBT as larval fishes from marine environments (Bushong et al., 1988; Pinkney et al., 1988), but more sensitive than rainbow trout yolk sac fry (Seinen et al., 1981). The toxicity of TPT is similar to that in fathead minnow (*Pimephales prome-las*), and rainbow trout (*Oncorhynchus mykiss*) larvae (deVries et al., 1991). For TPT compounds, the No Observable Effect Concentration (NOEC) was 0.05 µg/L and the LC_{50} (96 h) was 7.1 µg/L; reduced survival occurred at 2 µg/L, and reduced growth was seen at 0.23 µg/L after 30 days of exposure in fathead minnows larvae (Jarvinen et al., 1988). These data show that TBT and TPT are very toxic to the early life stages of fish.

Histopathology

In fish, general toxicity parameters such as hatching time and success, and survival were found to be less sensitive than were morphological and histological parameters. Histopathology has been investigated in studies on

Table 5. Toxicity of organotins to fish

Fish	Parameter	Compound	Conc. (µg/L)	Reference
Oncorhynchus mykiss	acute, LC_{50}	TBTO	3.5	Martin et al. (1989)
	chronic, immune system	TBTO	0.6	Schwaiger et al. (1992)
Larvae	chronic, LC_{100}, 12 d	TBTO	1.8	Seinen et al. (1981)
	chronic, 10 d	TBTO	0.2	
	chronic, 110 d	TBTCl	0.2	de Vries et al. (1991)
	chronic, 110 d	DBTCl, DPTCl	200	
	acute, 14 d	TPTCl	5.8	
	chronic, 110 d	TPTCl, TcHT	0.2	
	NOEC	TPTCl, TcHT	0.05	
Phoxinus phoxinus, larvae	acute, LC_{100}, 8 d	TBTC (TBTCl)	3.5	Fent and Meier (1992)
	histologic alterations	TBTCl	0.7	Fent and Meier (1992)
	acute, LC_{100}, 9 d	TBTCl	3.9	Fent and Meier (1994)
	histologic alterations	TPTCl	1.8	
Cyprinodon variegatus	chronic, LC_{50}, 14 d, 21 d	TBTO	0.4	Ward et al. (1981)
Leuresthes tenius, embryo	chronic, no embryotoxicity	TBT	1.72	Newton et al. (1985)
Lepistes reticulata	chronic (immune system)		0.4	Wester and Canton (1987)
Oncorhynchus tshawytscha	acute, LC_{50}, 96 h	TBTO	0.8	Short and Thrower (1986)
Menidia beryllina, larvae	acute, LC_{50}, 96 h	TBTCl	3.0	Bushong et al. (1988)
	chronic (growth), 28 d	TBT	0.09–0.5	Hall Jr. et al. (1988b)
Pimephales promelas, larvae	acute, LC_{50}, 96 h	TPTOH	7.1	Jarvinen et al. (1988)
	chronic, reduced growth		0.2	

DBTCl: dibutyltin chloride.
DPTCl: diphenyltin chloride.
TBTCl: tributyltin chloride.
TBTO: bis(tributyltin)oxide.
TcHT: tricyclohexyltin.
TPTCl: triphenyltin chloride.
TPTOH: triphenyltin hydroxide.

juvenile or adult fish exposed to TBT (Chliamovitch and Kuhn, 1977; Holm et al., 1991; Schwaiger et al., 1992; Wester and Canton, 1987; Wester et al., 1990), but only rarely applied to fish early life stages.

Significant histopathological effects of TBT (Fent and Meier, 1992) and TPT (Fent and Meier, 1994) were observed in minnow larvae. Various tissues of *Phoxinus phoxinus* were affected at aqueous concentrations of >0.7 µg/L TBT, and >1.8 µg/L TPT. Histological alterations include degenerative cellular changes in skeletal muscles, skin, kidneys, corneal epithelium, lens, pigment layer of the retina and choroid, retina, and CNS. These dose-related effects were marked at ≥3.5 µg/L TBT and at ≥3.9 µg/L TPT (Tab. 6).

The toxicities of TBT and TPT were essentially similar, but TPT acted more selectively on the CNS and spinal cord, retina, and eye lens, whereas effects on skin, skeletal muscles, and kidneys were less marked than with TBT. The following alterations were found both in TBT- and TPT-exposed larvae. Degenerative changes in the skin were dose-dependent. Alterations were minor at 0.73, moderate at 3.5, and marked at 8.0 µg/L TBT. In more pronounced cases, irreversible nuclear alterations such as pycnosis, karyorrhexis and karyolysis, and erosion or even ulceration of the epithelium were also evident (Fig. 13). The erosion of the body surface epithelia of *Phoxinus* yolk sac fry led to a subsequent loss of the epithelial barrier. As a consequence, osmoregulatory problems arose which led to hydropic cell alterations in the whole body of the larvae. It is hypothesized that these effects may be a consequence of the degenerative effects on biomembranes including the inhibition of ATPases, but also the perturbation of calcium homeostasis.

Severe alterations in muscle fibers were reflected in a spectrum of degenerative changes ranging from hydropic swelling to lysis of myocytes. Myocytolysis led to a disordering of the myocytes in the myomers (Fig. 13). Alterations were dose related and ranged from splitting of muscle fibers, broken and contorted myomers, to dislocation of the myosepta. Renal changes consisted predominantly of tubular epithelial lesions such as hydropic vacuolation and nuclear alterations (Fig. 13). In the nuclei, pycnosis, karyorrhexis or even karyolysis were observed. All these alterations were observed at 3.5 µg/L TBT, but were most evident at 8.0 µg/L. At 0.73 µg/L TBT, only hydropic vacuolation was evident. Tubulonephrosis in the kidney was found at ≥ 5.1 µg/L TPT.

In TBT and TPT exposed larvae, the corneal epithelium, lens, retina and pigment layer of the retina and choroid showed dose related degenerative changes. The external epithelial layer of the cornea was eroded (Fig. 14). The normal arrangement of retinal layers was disrupted, and single-cell necrosis was observed. Vacuolation and disintegration of collagen fibers were seen in the lens. The quantity of pigment in the pigment layer of the retina and the choroid was reduced depending on the organotin dose. These changes resulted in the opaque color of the eyes. Alterations in cel-

Table 6. Histological alterations in minnow (*Phoxinus phoxinus*) larvae after embryonic-larval and larval exposure to tributyltin chloride for 4 to 10 days[a]

Exposure	TBT (μg/L)[b]	Skin	Muscle	Kidney	Cornea	Lens	Retina, CNS	Pigment
Embryonic-larval, 16°C	0	–[c]	–	–	–	–	–	–
	0.7	+	+	+	– to +	– to +	–	–
	4.1	++	++	++	++	+	+	+
	8.6	+++	+++	+++	+++	+++	++	+++
Embryonic-larval, 21°C	0	–	–	–	–	–	–	–
	0.8	++	+ to ++	++	+	+ to ++	+	+
	7.9	+++	+++	+++	+++	+++	+++	+++
Larval, 21°C	0	–	–	–	–	–	–	–
	0.8	–	–	–	–	–	–	–
	3.5	++	+	–	+	–	–	+
	7.0	+++	+++	+	n.d.	n.d.	+ to ++	+++
	18.2	+++	+++	+++	+++	+++	+++	+++

[a] from Fent and Meier (1992).

[b] Mean of measured initial TBT concentrations and prior water renewal after 24 or 48 h.

[c] Rating values representing averages of all larvae in each group, and average degree of severity in all larvae. –, no alterations; + slight alteration; ++, moderate alteration; +++, marked alteration.

n.d. not determined.

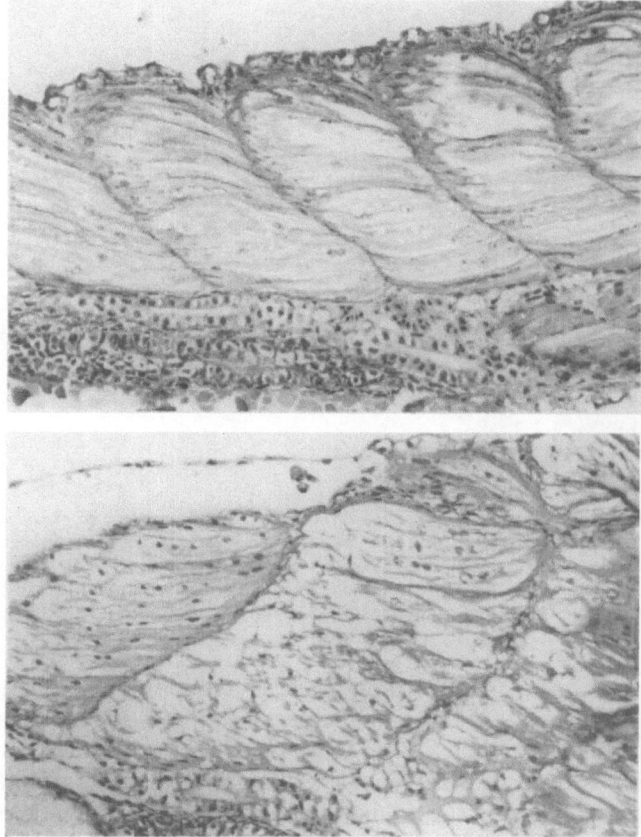

Figure 13. Histological effects on skin, skeletal muscle and posterior kidney of minnow larvae. Control (top) and fish exposed to 8.6 µg/L tributyltin chloride (below), in which hydropic vacuolation and erosion of the skin, myocytolysis and contorted myomers and dislocated myosepta are observed. The kidney epithelial cells of the tubules are vacuolated and have pyknotic or destroyed nuclei. From Fent and Meier (1992).

lular and fibrillar structures of the brain and spinal cord were pronounced in larvae exposed to 8.0 µg/L TBT. Severe, diffuse hydropic vacuolation occurred. In addition, multifocal nuclear alterations (e.g., pycnosis) were evident in severe cases. Alterations in cellular structures were more pronounced with TPT than with TBT.

Target organs and tissues affected by bis(tributyltin)oxide (TBTO) in adult rainbow trout included skin, cornea, neural retina layers, pigment layer of the retina, and kidney tubular cells (Chliamovitch and Kuhn, 1977). Very high concentrations of TBTO (nominal 5.85 mg/L) also had effects on red blood cells, gills, and liver (epithelial cells of bile duct). TBTO also induced neurotoxicity (Triebskorn et al., 1994). Histopathological effects in various organs were also observed after exposure to nomi-

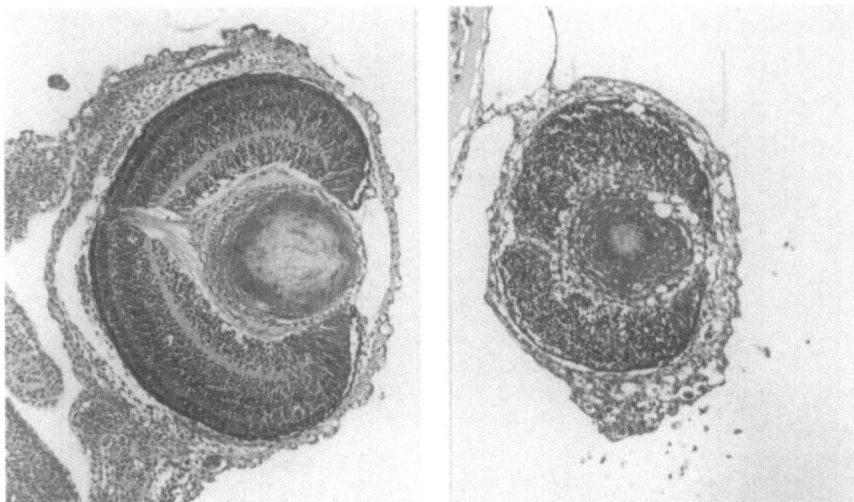

Figure 14. Histological effects on eyes of three days old minnow larvae exposed to 9.1 μg/L triphenyltin chloride (700 ×). Control eye (left), and eye of triphenyltin exposed larvae (right) that shows edematic vacuolization and necrosis of individual cells within the surface epithelium of the cornea, marked hydropic degenerative alteration in the lens, loss of architecture in the retinal part of the eye and widespread depigmentation. From Fent and Meier (1994).

nal concentrations of 0.32 to 10 μg/L TBTO in guppies and Japanese medaka (Wester et al., 1990). Effects on the kidneys were described after long-term exposure to nominal 3.2 up to 10 μg/L TBTO in guppy, and tubulonephrosis was observed in medaka. Gills are frequently affected by irritating compounds, as was found in rainbow trout, but not in guppies (Wester and Canton, 1987). Desquamation in skin and cornea can be attributed to the irritative property of TBT. Lesions of the cornea similar to those observed in minnows larvae were found in rainbow trout. Corneal keratitis and increased vacuolation as well as dermal hyperplasia occurred in guppies and medaka.

Histological alterations occurred at TBT concentrations as low as 0.7 μg/L in minnows larvae (Fent and Meier, 1992). Long-term exposure of rainbow trout yolk sac fry to nominal concentrations of 1 μg/L TBT resulted in a decrease in the number of red blood cells, increase in liver weight, and hyperplasia of liver cells (Seinen et al., 1981). Resistance of yolk sac fry to bacterial challenge was found to be decreased at 0.05 μg/L TBTCl and TPTCl, and 200 μg/L DBTCl (de Vries et al., 1991). The No Observed Effect Concentration (NOEC) in guppy was 0.01 μg/L, based on thymic atrophy, liver vacuolization or hyperplasia of the hemopoietic tissue (Wester and Canton, 1987), and in rainbow trout yolk sac fry, the NOEC was determined to be 0.05 μg/L for TBTCl and TPTCl (Tab. 5). These findings indicate that fish early life stages are very susceptible to triorganotins, and hence, they were negatively affected at polluted sites.

The histopathological effects of TBT and TPT on fish are based on their eye- and skin-irritating activity, and on their activity against the immune system, the liver, as well as renal and neuronal tissues. Triphenyltin has a lower irritating property than TBT, but seems to be more neurotoxic. Our study in minnow larvae and histopathological investigations in juvenile and adult fish showed that target organs and tissues in fish are for the most part identical to those in mammals, except for skeletal muscle and neuronal tissues of early life stages. The effects on hepatic and renal tissues correlate with the high TBT and TPT residues accumulated in these tissues (see section on Bioavailability and bioconcentration, p. 265), which may also be the case for the neuronal tissue, due to the lipophilic properties of these compounds. Not only the partitioning, but also the toxicokinetics may explain this observation. As the concentration at the target site governs the toxicity of chemicals, the effects on the skin, gills, eyes and the tissues with high residues (liver, gall bladder, kidney) can be understood by the irritating property of TBT, but also by its high binding affinity for sulfhydryl groups of proteins. Both the inhibition of oxidative phosphorylation and, particularly, the perturbation of calcium homeostasis result in cell necroses observed in many tissues. The effects on the immune system are also mainly based on these processes.

Immunotoxicity and hematological changes

The immunotoxicity of organotin compounds is only partially understood in fish, but seems to be essentially similar to that identified in rodents. Tributyltin oxide (TBTO) and dibutyltin (DBT) were shown in guppy (*Poecilia reticulata*) to induce thymic atrophy, hyperplasia of the hemopoietic interstitial tissue of kidneys, and an increase in circulating neutrophilic granulocytes after long-term exposure (Wester and Canton, 1987). Effects of TBTO occurred at concentrations that were almost three orders of magnitude lower than that of DBT; thymic atrophy was observed after exposure to aqueous concentrations of 0.32 µg/L TBTO after 3 months, and 320 µg/L DBT after 1 month. Similar to rats, specific depletion of lymphocytes in the cortex of the thymus with subsequent atrophy were observed in this fish.

The thymotoxic effects of TBTO in fish appear to be species- and age-specific, but difficulties in the identification of the organs may be one of the reasons for the differences. Contrary to adult guppies, no thymus atrophy was detected in rainbow trout yolk sac fry (deVries et al., 1991; Seinen et al., 1981) or Japanese medaka (Wester et al., 1990). In guppy, the spleen showed no lymphocyte depletion, whereas this was observed after 14 to 28 days exposure to 0.6 and 6.0 µg/L TBTO in juvenile rainbow trout (Schwaiger et al., 1992). Significant alterations in the spleen were associated with dose-related lymphocytic depletion, a decrease in circulating lymphocytes, marked proliferation of reticuloendothelial cells, and increased erythrophagia. Hematological changes were also characterized by an increase of monocytes and neutrophilic granulocytes and a reduction in the total number of leucocytes at ≥ 2.0 μg/L TBTO. The mononuclear phagocytic system was also affected (suppressive effect on phagocytic activity *in vitro*). Phagocytic activity of kidney macrophages was reduced in marine fish (oyster toadfish, hogchoker, Atlantic croaker) with increasing concentrations of TBT (Wishkovsky et al., 1989). Peripheral blood neutrophils and lymphocytes were suppressed in channel catfish after a single injection of 1 mg/kg TBTCl (Rice et al., 1995). Moreover, a significant suppression of the humoral immune response and a dose-dependent suppression of macrophage activation were observed.

Decreased resistance to bacterial challenge was found in rainbow trout yolk sac fry after 133 days of exposure to 0.2 μg/L TPT, at a concentration where growth and survival were also negatively affected (de Vries et al., 1991). Marked and similar effects as with TBT on blood parameters and histopathological effects were observed in juvenile rainbow trout exposed for 28 days to triphenyltin acetate (Schwaiger et al., 1996). Hematological findings included an increase in the total number of erythrocytes and a decrease in the total number of leukocytes at 4 and 6 μg/L, as well as a decrease in the percentage of lymphocytes. A dose related lymphocytic depletion of the spleen, accompanied by a proliferation of reticuloendo-thelial cells, and an increased erythrophagia, even at 1 μg/L, was observed. It seems possible that the processes leading to immunotoxicity are based on the alteration of calcium homeostasis in various types of immune cells (e.g., apoptotic thymocyte cell death causing thymus atrophy).

Ecotoxicological effects in polluted environments: consequences for fish populations?

The ecotoxicity of organotins in aquatic systems is mainly based on the concentrations of TBT and TPT compounds having similar toxicity. Maximal TBT concentrations found in late spring in harbors of Lake Lucerne, Switzerland, were below the LC_{50} value of 3.5 μg/L for rainbow trout, and below the 48 h static LC_{50} value for *Daphnia magna* of 1.7 μg/L (Cardwell, 1986). However, yolk sac fry of minnows (*Phoxinus phoxinus*), which could inhabit such environments showed alterations in skin, skeletal muscles and kidneys at concentrations of 0.7 μg/L TBT, which is in the range of the peak concentration measured in boat harbors (Fent and Hunn, 1991). Hence, organotin concentrations are primarily important in terms of chronic toxicity. Sublethal effects of TBT to fish larvae (decreased immune resistance, growth and survival) were reported at a similar range. Chronic toxicity towards the fish immune system was observed at 0.3–0.6 μg/L TBT (Schwaiger et al., 1992; Wester and Canton, 1987). Reduced immunological performance with increased susceptibility to infections and irritations of skin and eyes are important effects to be expected not only in fish early life stages, but in adults as well. Most of our knowledge about the ecotoxicity of organotins refers to marine organisms. It remains to be shown that in freshwater similar highly susceptible organisms exist as in the marine environments.

Sublethal changes may have important ecotoxicological consequences, such as impaired predator avoidance and reduced foraging and feeding activity, which may negatively affect fish populations in polluted environments. The eye irritating property of TBT and its negative effects on the cornea may result in such effects. Due to the possible effects on larvae, reduced reproductive success is expected in sensitive fish in heavily impacted areas. This ultimately may lead to a decline of fish populations. So far, no reports are available in which effects on fish populations were investigated. Studies in tadpoles of common European frogs demonstrate negative effects of TPT on survival and growth rate and increased time to metamorphosis. Because TPT most strongly affect the larval period, exposure to TPT could decrease survival in natural ponds because of the increased risk of predation in permanent ponds or desiccation in drying ponds, as well as the increased risk of additional chemical exposure (Fioramonti et al., 1997).

Consequences of cytochrome P450 inhibition

Our studies indicate important biochemical effects of TBT and TPT in fish liver, and suggest that exposure to these organotins may alter both cytochrome P450-dependent (xenobiotic) metabolism and the induction response to other environmental pollutants. Cytochromes P450 are also involved in the metabolism of endogenous substrates including steroids, arachidonic acid, prostaglandines, and others. Altered metabolism of these substrates may affect important biological processes such as the immune response, reproduction and development. As the effects on the monooxygenase system were observed at high organotin concentrations, it remains unclear, whether these inhibitory effects are causatively related to toxicity, or whether they are a consequence of a more general toxicity, which affects numerous sulfhydryl-containing proteins.

The interference of organotins with cytochrome P450 may indirectly explain alterations of macrophage function (Rice et al., 1995), because these cells have a considerable CYP-activity. It should be noted, however, that the inhibition of the hepatic monooxygenase system is in concert with additional biochemical effects of organotins, such as the inhibition of the glutathione S-aryltransferases (George and Buchanan, 1990), and interference with heme metabolism. High concentrations of TBT have also been shown to decrease the protein content of CYP3A, the major contributor to microsomal testosterone 6β-hydroxylase activity, and CYP2B, which oxidizes testosterone at several sites including the 15α-position after in vivo exposure.

Interference with steroid metabolism and prostaglandin synthesis should also be considered. Inhibition of CYP isozymes involved in steroid metabolism may have consequences for the sex hormone system. Among the most susceptible organisms affected by organotins are stenoglossan gastropod molluscs. Today, over 100 species are known to be affected by TBT (Bettin et al., 1996). The females are reproductively anomalous, in that they develop male reproduction organs at trace concentrations of a few ng/L of TBT, a phenomenon called imposex (Bryan and Gibbs, 1991; Bryan et al., 1986). In female dogwhelks (*Nucella lapillus*), an increase in penis length and increased levels of testosterone were found after exposure to TBT.

Imposex is mediated by the androgen receptor and is a consequence of a high testosterone titer. An increase in testosterone can be induced by inhibition of the cytochrome P450 dependent aromatase, which is responsible for the conversion of testosterone to estradiol-17β. Recent studies indicate that TBT disturbs steroid metabolism, namely the aromatization of androgens to estrogens in dogwhelks (Bettin et al., 1996; Oehlmann et al., 1993). Imposex is, therefore, an indirect effect of TBT, as this condition is related to and induced by an increased titer of testosterone (Oehlmann, 1994; Spooner et al., 1991). In dogwhelks experimentally exposed to a specific inhibitor of aromatase, imposex is also induced. Either the aromatase is destroyed or non-competitively inhibited, as is the case in fish for hepatic CYP enzymes (Brüschweiler et al., 1996a; Fent and Bucheli, 1994; Fent and Stegeman, 1993), or TBT acts as a competitive inhibitor of this enzyme. The inhibition of CYP enzymes by organotins, including the aromatase, may result in the observed disturbances in steroid homeostasis, ultimately resulting in the masculinization of females.

The observed loss of CYP3A- and CYP2B-like proteins in fish exposed to TBT (Fent and Stegeman, 1993) may also result in sex hormone imbalances. Hence, these findings may not only have implications for the understanding of the masculinization of females in neogastropods, but also for fish hormone disruption. It remains to be shown, however, whether similar effects on CYP occur with long-term exposure to low TBT concentrations.

The observed effects of organotins have also implications for the use of CYP1A as a biomarker in environmental monitoring (Bucheli and Fent, 1995; Stegeman et al., 1992). Elevated concentrations of hydrocarbon-inducible P450 and high rates of EROD and aryl hydrocarbon hydroxylase activity (both catalyzed by CYP1A) have been observed in fish from many sites, and linked to the exposure of fish to certain environmental organic pollutants (Stegeman and Hahn, 1994; Stegeman and Kloepper-Sams, 1987). Thus far, little attention has been paid to the inhibitory action of environmental chemicals on this cytochrome. The strong actions of TBT and TPT on CYP1A content and catalytic activity indicates that exposure to these environmental pollutants may result in a reduced induction response to other polycyclic and polychlorinated aromatic hydrocarbons. Modulation of the induction response by organic environmental pollutants has not been widely considered in the evaluation of this biochemical response as a biomarker. Given the widespread contamination of aquatic environments by organotins, these substances may mask the contamination with polycyclic and polychlorinated aromatic hydrocarbons in these areas, if only CYP1A catalytic activity is monitored as biomarker. CYP1A protein content is less influenced by organotins than is EROD activity. For that reason, immunochemical determination of CYP1A protein provides important complementary information. It will be important to investigate the dose-response relationship and time-course behavior of CYP1A induction *in vivo* when organisms are simultaneously exposed to inducing and inhibitory compounds at relevant environmental concentrations for an extended period of time.

Conclusions

Organotin contamination of marine and freshwater environments is still a global pollution problem. Although regulations in many industrialized countries resulted in decreased contamination, residues persist at ecotoxicologically relevant concentrations. Recent data illustrate considerable contamination in developing countries in all aquatic compartments. As trisubstituted organotins belong to the most toxic compounds for aquatic environments known thus far, this problem remains important not only with respect to fish, but to the ecosystem as a whole.

The various biochemical and molecular effects of TBT and TPT result in pronounced negative effects at the tissue and organism level. Maximal TBT concentrations measured in harbor environments after regulations of organotin-containing antifouling paints may still reach chronic values for the most susceptible aquatic species. The limited knowledge on the impact on fish populations and on other aquatic and benthic organisms indicate that further studies are needed to assess the ecotoxicological potential and hazard related to organotin contamination. In particular, the hazards associated with contaminated sediments should be investigated. This aspect is particularly important because harbor sediments are regularly dredged, and it remains unclear at present how and where such contaminated sediments should be treated and disposed. Hence, forthcoming research on tributyltin and triphenyltin compounds should focus on the following aspects:

1. The development of tools, preferably *in vitro* systems, for the evaluation of contaminated environmental compartments (in particular sediments).
2. The analysis of food chain biomagnification.
3. The study of subtle chronic effects on aquatic organisms, including fish, with respect to ecological and population consequences.

References

Aldridge, W.N. (1958) The biochemistry of organotin compounds. Trialkyltins and oxidation phosphorylation. *Biochem. J.* 69:367–376.

Aldridge, W.N. (1976) The influence of organotin compounds on mitochondrial function. *Adv. Chem. Ser.* 157:186–196.

Aldridge, W.N. and Rose, M.S. (1969) The mechanism of oxidative phosphorylation. A hypothesis derived from studies of trimethyltin and triethyltin compounds. *FEBS lett.* 4:61–68.

Aldridge, W.N. and Street, B.W. (1964) Oxidative phosphorylation: Biochemical effects and properties of trialkyltins. *Biochem. J.* 91:287–297.

Alzieu, C.,Sanjuan, J., Deltreil, J.P. and Borel, M. (1986) Tin contamination in Arcachon Bay: Effects on oyster shell anomalies. *Mar. Pollut. Bull.* 17:494–498.

Alzieu, C., Michel, P., Tolosa, I., Bacci, E., Mee, L.D. and Readman, J.W. (1991) Organotin compounds in the Mediterranean: A continuing cause for concern. *Mar. Environ. Res.* 32:261–270.

Arnold, C.G., Weidenhaupt, A., David, M.M., Müller, S.R., Haderlein, S.B. and Schwarzenbach, R.P. (1997) Aqueous speciation and 1-octanol-water partitioning of tributyltin- and triphenyltin: effect of pH and ion composition. *Environ. Sci. Technol.* 31:2596–2602.

Aw, T.Y., Nicotera, P., Manzo, L. and Orrenius, S. (1990) Tributyltin stimulates apoptosis in rat thymocytes. *Arch. Biochem. Biophys.* 283:46–50.

Beaumont, A.R. and Budd, M.D. (1984) Heigh mortality of the larvae of the common mussel at low concentrations of tributyltin. *Mar. Pollut. Bull.* 15:402–405.

Beaumont, A.R. and Newman, P.B. (1986) Low levels of tributyltin reduce growth in marine micro-algae. *Mar. Poll. Bull.* 17:457–461.

Becker-van Slooten, K. and Studer, C. (1993) Auch nach dem Verbot: Weiterhin hohe Organozinnbelastung in Häfen. *BUWAL-Bulletin* 2/93:10–16.

Becker-van Slooten, K., Merlini, L., Stegmueller, A.M., deAlencastro, L.F. and Tarradellas, J. (1994) Contamination des boues des stations d'épuration suisses par les organoétains. *Gas-Wasser-Abwasser* 74:104–110.

Bettin, C., Oehlmann, J. and Stroben, E. (1996) TBT-induced imposex in marine neogastropods is mediated by an increasing androgen level. *Helgol. Meeresunters.* 50:299–317.

Brüschweiler, B.J., Würgler, F.E. and Fent, K. (1995) Cytotoxicity *in vitro* of organotin compounds to fish hepatoma cells PLHC-I (*Poeciliopsis lucida*). *Aquat. Toxicol.* 32:143–160.

Brüschweiler, B.J., Würgler, F.E. and Fent, K. (1996a) Inhibition of cytochrome P4501A by organotins in fish hepatoma cells PLHC-1. *Environ. Toxicol. Chem.* 15:728–735.

Brüschweiler, B.J., Würgler, F.E. and Fent, K. (1996b) Inhibitory effects of heavy metals on cytochrome P4501A induction in permanent fish hepatoma cells. *Arch. Environ. Contam. Toxicol.* 31:475–482.

Bryan, G.W., Burt, G.R., Gibbs, P.E. and Pascoe, P.L. (1993) *Nassarius reticulatus* (Nassariidae: Gastropoda) as an indicator of tributyltin pollution before and after TBT restrictions. *J. Mar. Biol. Ass. U.K.* 73:913–929.

Bryan, G.W. and Gibbs, P.E. (1991) Impact of low concentrations of tributyltin (TBT) on marine organisms: a review. *In:* Newman, M.C. and McIntosh, A.W. (Eds) *Metal Ecotoxicology: concepts and applications.* Lewis Publishers, Boca Raton, Boston, pp 323–361.

Bryan, G.W., Gibbs, P.E., Hummerstone, L.G. and Burt, G.R. (1986) The decline of the gastropod *Nucella lapillus* around south-west England: Evidence for the effect of tributyltin from antifouling paints. *J. Mar. Biol. Ass. UK* 66:611–640.

Bucheli, T.D. and Fent, K. (1995) Induction of cytochrome P450 in fish as a biomarker for environmental contamination in aquatic ecosystems. *Crit. Rev. Environ. Sci. Technol.* 25:200–268.

Bushong, S.J., Hall, L.W.J., Hall, W.S., Johnson, W.E. and Herman, R.L. (1988) Acute toxicity of tributyltin to selected Chesapeake Bay fish and invertebrates. *Water Res.* 22:1027–1032.

Bushong, S.J., Ziegenfuss, M.C., Unger, M.A. and Hall Jr., L.W. (1990) Chronic tributyltin toxicity experiments with the Chesapeake Bay copepod *Acartia tonsa*. *Environ. Toxicol. Chem.* 9:359–366.

Cameron, J.A., Kodavanti, P.R.S., Pentyala, S.N. and Desaiah, D. (1991) Triorganotin inhibition of rat cardiac adenosine triphosphatases and catecholamine binding. *J. Appl. Toxicol.* 11:403–409.

Cardwell, R.D. (1986) A risk assessment concerning the fate and effects of tributyltins in the aquatic environment. In: *Proceedings of the organotin conference Oceans '86. Washington, D.C., USA*. The Institute of Electrical and Electronics Engineers and Marine Technology Society, New York, pp 1117–1129.

Chau, Y.K., Zhang, S. and Maguire, R.J. (1992) Occurrence of butyltin species in sewage and sludge in Canada. *Sci. Total Environ.* 121:271–281.

Chau, Y.K., Maguire, R.J., Brown, M., Yang, F. and Batchelor, S.P. (1997) Occurrence of organotin compounds in the Canadian aquatic environment five years after the regulation of antifouling uses of tributyltin. *Water Qual. Res. J. Can.* 32:453–521.

Chliamovitch, Y.P. and Kuhn, C. (1977) Behavioral, hematological and histological studies on acute toxicity of bis(tri-*n*-butyltin) oxide on *Salmo gairdneri* Richardson and *Tilapia rendalli* Boulenger. *J. Fish. Biol.* 10:575–585.

Chow, S.C., Kass, G.E.N., McCabe, M.J. and Orrenius, S. (1992) Tributyltin increases cytosolic free Ca^{2+} concentration in thymocytes by mobilizing intracellular Ca^{2+} activating a Ca^{2+} entry pathway, and inhibiting Ca^{2+} efflux. *Arch. Biochem. Biophys.* 298:143–149.

Connerton, I.F. and Griffith, D.E. (1989) Organotin compounds as energy-potentiated uncouplers of rat liver mitochondria. *Appl. Organomet. Chem.* 3:545–551.

deVries, H., Penninks, A.H., Snoeij, N.J. and Seinen, W. (1991) Comparative toxicity of organotin compounds to rainbow trout (*Oncorhynchus mykiss*) yolk sac fry. *Sci. Total Environ.* 103:229–243.

Fent, K. (1991) Bioconcentration and elimination of tributyltin chloride by embryos and larvae of minnows *Phoxinus phoxinus*. *Aquat. Toxicol.* 20:147–158.

Fent, K. (1992) Embryotoxic effects of tributyltin on the minnow *Phoxinus phoxinus*. *Environ. Pollut.* 76:187–194.

Fent, K. (1996a) Ecotoxicology of organotin compounds. *CRC Crit. Rev. Toxicol.* 26:1–117.

Fent, K. (1996b) Organotin compounds in municipal waste water and sewage sludge: contamination, fate in treatment process and ecotoxicological consequences. *Sci. Total. Environ.* 185:151–159.

Fent, K. (1998) Ökotoxikologie. Thieme-Verlag, Stuttgart, p. 1–288.

Fent, K. and Bucheli, T.D. (1994) Inhibition of hepatic microsomal monooxygenase system by organotins *in vitro* in freshwater fish. *Aquat. Toxicol.* 28:107–126.

Fent, K. and Hunn, J. (1991) Phenyltins in water, sediment, and biota of freshwater marinas. *Environ. Sci. Technol.* 25:956–963.

Fent, K. and Hunn, J. (1993) Uptake and elimination of tributyltin in fish yolk-sac larvae. *Mar. Environ. Res.* 35:65–71.

Fent, K. and Hunn, J. (1995) Organotins in freshwater harbors and rivers: temporal distribution, annual trend and fate. *Environ. Toxicol. Chem.* 14:1123–1132.

Fent, K. and Hunn, J. (1996) Cytotoxicity of organic environmental chemicals to fish liver cells (PLHC-1). *Mar. Environ. Res.* 42:377–382.

Fent, K. and Looser, P.W. (1995) Bioaccumulation and bioavailability of tributyltin chloride: influence of pH and humic acids. *Water Res.* 29:1631–1637.

Fent, K. and Meier, W. (1992) Tributyltin-induced effects on early life stages of minnows *Phoxinus phoxinus*. *Arch. Environ. Contam. Toxicol.* 22:428–438.

Fent, K. and Meier, W. (1994) Effects of triphenyltin on fish early life stages. *Arch. Environ. Contam. Toxicol.* 27:224–231.

Fent, K. and Müller, M.D. (1991) Occurrence of organotins in municipal waste water and sewage sludge and behavior in a treatment plant. *Environ. Sci. Technol.* 25:489–493.

Fent, K. and Stegeman, J.J. (1991) Effects of tributyltin chloride on the hepatic microsomal monooxygenase system in the fish *Stenotomus chrysops*. *Aquat. Toxicol.* 20:159–168.

Fent, K. and Stegeman, J.J. (1993) Effects of tributyltin in vivo on hepatic cytochrome P450 forms in marine fish. *Aquat. Toxicol.* 24:219–240.

Fent, K., Hunn, J. and Sturm, M. (1991a) Organotins in lake sediments. *Naturwissenschaften* 78:219–221.

Fent, K., Lovas, R. and Hunn, J. (1991b) Bioaccumulation, elimination and metabolism of triphenyltin chloride by early life stages of minnows *Phoxinus phoxinus*. *Naturwissenschaften* 78:125–127.

Fent, K., Woodin, B.R. and Stegeman, J.J. (1998) Effects of triphenyltin and other organotins on hepatic monooxygenase system in fish. *Comp. Biochem. Physiol.*, in press.

Fioramonti, E., Semlitsch, R.D., Reyer, H.-U. and Fent, K. (1997) Effects of triphenyltin and pH on the growth and development of *Rana lessonae* and *Rana esculenta* tadpoles. *Environ. Toxicol. Chem.* 16:1940–1947.

George, S.G. and Buchanan, G. (1990) Isolation, properties and induction of plaice liver cytosolic glutathione-S-transferases. *Fish Physiol. Biochem.* 8:437–449.

Gibbs, P.E., Pascoe P.L., and Burt, G.R. (1988) Sex change in the female dogwhelk, *Nucella lapillus*, induced by tributyltin from antifouling paints. *J. Mar. Biol. Assoc. UK* 68:715–731.

Gray, B.H., Porvaznik, M., Flemming, C. and Lee, L.H. (1987) Tri-*n*-butyltin: a membrane toxicant. *Toxicology* 47:35–54.

Hahn, M.E., Lamb, M., Schultz, M.E., Smolowitz, R.M. and Stegeman, J.J. (1993) Cytochrome P4501A induction and inhibition by 3,3′,4,4′-tetrachlorobiphenyl in an Ah receptor-containing fish hepatoma cell line (PLHC-1). *Aquat. Toxicol.* 26:185208.

Hall Jr., L.W., Bushong, S.J., Hall, W.S. and Johnson, W.E. (1988a) Acute and chronic effects of tributyltin on a Chesapeake Bay copepod. *Environ. Toxicol. Chem.* 7:41–46.

Hall Jr., L.W., Bushong, S.J., Ziegenfuss, M.C., Johnson, W.E., Herman, R.L. and Wright, D.A. (1988b) Chronic toxicity of tributyltin to Chesapeake Bay biota. *Water Air Soil Pollut.* 39:365–376.

Hasan, M.A. and Juma, H.A. (1992) Assessment of tributyltin in the marine environment of Bahrain. *Mar. Pollut. Bull.* 24:408–410.

Hightower, L.E. and Renfro, J.L. (1988) Recent applications of fish cell culture to biomedical research. *J. Exp. Zool.* 248:290–302.

His, E. and Robert, R. (1980) Action d'un sel organometallique, l'acetate de tributylétain sur les oeufs et les larves de *Crassostrea gigas* (Thunberg). International Council for the Exploration of the Sea (ICES) Mariculture Commission, Copenhagen.

His, E. and Robert, R. (1985) Développement des véligeres de *Crassostrea gigas* dans le Bassin d'Areachon, études sur les mortalités larvaires. *Rev. Trav. Inst. Peches Marit.* 47:63–88.

Holm, G., Norrgren, L. and Linden, O. (1991) Reproductive and histopathological effects of long-term experimental exposure to bis(tributyltin)oxide (TBTO) on the three-spined stickleback, *Gasterosteus aculeatus* Linnaeus. *J. Fish. Biol.* 38:373–386.

Huggett, R.J., Unger, R.J., Seligman, P.F. and Valkirs, A.O. (1992) The marine biocide tributyltin. Assessing and managing the environmental risks. *Environ. Sci. Technol.* 26:232–237.

Ishizaka, T., Nemoto, S., Sasaki, K., Suzuki, T. and Saito, Y. (1989) Simultaneous determination of tri-*n*-butyltin, di-*n*-butyltin, and triphenyltin compounds in marine products. *J. Agric. Food Chem.* 37:1523–1527.

Iwata, H., Tanabe, S., Miyazaki, N. and Tatsukawa, R. (1994) Detection of butyltin compound residues in the blubber of marine mammals. *Mar. Pollut. Bull.* 28:607612.

Iwata, H., Tanabe, S., Mizuno, T. and Tatsukawa, R. (1995) High accumulation of toxic butyltins in marine mammals form Japanese coastal waters. *Environ. Sci. Technol.* 29:2959–2962.

Jantzen, E. and Wilken, R.-D. (1991) Organotin compounds in harbor sediments – analysis and critical examination. *Vom Wasser* 76:1–11.

Jarvinen, A.W., Tanner, D.K., Kline, E.R. and Knuth, M.L. (1988) Acute and chronic toxicity of triphenyltin hydroxide to fathead minnows (*Pimephales promelas*) following brief or continuous exposure. *Environ. Pollut.* 52:289–301.

Kannan, K., Tanabe, S., Iwata, H. and Tatsukawa, R. (1995) Butyltins in muscle and liver of fish collected from certain asian and oceanian countries. *Environ. Pollut.* 90:279–290.

Kannan, K., Senthilkumar, K., Loganathan, B.G., Takahashi, S., Odell, D.K. and S. Tanabe, (1997) Elevated accumulation of tributyltin and its breakdown products in bottlenose dolphins (*Tursiops truncatus*) found stranded along the U.S. atlantic and gulf coasts. *Environ. Sci. Technol.* 31:296–301.

Kannan, K., Guruge, K.S., Thomas, N.J., Tanabe, S. and Griesy, J. (1998) Butyltinresidues in southern sea otters (Enhydra lutris nereis) found dead along California coastal waters. *Envir. Sci. Technol.* 32:1169–1175.

Kim, G.B., Tanabe, S., Iwakiri, R., Tatsukawa, R., Amano, M., Miyazaki, N. and Tanaka, H. (1996) Accumulation of butyltin compounds in Risso's dolphin (*Grampus griseus*) from the pacific coast of Japan: comparison with organochlorine residue pattern. *Environ. Sci. Technol.* 30:2620–2625.

Lapota, D., Rosenberger, D.E., Platter-Rieger, M.F. and P.F. Seligman, (1993) Growth and survival of *Mytilus edulis* larvae exposed to low levels of dibutyltin and tributyltin. *Mar. Biol.* 115:413–419.

Law, R.J., Waldock, M.J., Allchin, C.R., Laslett, R.E. and Bailey, K.J. (1994) Contaminants in seawater around England and Wales: Results from monitoring surveys, 1990–1992. *Mar. Pollut. Bull.* 28:668–675.

Lawler, I.F. and Aldrich, J.C. (1987) Sublethal effects of bis(tri-*n*-butyltin) oxide on *Crassostrea gigas* spat. *Mar. Pollut. Bull.* 18:274–278.

Lee, R.F. (1986) Metabolism of bis(tributyltin)oxide by estuarine animals. *In: Proceedings of the Organotin Symposium Oceans '86, Washington, D.C.* Institute of Electrical and Electronics Engineers and Marine Technology Society, New York, pp 1182–1188.

Lee, R.F. (1991) Metabolism of tributyltin by marine animals and possible linkages to effects. *Mar. Environ. Res.* 32:29–35.

Lee, R.F. (1993) Passage of xenobiotics and their metabolites from hepatopancreas into ovary and oocytes of blue crabs, *Callinectes sapidus*: possible implications for vitellogenesis. *Mar. Environ. Res.* 35:181–187.

Lee, R.F., Valkirs, A.O. and Seligman, P.F. (1989) Importance of microalgae in the biodegradation of tributyltin in estuarine waters. *Environ. Sci. Technol.* 23:1515–1518.

Looser, P.W., Bertschi. S. and Fent, K. (1998) Bioconcentration and bioavailability of organotin compounds: influence of pH and humic substances. *Appl. Organomet. Chem.*: 12:1–11.

Maguire, R.J., (1987) Review: Environmental aspects of tributyltin. *Appl. Organomet. Chem.* 1:475–498.

Maguire, R.J. (1991) Aquatic environmental aspects of non-pesticidal organotin compounds. *Water Poll. Res. J. Can.* 26:243–360.

Maguire, R.J., Tkacz, R.J., Chau, Y.K., Bengert, G.A. and Wong, P.T.S. (1986) Occurrence of organotin compounds in water and sediments in Canada. *Chemosphere* 15:253–274.

Martin, R.C., Dixon, D.G., Maguire, R.J., Hodson P.V. and Tkacz, R.J. (1989) Acute toxicity, uptake, depuration and tissue distribution of tri-*n*-butyltin in rainbow trout, *Salmo gairdneri*. *Aquat. Toxicol.* 15:37–52.

Meador, J.P. (1997) Comparative toxicokinetics of tributyltin in five marine species and its utility in predicting bioaccumulation and acute toxicity. *Aquat. Toxicol.* 37:307–326.

Miura, Y. and Matsui, H. (1991) Inhibitory effects of phenyltin compounds on stimulus-induced changes in cytosolic free calcium and plasma membrane potential of human neutrophils. *Arch. Toxicol.* 65:562–569.

Morcillo, Y. and Porte, C. (1997) Interaction of tributyl- and triphenyltin with the microsomal monooxygenase system in molluscs and fish from the western Mediterranean. *Aquat. Toxicol.* 38:35–46.

Müller, M.D. (1987) Comprehensive trace level determination of organotin compounds in environmental samples using high-resolution gas chromatography with flame photometric detection. *Anal. Chem.* 59:617–623.

Nagase, H., Hamasaki, T., Takahiko, S., Hideaki, K., Yoshioka, Y. and Ose, Y. (1991) Structure-activity relationships for organotin compounds on the red killifish *Oryzias latipes*. *Appl. Organomet. Chem.* 3:545–551.

Newton, F., Thum, A., Davidson, B., Valkirs, A. and Seligman, P.F. (1985) Effects on the growth and survival of eggs and embryos of the California grunion (*Leuresthes tenuis*) exposed to trace levels of tributyltin. Naval Ocean Center, San Diego, CA.

Oehlmann, J. (1994) *Imposex bei Muriciden (Gastropoda, Prosobranchia) – eine ökotoxikologische Untersuchung zu TBT-Effekten*. Ph.D. thesis, University of Münster, Germany.

Oehlmann, J., Stroben, E., Bettin, C. and P. Fioroni, (1993) Hormonal disorders and tributyltin-induced imposex in marine snails. *In: Twenty-seventh European Marine Biology Symposium*, Dublin, Ireland. JAPAGA, Ashford, pp 301–305.

Oshima, Y., Nirmala, K., Go, J., Yokota, Y., Koyama, J., Imada, N., Honjo, T. and Kobayashi, K. (1997) High accumulation of tributyltin in blood among the tissues of fish and applicability to the environmental monitoring. *Environ. Toxicol. Chem.* 16:1515–1517.

Penninks, A.H. and Seinen, W. (1980) Toxicity of organotin compounds. Impairment of energy metabolism of rat thymocytes by various dialkyltin compounds. *Toxicol. Appl. Pharmacol.* 56:221–231.

Pieters, R.H., Kampinga, J., Bol-Schoenmakers, M., Lamb, B.W., Pennkins, A.H. and Seinen, W. (1989) Organotin-induced thymus atrophy concerns the OX-44+ immature thymocytes: Relation to the interaction between early thymocytes and thymic epithelial cells? *Thymus.* 14:79–88.

Pinkney, A.E., Matteson, L.L. and Wright, D.A. (1988) Effect of tributyltin on survival, growth, morphometry and RNA-DNA ratio of larval striped bass, *Morone saxatilis*. *In: Proceedings of the Organotin Symposium Oceans '88. Baltimore, Maryland*. The Institute of Electrical and Electronics Engineers and Marine Technology Society, New York, pp 987–991.

Pinkney, A.E., Wright, D.A., Jepson, M.A. and Towle, D.W. (1989) Effects of tributyltin compounds on ionic regulation and gill ATPase activity in estuarine fish. *Comp. Biochem. Physiol.* 92C:125–129.

Polster, M. and Halacka, K. (1971) Beitrag zur hygienisch-toxikologischen Problematik einiger antimikrobiell gebrauchter Organozinnverbindungen. Ernährungsforsch. 16:527–535.

Raffray, M. and Cohen, G.M. (1991) Bis(tri-*n*-butyltin)oxide induces programmed cell death (apoptosis) in immature rat thymocytes. *Arch. Toxicol.* 65:135–139.

Reader, S., Marion, M. and Denizeau, F. (1993) Flow cytometric analysis of the effects of tri-*n*-butyltin chloride on cytosolic free calcium and thiol levels in isolated rainbow trout hepatocytes. *Toxicology* 80:117–129.

Rice, C.D., Banes, M.M. and Ardelt, T.C. (1995) Immunotoxicity in channel catfish, *Ictalurus punctatus*, following acute exposure to tributyltin. *Arch. Environ. Contain. Toxicol.* 28:464–470.

Roberts, M.H. (1987) Acute toxicity of tributyltin chloride to embryos and larvae of two bivalve mollusks, *Crassostrea virginica* and *Mercenaria mercenaria*. *Bull. Environ. Contam. Toxicol.* 39:1012–1019.

Rosenberg, D.W., Drummond, G.S. and Kappas, A. (1981) The influence of organometals on heme metabolism. *In vivo* and *in vitro* studies with organotins. *Mol. Pharmacol.* 21:150–158.

Schebek, L., Andreae, M.O. and Tobschall, H.J. (1991) Methyl- and butyltin compounds in water and sediments of the Rhine river. *Environ. Sci. Technol.* 25:87 1–878.

Schwaiger, J., Bucher, F., Ferling, H., Kalbfus, W. and Negele, R.D. (1992) A prolonged toxicity study on the effects of sublethal concentrations of bis(tri-*n*-butyltin)oxide (TBTO): histopathological and histochemical findings in rainbow trout (*Oncorhynchus mykiss*). *Aquat. Toxicol.* 23:31–48.

Schwaiger, J., Fent, K., Stecher, H., Ferling, H. and Negele, R.D. (1996) Effects of sublethal concentrations of triphenyltin acetate on rainbow trout (*Oncorhynchus mykiss*). *Arch. Environ. Contam. Toxicol.* 30:327–334.

Seinen, W., Helder, T., Vernij, H., Penninks, A. and Leeuwangh, P. (1981) Short term toxicity of tri-*n*-butyltin chloride in rainbow trout (*Salmo gairdneri Richardson*) yolk sac fry. *Sci. Total Environ.* 19:155–166.

Seligman, P.F., Valkirs, A.O. and Lee, R.F. (1986) Degradation of tributyltin in San Diego Bay, California, waters. *Environ. Sci. Technol.* 20:1229–1235.

Seligman, P.F., Grovhoug, J.G., Valkirs, A.O., Stang, P.M., Fransham, R., Stallard, M.O., Davidson, B. and Lee, R.F. (1989) Distribution and fate of tributyltin in the United States marine environment. *Appl. Organomet. Chem.* 3:31–47.

Selwyn, M.J. (1978) Triorganotin compounds as ionophores and inhibitors of translocating ATPases. *In:* Zuckerman, J.J. (Ed) Organotin compounds: New chemistry and applications. American Chemical Society, Washington, D.C., pp 204–226.

Selwyn, A., Dawson, A.P., Stockdale, M. and Gains, N. (1970) Chloride-hydroxide exchange across mitochondrial erythrocyte and artificial lipid membranes mediated by trialkyl- and triphenyltin compounds. *Eur. J. Biochem.* 14:120–126.

Semlitsch, R.D., Foglia, M., Müller, A., Steinert, I., Fioramonti E., and Fent, K. (1995) Short-term exposure to triphenyltin affects the swimming and feeding behavior of tadpoles. *Environ. Toxicol. Chem.* 14:1419–1423.

Short, J.W. and Thrower, F.P. (1986) Tri-*n*-butyltin caused mortality of chinook salmon, *Oncorhynchus tshawytscha*, on transfer to a TBT-treated marine net pen. *In: Proceedings of the Organotin Symposium. Oceans '86, Washington, D.C.* The Institute of Electrical and Electronics Engineers and Marine Technology Society, New York, pp 1202–1205.

Shoukry, M.M. (1993) Equilibrium study of tributyltin(IV) complexes with amino acids and related compounds. *Bull. Soc. Chim. Fr.* 130:117–120.

Snoeij, N.J., van Iersel, A.A.J., Perminks, A.H. and Seinen, W. (1985) Toxicity of triorganotin compounds: Comparative *in vivo* studies with a series of trialkyl compounds and triphenyltin chloride in male rats. *Toxicol. Appl. Pharmacol.* 81:274–286.

Spooner, N., Gibbs, P.E., Bryan, G.W. and Goad, L.J. (1991) The effect of tributyltin upon steroid titres in the female dogwhelk, *Nucella lapillus*, and the development of imposex. *Mar. Environ. Res.* 32:37–49.

Stäb, J.A., Rozing, M.J.M., van Hattum, B., Cofino, W.P. and Brinkman, U.A.T. (1992) Normal-phase high-performance liquid chromatography with UV irradiation, morin complexation and fluorescence detection for the determination of organotin pesticides. *J. Chromat.* 609:195–203.

Stäb, J.A., Cofino, W.P., van Hattum, B. and Brinkman, U.A.T. (1994) Assessment of transport routes of triphenyltin used in potato culture in the Netherlands. *Anal. Chim. Acta* 286:335–341.

Stäb, J.A., Traas, T.P., Stroomberg, G., van Kesteren, J., Leonards, P., van Hattum, B., Brinkman, U.A.T. and W.P. Cofino, (1996) Determination of organotin compounds in the foodweb of a shallow freshwater lake in the Netherlands. *Arch. Environ. Contam. Toxicol.* 31:319–328.

Stegeman, J.J. and Hahn, M.E. (1994) Biochemistry and molecular biology of monooxygenases: Current perspectives on forms, functions, and regulation of cytochrome P450 in aquatic species. *In:* Malins, D.C. and Ostrander, G.K. (Eds) *Aquatic toxicology: molecular, biochemical, and cellular perspectives.* Lewis Publishers, Boca Raton, Ann Arbor, London, Tokyo, pp 87–206.

Stegeman, J.J. and Kloepper-Sams, P.J. (1987) Cytochrome P-450 isozymes and monooxygenase activity in aquatic animals. *Environ. Health Perspect.* 71:87–95.

Stegeman, J.J., Brouwer, M., Di Giulio, R.T., Förlin, L., Fowler, B.A., Sanders, B.M. and van Veld, P.A. (1992) Enzyme and protein synthesis as indicators of contaminant exposure and effect. *In:* Huggett, R.J., Kimerle, R.A., Mehrle, P.M., and Bergman, H.L. (Eds) *Biomarkers. Biochemical, physiological, and histological markers of anthropogenic stress.* Lewis Publishers, Boca Raton, Ann Arbor, London, Tokyo 235–335.

Stromgren, T. and Bongard, T. (1987) The effect of tributyltin oxide on growth of *Mytilus edulis. Mar. Pollut. Bull.* 18:30–31.

Tanabe, S., Prudente, M., Mizuno, T., Hasegawa, J., Iwata, H. and Miyazaki, N. (1998) Butyltin contamination in marine mammals from North Pacific and Asian coastal waters. *Environ. Sci. Technol.* 32:193–198.

Tas, W.J. (1989) Bioconcentration and elimination of triphenyltin hydroxide in fish. *Mar. Environ. Res.* 28:215–218.

Thain, J.E. (1986) Toxicity of TBT to bivalves: Effects on reproduction, growth and survival. *In: Proceedings of the Organotin Symposium Oceans '86, Washington, D.C.* The Institute of Electrical and Electronics Engineers and Marine Technology Society, New York, pp 1306–1313.

Triebskorn, R., Köhler, H.-R., Flemming, J., Braunbeck, T., Negele, R.D. and Rahmann, H. (1994) Evaluation of bis (tri-*n*-butyltin) oxide (TBTO) neurotoxicity in rainbow trout (*Oncorhynchus mykiss*). 1. Behaviour, weight increase, and tin content. *Aquat. Toxicol.* 30:189–197.

Tsuda, T., Nakanishi, H., Aoki, S. and Takebayashi, J. (1986) Bioconcentration of butyltin compounds by round crucian carp. *Toxicol. Environ. Chem.* 12:137–143.

Tsuda, T., Nakanishi, H., Aoki, S. and Takebayashi, J. (1987) Bioconcentration and metabolism of phenyltin chlorides in carp. *Water Res.* 21:949–953.

Tsuda, T., Nakanishi, H., Aoki, S. and Takebayashi, J. (1988) Bioconcentration and metabolism of butyltin compounds in carp. *Water Res.* 22:647–651.

Tsuda, T., Aoki, S., Kojima, M. and Harada, H. (1990a) Differences between freshwater and seawater-acclimated guppies in the accumulation and excretion of tri-*n*-butyltin chloride and triphenyltin chloride. *Water Res.* 24:1373–1376.

Tsuda, T., Aoki, S., Kojima, M. and Harada, H. (1990b) The influence of pH on the accumulation of tri-*n*-butyltin chloride and triphenyltin chloride in carp. *Comp. Biochem. Physiol.* 95C:151–153.

Tsuda, T., Aoki, S., Kojima, M. and Harada, H. (1991) Accumulation of tri-*n*-butyltin chloride and triphenyltin chloride by oral and *via* gill intake of goldfish (*Carassius auratus*). *Comp. Biochem. Physiol.* 99C:69–72.

U'ren, S.C. (1983) Acute toxicity of bis(tributyltin)oxide to a marine copepod. *Mar. Pollut. Bull.* 14:303–306.

Valkirs, A.O., Davidson, B.M. and Seligman, P.F. (1987) Sublethal growth effects and mortality to marine bivalves from long-term exposure to tributyltin. *Chemosphere* 16:201–220.

Virkki, L. and Nikinmaa, M. (1993) Tributyltin inhibition of adrenergically activated sodium/proton exchange in erythrocytes of rainbow trout (*Oncorhynchus mykiss*). *Aquat. Toxicol.* 25:139–146.

Wade, T.L., Garcia-Romero, B. and Brooks, J.M. (1988) Tributyltin contamination in bivalves from United States coastal estuaries. *Environ. Sci. Technol.* 22:1488–1493.

Waldock, M.J., Waite, M.E. and Thain, J.E. (1988) Inputs of TBT to the marine environment from shipping activity in the U.K. *Environ. Technol. Lett.* 9:999–1010.

Walsh, G.E., L.L. McLaughlan, E.M. Lores, M.K. Louie and Deans, C.H. (1985) Effects of organotin on growth and survival of two marine diatoms, *Skeletoneina costatum* and *Thalassiosira pseudonata. Chemosphere* 14:383–392.

Ward, G.S., Cramm, G.C., Parrish, P.R., Trachman, H. and Slesinger, A. (1981) Bioaccumulation and chronic toxicity of bis(tributyltin)oxide (TBTO): tests with a saltwater fish. *In:* Branson, D.R. and Dickson, K.L. (Eds) *Aquatic toxicology and hazard assessment.* Americal Society for Testing and Materials, Philadelphia, pp 183–200.

Watling-Payne, A.S. and Selwyn, M.J. (1974) Inhibition and uncoupling of photophosphorylation in isolated chloroplasts by organotin, organomercury and diphenyleneiodonium compounds. *Biochem. J.* 142:65–74.

Wester, P.W. and Canton, J.H. (1987) Histopathological study of *Poecilia reticulata* (guppy) after long-term exposure to bis(tri-*n*-butyltin)oxide (TBTO) and di-*n*-butyltin dichloride (DBTCl). *Aquat. Toxicol.* 10:143–165.

Wester, P.W., Canton, J.H., van Iersel, A.A.J., Krajnc, E.I. and Vaessen, H.A.M.G. (1990) The toxicity of bis(tri-*n*-butyltin)oxide (TBTO) and di-*n*-butyltin dichloride (DBTCl) in the small fish species *Oryzias latipes* (medaka) and *Poecilia reticulata* (guppy). *Aquat. Toxicol.* 16:53–72.

WHO (1990) Tributyltin compounds. World Health Organization, Geneva, Switzerland.

Wishkovsky, A., Mathews, E.S. and Weeks, B.A. (1989) Effect of tributyltin on the chemiluminescent response of phagocytes from three species of estuarine fish. *Arch. Environ. Contam. Toxicol.* 18:826–831.

Wulf, R.G. and Byington, K.H. (1975) On the structure-activity relationships and mechanism of organotin induced non-energy dependent swelling of liver mitochondria. *Arch. Biochem. Biophys.* 167:176–185.

Yamada, H. and Takayanagi, K. (1992) Bioconcentration and elimination of bis(tributyltin)-oxide (TBTO) and triphenyltin chloride (TPTCl) in several marine fish species. *Water Res.* 26:1589–1595.

Zucker, R.M., Elstein, K.H., Easterling, R.E., Ting-Beall, H.P., Allis, J.W. and Massaro, E.J. (1988) Effects of tributyltin on biomembranes: Alteration of flow cytometric parameters in inhibition of Na$^+$, K$^+$-ATPase two-dimensional crystallization. *Toxicol. Appl. Pharmacol.* 96:393–403.

Fish Ecotoxicology
ed. by T. Braunbeck, D. E. Hinton and B. Streit
© 1998 Birkhäuser Verlag Basel/Switzerland

Multiple stressors in the Sacramento River watershed

David E. Hinton

Laboratory of Aquatic Toxicology, Department of Anatomy, Physiology and Cell Biology, School of Veterinary Medicine, University of California-Davis, Davis, CA 95616, USA

Summary. Aquatic biota in the Sacramento River watershed are stressed by diversion of river flows, by historical mining resulting in cadmium, copper, zinc, and mercury, and, more recently, contamination by agricultural and urban chemical runoff. In addition, the proposed redirection of drainage of saline waters – containing selenium – from the western slope of the San Joaquin River into the Delta formed by the confluence of the Sacramento and San Joaquin Rivers could add to the stress on resident organisms. These combined stressors have led to deterioration in surface water quality and the aquatic habitat. The potential interaction of these stressors, coupled with invasions of foreign species and the export of juvenile fish into aqueducts, has driven several species of fish to near extinction in the system. Effects of historical contamination by heavy metals are potentially exacerbated by presence of organophosphate pesticides, at concentrations exceeding National Academy of Sciences recommendations, throughout the lower watershed and the San Francisco Bay.

The Asian clam, *Potamocorbula amurensis*, an introduced non-indigenous species has apparently become a preferred food item of the sturgeon, *Accipenser transmontanus*, an important sport and aquaculture species. Since this introduction, sturgeon body burdens for selenium have increased dramatically and analytical chemistry of *P. amurensis* indicates that these organisms are effective bioaccumulators of selenium.

This review examines potential ecotoxicity associated with multiple stressors in the watershed. Data from field monitoring, laboratory toxicity assays with ambient water, and ecotoxicologic investigations are reviewed. Potential designs for multiple stressor investigations are discussed. The information presented on this watershed illustrates the challenge to investigators seeking to evaluate multiple stressor effects on riverine and estuarine organisms.

Introduction

The Sacramento River watershed includes all of Northern California except far western and eastern margins. The Sacramento River catchment includes the entire western slope of the Sierra Nevada Mountains to the east and the Coast Range to the west, and at the confluence with the San Joaquin River, forms the Delta, waters of which flow into the upper San Francisco Bay. The Delta and San Francisco Bay receive at least 80% of their fresh water from the Sacramento River, and its water is widely distributed to municipalities, to agriculture and for other uses throughout the system. The main features of the San Francisco Bay were reviewed by Bingham (1996) as well as Wright and Phillips (1988), and Bingham (1996) has provided a succinct account of human management and development of the Sacramento-San Joaquin Delta.

The Delta, 1127 km of interconnected waterways and encompassing 2987 km^2, was historically rich in wildlife resources, however, most of the marsh lands have been developed for agricultural (92%) and urban uses. Inherent modifications of water flow patterns to facilitate water exports and control flooding have dramatically altered the morphology and likely the carrying capacity of the Delta. Precipitous declines have affected the 29 indigenous fish species to the extent that 12 have been eliminated or are threatened with extinction (Herbold et al., 1992). Features of the watershed are shown in the accompanying maps (Figs. 1 and 2).

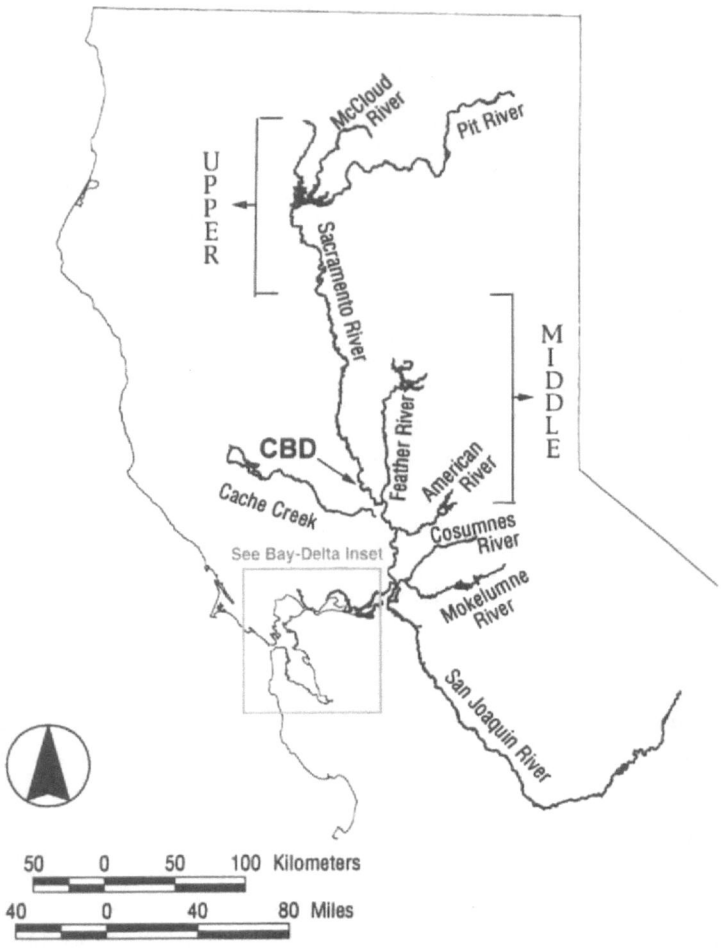

Map provided by the Information Center for the Environment, University of California, Davis
Data provided by Teale Data Center, US Environmental Protection Agency, and California Department of Fish and Game

Figure 1. Map showing upper and middle portions of the Sacramento River and selected tributaries. Approximate location of Colusa Basin Drain (CBD) is shown. Delta (inset) is formed by junction of Sacramento and San Joaquin Rivers and connects with the upper portion of the San Francisco Bay (right portion of watershed in box). Arrow in circle indicates direction (North).

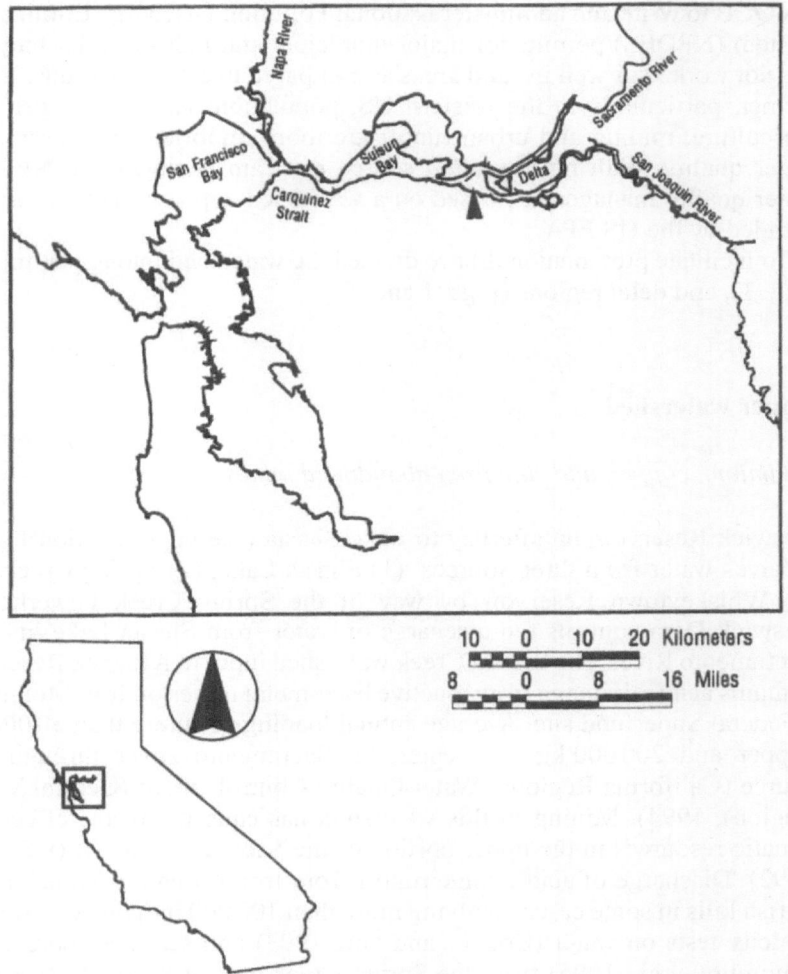

Map provided by the Information Center for the Environment, University of California, Davis
Data provided by Teale Data Center, US Environmental Protection Agency, and California Department of Fish and Game

Figure 2. Enlargement of region inside box (Fig. 1). Arrow head in figure indicates approximate site of entry of agricultural drain carrying selenium-laden water into Delta. Only northern reaches of San Francisco Bay are labeled. Arrow in circle indicates direction (North).

Water management in the Sacramento River watershed is among the most controversial environmental issues on the western US coast. The central question is how much water can be exported to southern California without causing unacceptable aquatic habitat alteration in the Delta region. The United States Environmental Protection Agency (US EPA) and the State of California Regional Water Quality Control Boards (RWQCBs) are changing their water management strategy. The old program, relying upon

RWQCB to write and administer National Pollution Discharge Elimination System (NPDES) permits for major municipal and industrial dischargers, has not worked as well in rural areas as compared to urban locations. In the former, particularly in the western US, population density is lower, and agriculture, mining and urban runoff are more important in determining water quality. With no *non-point* source program analogous to NPDES, water quality management based on a watershed approach is now recommended by the US EPA.

To facilitate presentation, I have divided the watershed into upper, middle (Fig. 1), and delta regions (Figs. 1 and 2).

Upper watershed

Cadmium, copper, and zinc from abandoned mines

Keswick Reservoir, an afterbay to Lake Shasta (see upper portion Fig. 1) receives water from three sources: (1) Shasta Lake, (2) Spring Creek and (3) Whiskeytown Reservoir by way of the Spring Creek Powerhouse. Keswick Dam controls the discharge of water from Shasta Lake into the Sacramento River. The Spring Creek watershed input to Keswick Reservoir contains acidic drainage from inactive base-metal mines on Iron Mountain, a Federal Superfund site. Average annual loadings of more than 50 000 kg copper and 200 000 kg zinc enter the Sacramento River through this source (California Regional Water Quality Control Board (Central Valley Region), 1994). Mining in this watershed has caused adverse effects on aquatic resources in the upper portion of the Sacramento River (US EPA, 1992). Discharge of acidic mine runoff from Iron Mountain Mine has led to fish kills in some cases involving more than 100 000 individuals. Aquatic toxicity tests on water (Connor and Foe, 1993) and sediment pore water (Fujimura et al., 1995) from the Spring Creek arm of Keswick Reservoir have shown acute toxicity to invertebrates, and to larval rainbow trout (*Oncorhynchus mykiss*). The long-term consequences to fish populations are not known, but the proximity of this input to spawning and rearing habitat for anadromous fish (1.5 km downstream) underscores the importance of ecotoxicologic investigations.

Cadmium, copper and zinc may adversely affect reproduction and development. Cadmium is a non-essential metal, and there is a literature related to developmental effects in fishes of cadmium exposure. Decreased growth of larvae following cadmium exposure may be due to decreased enzyme activity. Birge et al. (1991) exposed fathead minnows from immediately after fertilization until 4 days after hatching and detected developmental, primarily skeletal, defects. Other developmental defects have been reviewed by Weis and Weis (1989).

Copper is an essential micronutrient, but when embryos are exposed to excesses of this metal, even at low concentrations, toxic effects are made manifest. Herring larvae showed more toxicity, if exposure was initiated at an earlier embryologic stage (Blaxter, 1977). Larvae proved to be 30 times less susceptible than embryos. Interestingly, the reverse was true, when eight species of freshwater fishes were exposed as embryos or larvae by McKim and co-workers (1978). Prior embryo exposure may render larvae less resistant to copper exposure (Weis and Weis, 1991).

Zinc is another required micronutrient that, at higher concentrations, induces developmental toxicity. The vertebral column is one target often reported. The chorion may be altered as well, since reduced chorionic strength was observed following exposure of brook trout embryos to 1.4 mg/L zinc (Holcombe et al., 1979). Zebrafish adults exposed to zinc produced offspring with developmental anomalies (Speranza et al., 1977).

Copper from algal control in rice fields

One other source of copper to the watershed may be important, but has not received the attention it likely deserves. The California Department of Pesticide Regulation (1990) has recorded the application of 1 000 000 kg of copper-bearing compounds to rice fields for control of algae during 1990. Much of the rice production is in the Colusa Basin (area above Cache Creek on western side of Sacramento River; Fig. 1). Fate of this copper may include retention in the fields by sorption to minerals or to organic substances; however, no systematic study of mobility has been reported.

Mercury mining and the gold rush

Historically, mercury was mined intensively in the Coast Range and transported across the Central Valley for use in placer gold mining operations in the Sierra Nevada, particularly during the gold rush of the 19th century. With this usage pattern, the old cinnabar mining sites, sites of spillage during transport, and sediment in streams, in which gold ore was amalgamated to mercury in the mining process, are all of potential interest toxicologically. Biotic and abiotic methylation of the inorganic mercury produces methyl mercury, a neurotoxin and the most toxic chemical form of the metal (Porcella, 1994). Slotton and colleagues (1995) examined gold mining impacts on aquatic food chain mercury. They reported that the highly alkaline waters of the western US make atmospheric deposition, a major source in other regions of the globe, of less importance. They characterized the northern California mercury source as large-scale, bulk contamination. From both the mercury ore (cinnabar) mining sites in the

Coastal Range and from the gold mining sites on the Sierra Nevada streams, mercury has moved into downstream rivers and lakes, foothill reservoirs and into the Delta and San Francisco Bay. From estimates by the Central Valley Regional Water Quality Control Board (1987), over 3 million kg of mercury were deposited into Sierra Nevada streams during the California Gold Rush. As a way of determining mercury bioavailability, Slotton et al. (1995) measured metal in stream aquatic insects and in rainbow trout. These organisms integrated conditions over time and their bioaccumulation was an indicator of the bioavailable fraction of mercury. To date, the possible adverse effects on biota have not been examined.

The developmental toxicity of embryo and larval fishes exposed to methyl mercury has been reviewed (Weis and Weis, 1991). Defects were seen in optic cups, and in craniofacial development. Additionally, cardiovascular anomalies were observed. Salmonid reproduction may be sensitive to mercury. Adult female brook trout transferred Hg from adult tissues to developing oocytes resulting in some malformations (McKim et al., 1979). Birge et al. (1979) questioned the freshwater criterion of 0.05 µg/L mercury, after adults chronically exposed to 0.2 or 0.7 µg/L mercury produced larvae, 50% of which died at 4 days after hatching.

Middle river and delta

Agriculture in the Central Valley of California accounts for 10% of the annual pesticide usage for the entire nation (Kuivila and Foe, 1995). Pesticide usage data and the crops with which these are associated are available on a county by county basis from the Department of Pesticide Regulation. These data help to establish local usage patterns, and with a Geographical Information System they can be used to construct quantitative information on a watershed basis. For example, 100 000 kg of carbofuran were applied in 1990, primarily on rice, alfalfa, and grapes (Zalom, 1994); the active ingredient of diazinon applied in 1990 totaled 295 455 kg, 40% of which went to almonds, and another 15% for structural pest control and landscape maintenance (Sheipline, 1993). Together, this type of data organization has been used to demonstrate distinct patterns of exposure characterized with respect to season and location (Kuivila and Foe, 1995).

Our laboratory has contributed to procedures for identifying pesticide toxicity in ambient waters (Bailey et al., 1996). In this study, we investigated responses of *Ceriodaphnia dubia* to three of the pesticides which are commonly encountered in this portion of the watershed. Carbofuran, diazinon, or chlorpyrifos under acute Toxicity Identification Evaluation (TIE) procedures were used to characterize properties of each compound. Having a TIE "fingerprint" for specific pesticides could be very useful, because many of the current analytical methods may not provide the sensitivity necessary to detect biologically active concentrations of highly toxic chemicals. For example, EPA Method 614 for organophosphate pesticides has detection limits of 0.5 µg/L for both diazinon and chlorpyrifos (US EPA, 1982). The limits exceed LC_{50}s for *Ceriodaphnia dubia* meaning that toxicity could be present with no analytical confirmation. Our studies showed that diazinon degraded under acid, while carbofuran degraded under basic conditions with loss of toxicity within 6 h. Neither of the above conditions reduced chlorpyrifos. Solid phase extraction proved 95% effective with all but chlorpyrifos. All three pesticides eluted separately in methanol/water fractions. Piperonyl butoxide, a P450 inhibitor, ameliorated toxicity of the two organophosphates, but not carbofuran. Such use of TIE procedures enables sorting of possible candidates to better determine agent(s) responsible for toxicity.

Monitoring by Regional and State Water Quality Boards has centered around US EPA toxicity tests conducted on water samples from the river and from agricultural drains (cf. review by Bailey et al., 1995). When specific samples proved toxic, follow-up investigations using TIEs were employed and three major watershed problem areas were identified: (1) agricultural discharge in the Colusa Basin Drain (middle river – Fig. 1; Bailey et al., 1994), (2) urban storm water runoff, and (3) dormant spray applications to almonds and fruit (*Prunus* spp.) orchards in the Central Valley especially near the Delta (Kuivila and Foe, 1995). Some specific examples where these approaches were applied are given below for problem areas 1 and 3.

The striped bass (*Morone saxatilis*) is one of the most important sport fishes in California and the most important sport fish in the Delta (IESP, 1987; Moyle, 1976). Dramatic population decline since the mid-1970s has been largely attributed to failure of recruitment (Chadwick et al., 1977; Stevens et al., 1985). This decline has been attributed to factors such as water diversions, pollution and reduced abundance of food organisms. Because discharge from the Colusa Basin brings water from over 60 700 hectares, it can account for over 20% of the flow of the Sacramento River. Rice fields in this area drain directly into striped bass spawning grounds at the precise time of spawning. Early life stages could be adversely affected. Our laboratory conducted investigations over a 3-year period, modeling pesticide use and conducting toxicity studies with striped bass larvae and embryos and a prey species, the opossum shrimp, *Neomysis mercedis*, (Bailey et al., 1994). Toxicity to each species was demonstrated using 96 h bioassay with grab samples from the Colusa Basin Drain (Fig. 1). Histopathologic analysis of striped bass larvae surviving acute toxicity tests showed necrosis in brain stem and rostral spinal cord further suggesting that exposed larvae would be less likely to survive in their normal habitat. In a limited number of cases, exposure of striped bass larvae to river water, from sites above the level of the Colusa Basin Drain, has led to similar histologic central nervous system alterations. These findings suggest that compounds in addition to rice field pesticides may be producing these lesions as well.

From 1987 to 1992, striped bass and other fish populations with pelagic larvae declined markedly concurrent with dramatic reductions in phytoplankton and zooplankton. These patterns suggested that pelagic food was limited during times of low freshwater input and that larval fish starvation might be a cause of diminished recruitment. However, toxic compounds in agricultural runoff are less diluted in these same years of low-outflow, and this factor could enhance the potential for toxic impact. An integrated interdisciplinary study (Bennett et al., 1995) was conducted, and histopathology then enabled us to identify possible toxic effects. First, bass larvae were starved under laboratory-controlled conditions, and after 2 days of this treatment, indices of larval morphology and eye and liver tissue condition reflected food deprivation. In specimens caught in the field (1988–1991), however, > 90% were classified as feeding larvae having food in their guts and lacking tissue alterations consistent with starvation. Unexpectedly, 26–30% of field-caught larvae from 1988–1990 showed liver alterations consistent with exposure to toxic substances. Prevalence of livers with toxic alterations dropped to 15% in 1991 suggesting improvement, however, larval survival for 1991 did not increase (as established from tow surveys for 38 mm, young-of-the-year fish). The study indicates the need for interdisciplinary approaches to distinguish anthropogenic intervention from estuarine food-web processes and indicates the multifactorial considerations necessary to protect the resource.

State and federal water pumping facilities regularly divert 60–80% of the freshwater outflow to the San Francisco Bay (Herbold et al., 1992), altering water flow by drawing freshwater south across the Delta (Fig. 2). This reduced outflow has been associated with a decline in the abundance of fishes and invertebrates in the estuary (Herbold et al., 1992; Moyle et al., 1992). Not only is there a complex combination of habitat alteration and entrainment by water diversions (Bennett and Kennedy, 1993; Stevens et al., 1985) in the Sacramento River watershed, but effects of various physical and anthropogenic processes may co-vary with freshwater input to obscure the importance of a single factor on the recruitment process. However, flow and diversion are major factors affecting recruitment of young-of-the-year fishes (Stevens et al., 1985).

Kuivila and Foe (1995) coupled analytical chemistry and toxicity testing of water samples from the lower Sacramento- and upper San Joaquin-Rivers, the Delta and upper San Francisco Bay (Fig. 2) during and after dormant spray pesticide applications in January and February of 1993. Dormant spray pesticides including diazinon, methidathion, chlorpyrifos and malathion are typically applied to stone fruit orchards in the Central Valley during these months. Bioassays with *Ceriodaphnia dubia* (EPA, 1989) were performed, since these organophosphate insecticides are acetylcholinesterase inhibitors and are most toxic to zooplankton (Sittig,

1981; Verschueren, 1983). Following rainfall events, distinct pulses of the three compounds were detected in the San Joaquin River in January and February and in the Sacramento River in February. Higher pesticide loads in the latter were thought to be due to greater amounts of rainfall in the Sacramento Valley. Use patterns and differential water solubility accounted for observed temporal and spatial distributions in the two rivers. Importantly, Sacramento River water at Rio Vista, California (Fig. 2) proved acutely toxic to *Ceriodaphnia dubia* for three consecutive days, and San Joaquin River water at Vernalis, California (Fig. 2), was toxic for 12 consecutive days. All the above water samples contained diazinon at concentrations which could account for most, but not all of the toxicity. Interestingly, unpublished observations from this laboratory showed that following flooding of the Sacramento River in January 1997, river water samples from a large expanse of continuous floodwater proved acutely toxic in *Ceriodaphnia dubia* tests for 10 consecutive days. These landscape-scale events could cause both short- and long-term effects on resident organisms. However, assessments of effects on resident organisms have received little attention.

Selenium

California in general is heavily influenced by agricultural mobilization of selenium. Skorupa and co-workers (1996) reviewed the relatively extensive literature on selenium and provided guidelines for interpreting selenium exposures in fish and wildlife. Three points from that review are particularly relevant here. First, selenium is much less toxic to most plants and invertebrates than to vertebrates. It follows, therefore, that food web interactions, trophic transfer and biomagnification must be considered in evaluating the effects of this metal on resident vertebrates. Second, among oviparous vertebrates reproduction is especially sensitive to selenium. This could lead to reproductive failure and to population level effects. Third, metabolic stress, i.e., winter weather, may increase susceptibility of fish to selenium poisoning (Lemly, 1993). This underscores the need for ecotoxicological studies to incorporate multiple stressors in the watershed.

Kesterson Reservoir, an evaporation basin for irrigation drainage water on the western San Joaquin River valley, and managed as a series of ponds contains water with elevated (15–350 µg/L) selenium (for review, see Skorupa et al., 1996). Mosquitofish (*Gambusia affinis*) body burdens averaged 170 µg/kg. Studies with these fishes were performed using resident individuals inhabiting waters of the San Luis Drain, the source of Kesterson's drain Water (Saiki and Ogle, 1995; Zahm, 1986). In these organisms, 12 fry per brood were observed, and this was compared to a geographically relevant reference value of 25. To drain the selenium-laden Kesterson waters and to protect migratory waterfowl and fishes, various proposals have been made (e.g., Ohlendorf et al., 1988). The latest plan is to move water north into the Delta (see Fig. 2). This plan has concerned biologists, and a few of the reasons for that concern are presented below.

Agricultural translocation of selenium is not the only source of selenium to the Delta and upper San Francisco Bay. Another major source of selenium to these parts of the Sacramento River Watershed is refinery process wastewater (Nichols et al., 1986; Skorupa et al., 1996; Wright and Phillips, 1988). Brown and Luoma (1995) reported that selenium IV was accumulated efficiently by phytoplankton and converted to organic selenium. In this form, the metal was taken up by clams, when they consumed the phytoplankton (Luoma et al., 1992). It is also possible that phytoplankton may be a source of selenium to larval fish, but, this possibility has not been investigated to date.

Exotic species introduction

Release of seawater ballast from cargo vessels is believed to have been responsible for introduction of an Asian clam species (*Potamocorbula amurensis*; Carlton et al., 1990). First recorded in 1986, this clam has shown explosive population growth and has spread throughout northern San Francisco Bay (Nichols et al., 1990). The latter report indicates that the growth has led to the displacement of a former benthic community.

Effects of the invasion known to have impacted the lower watershed are: reduction of phytoplanktonic food sources for infaunal larval fishes, bioaccumulation of selenium by *Potamocorbula amurensis*, and apparent trophic transfer of selenium to ducks (scoters and scaup; Brown and Luoma, 1995) and to sturgeon (personal communications by Dr. David Kohlhorst, California Department of Fish and Game, Stockton, CA). Necropsy of sturgeon has shown their digestive tracts to contain numerous *Potamocorbula amurensis*, indicating that these organisms are a major food item. Interestingly, over the period of time during which the *Potamocorbula amurensis* growth has been documented, the body burdens for selenium in sturgeon, as established by the California Department of Fish and Game, have increased sharply (personal communications by Mary Dunne, California Department of Fish and Game, Stockton, CA). Concern exists for the possibility of reproductive toxicity in these sturgeon, an important game and aquacultural species for California. For example, mean selenium dry weight values in skeletal muscle of 30 sturgeon from Suisun Bay in the upper San Francisco Bay were 17.51 mg/kg. Selenium levels between 20 and 40 mg/kg dry weight in muscle have been shown to be associated with reproductive failure (Lemly, 1985a, b), while lower values (7–9 mg/kg) were associated with abnormal ovarian tissue (Sorenson et al., 1984).

Discussion

While the studies discussed above represent only a portion of the work that has been performed on the Sacramento River watershed, they are representative of the approaches taken and in some instances I have reviewed most of the major work with resident fishes and a given pollutant, e.g., selenium. It is obvious that by far the most common single approach has been to employ US EPA three species toxicity tests in the laboratory. However, the environmental relevance of the tests was enhanced by use of "real world" surface water samples. Although the grab samples represented what was present at the time of sampling, daily repeated sampling coupled with bioassays demonstrating acute toxicity over consecutive days (Kuivila and Foe, 1995, as well as unpublished results from our laboratory in January 1997) proved a significant finding. Spatially, this means that large volumes of river water contained sufficient organophosphates to result in toxicity to cladocerans and other zooplankters (Evans, 1988; Norberg-King et al., 1991). Temporally, results over consecutive days suggested that the toxic conditions were present over duration sufficient to be classed as of intermediate severity. Furthermore, if zooplankton food sources are reduced from a stretch of the watershed, growth of fish larvae may be retarded leaving them more susceptible to predation (Bennett et al., 1995). Individual-based modeling of larval striped bass (Rose et al., 1993) indicates that reduced growth rate could reduce larval survival by 40–45%.

Bailey and colleagues (1994) as well as Bennett et al. (1995) based their studies on larvae of feral striped bass. Since maternal brood stock, captured from the watershed, were used for egg and larval production, it is possible that the full extent of larval toxicity has not been detected. In similar studies with lake trout (Mac and Edsall, 1991), contaminants in yolk, derived from the mother, led to reproductive failure. Despite an extensive review, I was not able to find contaminant egg burdens for Sacramento River watershed striped bass. It is possible that larvae, by utilization of yolk, become exposed to maternally-derived contaminants, and this may influence their subsequent response, when used as subjects in acute toxicity tests. From the review above, laboratory studies have shown that for all of the metals which are present as significant contaminants in this watershed, embryo-larval (F_1) toxicity follows exposure of the adults.

Finally, the ecotoxicological significance of exposure to suites of stressors has not been addressed. What is the effect of multiple stressors on the ability of the individual to capture food, avoid predation, swim against currents, and withstand temperature elevation and salinity stress?

Simulation of multiple stressor exposure could be performed in the laboratory and provide useful information. For example, the exposure pattern for out-migrating fish could be simulated by initial exposure to low winter temperature, then to cadmium, copper and zinc followed by a recovery period, which mimics transit time downstream to areas where agricultural returns

Table 1. Selected multiple stressors in the Sacramento River watershed

Cd^{2+}, Cu^{2+}, Zn^{2+}	Temperature
Hg^{2+}	Diversions
Se^{2+}	Exports (pumps)
Agricultural returns	Exotic species
• Herbicides	Salinity
• Pesticides	

predominate. This could be followed by swimming challenges, exposure to increased temperature and to salinity fluxes. It is likely that these ecotoxicologic simulations would better reflect the composite response from exposure to a suite of stressors (Tab. 1). However, if tolerance to one or more stressors has occurred, the use of local organisms (i.e., those with a history of responding to selection pressure of their contaminated environment) may result in data which would not be indicative of the sensitivity demonstrated by conspecifics from a less-polluted river (Luoma, 1977).

Improved methods to demonstrate effects of these stressors have not been employed with resident fish populations. A major target of organophosphate and carbamate pesticides are the cholinesterases of the blood, muscle and nervous systems. Although toxicity is considered due to their inhibition of cholinergic synapses, myopathies are induced by excess acetylcholine at neuro-musclular junctions. Long-term delayed neuropathies due to selected organophosphates such as chlorpyrifos, but not diazinon and parathion, are well known (Cherniack, 1988; Dettbarn, 1984). Such responses could be highly chemical-specific. Inhibitions of specific acetylcholinesterases and non-specific cholinesterases of vertebrates and invertebrates are useful as biomarkers of exposure and of effect (e.g., Wilson and Henderson, 1992; Wilson et al., 1996). Reduced fecundity and developmental abnormalities in fishes have been observed after exposure to a single part per million (Davis, 1995). When salmonids were exposed to cholinesterase-inhibiting compounds, this form of toxicity altered respiration (Klaverkamp et al., 1977), swimming (Matton and Lattam, 1969; Post and Leisure, 1974), feeding (Bull and McInerney, 1974; Wildish and Lister, 1973), and social interactions (Symons, 1973); however, biomarker-based field studies focused on acetylcholinesterase or cholinesterase inhibition in fishes are rare and have not been performed in this watershed.

Studies in the Sacramento River watershed should employ better methods to detect exposure and deleterious effects of individual as well as combined stressors. To date, there is little reason to doubt that various life stages of fish will encounter localized conditions which are either directly toxic or adversely affect availability of food. Future investigations need to focus on resident organisms and to more closely mimic the exposures to multiple stressors. From a watershed perspective, all potential stressors need to be considered and efforts to mitigate exposures should be organized to consider the system as a whole.

Acknowledgements
Funded in part by US Public Health Service Grants CA-45131 from the National Cancer Institute, ES-04699 from the Superfund Basic Science Research Project, CR8191658-010 from the U.C. Davis, US EPA Center for Ecological Health Research and U.S. E.P.A. #R823297

References

Bailey, H.C., Alexander, C., DiGiorgio, C., Miller, M., Doroshov, S.I. and Hinton, D.E. (1994) The effect of agricultural discharge on striped bass (*Morone saxatilis*). *Ecotoxicol.* 3:123–142.

Bailey, H.C., Clark, S. and Davis, J. (1995) *The effects of toxic contaminants in waters of the San Francisco Bay and Delta.* Department of Water Resources, Contract No. B-59645.

Bailey, H.C., DiGiorgio, C., Kroll, K., Miller, J.L., Hinton, D.E. and Starrett, G. (1996) Development of procedures for identifying pesticide toxicity in ambient waters: carbofuran, diazinon, chlorpyrifos. *Environ. Toxicol. Chem.* 15:837–845.

Bennett, W.A. and Kennedy, T.B. (1993) Environmental Impact Assessment of an Aqueduct on Larval Striped Bass in the Sacramento-San Joaquin Delta, California. Chapter 3 in Dissertation *Interaction of food limitation, predation, and anthropogenic intervention on larval striped bass in the San Francisco Bay Estuary*. University of California, Davis.

Bennett, W.A., Ostrach, D.J. and Hinton, D.E. (1995) Larval striped bass condition in a drought-stricken estuary: evaluating pelagic food-web limitation. *Ecol. Appl.* 5:680–692.

Bingham, N. (1996) *Human management and development of the Sacramento-San Joaquin Delta: a historical perspective.* Interagency Ecological Program for the Sacramento-San Joaquin Estuary. 8:2, 13–18. Spring 1996.

Birge, W.J., Black, J.A., Hudson, J.E. and Bruser, D.M. (1979) Embryo-larval toxicity tests with organic compounds. *In:* Marking, L. and Kimmerle, R. (Eds) *Aquatic Toxicology*, ASTM S.T.P. 667. American Society for Testing Materials., Philadelphia, PA, pp 131–147.

Birge, W.J., Black, J.A. and Westerman, A.G. (1985) Short-term fish and amphibian embryo-larval tests for determining the effects of toxicant stress on early life stages and estimating chronic values for single compounds of complex effluents. *Environ. Toxicol. Chem.* 4:807–821.

Blaxter, J.H.S. (1977) The effect of copper on the eggs and larvae of plaice and herring. *J. Mar. Biol. Assoc. U.K.* 57:849–858.

Brown, C.L. and Luoma, S.N. (1995) Use of the euryhaline bivalve *Potamocorbula amurensis* as a trace metal contamination in San Francisco Bay. *Mar. Ecol. Prog. Ser.* 124:129–142.

Bull, C.J. and McInerney, J.E. (1974) Behavior of juvenile coho salmon exposed to sumithion, an organophosphate insecticide. *J. Fish. Res. Bd. Can.* 31:1867–1872.

California Department of Pesticide Regulation (1990) *Computer tapes of pesticide use records.*

California Regional Water Quality Control Board (Central Valley Region; 1994) *Monitoring results of the Iron Mountain mine complex and receiving water*. Supplemental Staff Report, 40 pp.

Carlton, J.T., Thompson, J.K., Schemel, L.E. and Nichols, F.H. (1990) Remarkable invasion of San Francisco Bay (California, USA) by the Asian clam *Potamocorbula amurensis*. I. Introduction and dispersal. *Mar. Ecol. Prog. Ser.* 66:81–94.

Chadwick, H.K., Stevens, D.E. and Miller, L.W. (1977) *Some factors regulating the striped bass population in the Sacramento-San Joaquin Estuary.* Proceedings of the Conference on Assessing the Effects of Power-plant induced Mortality on Fish Populations. Pergamon Press, New York, pp 18–35.

Cherniack, M.G. (1988) Toxicological screening for organophosphorus-induced delayed neurotoxicity: complications in toxicity testing. *Neurotoxicol.* Summer 9:249–271.

Connor, V. and Foe, C. (1993) *Sacramento River basin biotoxicity survey*. 1988–1990. California Regional water Quality Control Board, Sacramento, CA, USA, 105 pp.

Davis, J. (1995) Flea killers in the estuary. *Regional Monitoring News* 2 (1) Winter 1995.

Dettbarn, W.D. (1984) Pesticide induced muscle necrosis: mechanisms and prevention. *Fund. Appl. Toxicol.* 4:18–26.

EPA (1989) *Short-term methods for estimating the chronic toxicity of effluents and receiving water to freshwater organisms.* Environmental Monitoring and Support Laboratory, 2nd ed. Cincinnati, Ohio. EPA/600/4–89/001.

Evans, M.S. (1988) The effect of toxic substances on the structural and functional characteristics of zooplankton populations: a Great Lakes perspective. *In:* Evans, M.S. (Ed) *Toxic Contaminants and Ecosystem Health: a Great Lakes Focus.* Wiley, New York, pp 53–76.

Fujimura, R.W., Huang, C. and Finlayson, B. (1995) *Chemical and toxicological characterization of Keswick Reservoir sediments.* California Department of Fish and Game, Environmental Services Division, Elk Grove, CA. IA, 2-107-250-0. Final Report. February 1, 1995.

Herbold, B., Jassby, A.D. and Moyle, P.B. (1992) *Status and trends report on aquatic resources in the San Francisco estuary.* San Francisco Estuary Project, U.S. Environmental Protection Agency, Oakland, California.

Holcombe, G., Benoit, D. and Leonard, E. (1979) Long term effects of zinc exposure on brook trout (*Salvelinus fontinalis*). *Trans. Am. Fish. Soc.* 108:76–87.

IESP (1987) *Factors affecting striped bass abundance in the Sacramento-San Joaquin River System.* Sacramento, CA: Interagency Ecological Study Program for the Sacramento-San Joaquin Estuary. California Department of Fish and Game.

Klaverkamp, J.F., Duangsawadsi, M., MacDonald, W.A. and Majewski, H.S. (1977) An evaluation of fenitrothion toxicity in four life stages of rainbow trout, *Salmo gairdneri. In:* Mayer, F.L. and Hamelink, J.L. (Eds) *Aquatic Toxicology and Hazard Evaluation.* ASTM STP 634, American Society for Testing Materials, Philadelphia, PA, pp 231–240.

Kuivila, K.M. and Foe, C.G. (1995) Concentrations, transport and biological effects of dormant spray pesticides in the San Francisco Estuary, California. *Environ. Toxicol. Chem.* 14:1141–1150.

Lemly, A.D. (1985a) Ecological basis for regulating aquatic emissions from the Power Industry; the case with selenium. *Reg. Toxicol. Pharmacol.* 5:465–486.

Lemly, A.D. (1985b) Toxicology of selenium in a freshwater reservoir: implications for environmental hazard evaluation and safety. *Ecotox. Environ. Safety* 10:314–338.

Lemly, A.D. (1993) Metabolic stress during winter increases the toxicity of selenium to fish. *Aquat. Toxicol.* 27:133–158.

Luoma, S.N. (1977) Detection of trace contamination effects in aquatic ecosystems. *J. Fish. Res. Bd. Can.* 34:436–439.

Luoma, S.N., Johns, C., Fisher, N.S., Steinberg, N.A., Oremland, R.S. and Reinfelder, J.R. (1992) Determination of selenium bioavailability to a benthic bivalve from particulate and solute pathways. *Environ. Sci. Tech.* 26:485–491.

Mac, M.J. and Edsall, C.C. (1991) Environmental contaminants and the reproductive success of lake trout in the Great Lakes: an epidemiological approach. *J. Toxicol. Environ. Health* 33:375–394.

Matton, R. and Lattam, Q.N. (1969) Effect of organophosphate dylox on rainbow trout larvae. *J. Fish. Res. Bd. Can.* 26, 2193–2200.

McKim, J.M., Eaton, J.G. and Holcombe, G.W. (1978) Metal toxicity to embryos and larvae of eight species of freshwater fish. II. *Bull. Environ. Contam. Toxicol.* 19:608–616.

McKim, J.M., Olson, G.F., Holcombe, G.W. and Hunt, E.P. (1979) Long term effects of methylmercuric chloride on three generations of brook trout (*Salvelinus fontinalis*): toxicity, accumulation, distribution and elimination. *J. Fish. Res. Bd. Can.* 33:2726–2739.

Moyle, P.B. (1976) *Inland Fishes of California.* California University Press, Berkeley, CA, 405 pp.

Moyle, P.B., Herbold, B., Stevens, D.E. and Miller, L.W. (1992) Life history and status of the delta smelt in the Sacramento-San Joaquin Estuary, California. *Trans. Amer. Fish. Soc.* 121:67–77.

Nichols, F.H., Cloern, J.E., Luoma, S.N. and Peterson, D.H. (1986) The modification of an estuary. *Science* 231:567–573.

Nichols, F.H., Thompson, J.K. and Schemel, L.E. (1990) Remarkable invasion of San Francisco Bay (California, USA) by the Asian clam *Potamocorbula amurensis.* II. Displacement of a former community. *Mar. Ecol. Prog. Ser.* 66:95–101.

Norberg-King, T.J., Durhan, E.J., Ankley, G.T. and Robert, G. (1991) Application of toxicity identification evaluation procedures to the ambient waters of the Colusa Basin Drain, California. *Environ. Toxicol. Chem.* 10:891–900.

Ohlendorf, H.M., Kilness, A.W., Simmons, J.L., Stroud, R.K., Hoffman, D.J. and Moore, J.F. (1988) Selenium toxicosis in wild aquatic birds. *J. Toxicol. Environ. Health* 24:67–92.

Porcella, D.B. (1994) Mercury in the environment: biogeochemistry. *In:* Watras, C.J. and Huckabee, J.W. (Eds) *Mercury Pollution – Integration and Synthesis.* CRC Press, Boca Raton, pp 3–19.

Post, G. and Leisure, R.A. (1974) Sublethal effect of malathion to three salmonid species. *Bull. Environ. Contam. Toxicol.* 12:312–319.

Rose, K.A., Cowan Jr., J.H., Miller, L.W. and Stevens, D.E. (1993) Individual-based modeling of environmental quality effects on early life stages of fisheries: a case study using striped bass. *Amer. Fish. Soc. Symp.* 14:125–145.

Saiki, M.K. and Ogle, R.S. (1995) Evidence of impaired reproduction by Western mosquitofish inhabiting seleniferous agricultural drainwater. *Trans. Am. Fish. Soc.* 124:578–587.

Herbold, B., Jassby, A.D. and Moyle, P.B. (1992) *Status and Trends report on aquatic resources in the San Francisco Estuary*. San Francisco Estuary Project, Oakland, CA, 257 pp.

Sheipline, R. (1993) *Background information on nine selected pesticides*. California Regional Water Quality Control Board, Sacramento, CA, USA, 150 pp.

Sittig, M. (1981) *Handbook of Toxic and Hazardous Chemicals*. Noyes Publications, Park Ridge, NJ, 729 pp.

Skorupa, J.P., Morman, S.P. and Sefchick-Edwards, J.S. (1996) *Guidelines for interpreting selenium exposures of biota associated with nonmarine aquatic habitats*. U.S. Fish and Wildlife Service, Sacramento. March, 1996, 74 pp.

Slotton, D.G., Ayers, S.M., Reuter, J.E. and Goldman, C.R. (1995) *Gold mining impacts on food chain mercury in northwestern Sierra Nevada streams*. Technical Completion Report, Project UCAL-WRC-W-816. University of California Water Resources Center, Davis, CA, August 1995, 46 pp.

Sorenson, E.M.B., Cumbie, P.M., Bauer, T.C., Bell, J.S. and Harlan, C.W. (1984) Histopathological, hematological, condition factor, and organ weight changes associated with selenium accumulation in fish from Belews Lake, North Carolina. *Arch. Environ. Contam. Toxicol.* 13:153–162.

Speranza, A.E., Seeley, R.J., Seeley, V.A. and Perlmutter, A. (1977) The effects of sublethal concentrations of zinc on reproduction in the zebrafish, Brachydanio rerio (*Hamilton-Buchanan*). *Environ. Pollut.* 12:217–222.

Zalom, F.G. (1994) *Statewide Integrated Pest Management Project Database*. University of California, Division of Agriculture and Natural Resources, Davis, CA, USA.

Stevens, D.E., Kohlhorst, D.W. and Miller, L.W. (1985) The decline of striped bass in the Sacramento-San Joaquin Estuary, California. *Trans. Am. Fish. Soc.* 114:12–30.

Symons, P.E.K. (1973) Behavior of young Atlantic salmon exposed to or force-fed fenitrothion, an organophosphate insecticide. *J. Fish. Res. Bd. Can.* 30:651–655.

US EPA (1982) *Determination of organophosphorus pesticides in industrial and municipal waste water*. EPA-600/4-82-004. Environmental and Monitoring Support Laboratory, Cincinnati, OH.

US EPA (1992) *Public Comment Remedial Investigation Report, Boulder Creek Operable Unit, Iron Mountain Mine, Redding, California*. EPA WA No. 31-01-9N17, May, 1992.

Verschueren, K. (1983) *Handbook of environmental data on organic chemicals*. 2nd ed., Van Nostrand Reinhold, New York, 1310 pp.

Weis, J.S. and Weis, P. (1989) Effects of environmental pollutants on early fish development. *Aquat. Sci.* 1:45–73.

Weis, P. and Weis, J.S. (1991) The developmental toxicity of metals and metalloids in fish. *In:* Newman, M.C. and McIntosh, A.W. (Eds) *Metal Ecotoxicology – Concepts & Applications*. Lewis Publishers, Chelsea, pp 145–169.

Wildish, D.J. and Lister, N.A. (1973) Biological effects of fenitrothion in the diet of brook trout. *Bull. Environ. Contam. Toxicol.* 10:333–339.

Wilson, B.W. and Henderson, J.D. (1992) Blood esterase determinations as markers of exposure. *Rev. Environ. Contam. Toxicol.* 128:55–69.

Wilson, B.W., Padilla, S. and Henderson, J.D. (1996) Factors in standardizing automated cholinesterase assays. *J. Toxicol. Environ. Health* 48:187–195.

Wright, D.A. and Phillips, D.J.H. (1988) Chesapeake and San Francisco Bays -A study in contrasts and parallels. *Mar. Pollut. Bull.* 19:405–413.

Zahm, G.R. (1986) Kesterson Reservoir and Kesterson National Wildlife Refuge: history, current problems, and management alternatives. *Trans. Am. Wildl. Nat. Resour. Conf.* 51:324–329.

Fish Ecotoxicology
ed. by T. Braunbeck, D.E. Hinton and B. Streit
© 1998 Birkhäuser Verlag Basel/Switzerland

Effects of estrogenic substances in the aquatic environment *

Peter Matthiessen[1] and John P. Sumpter[2]

[1] Centre for Environment, Fisheries and Aquaculture Science, Remembrance Avenue, Burnham-on-Crouch, Essex CM0 8HA, UK
[2] Department of Biology and Biochemistry, Brunel University, Uxbridge, Middlesex UB8 3PH, UK

Summary. This review describes the research that has been carried out into estrogenic effects occurring in aquatic environments, both freshwater and marine, and the substances found to be responsible. In summary, estrogenic (and probably some anti-androgenic) activity has mainly been detected in a variety of treated sewage and other effluents, but also as a result of certain chemical spills and deliberate applications. This activity has resulted in a number of effects in vertebrate wildlife that can best be described as feminization, although the severity of these effects ranges from biomarkers of exposure such as vitellogenin induction in males through to morphological changes in sex organs and complete sex reversal. The implications of these changes for the future of aquatic wildlife populations have not yet been thoroughly explored. It is unlikely that all the causative substances have yet been discovered, but those which have been positively identified include natural and synthetic estrogenic hormones, natural plant sterols, synthetic alkylphenols, and certain organochlorine substances. The review concludes that there is now a need to investigate the consequences for wildlife populations of exposure to these materials, by means of a variety of field experiments and investigations.

Introduction

There is much interest at present in the effects on wildlife and humans of endocrine disruptors in general, and estrogenic substances in particular (Colborn et al., 1993; IEH, 1995; Kavlock et al., 1996; Toppari et al., 1995). Endocrine-disrupting effects of contaminants have been observed or suspected in most classes of organisms from gastropod molluscs to beluga whales, and calls are already being made for the withdrawal from use of many anthropogenic substances which could be causing such changes in humans (Colborn et al., 1996).

A common factor between the many and varied examples of endocrine disruption is that most of the affected species are part of aquatic eco-systems, the pollutants in question often being organochlorine substances biomagnified from contaminated waters into piscivorus top-predators. Until recently, little was known about the effects of endocrine disruptors on fish themselves, although there are good reasons for believing that they may also be susceptible to similar types of interference. In particular,

---

steroid hormone structure and function has been highly conserved during the evolution of the vertebrates, such that there are many similarities between the endocrine systems in fish and mammals. As fish are directly exposed to water-borne contaminants *via* their gills, as well as to contaminants in their food and in the yolk they inherit from their mothers, it was perhaps to be expected that they, as well as their predators, would respond to anthropogenic substances that interact with the steroid hormone system. It should, of course, be remembered that the reproductive systems of fish and other organisms are also susceptible to a range of deleterious contaminant-related effects that cannot be classified as endocrine disruption (Kime, 1995), although the distinction between endocrine and non-endocrine effects is not always clear.

Endocrine systems are extremely complex and can be perturbed by many exogenous chemical influences. These can be crudely categorized as hormone receptor agonists, receptor antagonists, hormone synthesis/metabolism stimulants or blockers, and a range of others with less specific effects. This paper is mainly concerned with a group which has perhaps received the greatest study to date, the estrogenic substances which to some extent mimic the action of steroid hormones such as estradiol, thereby acting as partial ("weak") or complete estrogen agonists. They therefore have a highly specific action, triggering the cascade of events which usually flows from estradiol exposure. Under normal circumstances, these events range from uterine growth stimulation in female mammals to synthesis of the yolk precursor protein vitellogenin in female fish, but when they occur at the wrong stage of the life cycle, or in males, there is obvious potential for damage to an organism's reproductive success.

Another type of endocrine disruption concerns those hormone antagonists which bind more or less irreversibly with androgen receptors and prevent the true agonists (e.g., testosterone) from triggering their intended effects. Although weak antagonists often simultaneously have weak agonistic action, many of the results of weak androgen blockers are phenotypically similar to, or even identical with, the effects of estrogens with which they can be confused.

The purpose of this review is to describe some examples of reproductive interference in aquatic life which have been attributed to exposure to estrogen mimics and androgen blockers, to evaluate their environmental significance, and to recommend new research directions.

Estrogenic and anti-androgenic effects of pesticides

At least 18 pesticides have been reported as showing weak estrogenic or anti-androgenic activity (Lyons, 1996), although few examples have explicitly involved aquatic organisms. One early recorded case of a probable estrogenic response in fish concerns the cyclodiene organochlorine insecticide endosulfan, which is now suspected by some to be a weak estrogen-mimic (ATSDR, 1990; Soto et al., 1994). It was sprayed from the air to kill tsetse flies in parts of southern Africa, and its aquatic environmental effects were extensively studied in the Okavango Delta of Botswana in the late 1970's (Douthwaite et al., 1981 and 1983). It was observed that the nest-building activities of the cichlid fish *Tilapia rendalli* were interrupted during endosulfan spraying operations that contaminated shallow swamp breeding areas, and reproduction did not resume until the following season. Nest-building is a vital part of reproduction in many cichlids, and is mediated by the use of elaborate body patterns and coloration. Laboratory experiments with a closely-related cichlid *Oreochromis mossambicus* (Matthiessen and Logan, 1984) showed that the concentrations of endo-sulfan which had been measured in Okavango waters soon after spraying were indeed able to interfere with the striking body patterns and nesting behavior of male fish, although the mode of action of this effect was never elucidated.

However, undoubtedly the best pesticide-related example of phenotypic estrogenic changes in an aquatic species concerns the American alligator (*Alligator mississippiensis*) population in Lake Apopka, Florida. This population of fish-eating reptiles was reported by Woodward et al. (1993) to have shown a drastic decline since 1980 in numbers of juveniles, accom-panied by poor egg viability. Subsequent studies by Guillette et al. (1994, 1995, 1996) demonstrated that juvenile Lake Apopka alligator females dis-played elevated plasma estradiol levels and abnormal ovarian morphology. Male juveniles showed depressed plasma testosterone, abnormal testes, and small phalli. Furthermore, male testes *in vitro* secreted elevated levels of estradiol. Guillette et al. (1994) interpret these abnormalities as a per-manent change triggered at the embryonic stage which had led to altered steroidogenesis, and suggest that the effect was caused by exposure to the estrogenic effects of dicofol and DDT (plus its metabolites). Heinz et al. (1991) have shown that eggs collected from Lake Apopka in 1984–85 con-tained the DDT metabolites *p,p'*-DDE and *p,p'*-DDD at mean concentra-tions of about 5.8 and 0.8 mg/kg wet wt., respectively, and it is assumed that these residues largely derive from a spill at the Tower Chemical Com-pany in 1980 of dicofol together with DDT and its metabolites.

Although *o,p'*-DDT and dicofol undoubtedly act as weak estrogen recep-tor agonists (Forster et al., 1975; McLachlan et al., 1992), they degrade fair-ly rapidly in the environment and have not been found at high concentra-tions in the water or biota of Lake Apopka. On the other hand, the *p,p'* iso-mer of DDT and its metabolites (including *p,p'*-DDE, which was the most significant contaminant in Lake Apopka alligator eggs) do not interact significantly with the estrogen receptor (Forster et al., 1975). However, it has recently been shown (Kelce et al., 1995) that *p,p'*-DDE antagonistical-ly blocks binding of testosterone to the androgen receptor, and thereby prevents androgenic action in male rats. When *p,p'*-DDE is fed to pregnant rats, it produces a variety of phenotypic changes in the male offspring which appear similar to those induced by estradiol exposure.

The situation in Lake Apopka is therefore multifactorial, with estrogenic and anti-androgenic activity probably taking place simultaneously. It may, indeed, be further complicated by the fact that Lake Apopka alligator eggs also contain about 0.1 mg/kg wet wt. of PCBs (Heinz et al., 1991). While some of the 209 PCB congeners are known to be weak estrogen mimics, others act as anti-estrogens by binding to the aryl hydrocarbon (Ah) receptor (Jansen et al., 1993; Krishnan and Safe, 1993). Of particular relevance to the alligators in Lake Apopka is some elegant experimental work by Bergeron et al. (1994) who have shown that two PCB metabolites (trichloro- and tetrachloro-hydroxybiphenyls) are together able to override temperature-dependent male sex determination in turtle eggs at a combined concentration of <1 mg/kg, thereby producing female hatchlings at a male-inducing temperature. Temperature-dependent sex determination also occurs in alligators, so it seems possible that interference with this mechanism could also be occurring in Lake Apopka.

Estrogenic effects of sewage effluents

Undiluted effluents

In the late 1970s, anglers started reporting the presence of abnormal fish (roach, *Rutilus rutilus*) in sewage effluent settlement lagoons discharging to the River Lea in north London. Investigations by the Thames Water Authority (now United Kingdom Environment Agency – unpublished internal report entitled *Hermaphrodite roach in the River Lee*) in 1981 showed that intersex or hermaphrodite fish were indeed present in the Lea catchment, although the incidence was low and no immediate action was taken. In 1986/87, research by Brunel University and the UK Ministry of Agriculture, Fisheries and Food (MAFF Directorate of Fisheries Research, now, Centre for Environment, Fisheries and Aquaculture Science) into the reproductive physiology of rainbow trout at the MAFF Experimental Trout Culture Unit in East Anglia had discovered that some male fish contained measurable levels of the female yolk protein vitellogenin in their blood plasma (Purdom et al., 1994). This had been tentatively ascribed to unknown contaminants derived from a treated sewage discharge upstream from the unit's water intake. These unexpected and possibly unrelated observations were, however, sufficiently curious to trigger an intensive investigation of the estrogenic properties of sewage effluent in the UK.

The hypothesis that treated domestic and/or industrial sewage effluents may have endocrine disrupting effects, and more specifically might contain estrogen-mimicking substances, was tested by Purdom et al. (1994). They placed caged adult rainbow trout (*Oncorhynchus mykiss*) in undiluted sewage discharges at various English locations in 1986–89, and measured the induction of the phospholipoprotein vitellogenin after 3 weeks exposure. Vitellogenin is a yolk-precursor normally only found in the blood plasma and ovaries of female egg-laying vertebrates. It is synthesized in the liver under the control of estradiol originating in the ovary (which in turn is stimulated by pituitary gonadotropin hormone), but can also be induced by exogenous estrogen. Purdom et al. (1994) measured vitellogenin concentrations by means of radio-immunoassay (Sumpter, 1985), although it is of interest to note that vitellogenin gene expression can also be measured directly (Guellec et al., 1988; Ren et al., 1996). Vitellogenin synthesis is a highly specific marker of exposure to estrogenic substances; indeed, vitellogenin production by estradiol-exposed liver cells is depressed by simultaneous exposure to a variety of cytochrome P4501A1-inducing polycyclic and halogenated aromatic hydrocarbons (Anderson et al., 1996), and nonspecific stress also diminishes the vitellogenin response (Carragher et al., 1989). Furthermore, *in vitro* work has shown that vitellogenin production by rainbow trout liver cell cultures exposed to exogenous estrogen-mimics is inhibited by the estrogen receptor-blocker tamoxifen (Harries et al., 1995; Jobling and Sumpter, 1993).

As one might expect, Purdom et al. (1994) found that some of the effluents were rapidly lethal to rainbow trout, but data from 9 sites showed that the remaining effluents were all strongly estrogenic, giving rise to plasma vitellogenin concentrations of up to 147 000 µg/mL, which are equivalent to those seen in normal pre-spawning females. Some vitellogenin synthesis was also seen in caged immature carp (*Cyprinus carpio*), although they did not respond so strongly as rainbow trout.

These results suggested that receiving waters downstream of estrogenic effluents might also show estrogenic activity. Although a limited investigation (Purdom et al., 1994) showed that the few downstream locations where fish cages were placed were *not* estrogenically active, it was decided to conduct a more comprehensive survey.

Surveys of rivers and reservoirs

An initial investigation (Harries et al., 1995, 1996) of the River Lea near London was made in the summer of 1992 using a similar caged rainbow trout technique to that of Purdom et al. (1994). This small (80 km long) river was chosen partly because it receives treated effluent from five sewage treatment works (mixed domestic and industrial) along its length, and these contribute up to 82% of the total flow in summer on the upper Lea. The R. Lea was therefore expected to show some estrogenic activity, and of course it was already known that intersex roach *R. rutilus* had been found there. Nineteen sites on the Lea were surveyed with caged male rainbow trout exposed for 3 weeks, and most fish showed some vitellogenesis (up to 576 µg/mL in plasma), although not all the responses were statistically significant. Highest responses were seen immediately downstream of sewage treatment works' discharges, and these then diminished rapidly with distance, although statistically significant effects were seen up to 4.5 km downstream of discharges. Simultaneous measurements of testicular growth (gonadosomatic index, or GSI) showed that testes had grown more slowly than the laboratory controls at seven sites, but 11 sites showed no testicular retardation, and indeed one showed excess testicular growth. Limited measurements of vitellogenesis in wild roach from the Lea system (Harries et al., 1995) suggest that they were also receiving exposure to estrogenic material, but more research is required to confirm this. It is additionally worth noting that similar elevations in vitellogenin have now been observed in feral male carp *Cyprinus carpio* in the vicinity of a sewage treatment works in the United States, accompanied by reduced serum testosterone concentrations (Folmar et al., 1996).

A more detailed survey downstream of one of the most estrogenically active discharges (Harpenden sewage treatment works) was conducted in the winter of 1992 (Harries et al., 1995, 1996) and demonstrated that the effects, although still present, were diminished by up to two orders of magnitude compared to the summer survey. It was calculated that sewage treatment works discharges to the Lea in winter were receiving an extra 36% dilution compared with summer, and it was hypothesized that the reduced vitellogenin concentrations might be a response to this additional dilution effect. Although seasonal variations in ability to produce vitellogenin could not be ruled out, it seemed likely that dilution was the deciding factor.

This led to an experiment at Harpenden sewage treatment works in late 1994 (Harries et al., 1995) in which male rainbow trout in tanks were exposed to treated effluent continuously diluted with tap water. This clearly showed that although the undiluted effluent was still strongly estrogenic (mean plasma vitellogenin concentration was 36 500 µg/mL), increasing dilutions produced a rapidly diminishing response, such that a 1 : 1 dilution gave a plasma vitellogenin concentration of only 1 µg/mL, and a 1 : 3 dilution gave no response at all. The reasons for such a steep and apparently non-linear response are unknown. This fast reduction in effect with increasing dilution suggests that most of the diminution in effects downstream of discharges is indeed the result of dilution and dispersion, although other processes such as adsorption and degradation are probably also occurring. The seasonal increase in dilution in the winter is probably also sufficient to explain the smaller responses at that time of year.

Further investigations similar to those on the Lea were conducted on five other English rivers in the summer of 1994 (Harries et al., 1995, 1997), downstream of sewage treatment works discharges handling largely domestic effluent, but with a small proportion of industrial effluent. The rivers were the Arun and Kent Stour in southern England, the Chelmer and Essex Stour in eastern England, and the Aire in northern England. A small response (2 µg/mL) was seen in rainbow trout held 1.5 km downstream of the sewage treatment works which was studied on the Arun, but no responses were seen on the surveyed stretches on either of the Stours, or on the Chelmer. Indeed, the sewage treatment works discharge on the Essex Stour (Bures sewage treatment works) was the only one investigated that was not estrogenic to rainbow trout when undiluted. It is a small (about 1200 people) rural discharge with no industrial inputs, but it is not known whether this absence of industrial activity is significant.

The most marked effects seen at any site were on the River Aire in Yorkshire downstream of Marley sewage treatment works, which takes influent *inter alia* from wool-scouring plants as well as from domestic sources. At all stations on a 5 km stretch below the discharge, plasma vitellogenin concentrations after 3 weeks exposure were in the range of 25 000 to 52 000 μg/mL and there were only weak signs of a diminution of effect over this distance. Simultaneous measurements of liver (hepatosomatic index – HSI) and testis growth (GSI) showed that liver weights were significantly elevated with respect to controls over the whole surveyed stretch, and testis weights were all significantly depressed. In particular, mean GSIs after 3 weeks ranged from 0.31 to 0.49, in comparison with a control GSI of 1.27. The elevated HSIs were not unexpected, given that the livers of these male fish had been induced to synthesize large amounts of vitellogenin. The very poor testis growth implied that fish in the Aire may be at a significant reproductive disadvantage, and unpublished information indeed suggests that the coarse angling fishery on this river is impoverished. Unfortunately, such field data cannot be taken as proof that estrogenic compounds in complex discharges are causing reproductive inhibition, because there are probably many other pollutants present which may damage reproduction without directly influencing the steroid hormone system.

These data raised concerns about possible human exposure because raw water is abstracted from rivers prior to extensive treatment and ultimate supply as drinking water. The caged rainbow trout technique was therefore used to survey 15 raw water reservoirs in the southeast of England in summer 1993 (Harries et al., 1995, 1996). In order to ensure that even small effects could be detected, the vitellogenin radio immunoassay was made about 10 times more sensitive, and the fish were exposed for both 3 and 6 weeks. The results showed clearly that there was no detectable estrogenic activity in any of the reservoirs, even though positive controls placed in undiluted sewage treatment works effluent upstream showed strong effects, indicating that the fish were fully responsive. On the basis of data described above, it seems likely that this lack of response largely resulted from the significant dilution which sewage effluents receive in rivers.

Causative substances in sewage

The first attempt to identify the estrogenic materials in English sewage effluent (Purdom et al., 1994) was based largely on guesswork, and focused on the synthetic estrogen 17α-ethinylestradiol (EE2) which is a component of many contraceptive pills. Simple calculations based on known excretion rates etc. suggested that sewage effluent might contain 0.1– 1.0 ng/L of this substance, although it is excreted by women primarily as a conjugate with glucuronic acid (while a little is unchanged, and some is sulfated or hydroxylated). Laboratory experiments (unpublished) showed that the conjugate is not estrogenic to fish, but the original compound is highly active, with a no-observed-effect-concentration (NOEC) for vitellogenesis in chronic tests with rainbow trout of 0.1–0.3 ng/L (Sheahan et al., 1994). It is probable that the conjugate is enzymatically cleaved in sewage to re-create the parent material, but EE2 is very difficult to quantify in effluent or river water. This may be due to the fact that traditional analytical techniques for EE2 in "dirty" water are at their detection limits, although Aherne and Briggs (1989) have indeed reported EE2 concentrations in English rivers in the range of <5 to 15 ng/L.

Natural estrogenic hormones, such as estradiol, estrone and estriol, are also potential candidates for the estrogenic substances in effluents. These natural estrogens can originate from both humans (particularly pregnant women) and farm animals such as chickens, pigs and cows, which excrete considerable amounts of estrogens into the environment (Knights, 1980; Shore et al., 1988). Work conducted in Israel, a country which often has relatively little rainfall, by Shore and colleagues, has shown that concentrations of estrogen (the technique used did not distinguish between estradiol and oestrone) in sewage treatment works effluent ranged from 7 ng/L after a period of heavy rainfall up to 60 ng/L during a period of severe drought (Shore et al., 1993). Estrogen was also detectable in lake water, at concentrations ranging from 6 to 25 ng/L. The highest concentrations measured (up to 350 ng/L) were present in treated effluents, originating from small farms and municipal sewage treatment works, used for irrigation. In these studies free estrogens, rather than conjugated estrogens, were quantified; thus de-conjugation of the excreted form of estrogens presumably occurred during treatment of the sewage.

More recently, other suspect chemicals have to some extent shifted attention away from synthetic and natural hormones. Work by Soto et al. (1991) showed that the common industrial chemical (and breakdown product of certain surfactants) p-nonylphenol was estrogenic in breast cancer cell cultures, and subsequent studies in the UK (Jobling and Sumpter, 1993) demonstrated that several alkylphenolic chemicals induce vitellogenesis in rainbow trout hepatocyte culture. Further investigations with rainbow trout *in vivo* (Jobling et al., 1996) confirmed that alkylphenols in the ambient water were able to induce vitellogenesis, although only weakly by comparison with, for example, EE2. This work established a 3-week NOEC for vitellogenesis of about 10 µg/L nonylphenol, and an equivalent value for octylphenol of 3 µg/L. Short-chain alkylphenol ethoxylates (intermediate degradation products of the long-chain alkylphenol ethoxylates, or APEs, used as industrial surfactants) also had some estrogenic activity *in vivo*. Perhaps of greater environmental significance was that 30–54 µg/L nonylphenol was sufficient to retard testicular growth significantly over a 3 week exposure, and this was accompanied by some interference with spermatogenesis and relative frequency of testicular cell types.

The wool scouring plants in the River Aire catchment used APE's in large amounts to wash the grease from fleeces and discharged the spent washing liquor *inter alia via* Marley sewage treatment works. Water from the Aire was therefore analyzed for nonyl- and octylphenol (Blackburn and Waldock, 1995), and although little octylphenol was found (< 1 µg/L), 24–53 µg/L of dissolved nonylphenol was present in the stretch of river where maximal vitellogenesis and retarded testicular growth were seen in caged fish. When these data are compared with the work of Jobling et al. (1996), it becomes apparent that most of the estrogenic activity in the Aire can be plausibly attributed to nonylphenol. As a result, the UK textile industry agreed to phase out the use of nonylphenol ethoxylate (NPEO) surfactants by the end of 1996. It should also be remembered, however, that nonylphenol is not the only degradation product of the NPEs. As the short-chain NPEOs and nonylphenol carboxylates (NPECs) also have some estrogenic activity (Jobling and Sumpter, 1993), it is probable that they too are contributing to the overall effect seen in the River Aire and elsewhere.

It should be noted that Blackburn and Waldock (1995) only reported the presence of low concentrations of alkylphenols in the Lea (0.2–9.0 µg/L nonylphenol; 0.4 µg/L octylphenol), although again it is worth remembering that short-chain NPEOs and NPECs are also likely to have been present at concentrations probably exceeding those of nonylphenol. Octylphenol and nonylphenol were not thought to be responsible for all the observed effects in that river because they were below their respective NOECs, and were therefore likely to be acting in combination with NPEOs, NPECs, and with other chemically unrelated estrogen receptor agonists, as has been seen in fish hepatocyte culture (Sumpter and Jobling, 1995).

Due to the limitations of the approaches described above, a successful new strategy based on Toxicity Identification Evaluation (TIE) procedures was developed. In essence, estrogenic effluent from seven sewage treatment works where alkylphenols are not present in large quantities were fractionated, and the estrogenic fractions identified using a rapid estrogen screen based on a recombinant yeast cell line containing the human estrogen receptor gene and a reporter gene which ultimately produces a color change in the test medium (Routledge and Sumpter, 1996). Estrogenic fractions, which were largely confined to the dissolved phase, were successively sub-fractionated using solid-phase C18 columns with methanol/water elution, thus producing progressively simpler fractions which were eventually separated by HPLC. More traditional analytical procedures were then used to identify the substances in what proved to be the only estrogenic HPLC fraction. Recently published results of this work (Desbrow et al.,

1998) show that the estrogenically-active substances in these seven sewage treatment works effluents were the natural estrogenic hormones estrone and 17β-estradiol, and the synthetic estrogen ethinylestradiol. It therefore seems likely that the most common causes of estrogenicity in sewage effluents are the estrogen hormones and their analogs.

Pulpmill effluents

In 1980, Howell et al. reported the presence of masculinized female mosquito fish (*Gambusia affinis*) in a small creek downstream of a paper mill discharging bleached kraft mill effluent. Shortly thereafter, a second population of masculinized fish was discovered in another stream (Bortone and Drysdale, 1981), also downstream of where bleached kraft mill effluent was discharged. Masculinized females of other (but not all) species were also found. Subsequent laboratory work showed that exposure to bleached kraft mill effluent induced male secondary sexual characteristics (Bortone et al., 1989; Drysdale and Bortone, 1989). Exposure of female mosquito fish, and other poeciliids, to androgens, such as methyl testosterone and androstenedione, produced the same phenotypic masculinization as was seen in fish exposed to bleached kraft mill effluent; thus, it appears that bleached kraft mill effluent contains an androgenic chemical, or mixture of chemicals. Although the chemical (or chemicals) responsible for the observed effects has not been identified, it has been suggested that phytosterols might be responsible. Exposure of mosquito fish to some phytosterols (such as sitosterol and stigmastanol) resulted in masculinization of females (Howell and Denton, 1989). A comprehensive review of all the work on endocrine disruption of mosquito fish can be found in Bortone and Davies (1994).

Recent work on the effects of bleached kraft mill effluent has tended to focus more on the mechanisms whereby bleached kraft mill effluent disrupts reproduction. Exposure to bleached kraft mill effluent has been shown to affect reproduction adversely by reducing plasma sex steroid hormone levels, delaying sexual maturity, and reducing gonad and egg size (Gagnon et al., 1995; Munkittrick et al., 1994; Van der Kraak et al., 1992). These multiple effects, which are undoubtedly effects of bleached kraft mill effluent on the endocrine system, are difficult, if not impossible, to classify as primarily androgenic or estrogenic. This is unsurprising when one considers that bleached kraft mill effluent is a highly complex mixture of natural and anthropogenic substances, many of which have been shown to alter the reproductive endocrine system. For example, synthetic polychlorinated aromatic compounds formed during the bleaching process, such as dioxins and furans, can act as endocrine disruptors (Sumpter et al., 1996). Additionally, bleached kraft mill effluent contains a wide range of natural compounds from trees, some at high concentrations. One of these chemicals, the sterol β-sitosterol, which can be present in bleached kraft mill effluent at concentrations greater than 1 mg/L, alters the reproductive status of fish (MacLatchy and Van der Kraak, 1995), possibly because it is weakly estrogenic (Mellanen et al., 1996), although other mechanisms of action are likely. Interestingly, it has been proposed that it is the degradation products of sitosterol and related phytosterols such as stigmastanol that are the active agents in masculinizing female mosquito fish exposed to bleached kraft mill effluent (Howell and Denton, 1989).

Estrogenic effects in the marine environment

Little research has been conducted on such effects in marine fauna, although there is no doubt that estuaries and the sea receive inputs of sewage and other estrogenic wastes (e.g., nonylphenol, Blackburn and Waldock, 1995). There have been reports of intersexuality in estuarine crustaceans which might be attributable to sewage discharges (Moore and Stevenson, 1991). Furthermore, Pereira et al. (1992), working solely with female winter flounder (*Pleuronectes americanus*) from Long Island Sound and Boston Harbor, have shown that fish from contaminated areas can have an elevated blood vitellogenin concentration (as measured by alkali-labile phosphate). However, they attribute this to impaired vitellogenin uptake by the ovary, rather than to exposure to estrogenic xenobiotics.

The few remaining examples of suspected estrogenic effects in marine organisms all involve fish. Lye et al. (1997) have recently studied vitellogenesis in wild flounder (*Platichthys flesus*) caught at two locations in the River Tyne estuary near a sewage treatment works discharge, and at a less-contaminated site on the Solway Firth in northern England. Using polyacrylamide gel electrophoresis to identify vitellogenin, they detected elevated vitellogenin levels in male flounder from the two sites on the Tyne. Small amounts of vitellogenin were also detected in two of the male Solway fish. Vitellogenesis was associated with gross malformations of the testis in up to 53% of the Tyne fish, although histological evidence was not presented. No testicular malformations were seen in the Solway sample. The authors conclude that the reproductive health of Tyne flounder appears to have been affected by sewage effluent, but provide no evidence as to the causative substances.

Further work on wild flounder (Allen et al., 1997) in which VTG was measured by a specific radioimmunoassay, has confirmed that several UK estuaries (especially the Tyne and the Mersey) are severely contaminated with estrogens. In the case of the Mersey in 1996, approximately 20% of male flounder were also shown to contain primary oocytes in their testes, a sign that they had been exposed to oestrogenic materials at the stage of gonadogenesis. Intersex conditions of this type were not seen in the other four estuaries sampled (Tyne, Thames, Crouch and Alde), perhaps because the sensitive larval stage of the flounder life cycle is spent at sea where contaminant levels are generally lower than in estuaries.

Recently, results from a long-term experiment in which flounders (*Platichthys flesus*) were exposed to organic pollutants have been reported (Janssen et al., 1997). To study the chronic effects of environmental pollution, fish were held in large mesocosms; a reference mesocosm contained relatively clean sediment and water from the Dutch Wadden Sea, another mesocosm contained contaminated dredged sediment from Rotterdam harbor, which is known to contain organic chemical pollutants and heavy metals, and an indirectly contaminated mesocosm, which contained relatively clean sediment, but received effluent water from the mesocosm containing contaminated sediment. Fish were maintained in these mesocosms for 2.5 to 3 years before being analyzed. In general, the flounders appeared relatively unaffected by the (presumed) exposure to environmental pollutants. Testicular development was not affected by exposure to pollutants, but ovarian development was advanced. This was accompanied by premature vitellogenin synthesis, probably resulting from an elevated estradiol level in the blood. However, male flounders exposed to contaminated sediment and water did not show any evidence of vitellogenin synthesis. This makes it unlikely that xenoestrogens (in the sediment and/or water) were responsible for the advanced vitellogenesis seen in female flounders. Instead, exposure appeared to affect steroidogenesis in some undefined way, leading to elevated 17β-estradiol levels, and hence increased vitellogenin synthesis.

Although the approach taken by Janssen et al. (1997) was admirable, in that the flounders were exposed for a long time to realistic environmental conditions, their studies illustrate the difficulties associated with unequivocally demonstrating endocrine disruption in the "real world", and linking it to exposure to known concentrations of identified chemicals.

It is of passing interest to record that unusual sex ratios have been observed in a North Sea flatfish, the dab *Limanda limanda* (Lang et al., 1995). Over the whole North Sea, there is a slight but significant downward temporal trend in the proportion of females. This is largely due to changes in central and northwestern areas, and is almost balanced by an upward trend in the southeast. There are several possible explanations of these observations, however, and it is by no means clear whether anthropogenic substances are responsible.

Finally, Waring et al. (1996) have conducted some interesting experiments in which they exposed pre-spawning sand gobies (*Pomatoschistus minutus*) to 0.1% sewage sludge for 19 weeks. This work is environmentally relevant, not just because sewage (treated and untreated) is discharged to estuarine and coastal waters, but because ocean disposal of sewage sludge continues in several countries. The experimental exposures had no major effects on gonadosomatic indices, androgen levels, fecundity or numbers of fertile eggs produced by spawning females, but a high proportion (about 50%) of exposed females failed to spawn, even though they contained eggs and had a normal GSI. This produced a major reduction in the numbers of larvae produced by the population as a whole. The authors speculate that these effects could have been caused by several mechanisms, including interference with final oocyte maturation and abnormal male breeding behaviour (the latter perhaps being caused by exposure to estrogenic substances), but conclusive evidence was not obtained.

Conclusions

Although it has been conclusively shown that estrogen-mimicking chemicals, and indeed estrogenic hormones themselves, are present at some locations in the freshwater and marine environments, it is apparent that they are almost invariably accompanied by other substances which can disrupt the endocrine system of aquatic fauna by alternative means. For example, effects which are phenotypically similar to those caused by estrogens can be induced by anti-androgens, and the effects of estrogens can be blocked by anti-estrogens. Disentangling these competing processes is difficult, especially with regard to such complex effluents as treated sewage and pulp mill waste. However, it seems that the net effects of these two types of effluent are generally feminizing and masculinizing, respectively, although there will always be exceptions caused by the local predominance of particular substances.

One of the clear implications of this complexity is that chemical analysis on its own will rarely, if ever, allow a confident prediction of endocrine effects. This is a powerful argument in favour of using bioassays and biomarkers which are able to integrate the various endocrine-disrupting processes in order to achieve a more holistic picture of likely environmental impacts. Detection of effects can then be followed up with toxicity identification procedures, providing that suitable rapid screening bioassays are available. This approach has been successfully used on a research basis in the United Kingdom to identify the estrogenic components of certain sewage effluents, and it is recommended that a wider range of screening bioassays for other endocrine disruptors should therefore be developed. Such methodologies would find ready application in the suites of tests which are being increasingly used by regulatory authorities for the direct toxicity assessment of complex effluents. The development of relatively cheap bioassays for the detection of environmental endocrine disruptors is also a priority for chemical manufacturers, who may wish or be required to demonstrate that their products are benign in this regard. Aquatic tests, perhaps including one based on the induction of vitellogenesis in fish, are likely to play an important role for this purpose.

However, the arguments set out above presuppose that endocrine-disrupting substances and the effluents containing them are indeed causing present or future damage to aquatic ecosystems, or to human consumers of products derived from those systems. Such damage has not yet been clearly established. At one extreme, one could take the position that receptor-mediated effects such as these carry the potential for additivity or even synergism, so a strongly precautionary approach (e.g., preventing the discharge to water of all such substances) might be justified. At the other extreme, it could be argued that deleterious environmental effects of endocrine disruptors must be proven on a case-by-case basis before the costs of regulatory action can be justified. Ideally, pragmatic approaches will be taken at positions on the continuum between these two extremes.

This debate is hampered by a lack of information about the aquatic ecological impacts which environmental estrogens and other endocrine disrupting substances are already causing. In other words, are the known hazards being translated into unacceptable risks? We are not referring here to localized impacts which have undoubtedly been caused by accidental spills such as that on Lake Apopka, so much as to the risks that may be associated with routine industrial and domestic discharges. We therefore recommend the instigation of field and semi-field studies designed to establish the extent to which aquatic ecosystems are at risk of damage. These are by no means easy or cheap to execute, but a few are already underway, and we believe their cost is justified by the scale of the potential problem.

Acknowledgements
The authors thank all their co-workers at the Centre for Environment, Fisheries and Aquaculture Science at Lowestoft and Burnham-on-Crouch, and the Department of Biology and Biochemistry at Brunel University, Uxbridge, for their hard work and advice. Without the many fruitful discussions which have taken place with them on the topics covered by this review, it would not have been possible to write it.

References

Aherne, G.W. and Briggs, R. (1989) The relevance of the presence of certain synthetic steroids in the aquatic environment. *J. Pharm. Pharmacol.* 41:735–736.

Allen, Y., Thain, J., Matthiesen, P., Scott, S., Haworth, S. and Feist, S. (1997) A survey of oestrogenic activity in UK estuaries and its effects on gonadal development of the flounder *Platichthys flesus*. ICES CM 1997/U:01, International Council for the Exploration of the Sea, Copenhagen, 13 pp. + figs.

Anderson, M.J., Miller, M.R. and Hinton, D.E. (1996) *In vitro* modulation of 17β-estradiol-induced vitellogenin synthesis: effects of cytochrome P4501A1 inducing compounds on rainbow trout (*Oncorhynchus mykiss*) liver cells. *Aquat. Toxicol.* 34:327–350.

ATSDR (1990) *Toxicological profile for endosulfan, endosulfan alpha, endosulfan beta, endosulfan sulfate*. Agency for Toxic Substances and Disease Registry, Atlanta, USA.

Bergeron, J.M., Crews, D. and McLachlan, J.A. (1994) PCBs as environmental estrogens: turtle sex determination as a biomarker of environmental contamination. *Environ. Health Perspect.* 102:780–781.

Blackburn, M.A. and Waldock, M.J. (1995) Concentrations of alkylphenols in rivers and estuaries in England and Wales. *Water Res.* 29:1623–1629.

Bortone, S.A. and Davis, W.P. (1994) Fish intersexuality as an indicator of environmental stress. *Bioscience*, 44:165–172.

Bortone, S.A. and Drysdale, D.T. (1981) Additional evidence for environmentally-induced intersexuality in Poeciliid fishes. *Assoc. Southeastern Biologists Bull.* 28:67.

Bortone, S.A., Davis, W.P. and Bundrick, C.M. (1989) Morphological and behavioural characters in mosquito fish as potential bioindicators of exposure to kraft mill effluent. *Bull. Environ. Contam. Toxicol.* 43:370–377.

Carragher, J.F., Sumpter, J.P., Pottinger, T.G. and Pickering, A.D. (1989) The deleterious effects of cortisol implantation on reproductive function in two species of trout, *Salmo trutta* L. and *Salmo gairdneri* Richardson. *Gen. Comp. Endocrinol.* 76:310–321.

Colborn, T., Dumanoski, D. and Myers, J.P. (1996) *Our Stolen Future*. Dutton/Penguin Books USA, New York, 306 pp.

Colborn, T., vom Saal, F.S. and Soto, A.M. (1993) Developmental effects of endocrine-disrupting chemicals in wildlife and humans. *Environ. Health Perspect.* 101:378–384.

Desbrow, C., Routledge, E.J., Brighty, G.C., Sumpter, J.P. and Waldock, M. (1998) Identification of estrogenic chemicals in STW effluent. I. chemical fractionation and *in vitro* biological screening. *Environ. Sci. Technol.* 32:1549–1558.

Douthwaite, R.J., Fox, P.J., Matthiessen, P. and Russell-Smith, A. (1981) *Environmental Impact of Aerosols of Endosulfan, Applied for Tsetse Fly Control in the Okavango Delta, Botswana.* Final report of the Endosulfan Monitoring Project, Overseas Development Administration, London, 141 pp.

Douthwaite, R.J., Fox, P.J., Matthiessen, P. and Russell-Smith, A. (1983) Environmental impact of aerial spraying operations against tsetse fly in Botswana. *In: 17th Meeting of the International Scientific Council for Trypanosomiasis Research and Control.* Arusha, Tanzania, 1981. Organization of African Unity/International Scientific Council for Trypanosomiasis Research & Control, OAU/ISTRC Publication No. 112, pp 626–633, Eleza Services, Nairobi.

Drysdale, D.T. and Bortone, S.A. (1989) Laboratory induction of inter-sexuality in the mosquito-fish *Gambusia affinis*, using paper mill effluent. *Bull. Environ. Contam. Toxicol.* 43:611–617.

Folmar, L.C., Denslow, N.D., Rao, V., Chow, M., Crain, D.A., Enblom, J., Marcino, J. and Guillette, L.J. (1996) Vitellogenin induction and reduced serum testosterone concentrations in feral male carp (*Cyprinus carpio*) captured near a major metropolitan sewage treatment plant. *Environ. Health Perspect.* 104:1096–1101.

Forster, M.S., Wilder, E.L. and Heindrichs, W.L. (1975) Estrogenic behaviour of 2-(*o*-chlorophenyl)-2-(*p*-chlorophenyl)-1,1,1-trichloroethane and its homologues. *Biochem. Pharmacol.* 24:1777–1780.

Gagnon, M.M., Bussieres, D., Dodson, J.J. and Hodson, P.V. (1995) White sucker (*Catostomus commersoni*) growth and sexual maturation in pulp mill contaminated and reference rivers. *Environ. Toxicol. Chem.* 14:317–327.

Guellec, K.L., Lawless, K., Valotaire, Y., Kress, M. and Tenniswood, M. (1988) Vitellogenin gene expression in male rainbow trout (*Salmo gairdneri*) *J. Comp. Endocrinol.* 71:359–371.

Guillette, L.J., Gross, T.S., Gross, D.A., Rooney, A.A. and Percival, H.F. (1995) Gonadal steroidogenesis *in vitro* from juvenile alligators obtained from contaminated or control lakes. *Environ. Health Perspect.* 103 Suppl. 4:31–36.

Guillette, L.J., Gross, T.S., Masson, G.R., Matter, J.M., Percival, H.F. and Woodward, A.R. (1994) Developmental abnormalities of the gonad and abnormal sex hormone concentrations in juvenile alligators from contaminated and control lakes in Florida. *Environ. Health Perspect.* 102:680–688.

Guillette, L.J., Pickford, D.B., Crain, D.A., Rooney, A.A. and Percival, H.F. (1996) Reduction in penis size and plasma testosterone concentrations in juvenile alligators living in a contaminated environment. *Gen. Comp. Endocrinol.* 101:32–42.

Harries, J.E., Jobling, S., Matthiessen, P., Sheahan, D.A. and Sumpter, J.P. (1995) *Effects of Trace Organics on Fish – Phase 2.* Report to the Department of the Environment, Foundation for Water Research, Marlow, Report no. FR/D 0022, 90 pp.

Harries, J.E., Sheahan, D.A., Jobling, S., Matthiessen, P., Neall, P., Routledge, E., Rycroft, R., Sumpter, J.P. and Tylor, T. (1996) Survey of estrogenic activity in United Kingdom inland waters. *Environ. Toxicol. Chem.* 15:1993–2002.

Harries, J.E., Sheahan, D.A., Jobling, S., Matthiessen, P., Neall, P., Sumpter, J.P., Tylor, T. and Zaman, N. (1997) Estrogenic activity in five United Kingdom rivers detected by measurement of vitellogenesis in caged male trout. *Environ. Toxicol. Chem.* 16:534–542.

Heinz, G.H., Percival, H.F. and Jennings, M.L. (1991) Contaminants in American alligator eggs from Lakes Apopka, Griffin and Okeechobee, Florida. *Environ. Monit. Assess.* 16: 277–285.

Howell, W.M. and Denton, T.E. (1989) Gonopodial morphogenesis in female mosquitofish, *Gambusia affinis affinis*, masculinized by exposure to degradation products from plant sterols. *Environ. Biol. Fish* 24:43–51.

Howell, W.M., Black, D.A. and Bortone, S.A. (1980) Abnormal expression of secondary sex characters in a population of mosquitofish, *Gambusia affinis holbrooki*: evidence for environmentally-induced masculinization. *Copeia*, 1980:676–681.

IEH (1995) *Environmental Estrogens: Consequences to Human Health and Wildlife*. Institute for Environment and Health, Leicester, UK, 107 pp.

Janssen, P.A.H., Lambert, J.G.D., Vethaak, A.D. and Goos, H.J.Th. (1997) Environmental pollution caused elevated concentrations of oestradiol and vitellogenin in the female flounder, *Platichtys flesus* (L.). *Aquat. Toscicol.* 39:195–214.

Jansen, H.T., Cooke, P.S., Porcelli, J., Liu, T.C. and Hansen, L.G. (1993) Estrogenic and anti-estrogenic actions of PCBs in the female rat: *in vitro* and *in vivo* studies. *Reprod. Toxicol.* 7:237–248.

Jobling, S. and Sumpter, J.P. (1993) Detergent components in sewage effluent are weakly estrogenic to fish: an *in vitro* study using rainbow trout (*Oncorhynchus mykiss*) hepatocytes. *Aquat. Toxicol.* 27:361–372.

Jobling, S., Sheahan, D.A., Osborne, J.A., Matthiessen, P. and Sumpter, J.P. (1996) Inhibition of testicular growth in rainbow trout (*Oncorhynchus mykiss*) exposed to estrogenic alkylphenolic chemicals. *Environ. Toxicol. Chem.* 15:194–202.

Kavlock, R.J., Daston, G.P., DeRosa, C., Fenner-Crisp, P., Gray, L.E., Kaattari, S., Lucier, G., Luster, M., Mac, M.J., Maczka, C., Miller, R., Moore, J., Rolland, R., Scott, G., Sheehan, D.M., Sinks, T. and Tilson, H.A. (1996) Research needs for the risk assessment of health and environmental effects of endocrine disruptors: a report of the U.S. EPA-sponsored workshop. *Environ. Health Perspect.* 104:715–740.

Kelce, W.R., Stone, C.R., Laws, S.C., Gray, L.E., Kemppainen, J.A. and Wilson, E.M. (1995) Persistent DDT metabolite *p,p'*-DDE is a potent androgen receptor antagonist. *Nature* 375:581–585.

Kime, D.E. (1995) The effects of pollution on reproduction in fish. *Rev. Fish Biol. Fish.* 5:52–96.

Knights, W.M. (1980) Estrogens administered to food-producing animals: environmental considerations. *In*: McLachlan, J.A. (Ed) *Estrogens in the Environment*. Elsevier, North Holland, Amsterdam, pp 391–401.

Krishnan, V. and Safe, S. (1993) PCBs, PCDDs and PCDFs as antiestrogens in MCF-7 human breast cancer cells: quantitative structure-activity relationships. *Toxicol. Appl. Pharmacol.* 120:55–61.

Lang, T., Damm, U. and Dethlefsen, V. (1995) Changes in the sex ratio of North Sea dab (*Limanda limanda*) in the period 1981–1995. ICES CM 1995/G:25 Ref E, International Council for the Exploration of the Sea, Copenhagen, 11 pp.

Lye, C.M., Frid, C.J.J., Gill, M.E. and McCormick, D. (1997) Abnormalities in the reproductive health of flounder *Platichthys flesus* exposed to effluent from a sewage treatment works. *Mar. Pollut. Bull.* 34:34–41.

Lyons, G. (1996) *Pesticides posing hazards to reproduction*. Worldwide Fund for Nature, WWF-UK, Godalming, unpublished report, 67 pp.

MacLatchy, D.L. and Van Der Kraak, G.J. (1995) The phytoestrogen β-sitosterol alters the reproductive endocrine status of goldfish. *Toxicol. Appl. Pharmacol.* 134:305–312.

Matthiessen, P. and Logan, J.W.M. (1984) Low concentration effects of endosulfan insecticide on reproductive behaviour in the tropical cichlid fish *Sarotherodon mossambicus*. *Bull. Environ. Contam. Toxicol.* 33:575–583.

McLachlan, J.A., Newbold, R.R., Teng, C.T. and Korach, K.S. (1992) Environmental estrogens: orphan receptors and genetic imprinting. *In:* Colborn, T. and Clement, C. (Eds) *Chemically induced alterations in sexual and functional development: the wildlife/human connection*. Princeton Scientific Publishing, Princeton, pp 107–112.

Mellanen, P., Petanen, T., Lehtim, S., Bylund, G., Holmbom, B., Mannila, E., Oikari, A. and Santti, R. (1996) Wood-derived estrogens – studies *in vitro* with breast cancer cell lines and *in vivo* in trout. *Toxicol. Appl. Pharmacol.* 136:381–388.

Moore, C.G. and Stevenson, J.M. (1991) The occurrence of intersexuality in harpacticoid copepods and its relationship with pollution. *Mar. Pollut. Bull.* 22:72–74.

Munkittrick, K.R., Van Der Kraak, G.J., McMaster, M.E., Portt, C.B., van den Heuvel, M.R. and Servos, M.R. (1994) Survey of receiving water environmental impacts associated with discharges from pulp mills. 2. Gonad size, liver size, hepatic EROD activity and plasma sex steroid levels in white sucker. *Environ. Toxicol. Chem.* 13:1089–1101.

Pereira, J.J., Ziskowski, J., Mercaldo-Allen, R., Kuropat, C., Luedke, D. and Gould, E. (1992) Vitellogenin in winter flounder (*Pleuronectes americanus*) from Long Island Sound and Boston Harbor. *Estuaries* 15:289–297.

Purdom, C.E., Hardiman, P.A., Bye, V.J., Eno, N.C., Tyler, C.R. and Sumpter, J.P. (1994) Estrogenic effects of effluents from sewage treatment works. *Chemistry and Ecology* 8:275–285.

Ren, L., Lattier, D. and Lech, J.J. (1996) Estrogenic activity in rainbow trout determined with a new cDNA probe for vitellogenesis, pSG5Vg1.1. Bull. *Environ. Contam. Toxicol.* 56:287–294.

Routledge, E.J. and Sumpter, J.P. (1996) Estrogenic activity of surfactants and some of their degradation products assessed using a recombinant yeast screen. *Environ. Toxicol. Chem.* 15:241–248.

Sheahan, D.A., Bucke, D., Matthiessen, P., Sumpter, J.P., Kirby, M.F., Neall, P. and Waldock, M. (1994) The effects of low levels of 17α-ethynylestradiol upon plasma vitellogenin levels in male and female rainbow trout, *Oncorhynchus mykiss* held at two acclimation temperatures. *In:* Müller, R. & Lloyd, R. (Eds) *Sublethal and Chronic Effects of Pollutants on Freshwater Fish.,* Fishing News Books, Blackwell Science, Oxford, pp 99–112.

Shore, L.S., Shemesh, M. and Cohen, R. (1988) The role of estradiol and oestrone in chicken manure silage in hyperestrogenism in cattle. *Aust. Vet. J.* 65:67.

Shore, L.S., Gurevitz, M. and Shemesh, M. (1993) Estrogen as an environmental pollutant. *Bull. Environ. Contam. Toxicol.* 51:361–366.

Soto, A.M., Justicia, H., Wray, J.W. and Sonnenschein, C. (1991) p-nonyl-phenol: an estrogenic xenobiotic released from "modified" polystyrene. *Environ. Health Perspect.* 92:167–173.

Soto, A.M., Chung, K.L. and Sonnenschein, C. (1994) The pesticides endosulfan, toxaphene, and dieldrin have estrogenic effects on human estrogen-sensitive cells. *Environ. Health Perspect.* 102:380–383.

Sumpter, J.P. (1985) The purification, radioimmunoassay and plasma levels of vitellogenin from the rainbow trout *Salmo gairdneri. In:* Lofts, B. and Holmes, W.H. (Eds) *Trends in Comparative Endocrinology.* Hong Kong University Press, Hong Kong, pp 355–357.

Sumpter, J.P. and Jobling, S. (1995) Vitellogenesis as a biomarker of estrogenic contamination of the aquatic environment. *Environ. Health Perspect.* 103 suppl. 7:173–178.

Sumpter, J.P., Jobling, S. and Tyler, C.R. (1996) Estrogenic substances in the aquatic environment and their potential impact on animals, particularly fish. *In:* Taylor, E.W. (Ed) *Toxicology of aquatic pollution: physiological, molecular and cellular approaches.* Cambridge University Press, Cambridge, pp 205–24.

Toppari, J., Larsen, J.C., Christiansen, P., Giwercman, A., Grandjean, P., Guillette, L.J., Jégou, B., Jensen, T.K., Jouannet, P., Keiding, N., Leffers, H., McLachlan, J.A., Meyer, O., Müller, J., Rajpert-De Meyts, E., Scheike, T., Sharpe, R., Sumpter, J. and Skakkebæk, N.E. (1995) *Male Reproductive Health and Environmental Chemicals with Estrogenic Effects.* Danish Environmental Protection Agency, Copenhagen, Miljøprojekt no. 290, 166 pp.

Van Der Kraak, G.J., Munkittrick, K.R., McMaster, M.E., Portt, C.B. and Chang, J.P. (1992) Exposure to bleached kraft mill effluent disrupts the pituitary-gonadal axis of white sucker at multiple sites. *Toxicol. Appl. Pharmacol.* 115:224–233.

Waring, C.P., Stagg, R.M., Fretwell, K., McLay, H.A. and Costello, M.J. (1996) The impact of sewage sludge exposure on the reproduction of the sand goby, *Pomatoschistus minutus. Environ. Pollut.* 93:17–25.

Woodward, A.R., Jennings, M.L., Percival, H.F. and Moore, C.T. (1993) Low clutch viability of American alligators on Lake Apopka. *Florida Sci.* 56:52–63.

Fish Ecotoxicology
ed. by T. Braunbeck, D. E. Hinton and B. Streit
© 1998 Birkhäuser Verlag Basel/Switzerland

Testing of chemicals with fish – a critical evaluation of tests with special regard to zebrafish

Roland Nagel and Karla Isberner

Institute of Hydrobiology, Technical University of Dresden, D-01062 Dresden, Germany

Summary. In this review, different test systems with fish according to the German Chemicals and Plant Protection Acts are introduced and evaluated. On the basis of a critical consideration of these test systems, the following test concept with zebrafish (*Danio rerio*) is proposed:
   Base level: For ethical reasons, the conventional routine acute toxicity test with juvenile or adult fish should be replaced by a 48 h embryo test with zebrafish.
   Level 1: At level 1, the early life-stage test (OECD guideline 210) deserves priority over the prolonged fish test according to OECD guideline 204.
   Level 2: Only a complete life-cycle test fulfills the requirements of a chronic toxicity test. It can, therefore, not be substituted by an early life-stage test. Since extrapolation from acute-chronic ratio (ACR) data is not possible as well, there is no alternative to a complete life-cycle test.

Importance of fish

Fish are of particular importance for man: In many countries, fish have remained a major source of protein for large portions of the population, and commercial fisheries are not only important in industrial, but also in developing countries. Apart from the significance of fish as a direct food source for man, processed fish meal plays a prominent indirect role in human nutrition as an important feed in the production of meat.

As the only primary aquatic vertebrates, fish also deserve particular attention as monitor system in the surveillance of aquatic ecosystems. Thus, fish have become classical "test organisms". In the attempt to evaluate adverse effects of environmental chemicals on fish, however, it is essential to realize that the "fish" (species) as an archetype does not exist. *Bony fish* in terms of a zoological taxon (Osteichthyes) comprise the vast majority of fish and, among these, teleosts represent the dominant recent subclass. Estimates of teleost species numbers range between 20 000 and 30 000 (Groombridge, 1992; Nelson, 1994). In fact, even within teleost fishes, structural and biological diversity is immense. In almost any limnic and marine habitat, a huge variety of ecological niches has been occupied by teleost fishes. As primary, secondary, and tertiary consumers, they represent different trophic levels within aquatic food webs. It is, thus, important to give credit to the considerable morphological, physiological and ecological diversity of fish when potential interactions between environmental chemicals and fish are investigated.

Table 1. Test levels within the framework of the German Chemicals Act

Test level	Production of the chemical	Type of fish test
Base level	> 1 t/a	Fish test in the acute toxicity range (LC$_{50}$ over 96 h)
Level 1	> 100 t/a or 500 t in total	Prolonged fish test (14 and 28 days, respectively)
Level 2	1000 t/a or 5000 t in total	Long-term studies including reproduction

The role of fish within the framework of testing regulations according to the Chemicals Act and Plant Protection Act

Chemicals may cause biochemical, physiological, morphological, and/or genetic alterations in fish, and, at different levels of biological organization, these alterations can influence the specific performances of the organisms in, e.g., development, growth and reproduction.

Depending on the rate of chemical damage, these effects may be of sublethal nature or result in the death of the organism. In an attempt to account for these facts, toxicity tests with fish within the regulations for the estimation of chemical toxicity by the German Chemicals Act discriminate between three stages according to the production rate of the chemicals (Tab. 1). In contrast, there is no such rigid test scheme in the Plant Protection Act; yet, the acute fish test also represents the initial step in the test procedure.

Toxicity studies at the base level within the framework of the German Chemicals Act

Conventional acute toxicity tests with fish

Usually, the acute toxicity of chemicals to fish is determined as an LC$_{50}$ (96 h) value (e.g., according to OECD guideline 203), i.e., as the concentration of the test substance resulting in 50% mortality of the experimental fish over a period of 96 h. Over many years, an extensive data base has thus been accumulated for acute toxicity. Different fish species may vary by orders of magnitude with respect to their sensitivity in acute tests to environmental contaminants (Tab. 2). As a rule, salmonid fishes are considered to be more susceptible than Cypriniform and Cyprinodontiform species, respectively (Tab. 3).

Attempts to extrapolate from toxicity data determined for freshwater fish to toxicity in marine species has frequently been made and seems to be possible. For instance, Hutchinson and coworkers (1998) compared the acute

Table 2. A selection of LC$_{50}$-values from tests with different species

Chemical	Test fish	LC$_{50}$ (μmol/L)
Trifluralin	Rainbow trout (*Oncorhynchus mykiss*)	0.12
	Channel catfish (*Ictalurus punctatus*)	6.6
Cadmium chloride	Rainbow trout (*Oncorhynchus mykiss*)	0.08
	Guppy (*Poecilia reticulata*)	490
Mercuric chloride	Fathead minnow (*Pimephales promelas*9	0.2
	Fuppy (*Poecilia reticulata*)	52.5
Parathion	Guppy (*Poecilia reticulata*)	0.2
	Medaka (*Oryzias latipes*)	10
Malathion	Mosquito fish (*Gambusia affinis*)	0.01
	Fathead minnow (*Pimephales promelas*)	45.7
Benzene	Rainbow trout (*Oncorhynchus mykiss*)	85.1
	Mosquito fish (*Gambusia affines*)	5,012
1,1,2-Trichloroethylene	Medaka (*Oryzias latipes*)	14.5
	Golden ide (*Leuciscus idus melanotus*)	1,380
	Guppy (*Poecilia reticulata*)	1,380
Pyridine	Rainbow trout (*Oncorhynchus mykiss*)	57.5
	Guppy (*Poecilia reticulata*)	17,378
Allylamine	Fathead minnow (*Pimephales promelas*)	37.2
	Golden ide (*Leuciscus idus melanotus*)	1,096

Data for trifluralin cited from Johnson and Finley (1980); data for all other chemicals as given by Vaal et al. (1997).

Table 3. Sensitivity of the cypriniform golden ide (*Leuciscus idus melanotus*) and fathead minnow (*Pimephales promelas*), the cyprinodontiform medaka (*Oryzias latipes*) and guppy (*Poecilia reticulata*) as well as the salmonid rainbow trout (*Oncorhynchus mykiss*) in acute toxicity tests with 17 compounds (data cited from Vaal et al. 1997)

	Species	Average sensitivity LC$_{50}$ (μmol/L)
Cypriniformes	Golden ide (*Leuciscus idus melanotus*)	118.6
	Fathead minnow (*Pimephales promelas*)	74.3
Cyprinodontiformes	Medaka (*Oryzias latipes*)	76.8
	Guppy (*Poecilia reticulata*)	168.2
Salmoniformes	Rainbow trout (*Oncorhynchus mykiss*)	33.1

toxicity (EC$_{50}$) of 22 substances in both freshwater and marine species. In five cases, freshwater fish proved to be significantly more sensitive than marine fish (ratios freshwater:marine < 0.5). For six compounds, the EC$_{50}$ ratios ranged between 0.5 and 2.0. In 11 cases, freshwater fish were significantly less sensitive than saltwater fish (ratios freshwater:marine > 2.0). The sensitivity ratios between freshwater and marine fish species were within a factor of 10 for 91% of all substances tested by Hutchinson et al. Remarkable exceptions were chlopyrifos and benzene with ratios of 26.3 and 183, respectively.

In some cases, differences in toxicity could be explained by differences in metabolism. For instance, quantitative differences in the oxidative metabolism of diazinon between guppy and zebrafish resulted in different mechanisms of action and in selective toxicity in these two species. In the guppy, the relevant biologicaly active compound is a metabolite (probably diazoxon), whereas in the zebrafish it is the parent compound (Keizer et al., 1993).

As an addition to acute toxicity expressed as LC_{50} data, the slope of the dose-response relationship may be used as a source of important information. The ratio of LC_{50} (48 h) to LC_{50} (96 h) gives an indication whether time plays an important role in toxicity.

Critical assessment of conventional acute toxicity tests with fish and possible alternatives

Death as an endpoint in toxicological research represents an unambiguous parameter for the individual. However, unless detrimental mortal effects affect large parts of the entire population (as could be observed, for example, during the massive fish kill after the chemical spill into the Rhine river at Basel in November 1986), the significance of the death of individuals for the population should be discussed and depends on species, reproductive strategy and the actual structure of the population etc.

Moreover, the ecotoxicological relevance of death is limited by the fact that, at least in our opinion, extrapolation of chronic effects from acute toxicity data is, by principle, impossible (see also below), and, thus, do not allow for evaluation and prediction of chronic effects from LC_{50} data. However, it is chronic toxicity which represents the ultimately important endpoint in ecotoxicology.

In recent years, at least in developed countries acute toxicity tests with fish have also aroused considerable ethical concern (for a closer discussion, see the contributions by Segner as well as Braunbeck in this volume): Since acute toxicity to fish is determined in tests with juvenile or adult animals, intact fish are subjected to considerable pain and suffering, which is clearly in conflict with current Animal Rights Welfare legislation. Since results of LC_{50} tests are only of minor ecotoxicological significance, acute toxicity tests with fish should be replaced by an alternative method.

Possible alternatives to the acute fish test might be acute toxicity tests with embryo of zebrafish, *Danio rerio* (formerly *Brachydanio rerio*) and cytotoxicity tests with fish cells. The methodology of the zebrafish embryo test and preliminary results have been published by Schulte and Nagel (1994): In brief, fertilized zebrafish eggs are exposed individually to five concentrations of a test substance for 48 h. During the static exposure, the development of the embryos is examined with respect to 13 toxicological endpoints. With this method, sublethal and lethal effects can be determined.

In three laboratories, 21 reference chemicals were investigated with the zebrafish embryo test. Consistency of the results was confirmed by relatively low coefficients of variation between the laboratories (highest coefficient of variation: 36%). The average value from the inter-laboratory study and LC_{50} data for 16 additional chemicals were compared to LC_{50} values resulting from acute toxicity tests with zebrafish (*Danio rerio*) or, if not available, golden ide (*Leuciscus idus melanotus*). Regression analysis with data for these 37 chemicals revealed a good correlation with a slope of 0.81 and a regression coefficient of 0.87 (Schulte et al., 1996).

On the basis of these results, an international inter-laboratory study with 12 laboratories, funded by the German Environmental Protection Agency was initiated (Fußmann and Nagel, 1997). In this validation study with three chemicals, the variability of LC_{50} data within laboratories was very low. In contrast, variability between laboratories was higher for all three test substances. However, for a sub-set consisting of eight out of twelve laboratories, LC_{50} values were very homogenous. Deviations from the mean of all laboratories may be partly due to the fact that two laboratories used strains of *Danio rerio* different from those used by the other participants. A standard test protocol and a prediction model (fish embryo – fish) was developed, and an OECD guideline has been proposed and is now under discussion.

In the meantime, the database for the toxicity of chemicals to zebrafish embryos has been extended to 44 compounds. The correlation is given in Figure 1. These data corroborate the conclusion that toxicity estimates derived from the embryo test are in good accordance with data from corresponding acute toxicity tests with juvenile or adult fish.

In another comparative study, Lange and colleagues (1995) related results of the zebrafish embryo test to those of the cytotoxicity test with the permanent cell line RTG-2 derived from rainbow trout (*Oncorhynchus mykiss*) gonads for 10 selected compounds with different modes of action. It could be shown that in most cases the zebrafish embryo test was more sensitive than both the acute toxicity test with adult zebrafish and the RTG-2 cell test. In another study for 17 compounds, results of the embryo and the RTG-2 cytotoxicity tests were compared to LC_{50} data from tests with golden ide (F. Moldenhauer and H. Spielmann, personal communication). For either alternative test system, linear regression analysis documented satisfactory correlation. The correlation coefficient for the comparison of the zebrafish embryo test and juvenile golden ide fish tests proved to be slightly better (0.991) than the correlation between RTG-2 cytotoxicity and juvenile golden ide toxicity (0.796). A comparixon of y-axis intercepts revealed that the zebrafish embryo test (y = – 0.13) reflected the acute toxicity more accurately than the RTG-2 cytotoxicity test (– 0.89).

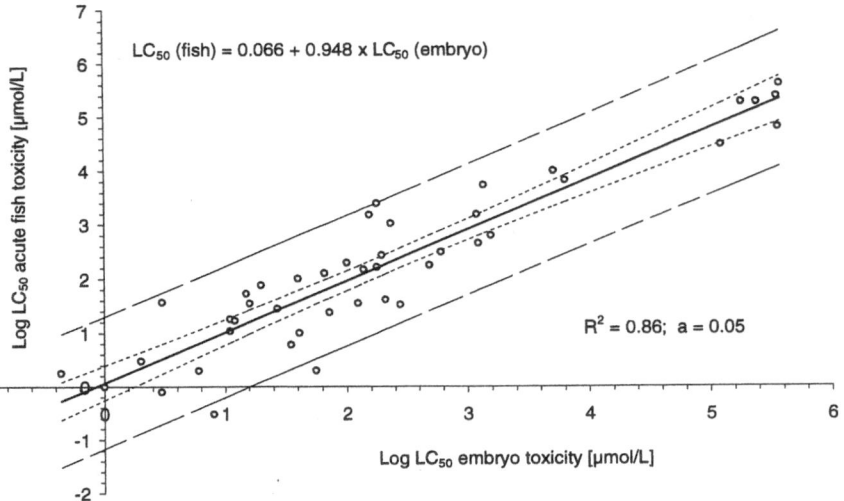

Figure 1. Correlation of LC_{50} acute fish toxicity and LC_{50} embryo toxicity. Data have been taken from Bachmann (1996), Maiwald (1997) as well as Schulte et al. (1996).

These data indicate that the embryo test is a particularly suitable candidate to replace the acute fish test at the base level. The zebrafish embryo test might also represent a promising alternative to acute toxicity tests with intact fish in routine waste water control. Friccius and coworkers (1995) tested 29 samples of industrial effluents from 11 sewage plants. The zebrafish embryo toxicity test was as sensitive as or even more sensitive than the conventional fish test according to the German DIN-Norm 38 412, L 31. At present, the alternative test with zebrafish embryos is being validated in a DIN working group.

Studies at level 1 within the framework of the German Chemicals Act

Prolonged fish toxicity tests

Apart from the determination of threshold concentrations for lethal effects, the intention of the prolonged fish test is to reveal other macroscopically overt signs of toxicity other than mortality, including parameters such as modified swimming behavior, changes in the external appearance of individuals or food uptake. In most cases, the prolonged fish test has proven to be more sensitive than the acute test (Rudolph and Boje, 1986). Unfortunately, however, in most studies *death* has represented the crucial test parameter for the determination of LOEC (lowest observed effect concentration) and NOEC (no observed effect concentration) as well. In fact, the prolonged fish test (OECD-guideline 204) has practically evolved into a lethal test over 14 and 28 days, respectively. One reason is certainly the fact that *death* – in contrast to many other parameters – is inambiguous and, thus, free of any discussion as to its toxicological significance.

Critical appraisal of the prolonged fish test

Although the prolonged fish test not only allows, but has indeed designated for the registration of other endpoints than lethality, death is frequently the sole endpoint measured in practice. Therefore, if compared to the acute toxicity test, the only additional information of the prolonged fish test is the impact or prolonged exposure (i.e., time) on the onset of lethality. Measurement of growth (draft: OECD-fish growth test), e.g., would additionally allow for the determination of the performance of individual organisms. The significance of this parameter at the level of the population, however, is still obscure. There is good evidence of considerable differences in the "grow out" of fish, and it has frequently been shown that fish are capable of rapidly compensating diet-dependent variations in body size. Whether this holds true for changes induced by exposure to xenobiotics remains to be elucidated.

However, the use of young fish during their growth phase in the prolonged fish test should enable us to incorporate growth as a sublethal test parameter, which can easily and unambiguously be identified and quantified. With regard to growth as a toxicological endpoint, rainbow trout and zebrafish have shown comparable susceptibility to, e.g., 3,4-dichloroaniline (Nagel, 1988.)

Early life-stage tests (ELS-tests)

The impact of chemicals on early life stages of fish may be tested according to the OECD guideline 210. An inter-laboratory validation study including eight laboratories on 3,4-dichloroaniline-dependent effects in zebrafish has documented that an early life-stage test with this fish species may yield comparable and conclusive results (Nagel et al., 1991). Whereas another early life-stage study on the effects of 3,4-dichloroaniline using fathead minnow (*Pimephales promelas*) has documented zebrafish to be apparently slightly less sensitive than the fathead minnow (Call et al., 1987), a comparative study of effects by 3,4-dichloroaniline in early life-stages of zebrafish, perch (*Perca fluviatilis*) and roach (*Rutilus rutilus*) revealed that all three species displayed similar susceptibility (Schäfers and Nagel, 1993). The results of the latter study indicate that, despite different water temperature and fish body size, zebrafish may be used as a generalized model to predict early life-stage toxicity of 3,4-dichloroaniline to fish species common to freshwaters in temperature European freshwater eco-systems.

Critical assessment of early life-stage tests with fish

Early life-stage tests with fish are used to record the impact of a substance on embryonic development, hatching, and larval development. Thus, a multitude of diverse processes in differentiation and development as well as the weaning period with the particularly important shift from utilization of endogenous reserves to external feeding are included. Numerous tests with a variety of fish species have documented the conclusiveness and relevance of this methodology (McKim, 1977; Woltering, 1984).

In a data set published by the German Federal Environmental Protection Agency, results for 40 chemicals from both early life-stage tests and the prolonged fish test (28 days) have been made available. For 29 out of 40 substances, the early life-stage test was found to be more sensitive than the prolonged fish test, with an average ratio of 28.6 between the NOECs of the prolonged fish test and the early life-stage test. In turn, for 11 substances, the ratio between the NOECs of the prolonged fish test and the early life-stage test averaged 0.3 (W. Heger, personal communication).

Likewise, from the registration procedures of pesticides, a considerable body of classified information exists, which also documents that for the majority of cases early life-stage tests are more sensitive than prolonged fish tests. The effects of a given fungicide on rainbow trout (*Oncorhynchus mykiss*) may serve as an example: The LC_{50} (96 h) was determined at 9.5 mg/L. Whereas the NOEC/LOEC values established with early life-stage tests are 0.00016 mg/L and 0.0008 mg/L, respectively, the NOEC/LOEC values from the prolonged fish test (21 d) as determined at 0.9 mg/L and 1.7 mg/L, respectively, were four orders of magnitude higher.

Therefore and due to its higher ecotoxicological relevance, the early life-stage test with fish according to the OECD guideline 210 is recommended as a sensitive and suitable method for toxicity testing at level 1 according to the German Chemicals Act. In this context, it should be noted that the test procedures "Fish toxicity test on egg and sac-fry stages" submitted as a draft to the OECD as well as "Water quality – Determination of embryo-larval toxicity to freshwater fish – Semistatic method" submitted as a draft to ISO/DIS 12890 do not represent tests which can be compared to the OECD guideline 210. According to these protocols, fish are exposed without feeding during the entire test procedures, and the most sensitive periods including the shift from utilization of endogenous reserves to external feeding have not been included. For these reason, this test design appears unacceptable, and it is still under debate as to at which test level these protocols may provide meaningful results.

Studies at level 2 within the framework of the German Chemicals Act

According to the German Chemicals Act, at level 2 long-term tests are required which should include investigations of the effects of chemicals on fish reproduction. In routine laboratory testing, this regulation can only be met by conducting either a partial or a complete life-cycle test. Whereas the abbreviated test is restricted to the determination of egg production followed by an early life-stage test, the complete life-cycle test comprises effects of permanent chemical exposure in embryos, larvae, juveniles, and adults in the F_I-generation as well as during embryo-larval development in the F_{II}-generation. Numerous processes such as development in the F_{II}-generation. Numerous processes such as development of gametes, fertilization, courtship, and mating can thus be studied. As a matter of course, within a reasonable period of time, complete life-cycle tests can only be carried out with small, rapidly growing warm water fish such as zebrafish (*Danio rerio*), an egg-laying species, which reaches sexual maturity after only 3 months (Laale 1977; Ekker and Akimenko, 1991), or other fish species of tropical origin. The test scheme of a complete life-cycle test with zebrafish is given in Figure 2. A draft of an OECD-guideline is under preparation.

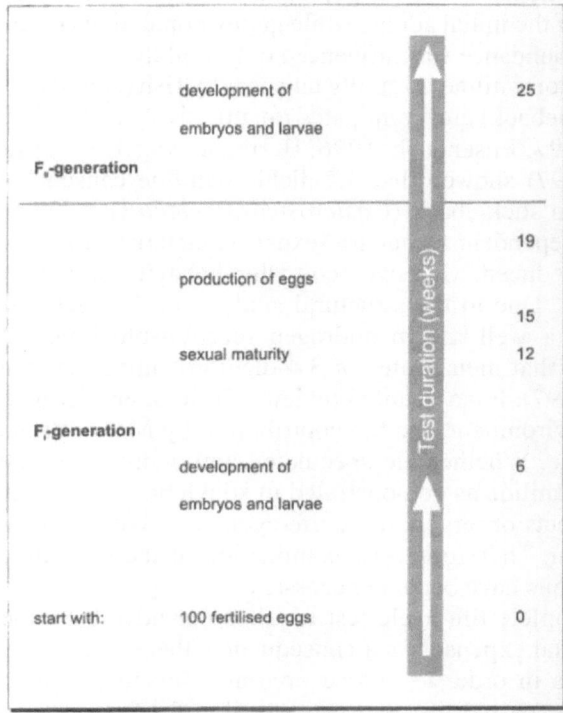

Figure 2. Test scheme of the complete life-cycle test with zebrafish (*Danio rerio*).

The difficulties in the extrapolation from one species to another, which are well-known for acute toxicity data, do not appear as problematic for data of chronic toxicity. As shown by Nagel (1993), zebrafish may serve as "model species" for other freshwater fish with comparable reproduction strategies. In contrast, comparison of effects by 3,4-dichloroaniline on zebrafish as a model r-strategist with results from experiments with the live-bearing guppy (*Poecilia reticulata*) as a model for a K-strategist documents that zebrafish *cannot* be used as a "model" for the guppy. Whereas in zebrafish larval survival in the F_{II}-generation was the most sensitive parameter to 3,4-dichloroaniline exposure, in guppies reproduction proved to be most susceptible to this compound (Schäfers and Nagel, 1991). The consequences of chemical exposure at the population level have been simulated in a stochastic individuum approach (Schäfers et al., 1993): At a given concentration that did not take influence on the endpoints measured in the zebrafish life-cycle test, reproductioin of the guppy was significantly reduced by 30%. These findings apparently correspond to the statement commonly made that K-strategists should react more sensitively than r-strategists (Reichholf, 1983). At the population level, however, this results in a decrease of the mean population abundance by only 12% in the computer simulations. On the other hand, zebrafish react more drastically than guppy: At 200 µg/L 3,4-dichloroaniline, e.g., experimental zebrafish populations were eventually eliminated due to the death of the initial adults, while guppy populations reacted more flexible and final abundance was influenced only slightly.

3,4-dichloroaniline is rapidly taken up by fish (zebrafish, guppy, rainbow trout, stickleback) and conjugated quantitatively to 3,4-dichloroacetanilide (Allner, 1997; Ensenbach, 1996; Hertl and Nagel, 1993; Zok et al., 1991). Allner (1997) showed that 3,4-dichloroaniline caused lowered androgen synthesis in stickleback (*Gasterosteus aculeatus*), a K-strategist, and that androgen dependent secondary sexual characteristics such as nuptial coloration were reduced. Moreover, courtship behavior of male sticklebacks was suppressed. Due to the structural analogy of 3,4-dichloroacetanilide and flutamide, a well-known androgen receptor-blocking substance, it was speculated that metabolites of 3,4-dichloroaniline acted as anti-androgen (Allner, 1997). For a detailed review of role of endocrine disruptors in the aquatic environment, see the contribution by Matthiessen and Sumpter in this volume. Whether the speculated anti-androgen properties of 3,4-dichloroacetanilide as demonstrated in stickleback can be made responsible for the effects observed in the life-cycle tests with zebrafish and guppy is still unclear. Therefore, a re-examination of the data and complementary investigations have become necessary.

The complete life-cycle test requires considerable expenditure of time and financial expense. As a consequence, there is a need for the search of alternatives in order to reduce time and financial expense. As potential alternatives, partial life-cycle tests, early life-stage tests and extrapolation from acute toxicity data to chronic toxicity data have been discussed.

Partial life-cycle tests as a replacement for complete life-cycle tests were performed by McKim and Benoit (1971, 1974), McKim et al. (1976), Holcombe et al. (1976) as well as Norberg-King (1989). In studies in our laboratory, however, results with 4-nitrophenol revealed the survival rate in the F_{II}-generation as the most sensitive endpoint. Therefore, a partial life-cycle test is less sensitive than the full life-cycle test starting with eggs and followed by an investigation of embryonic and larval development in the F_{II}-generation (Nagel, 1988).

Results by Nagel (1994) as well as Bresch and coworkers (1990) indicate that the replacement of the complete life-cycle test by an early life-stage test as advocated by McKim (1977, 1985) and Woltering (1984) is questionable. Both Nagel (1988) and Bresch and colleagues (1990), using zebrafish exposed to 4-nitrophenol and 3,4-dichloroaniline, and 4-chloroaniline, respectively, demonstrated that the early life-stages of the F_{II}-generation were more sensitive than those of the F_I-generation. These results are also supported by a literature review by Chorus (1987) and in line with drawn conclusions by Peter and Heger (1997) that early life-stage tests cannot substitute for truly chronic tests. In this context and in the interest of a fruitful discussion, it does not appear helpful to re-publish exactly the same set of data and the same conclusions ten years later (McKim, 1985, 1995).

Another method frequently discussed as an alternative to life-cycle experiments is the extrapolation of chronic toxicity from acute effects. One possibility to predict chronic toxicity is to determined the acute-to-chronic ratio (ACR; Holcombe et al., 1995; Kenaga, 1982), which has been defined as the quotient of the acute LC_{50} value and the mean between NOEC and LOEC (or, in some publications, the NOEC) derived from a chronic test, at best from a complete lifecycle test. Kenaga (1982) recommended ACR values to predict safe concentration limits for chronic toxicity derived from acute toxicity. Comparing ACR data of industrial chemicals with those of pesticides and heavy metals, Kenaga (1982) was able to show that high ACRs (above 25) were principally represented by metals and pesticides. From this, Kenaga concluded that the use of ACRs of 25 or less represented a good tool for the prediction of chronic toxicity from the acute toxicity for organic industrial chemicals. However, as documented in Table 4, results by several authors document that acute to chronic ratios may be considerably higher than 25.

Länge and coworkers (1998) calculated acute to chronic ratios for fish and invertebrates on the basis of acute EC_{50}s (or LC_{50}s) and (sub)chronic NOEC values from the ECETOC Aquatic Toxicity (EAT) data base (Solbé et al., 1998). Their main conclusions for risk assessment purposes were (1) that for all species and all substances, an ACR of approximately 73 would safely predict the chronic NOEC for an estimated 90% of substances, (2) that for general organic chemicals, an ACR of approximately 15–25 would safely predict the chronic NOEC for an estimated 90% of such substances,

Table 4. Selection of high acute–chronic ratios (ACR) for different chemicals

Chemical	ACR	Class of substance	Test fish	Reference
Propanil	18,100	Herbicide	Fathead minnow (*Pimephales promelas*)	Kenaga (1982)
Azinphos-methyl	5,555	Insecticide	Fathead minnow (*Pimephales promelas*)	Kenaga (1982)
3,4-dichloroaniline	4,000 > 4,000	Industrial organic chemical	Zebrafish (*Danio rerio*) Guppy (*Poecilia reticulata*)	Nagel (1994) Schäfers and Nagel (1991, 1994)
4-chloroaniline	> 1,000	Industrial organic chemical	Zebrafish (*Danio rerio*)	Bresch et al. (1990)
Diazinon	500–769	Insecticide	Fathead minnow (*Pimephales promelas*), brook trout (*Salvelinus fontinalis*)	Kenaga (1982)
Diuron	435	Herbicide	Fathead minnow (*Lepomis macrochirus*)	Kenaga (1982)
Atrazine	370	Herbicide	Fathead minnow (*Pimephales promelas*), bluegill sunfish (*Lepomis macrochirus*), lake trout (*Salvelinus namaycush*)	Kenaga (1982)
Cadmium (chloride)	312.5–625	Heavy Metal	Flagfish (*Jordanella floridae*)	Spehar (1976)
Cadmium (sulfate)	254–370	Heavy Metal	Fathead minnow (*Pimephales promelas*)	Kenaga (1982)

(3) that heavy metals, organometal compounds and ingredients active as pesticides may give very high ACRs, and for these substances consideration should be given on an individual basis rather using a default value, and (4) that ACRs derived for a given fish species might carefully be extrapolated to ACR values of a second fish species.

These conclusions appear worth of discussion: First of all, it should be re-called that, in contrast to conventional toxicology, ecotoxicology is focused on the protection of *populations* and not of individual fish. As a consequence, in chronic studies endpoints with relevance to the population, e.g., reproduction, need to be examined. Therefore, the definition of the term "chronic toxicity" as "the harmful properties of a substance which are demonstrated only after long-term exposure in relation to the life of the test organism (typically over one-third to sexual maturity) or the full life-cycle" by Länge and coworkers (1998) cannot be accepted. In this context, not duration of exposure is the relevant factor, but the intrinsic potentials of a compound to impair, e.g., fecundity. Moreover, such potentials of compounds to interfere with reproductive success are independent of the classification as pesticides or general organic compounds. 3,4-dichloroaniline might serve as an excellent example for this basic toxicological rule.

To the best of our knowledge, no (classical) toxicologist has ever made an attempt to estimate reproductive toxicity on the basis of acute toxicity data. Likewise, ecotoxicologists should also come to an end with the discussion whether ACRs represent a suitable tool to predict chronic toxicity from acute toxicity data or not. In agreement with conclusions drawn by other authors such as Chorus (1987), Suter and colleagues (1987) as well as Peter and Heger (1997), we are convinced that only a complete life-cycle test is able to elucidate the true chronic toxicity of chemical compounds to fish. Thus, if we are serious about the protection of fish populations, our sole option are full life-cycle tests.

Conclusions

A careful assessment of the data available for various test systems suggests that a modification of the test concept presently used in the Federal Republic of Germany is required. The restriction to zebrafish as a model species can be maintained on the basis of the present data from interspecies comparisons and even appears advantageous, since, within the test schedule proposed, experiments at higher levels can be consequently related to and based on results from studies at lower levels. Furthermore, zebrafish has undoubtedly become the most important model in developmental biology of vertebrates (Ekker and Akimenko, 1991; Nüsslein-Volhard, 1994; deMonte Westerfield, 1995). Following the regulations by the German Chemicals Act in its present form, the test strategy proposed is divided into investigations at three levels: (1) For ethical reasons, conventional acute toxicity tests with juvenile or adult fish at the *base level* should be replaced by the embryo test with zebrafish (48 h). (2) At *level 1*, early life-stage tests according to OECD guideline 210 deserve priority over the prolonged fish test according to OECD guideline 204 (3) At *level 2*, only complete life-cycle tests are fully compatible with the requirements of a chronic toxicity test. It cannot be substituted by early life-stage tests or extrapolation from ACR data.

Acknowledgement
The authors wish to express their thanks to Dr. G. Fußmann for reading the manuscript.

References

Allner, B. (1997) Toxikokinetik von 3,4-Dichloranilin beim Dreistachligen Stichling (*Gasterosteus aculeatus*) unter besonderer Berücksichtigung der Fortpflanzungsphysiologie. *Ph.D. thesis*, Johannes Gutenberg-University of Mainz, FRG.

Bachmann, J. (1996) Wirkung von Chemikalien auf die Embryonalentwicklung des Zebrabärblings (*Brachydanio rerio*). *M.Sc. Thesis*, Technical University of Dresden, FRG.

Bresch, H., Beck, H., Ehlermann, D., Schlaszus, H. and Urbanek, M. (1990) A long-term toxicity test comprising reproduction and growth of zebrafish with 4-chloroaniline. *Arch. Environ. Contam. Toxicol.* 19:419–427.

Call, D.J., Poirier, S.H., Knuth, M.L., Haring, S.L. and Lindberg, C.A. (1987) Toxicity of 3,4-dichloroaniline to fathead minnow *Pimephales promelas* in acute and early life-stage exposures. *Bull. Environ. Contam. Toxicol.* 38:352–358.

Chorus, I. (1987) Literaturrecherche und Auswertung zur Notwendigkeit chronischer Tests – insbesondere des Reproduktionstests – am Fisch für die Stufe II nach Chemikaliengesetz. *Federal Environmental Protection Agency of Germany*, Z.4.1.-97 316/7.

deMonte Westerfield, M. (1995) *The zebrafish book – a guide for the laboratory use of zebrafish (Danio rerio)*. 3rd ed. University of Oregon, Oregon, 334 pp.

Ekker, M. and Akimenko, M.A. (1991) Le poisson zèbre (*Danio rerio*), un modèle en biologie du developement. *Médicine/sciences* 7:553–560.

Ensenbach, U., Hryk, R. and Nagel, R. (1996) Kinetics of 3,4-dichloroaniline in several fish species exposed to different types of water. *Chemosphere* 32:1643–1654.

Friccius, T., Schulte, C., Ensenbach, U., Seel, P. and Nagel, R. (1995) Der Embryotest mit dem Zebrabärbling – eine neue Möglichkeit zur Prüfung und Bewertung der Toxizität von Abwasserproben. *Vom Wasser* 84:407–418.

Fußmann, G. and Nagel, R. (1997) Laborvergleichsversuch – Embryotest mit dem Zebrabärbling *Brachydanio rerio*. *Federal Environmental Protection Agency of Germany*, Report No. 106 03 908.

Groombridge, B. (1992) *Global biodiversity. Status of the earth's living resources*. World Conservation Monitoring Center. Chapman and Hall, London, 585 pp.

Hertl, J. and Nagel, R. (1993) Bioconcentration and metabolism of 3,4-dichloroaniline in different life stages of guppy and zebrafish. *Chemosphere* 27:2225–2234.

Holcombe, G.W., Benoit, D.A., Leonard, E.N. and McKim, J.M. (1976) Long-term effects of lead exposure on three generations of brook trout (*Salvelinus fontinalis*). *J. Fish. Res. Bd. Can.* 33:1731–1741.

Holcombe, G.W., Benoit, D.A., Hammermeister, D.E., Leonard, E.N. and Johnson, R.D. (1995) Acute and long-term effects of nine chemicals on the Japanese medaka (*Oryzias latipes*). *Arch. Environ. Contam. Toxicol.* 28:287–297.

Hutchinson, T.H., Scholz, N. and Guhl, W. (1998) Analysis of the ECETOC aquatic toxicity (EAT) database. IV. Comparative toxicity of chemical substances to freshwater versus saltwater organisms. *Chemosphere.* 36:143–154.

Johnson, D.W. and Finley, M.T. (1980) Handbook of acute toxicity of chemicals to fish and aquatic invertebrates. *U.S. Dept. Inter. Fish. Wildl. Serv. Res. Publ.* 137, 98 pp.

Keizer, J., D'Agostino, G., Nagel, R., Gramenzi, F. and Vittozzi, L. (1993) Comparative diazinon toxicity in guppy and zebrafish: different role of oxidative metabolism. *Environ. Toxicol. Chem.* 12:1243–1250.

Kenaga, E.E. (1982) Predictability of chronic toxicity from acute toxicity of chemicals in fish and aquatic invertebrates. *Environ. Toxicol. Chem.* 1:347–358.

Laale, H.W. (1977) The biology and use of zebrafish, *Brachydanio rerio*, in fisheries research. A literature review. *J. Fish Biol.* 10:121–173.

Lange, M., Gebauer, W., Markl, J. and Nagel, R. (1995) Comparison of testing acute toxicity on embryo of zebrafish, *Brachydanio rerio* and RTG-2 cytotoxicity as possible alternatives to the acute fish test. *Chemosphere* 30 (11):2087–2102.

Länge, R., Hutchinson, T.H., Scholz, N. and Solbé, J. (1998) Analysis of the ECETOC Aquatic Toxicity (EAT) database II – Comparison of acute to chronic ratios for various aquatic organisms and chemical substances. *Chemosphere* 36:115–127.

Maiwald, S. (1997) Wirkung von Lösungsvermittlern und lipophilen Substanzen auf die Embryonalentwicklung des Zebrabärblings (*Brachydanio rerio*). *M.Sc. thesis*, Technical University of Dresden, FRG.

McKim, J.M. (1977) Evaluation of tests with early life stage of fishes for predicting long-term toxicity. *J. Fish. Res. Bd. Can.* 34:1148–1154.

McKim, J.M. (1985) Early life stage toxicity tests. *In:* Rand, G.M. and Petrocelli, S.R. (Eds) *Fundamentals of aquatic toxicology.* Hemisphere Publishing, Washington, pp 58–95.

McKim, J.M. (1995) Early life stage toxicity tests. *In:* Rand, G.M. (Ed) *Fundamentals of aquatic toxicology.* Taylor & Francis, Washington, pp 974–1011.

McKim, J.M. and Benoit, D.A. (1971) Effects of long-term exposure to copper on the survival, growth and reproduction of brook trout. *J. Fish. Res. Bd. Can.* 28:655–662.

McKim, J.M. and Benoit, D.A. (1974) Duration of toxicity tests for establishing "no effect" concentration for copper with brook trout. *J. Fish. Res. Bd. Can.* 31:449–452.

McKim, J.M., Holcombe, G.W., Olson, G.F. and Hunt, E.P. (1976) Long-term effects of methyl mercuric chloride on three generations of trout: toxicity, accumulation, distribution and elimination. *J. Fish. Res. Bd. Can.* 33:2726–2739.

Nagel, R. (1988) Umweltchemikalien und Fische – Beiträge zu einer Bewertung. *Habilitation Thesis*, Johannes Gutenberg-University of Mainz, FRG.

Nagel, R. (1993) Fish and environmental chemicals – a critical evaluation of tests. *In:* Braunbeck, T., Hanke, W. and Segner, H. (Eds) *Ecotoxicology and Ecophysiology.* VCH, Weinheim, pp 174–176.

Nagel, R. (1994) Complete life-cycle test with zebrafish – a critical assessment of results. *In:* Müller, R. and Lloyd, R. (Eds) *Sublethal and chronic effects of pollutants on freshwater fish.* FAO, Fishing New Books, Blackwell Science, Oxford, pp 188–195.

Nagel, R., Bresch, H., Caspers, N., Hansen, P.D., Markert, M., Munk, R., Scholz, N. and ter-Höfte, B.B. (1991) Effect of 3,4-dichloroaniline on the early life stages of the zebrafish (*Brachydanio rerio*): Results of a comparative laboratory study. *Ecotox. Environ. Safety.* 21:157–164.

Nelson, J.S. (1994) *Fishes of the World.* 3rd ed., Wiley and Sons, New York, 624 pp.

Norberg-King, T.J. (1989) An evaluation of the fathead minnow seven-day subchronic test for estimating chronic toxicity. *Environ. Toxicol. Chem.* 8:1075–1089.

Nüsslein-Volhard, C. (1994) Of flies and fishes. *Science* 266 (5185):572–576.

Peter, H. and Heger, W. (1997) Long-term effects of chemicals in aquatic organisms. *In:* Schüürmann, G. and Markert, B. (Eds) *Ecotoxicology.* Wiley Sons, New York, and Spektrum Akademischer Verlag, Heidelberg, pp 571–586.

Reichholf, J. (1983) Populationsdynamik. *In:* Deutsche Forschungsgemeinschaft (Ed) *Ökosystemforschung als Beurteilung der Umweltwirksamkeit von Chemikalien.* Verlag Chemie, Weinheim, pp 30–31.

Rudolph, P. and Boje, R. (1986) Ökotoxikologie – Grundlagen für die ökotoxikologische Bewertung von Umweltchemikalien nach dem Chemikaliengesetz. Ecomed, Landsberg, 106 pp.

Schäfers, C. and Nagel, R. (1991) Effect of 3,4-dichloroaniline on fish populations. Comparison of r- and K-strategists. A complete life-cycle test with guppy. *Arch. Environ. Toxicol.* 21:297–302.

Schäfers, C. and Nagel, R. (1993) Toxicity of 3,4-dichloroaniline to perch (*Perca fluviatilis*) in acute and early life stage exposures. *Chemosphere.* 26 (9):1641–1651.

Schäfers, C. and Nagel, R. (1994) Fish toxicity and population dynamics: effects of 3,4-di-chloroaniline and the problems of extrapolation. *In:* Müller, R. and Lloyd, R. (Eds) *Sublethal and chronic effects of pollutants on freshwater fish.* FAO, Fishing New Books, Blackwell Science, Oxford, pp 229–238

Schäfers, C., Oertel, D. and Nagel, R. (1993) Effects of 3,4-dichloroaniline on fish populations with differing strategies of reproduction. *In:* Braunbeck, T., Hanke, W. and Segner, H. (Eds) *Exotocicology and ecophysiology.* VCH, Weinheim, pp 133–146.

Schulte, C. and Nagel, R. (1994) Testing acute toxicity in the embryo of zebrafish, *Brachydanio rerio*, as an alternative to the acute fish test: preliminary results. *ATLA* 22:12–19.

Schulte, C., Bachmann, J., Fliedner, A., Meinelt, T. and Nagel, R. (1996) Testing acute toxicity in the embryo of zebrafish, *Brachydanio rerio* – an alternative to the fish acute toxicity test. *Proceedings, 2nd World Congress – Alternatives & animal use in the life science.* Utrecht, NL.

Solbé, J., Mark, U., Guhl, W. Hutchinson, T., Kloepper-Sams, P., Länge, R., Munk, R., Scholz, N., Bontinck, W. and Niessen, H. (1998) Analysis of the ECETOC Aquatic Toxicity (EAT) database I – general introduction. *Chemosphere* 36:99–113.

Spehar, R.L. (1976) Cadmium and zinc toxicity to flagfish, *Jordanella floridae. J. Fish. Res. Bd. Can.* 33:1939–1945.

Suter, G., Rosen, A.E., Linder, E. and Parkhurst, P.V. (1987) Endpoints for responses of fish to chronic toxic exposure. *Environ. Toxicol. Chem.* 6:793–809.

Vaal, M., van der Wal, T.J., Hermens, J. and Hoekstra, J. (1997) Pattern analysis of the variation in the sensitivity of aquatic species to toxicants. *Chemosphere* 35:1291–1309.

Woltering, D.M. (1984) The growth response in fish chronic and early life stage toxicity tests: a critical review: *Aquat. Toxicol.* 5:1–21.

Zok, S., Görge, G., Kalsch, W. and Nagel, R. (1991) Bioconcentration, metabolism and toxicity of substituted anilines in the zebrafish (*Brachydanio rerio*) *Sci. Total Environ.* 109/110: 411–421.

Fish Ecotoxicology
ed. by T. Braunbeck, D.E. Hinton and B. Streit
© 1998 Birkhäuser Verlag Basel/Switzerland

Bioaccumulation of contaminants in fish

Bruno Streit

Department of Ecology and Evolution, Biological Sciences, J. W. Goethe-University of Frankfurt, D-60054 Frankfurt, Germany

Summary. The term bioaccumulation is defined as uptake, storage, and accumulation of organic and inorganic contaminants by organisms from their environment. Bioaccumulation therefore results from complex interactions between various routes of uptake, excretion, passive release, and metabolization. For fish, the bioaccumulation process includes two routes of uptake: aqueous uptake of water-borne chemicals, and dietary uptake by ingestion of contaminated food particles. The contribution to bioaccumulation that results from aqueous exposure and is taken up by the gills is called bioconcentration. The contribution to bioaccumulation resulting from dietary exposure *via* uptake by intestinal mucosa is termed biomagnification. In both cases, important co-determinants for bioaccumulation are the various elimination mechanisms. This chapter presents a short historical survey of the problem of bioaccumulation with particular reference to fish and of the various approaches to study bioaccumulation. This is followed by an overview of our present knowledge about basic physico-chemical determinants that either increase or reduce the bioaccumulation potential of various chemicals, and about the physiological basis of gills, blood circulation and intestines, as far as they are crucial for our understanding of uptake and accumulation. Finally, selected quantitative data and modelings of bioaccumulation in fish will be discussed, with regard to such problems as the relative importance of aqueous and dietary uptake.

Introduction

When aquatic ecosystems are polluted with organic or inorganic contaminants, fish will almost inevitably be contaminated. It was most likely first recognized in the 1960s that some contaminants were taken up and retained by fish to a substantial extent and finally led to considerably higher biotic than abiotic concentrations. The study of bioaccumulation of organic as well as inorganic chemicals turned out to become relevant to the twin goals of (1) protecting fish and the other organisms interacting with them in the ecosystem, and (2) protecting human health (Phillips, 1993).

There is some confusion about the use of the terms bioaccumulation and bioconcentration, as well as bioaccumulation factor and bioconcentration factor, respectively. In aquatic studies, the term bioaccumulation is mostly used as it is throughout the present chapter, i.e., with reference to uptake, storage, and accumulation of contaminants by organisms from their environment. For fish, this includes the consideration of at least two different routes of uptake: (1) aqueous uptake of water-borne chemicals, and (2) dietary uptake by ingestion of contaminated food particles. The result of the uptake from water is termed bioconcentration; the result of the combined effects is termed bioaccumulation. In some texts, however, e.g., from terrestrial ecotoxicologists, these and other terms are used differently. Plant ecologists frequently use the term accumulation to indicate direct uptake of contaminants by the roots into the plant, which would correspond to bioconcentration in the aquatic ecotoxicologist's nomenclature. On the other hand, the ratios of contaminant levels in organisms to those in the external reference compartment are frequently termed bioconcentration factors even by aquatic ecotoxicologists in cases when bioaccumulation factor should be used, which, however, is not as common a term as the term bioconcentration factor.

One of the reasons for the inconsistent use of the terms by different specialists (and by authors of secondary literature) is probably the lack of generally accepted ecotoxicological definitions and concepts for terrestrial and aquatic systems. Another reason is the difficulty in many experimental and field studies to differentiate between dietary and aqueous contribution to bioaccumulation. In the present review, the terms will be used as defined above (and explained in more detail in the following sections), since this is in line with the most conventional use among aquatic ecotoxicologists.

The most important and harmful contaminants for aquatic systems in the past and present have been persistent organic compounds such as chlorinated hydrocarbons and alkylated metals, which both show a high bioaccumulation potential. Bioaccumulation mechanisms of elements and isotopes are frequently based on mechanisms highly different to those revealed for organic contaminants and will be treated in this chapter only marginally. Fortunately, the use and release of many of these organic or inorganic contaminants into the environment has been drastically reduced or even banned from many industrial countries, but some are still widely used in other countries, and accidents and spills will continue to threaten aquatic ecosystems and fish communities.

The aims of this review are to provide an overview of concepts and principles of contaminant bioaccumulation by fish, and to give an outline of the determinants governing uptake, storage, and elimination of mainly organic contaminants. It will give precise definitions important to understanding the theory of bioaccumulation as well as the physico-chemical and physiological bases important for an adequate appreciation of the main processes. As predictive tools, a variety of models have been developed during the last years (MacKay, 1998), some of which will be mentioned, although none of them can be treated in detail. Problems associated with the use of test systems and with modeling by implementing the inherent complexity of natural aquatic communities, which greatly exceeds that of any test system, have been widely omitted; some aspects have recently been discussed elsewhere (Streit, 1998).

Concepts and approaches to bioaccumulation studies

Early studies of bioaccumulation in fish

Data on contaminant bioaccumulation in fish have been gathered from field as well as laboratory studies. As soon as the importance and extent of natural bioaccumulation for fish and fish-eating organisms at higher trophic levels such as birds, mammals and man had become known, detailed studies on the uptake process were initiated under laboratory conditions to better understand the kinetics of uptake and elimination as well as the determinants of bioaccumulation.

Figure 1 represents the result of such an early study with juvenile whitefish in a laboratory-scale study (Gunkel and Streit, 1980) using ^{14}C-labeled atrazine, a formerly widely-used herbicide of the s-triazine group, which is still applied in agricultural use in various countries. It is moderately lipophilic with a K_{OW} of 2.64 (for explanation of the term K_{OW}, see below). Individual fish had been brought from the field into the laboratory on the day of hatching. They grew up to one year, when they reached a wet weight of 2.5–5 g and a length of 7.5–9 cm. At this size, they were used for the experiments, which were performed in round flasks at 16 °C. Experimental contaminant concentrations in the water ranged from 50 to 253 µg/L (roughly

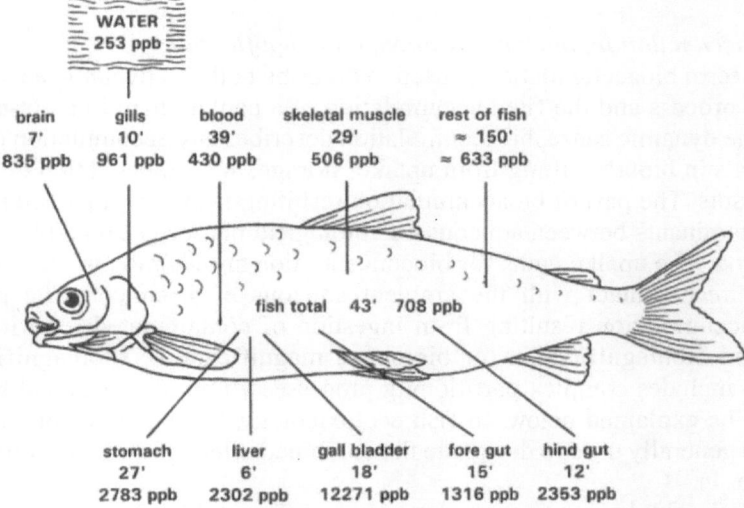

Figure 1. Uptake of water-borne atrazine by whitefish (*Coregonus fera*), showing (1) the time required for organs to reach the ambient concentration (minutes as indicated by a dash), (2) the final steady-state concentrations in various organs. Fish length ranged from 7.5–9 cm, fresh weight from 2.5–5 g. Figures indicate means for atrazine uptake (measured at 253 mg/L ambient concentration) and exchange experiments (measured at 50 mg/L ambient concentration); tissue concentrations were calculated for the 253 mg/L concentration. Adapted from Gunkel and Streit (1980).

0.2–1 µM). Uptake of atrazine was possible only from the water, since no food was provided. It took 6–150 min, until the concentration in various organs was equal to the concentration in the water (Fig. 1). The organs with highest lipid content, i.e., liver and brain, reached the equivalent to the ambient concentration fastest, even faster than the gills. The highest final concentrations, however, were measured in the gall bladder, most probably as a consequence of sequestration and biliary excretion. From additional studies with tritiated water, it was found that the rates of exchange of water and atrazine were almost proportional, and passive exchange and partitioning was the most probable determinant for the kinetics observed.

Various questions arose from this and related studies: What are the physico-chemical and physiological bases of the partitioning process? What about the concentrations in fish with different size or fat contents? How do other classes of organic as well as inorganic chemicals bioaccumulate? What are the determinants for uptake and elimination rates? What are final bioaccumulation levels under varying conditions? How would contaminated food further influence the concentrations in the fish? After the definitions and a short summary of the historical development of concepts, these questions, most of which are relatively well understood today, will be treated in the following sections.

Definitions of bioaccumulation and related terms

Bioaccumulation, bioconcentration, biomagnification

The term bioaccumulation is used to describe both the dynamic accumulation process and the final accumulation of a contaminant in an organism. In the dynamic sense, bioaccumulation describes any accumulation of chemicals in biota resulting from uptake, storage, and sequestration of contaminants. The part of bioaccumulation resulting from direct partitioning of contaminants between aqueous and biological phases is called bioconcentration. The uptake paths for bioconcentration are *via* gills or other tissues in direct contact with the ambient aqueous environment. The part of bioaccumulation resulting from ingestion of contaminated food items is called biomagnification (or biological magnification). Biomagnification also includes complex partitioning processes within the intestinal tract as will be explained below. In fish ecotoxicology, bioaccumulation is therefore generally used to designate the combined effect of these two processes (Tab. 1).

It should be mentioned that these different bioaccumulation determinants are not easily and fully separable, neither in field nor in experimental studies. There are complex equilibration processes in continuous operation between dissolved and bound contaminants anywhere at the fish/water interface and within the fish. For instance, contaminants taken up by digestion will enter the blood circulation and underlie the same

Table 1. Definitions of terms as most frequently used in fish ecotoxicology

Term	Definition
Bioaccumulation	The accumulation of a contaminant into an organism or a biological community, resulting either from direct uptake from the water (i.e., by bioconcentration) or from ingestion (i.e., by biomagnification)
Bioconcentration	The accumulation of a water-borne contaminant by a non-dietary route. In fish, the major route is via the gills. Bioconcentration results from the physico-chemical partitioning between ambient water and organism and the morphological and physiological features of the gill.
Biomagnification	The accumulation of a food-borne contaminant by oral uptake and digestion. Biomagnification of fishes is often measured or calculated as the increase in contaminant concentration in excess of bioconcentration. It results from the physico-chemical partitioning between intestinal fluids and organism and the special food uptake processes in the gut system.

equilibrium exchange with the water phase in the gills as is the case for the water-borne contaminant fraction. The same is true for the various elimination mechanisms, since in practice renal excretion may be difficult to distinguish from biliary excretion or loss *via* the gills, and contaminants taken up from either water or diet will basically behave identically, as soon as they are in the blood stream.

The complex interactions are a major reason for the inconsistent usage of the introduced terms. Differentiating between bioconcentration and biological magnification is to some extent artificial in fish experiments (Opperhuizen, 1991; Streit, 1979b, 1992), but will nevertheless be done in this review wherever applicable. In cases, when the route of contaminant uptake remains unclear or non-definable, the general term bioaccumulation will be used, as proposed by other authors as well (e.g., Connell, 1998).

The terms introduced can be applied to any kind of chemicals: (1) organic contaminants (e.g., PCB, DDT), (2) inorganic contaminants (e.g., Cd^{2+}), (3) organo-metallic contaminants (e.g., methyl-Hg), and (4) nutrients (e.g., Zn^{2+}, natural lipids). Moreover, the terms can be applied in cases where the uptake across biological membranes is the result of passive diffusion, active transport or facilitated uptake mechanisms. The more or less selective uptake and bioaccumulation of many metals is to some degree the result of a long-term evolutionary adaptation to natural concentrations. On the other hand, there is no protective anti-contaminant mechanism active against lipophilic chemicals, since most lipophilic chemicals did not occur at all in the past or at least not in concentrations as high as those encountered in present anthropogenically contaminated environments.

One potential further uptake route by fish has not yet been mentioned so far: drinking of ambient water. The osmotic values of teleost fish is intermediate between fresh and marine water. Freshwater teleosts live in a hypoosmotic environment and, thus, will usually not drink notable amounts of water, since the ambient water constantly invades the fish at quite high rates, basically across the surface of the gills. Freshwater fish thus face the problem of having to excrete constantly diluted urine to avoid swelling. For goldfish, *Carassius auratus*, a drinking rate of 1.2% wet weight per day and a urine flow of 35% per day were reported (Motais et al., 1969). In contrast, marine teleost fish will constantly lose water and therefore need to drink appreciable amounts of water. However, a special term for contaminant uptake as a result of drinking is not in use. Elasmobranchs (sharks, rays, etc.) are homoeoosmotic to their mostly marine environment, so that the problem of passive water loss or gain does not exist for these fish. In general, the total amount of contaminants taken up by drinking can be calculated to be marginal for most scenarios.

How to express concentrations and concentration factors?

Bioaccumulation and bioconcentration factors can be defined and measured under steady-state conditions, after an input-output equilibrium has been established. They will then correspond to a static bioaccumulation concept (BAF_s and BCF_s for steady-state bioaccumulation and bioconcentration factor). Likewise, both parameters can be defined and measured as a function of time, depending on actual input-output rates and the metabolic activities ($BAF(t)$, $BCF(t)$). If only the terms BAF and BCF are used, they usually stand for steady-state factors.

Bioaccumulation and bioconcentration factors in the static sense can be expressed either by designating the respective concentration terms (e.g., ng chemical per kg living mass of fish), or by relative figures expressing the concentration in the fish in relation to the environmental compartment (i.e., the ambient water or the food items). The latter is a unitless number between 0 and infinity.

Again, the use of these terms according to the definitions given has not been used very consistently in the literature, and many authors use the term bioconcentration factor for all cases. Yet, several recent reviews propagate the use of the term bioaccumulation factor as defined above (e.g., Barron, 1995), and we will adhere to this usage in the present chapter.

For expressing contaminant loads in rivers and lakes, the unit of mass in the whole ecosystem can be used (e.g., the load of a river may be 10 kg of a contaminant per day). For chemical reaction mechanisms, the preferred units in water and body fluids are generally molar units. The most widespread types of units for fish bioaccumulation studies are the mass per mass units, sometimes abbreviated by relative units, such as ppm, ppb and ppt. These units (ppm = parts per million = 10^{-6}, ppb = parts per billion = 10^{-9}, ppt = parts per trillion = 10^{-12}) should only be used, if they have been clearly defined with respect to the underlying calculation.

A correct basis for these units would be amount contaminant per kg water compared to amount contaminant per kg fresh or dry tissue, or

Table 2. Examples of acceptable units used for bioaccumulation and bioaccumulation factors; bioaccumulation factors (BAF) and bioconcentration factors (BCF) must always indicate the basis of calculation

Basis	Concentration in reference compartment	Concentration in fish (or any other aquatic organism)	Calculated BAF/BCF should indicate calculation basis
Volume basis	µg Contaminant/liter water (µg/L)	µg Contaminant/liter fish volume	Volume-based BAF/BCF
Wet weight basis	µg Contaminant/kg water (µm/kg)	µg Contaminant/kg fish wet weight	Wet weight-based BAF/BCF
Dry weight basis	µg Contaminant/kg water (µm/kg)	µg Contaminant/kg fish dry weight	Dry weight-based BAF/BCF
Lipid basis	µg Contaminant/kg water (µm/kg)	µg Contaminant/kg fish lipid mass	Lipid-based BAF/BCF

Note: For convenience, the term "kg water" is frequently measured as "liter water" and transformed by equaling 1 kg water and 1 liter water. The given examples can also be expressed as "ppb", but should then indicate the calculation basis.

amount contaminant per liter water compared to amount contaminant per "liter" of fish volume (Tab. 2). Most frequently the wet weight (correctly: wet mass) and dry weight (dry mass) bases are in use. The concentration in water is mostly expressed in mass per volume (e.g., µg chemical per liter), which is comparable to the µg chemical per kg biomass unit as used for living organisms, since specific density of tissues is close to 1.

It should be noted that published bioaccumulation and bioconcentration factors calculated on the bases of either wet weight, dry weight or lipid contents will be very different. Whereas the percentage of lipids may vary considerably (see below), the relationship between wet and dry weight is fairly constant in healthy fish, since the relative amount of water in a fish is mostly around 72% for both Osteichthyes and Chondrichthyes (Holmes and Donaldson, 1969). Therefore, bioaccumulation (and bioconcentration) factors for lipophilic contaminants calculated on the basis of dry weight, are approximately 3.6 times higher than those calculated on the basis of wet weight. Example: 1 kg of fish may contain 1 µg/kg wet weight of a contaminant; since the dry weight is only 280 g, the concentration on the basis of dry weight would be 1 µg/280 g or 3.6 µg/kg dry weight. The contaminant concentration based on the lipid fraction would be higher again by a factor defined as (total dry weight / lipid weight), which, however, is highly variable. The percentile lipid concentrations display interspecies variations from about 1% on a fresh weight basis in *Tilapia* to approximately 25% in European eel (cf. Geyer et al., 1994a, b). For comparative purposes, lipid-normalized concentrations have occasionally been calculated to reduce this variation.

Quantitative structure-activity relationships (QSAR) and lipophilicity approach

A major aim of ecotoxicology is to predict the impact of potentially harmful chemicals. However, a thorough investigation of all chemicals released into the environment, and of all possible derivatives that may arise by abiotic or biotic transformations, is not possible. It would be too time-consuming and too expensive to study these virtually hundreds of thousands of possible chemicals with respect to sublethal and lethal toxic effects in the various species of plants and animals, including aspects of bioaccumulation and the potential impact to human health.

There is, however, an approach available to predict at least to some extent certain properties on the basis of molecular features of the various chemical compounds. The different compounds of a chemical group can actually be shown to alter ecotoxicological behavior in a systematic way in relation to some chemical features or properties, which can be used as molecular descriptors. The utility of such structure-activity relationships (SAR) and property-activity relationships (PAR) to further our understanding of toxicological behavior has been demonstrated for more than one hundred years now, but their application has been extended to all aspects of the environmental behavior of organic chemicals. Frequently, the term SAR is used in the literature for PAR as well, and, as far as quantitative aspects are considered, the term QSAR (for quantitative structure-activity relationships) is used (Donkin, 1994).

With respect to bioaccumulation in fish, the most important molecular properties of organic and organo-metallic compounds are those related to solubility in aqueous and lipid solvents, since these influence uptake, internal transport and storage by fish. Other properties such as surface adsorption are also important. Adsorption is crucial, because this property allows the initial binding of molecules onto gill surfaces and then to transport proteins within the body. A great many studies on the relationship of lipophilicity and adsorption have shown that there is a positive correlation between these two properties, and that they usually both lead to increasing bioaccumulation. However, chemicals with a very strong lipophilic behavior (super-lipophilic compounds) and strong tendency to adsorption may show properties that are active against bioaccumulation for various reasons (see below).

Some chemicals have properties that favor evaporation, hydrolysis or photolysis in the aquatic environment, so that their total load in the ecosystem will successively be reduced. For instance, various low-molecular lipophilic xenobiotics (e.g., low-molecular chlorohydrocarbons) are not bioaccumulated, since they exit the aquatic system by evaporation, and long-chain alkanes or alkenes show little bioaccumulation, because they are unstable in natural water. In contrast, high stability and thus high persistence in the aquatic ecosystem including organisms has generally been observed for many chlorinated hydrocarbons. Finally, bioaccumulation varies a great deal due to highly different ecophysiological properties such as body size, longevity or trophic interrelationships with other organisms. Therefore, the use of QSARs provides only one of several approaches to estimate bioaccumulation.

Despite these limitations, various studies have shown that the mixing and separation of 1-octanol (= n-octanol) with water may provide a good solvent mix to simulate the partitioning of lipids (lipophilic compounds) between the aquatic environment and organismic tissues. Bioaccumulation of lipophilic substances in the 1-octanol phase was found to be fairly well correlated to bioconcentration factors observed in fish. The octanol-water partition coefficient, K_{OW} (sometimes also designated as P_{OW}) therefore became the most widely used molecular descriptor in quantitative structure-activity relationships for organic contaminants. The relationship between the steady state bioconcentration factor (BCF_s) and K_{OW} has been found repeatedly to follow a logarithmic regression of the following form:

$$\log BCF_s = a \times \log K_{OW} + b$$

Regression models of this kind have been published for various organisms, especially from aquatic environments. For fish, tables or figures covering between 8 and 122 individual points were presented by Banerjee and Baughman (1991), Chiou (1985), Kenaga and Goring (1980), MacKay (1982), Neely et al. (1974), Oliver and Niimi (1983), Streit (1990) as well as Veith and Kosian (1983). However, the relationship was also found valid for other constituents of the aquatic ecosystem, such as microorganisms (Baughman and Paris, 1982), molluscs (Geyer et al., 1982; Hawker and Connell, 1986; Ogata et al., 1984) and even sediments (Karickhoff et al., 1979).

It should be emphasized that published graphs of this kind sometimes include data that had not been empirically and independently elaborated, but that have been estimated from previously calculated correlations. A critical evaluation of published correlations with regard to the original data set therefore has to precede any evaluations of the accuracy of K_{OW} for bioaccumulation prediction.

There is also a relationship between water solubility and K_{OW}, so that a further linear relationship exists between water solubility and bioconcentration factors under steady state conditions is found, although with a lower predictability than the K_{OW}/bioconcentration relationship. This solubility/bioaccumulation relationship is presented in Figure 2 and illustrates the variability still observed, which must be attributed to other factors, such as steric influences in various molecules or physiological variability of the test fish specimens.

The increase in bioconcentration as a function of lipophilicity does not follow a straight line up to the highest K_{OW} values. Beyond log K_{OW} values

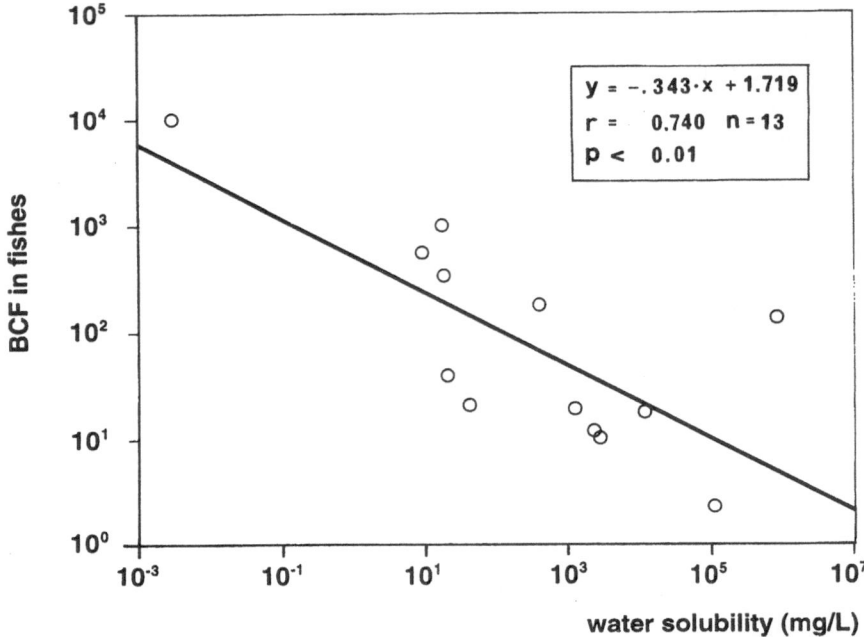

Figure 2. Correlation between water solubility of organic compounds and steady-state bioconcentration factor (BCf$_s$) in teleost fish. The correlation BCf$_s$ *vs* octanol/water partition coefficient (K$_{OW}$) is generally higher, but still shows a certain variability. Adapted from Streit (1990).

of approximately 6 (i.e., in super-lipophilic compounds), a lower increase or even a slight decrease has been observed. The exact curve and the mathematically best fit, however, are still matters of debate, as is the underlying theory explaining this phenomenon (Geyer et al., 1992, 1994b, Nendza, 1991).

Food-chain approach

In contrast to the QSAR approach, which was originally introduced by toxicologists, pharmacologists and physico-chemists, the significance of biomagnification by successive ingestion of contaminated food was propagated especially by biologists and ecologists. Food-borne contaminant uptake and bioaccumulation is possible for (almost) all aquatic animal species, but not, of course, for bacteria, algae, fungi, and higher plants. On the other hand, water-borne contaminant uptake and bioaccumulation is widely negligible for air-breathing aquatic animals such as whales, dolphins, ducks, penguins, adult aquatic beetles, etc., which do not expose any notable areas of permeable membranes to their aqueous environment.

The food-chain approach, responsible for the biomagnification process, implies the consecutive increase in concentration along a food-chain, e.g., from algae to zooplankton to juvenile fish and adult fish and finally to top carnivores like predatory fish or birds. Digestive uptake of contaminants is not fundamentally different from aqueous uptake, since partitioning occurs in a similar way as at the gill/water interface. Furthermore, the direct equilibrium exchange of the contaminants at the gill epithelium will also persist during digestive partitioning and uptake into the blood circulation. Thus, in a certain sense, dietary exposure is quantitatively rather different from aquatic exposure, and the two mechanisms are in equilibrium interactions within the blood system. In mammals, birds and many aquatic reptiles and insects, however, dietary uptake is definitely the major source of contamination. The lack of gills in these groups has very profound effects on elimination as well, since their internal contaminant concentration is not in equilibrium exchange at a permeable body/water interface. Elimination in these groups basically occurs by sequestration as well as renal, biliary or Malpighian excretion, and this is the major reason why aquatic mammals and birds have been reported in the past to suffer from very high concentrations of organochlorines and other organic chemicals. The differences in concentrations between the various "true" aquatic animals is frequently found to be not very high, and the accumulation by non-animal organisms is partly due to other mechanisms of uptake and storage (for reference, see Streit, 1992).

In any case, there is no "automatic" increase in concentrations with increasing trophic levels, as had occasionally been postulated in the past. Early protagonists of the dominant importance of biomagnification (e.g., Woodwell, 1967) neglected the frequently extensive equilibrium exchange observable in fish. The decline of the predominance of this approach for the aquatic community began in the mid 1970s, when it could be shown for lipophilic xenobiotics that they are taken up in aquatic invertebrates directly from water very rapidly, and can, at least to a large extent, be lost again quite quickly (Södergren and Svensson, 1973; Streit and Schwoerbel, 1976/77). For fish, until the 1980s several authors concluded that direct uptake from the water (*via* the gills) can account for only a few percent of the total uptake of hydrophobic chemicals such as PCBs and hexachlorobenzene, since the supply from the aqueous phase must be very limited, as observed concentrations (and frequently also the maximum possible solubility) are low (Weininger, 1978; Oliver and Niimi, 1983; Spigarelli et al., 1983; Thomann and Connolly, 1984). The relative importance of aqueous *vs.* dietary uptake, however, will always depend on the exposure scenario considered. For goldfish, bioaccumulation studies with differentially labeled DDT revealed that either direct uptake from water or uptake from food could be the predominant route, depending on prevailing exposure conditions (Barber et al., 1991).

Ecophysiological and compartmental approach

QSAR- and food-chain-based models both facilitate estimating exposure and bioaccumulation under various scenarios. They fail to provide predictions, however, for non-equilibrium conditions, and cannot be applied to all types of animals with their highly diverse physiology. Non-equilibrium is very frequent in nature, whenever ambient exposure concentrations are variable in time or when complex time-delayed uptake/elimination rates occur in the fish body. Furthermore, it has been shown for many aquatic animals that sequestration rates are small in relation to rates of uptake and depuration by diffusion. However, fish do have a multitude of metabolic capabilities for a wide range of organic compounds, which can be of importance for elimination of some chemicals. Non-equilibrium conditions can also be observed, if the time required to induce the maximum metabolic activity is shorter than the duration of bioconcentration experiments (Donkin, 1994). Specific ecophysiological and biochemical features must therefore be implemented into any detailed consideration and prediction of bioaccumulation in fish.

A fish individual can therefore be treated as a complex system with respect to uptake and release of contaminants and consists of a multitude of compartments (or subsystems) that are linked to each other in a correspondingly complex way. The resulting uptake, elimination, and allocations within the body will reveal multi-phase kinetics, although for practical purposes the fish system will be mostly assumed to represent a one- or two-compartment model. Compartmental modelings are a powerful tool to estimate bioaccumulation behavior. Many compartment models applied to fish in ecotoxicology have been derived from those used in human (or mammalian) pharmacology and toxicology and are basically pharmacokinetic models. Especially under variable environmental concentrations, as they can easily occur in running water ecosystems with a temporary load of chemicals, compartment modelings are essential. We will not go into any detail, but simply present the most basic approach for any compartmental analyses.

Compartment models are constructed on the assumption that different parts ("pools") of an ecological system (e.g., a fish and its surrounding water) contain different concentrations of a chemical, and that there are exchange rates ("fluxes") between these parts. The simplest compartment model consists of a single compartment with an input and an output flux (Fig. 3, top). The rate constant (or rate coefficient) for the input mass flow is usually designated as k_{in} or k_1; the rate constant of the output mass flow from compartment 1 is k_{out} or K_2. In more complex system structures, it is convenient to characterize each transfer rate by a double index: Biologists have frequently used the notation k_{ij} for the transfer rate from element i to element j; more commonly used in mathematical compartment theory for this case would be the notation k_{ji} (e.g., Anderson, 1983; Carson et al.,

1983; Cobelli and Goffolo, 1985; Covell et al., 1984). Whereas the first notation seems more logical at first sight, as index letters are written in the order of the material transfer, the second formally corresponds to the matrix notation (with i being the row number and j the column number) and is therefore more consistent with generalized m × n compartment models.

The second model in Figure 3 is a two-compartment model: C_1 represents the concentration in compartment 1 and may represent, e.g., the concentration in the whole body except a high-concentration compartment, where a concentration C_2 is found. The latter may represent concentrations in organs that are rich in lipids (essential for storage of lipophilic compounds) or in metal-binding proteins (such as metallothionein, essential for storage of certain metals). The third model in Figure 3, although of similar complexity, reveals a different topology. It represents a system that lacks a storage compartment, but instead has a metabolic compartment, where the

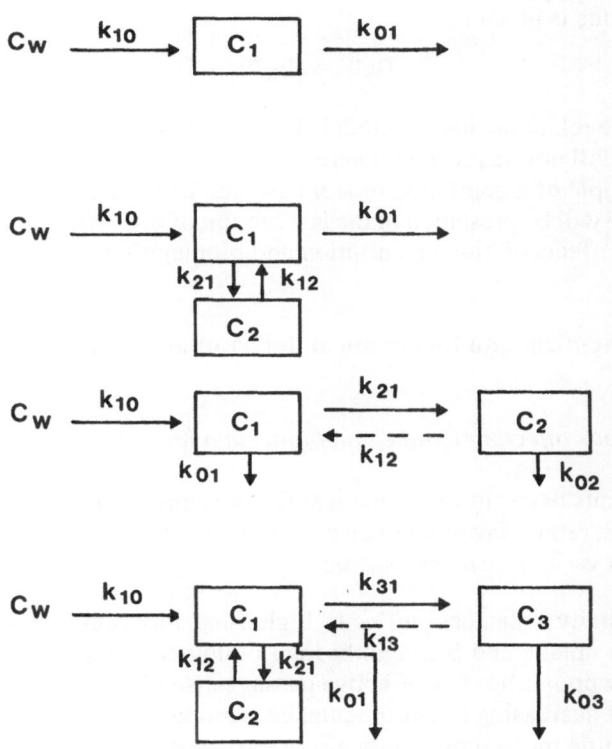

Figure 3. General compartmental concept of bioaccumulation in individuals, representing one- to three-compartmental systems, including one-way excretory metabolic pools in models III and IV; the environmental compartment in these representations is considered compartment zero. Notation of the fluxes is adapted from that used by mathematics and physicists, i.e., the first index of the constant rate (k_{ij}) refers to the destination compartment (i), the second to the source compartment (j). From Streit (1992).

chemical is transformed into an excretory product. If we assume the metabolization to be enzymatically controlled, then the flow is nearly unidirec-
tional and the backward reaction is negligible. The fourth model in Figure 3,
finally, combines the model versions II and III.

BCF_s (steady-state bioconcentration factor) values can now be defined
on the basis of the topology and the notations used in the models:

$$BCF_s = (C_1/C_w)_s$$

for the uppermost model, and

$$BCF_s = [(p_1 \times C_1) + (p_2 \times C_2)]/C_w$$

for the second to fourth model (p_i indicating volume or mass fraction of the
respective compartment sizes).

In the first (uppermost) model), the relationship between BCF_s and the
rate constants is given by

$$BCF_s = k_{10}/k_{01},$$

whereas the relationships for models II through IV are much more complicated and will not be presented here.

An example of a combined model based on compartment theory and on
physiology will be presented in the last section of this chapter to explain the
combined effects of bioconcentration and biomagnification in fish.

**Physico-chemical and biochemical determinants of bioaccumulation
in fish**

Determinants affecting uptake and bioaccumulation

Contaminants have physico-chemical features that either increase bioaccumulation or, rather, favor equal distributions between water and fish as well
as between various fish compartments:

- Lipophilicity enhances not only high final bioaccumulation, but also
  effective uptake and passage through biological membranes, which are
  basically lipophilic. The effective passage also allows an effective elimination at decreasing environmental contamination.
- Hydrophilic molecules cannot easily pass biological membranes, unless
  they are very small. Metal ions are transported by "ionic pumps" ("ionic
  channels"), filtration by hydrostatic or osmotic gradients, pinocytosis or
  phagocytosis at the gill surface or within the fish. Organometal complexes behave according to the chemical properties of the organic part of
  the molecule.

- Molecular size and shape can also affect membrane passage: Beyond a relative molecular weight of 500–1000 and/or a molecule diameter of about 1 nm, passages are strongly reduced. Also, 'steric hindrance' (Vieth et al., 1979) may limit the diffusability.
- Biochemical degradability co-determines the extent of bioaccumulation. Metabolic and excretory activities reduce bioaccumulation, but degradation rates will become noteable in those cases only where they approach the rates for uptake and diffusional elimination.
- Metabolized contaminants are usually more hydrophilic, which enhances their urinary excretion and limits bioaccumulation. If the metabolites are transported in the blood stream, they may also underlie dynamic equilibrium exchanges with the water across the gill membranes, which might also reduce bioaccumulation of the metabolites.
- Bioavailability may be affected by metal speciation, i.e., occurrence in different forms. Pure metals (as non-charged elements) are practically inexistent in natural aquatic environments. Metals mostly occur as positively charged ions, which are taken up effectively by the gills (Tjalve et al., 1988). In the water, pH and co-occurring metal ions can influence the uptake and thus the bioaccumulation potential for various heavy metal ions. For instance, high pH or high Ca^{2+} concentration will reduce Cd^{2+} uptake and bioaccumulation. A simplified scheme of some processes involved with metals during the passage from water into fish is given in Figure 4.

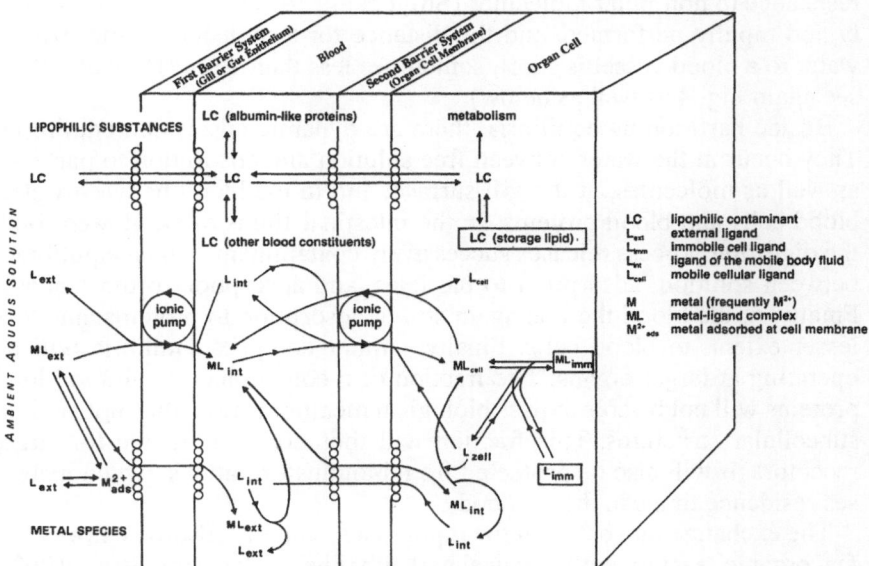

Figure 4. Diagrammatic representation of membrane passage mechanisms of lipophilic and metal contaminants. Adapted from Streit (1994).

– Organic metal complexes can form, wherever adequate organic ligands coincide with complex-forming metals such as copper. In water, ligands occur as natural compounds (e.g., humic acids) or as anthropogenic compounds (e.g., EDTA, NTA), but usually the metals will not be taken up in complex form, unless they are lipophilic (e.g., triphenyltin; Fent et al., 1991).

– Those metal ions that show strong chemical binding behavior to nitrogen or sulfur will easily bind to certain proteins or polypeptides such as metallothioneins (e.g., Cu^{2+}, Cd^{2+}). Those metals that are similar in behavior and ionic radius to calcium, may easily concentrate in bone (e.g., Sr^{2+}, Pb^{2+}). Alkali metals (Na^+, K^+, Cs^+ etc.), on the other hand, will usually not bioaccumulate. Generally, there are complex mechanisms operating in organisms, which maintain concentrations in body fluids within certain limits. The reason for a multitude of such homeostatic mechanisms is that appropriate concentrations of many metal ions are biologically essential and that, in contrast to anthropogenic lipophilic contaminants, long-time evolutionary adaptations have developed to keep internal concentrations within rather narrow limits despite natural variations in the external aquatic medium.

Determinants affecting internal transport and elimination

To enter the blood system of a fish from ambient water, a contaminant has to pass through several membranes, which exert only a limited diffusion resistance to non-polar molecules (Streit et al., 1991). The passage is easily and rapidly performed, and the distance for a molecule to move from water to a blood vessel is short, sometimes less than 1 μm (Hughes, 1984; see again Fig. 4 as well as below).

Beside partitioning equilibria, there are dynamic adsorption equilibria: They occur in the water between free solution and adsorption to particles as well as molecules, at the gill surface, and in the blood between water, blood cells and blood proteins. In the intestinal fluids of swallowed food, the digesting process releases successively contaminants which equilibrate between solution, adsorption to particles, and adsorption to the mucosa. Finally, in the blood there is again strong adsorption to proteins and, to a lesser extent, to blood cells. Finally, similar complex equilibria will be operating at target organs: The fraction of a contaminant bound to blood proteins will not be able to pass biological membranes and thus not get into subcellular structures. This fraction will thus not be available for target receptors. It will also be protected from biotransformations, which increases residence times in the fish body.

The exchange rate between the liquid phase and the adsorbate is a rather fast organic reaction with a typical half-life time of 20 ms (cf. Streit, 1992). This high rate explains the fast and effective reversible binding during both bioaccumulation and elimination processes, e.g., at the blood passage

Table 3. Relationship between log K_{OW} and mean residence times in a 750 g fish

Log K_{OW}	1.85	2.39	2.64	2.93	3.63	4.61
Mean residence time (h)	5.4	8.0	9.6	11.8	42.6	259

Log K_{OW} data have been taken empirically from literature, mean residence time (MRT) has been calculated. For details, see Streit and Siré (1993). It should be noted that perfusion of the gills and other parameters are highly variable in nature, so that model calculations of this kind can provide only tendencies.

through the gills. Mean residence times indicate the time a substance remains in the fish body and its circulation system without being exchanged with the environment. They have been studied theoretically (Covell et al., 1984) and with respect to the effects of lipophilic contaminants. Mean residence times of organic chemicals with variable K_{OW} values have been calculated by combining published experimental results with modeling studies, and reveal a correlation to log K_{OW} (Table 3; Streit and Siré, 1993).

Physiological basis of uptake and bioaccumulation in fish

Uptake from water via *gills and skin*

The gills are an organ which allows various functions: (1) respiratory gas exchange, (2) water ingress and egress by osmosis, (3) ionic and acid-base regulation, (4) excretion of nitrogen and other compounds including toxicants and heavy metals, and (5) passive uptake and elimination of natural and anthropogenic lipophilic compounds. The importance of the gills as an exchange organ has been intensively studied with respect to gas molecules by Piiper and Scheid (1984) and Piiper et al. (1986), and with respect to xenobiotic exchange by Barber et al. (1988), Bruggeman et al. (1981), Gobas and MacKay (1987), Opperhuizen et al. (1985), as well as Streit et al. (1991).

In teleost fishes, the gills consist of four arches at each side of the pharynx, which all bear a double row of gill filaments. The tips of the filaments of adjacent arches touch, so that water has to flow between adjacent filaments. Each filament bears, on its upper and lower surfaces, a row of closely spaced leaflets, the secondary lamellae. A secondary lamella can be likened to an envelope. Its two parallel faces are spaced apart and anchored by the pillar cells.

The key process at the gill surfaces is diffusion, which is important for most processes mentioned, but not, however, for the uptake and elimination of ionic chemicals. The diffusion law can be described by the relationship

$$k = D \times A \times \Delta p/d,$$

where k is the rate of diffusion, D is the diffusion constant, A is the area across which diffusion takes place, Δp is the difference in the partial pressure of the contaminant in the water, and d is the distance of diffusion.

Morphologically and physiologically, the basis of the gill function lies in (1) the large surface area, (2) countercurrent flows of water and blood, and (3) small diffusion distances between water and blood. A scheme of a secondary lamellar structure is shown in Figure 5: Blood flows between the walls of the lamellae by percolating around the pillar cells and along a marginal channel free of pillar cells, which encircles its outer edge. The pillar cells are shaped like spools and their widely spread flanges have their edges joined by desmosomes (Satchell, 1991). The two cell layers of the lamella wall constitute the water-blood barrier. They consist of the epithelial layer (which roughly makes some 90% of the total thickness), a basement membrane, and the pillar cell flange (Satchell, 1991).

The difference in partial pressure of water and blood solutes is maintained by a counter-current system. For contaminant uptake, this means that two phases of different concentrations are very close to each other, and exchange is very effective in both directions. Short distances at exchange sites are also characteristic for exchange in various tissues, such as highly vascularized tissues or storage tissues.

Morphological, physiological and biological features of the gill system vary according to the life modes of fish species (Hughes, 1966, 1972, 1984; Wootton, 1990): Fast-swimming, oceanic teleost fishes such as tunas have gills with a large surface area resulting from a high total filament length and a high number of lamellae. Such fish frequently ventilate the gills passively by ram ventilation, keeping the mouth open as they swim so that water enters the mouth, passes over the gills, and finally leaves by the opercular slits. In teleosts with moderate activity the gill arches tend to be

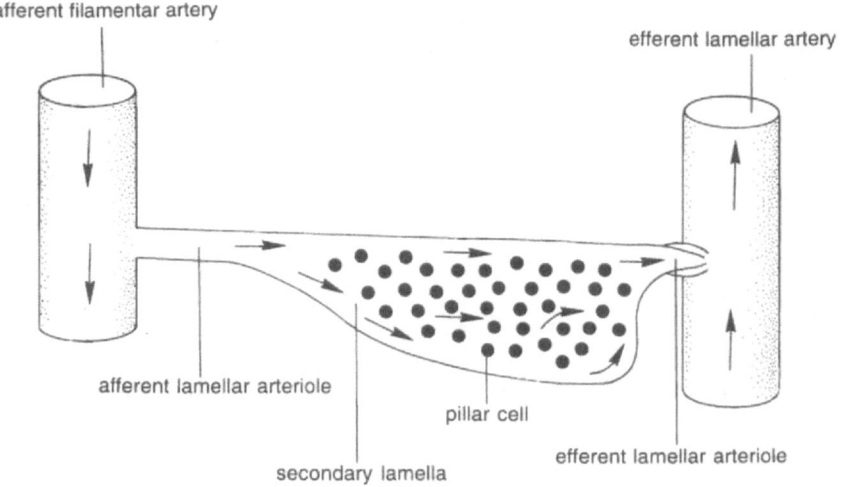

Figure 5. Secondary lamella and some main blood vessels of teleost gills. Highly simplified and partial view only. Redrawn in part after Boland and Olson (1979).

well developed with filaments of average length. Ventilation of their gills usually involves rhythmic respiratory movements, in which buccal and opercular suction pumps are equally important in maintaining a flow of water over the gills. In sluggish teleosts, the gills are more poorly developed with short filaments. Here, the opercular suction pump is often more important than the buccal pump.

Likewise, the thickness of the epithelium as a water-blood barrier varies considerably: In trout it has an overall thickness of ca. 6 μm, in fast-swimming pelagic fish only of about 0.6 μm. The diffusivity of the gill epithelium for lipophilic contaminants is generally assumed to approximate one half the diffusivity of interlamellar water or cytosol (Barber et al., 1991, based on Piiper et al., 1986 as well as Sidell and Hazel, 1987).

The gill area in a given fish species is a function of body size, following the allometric relationship

$$A = a \times W^b,$$

where A is gill area, W is body weight and a and b are parameters.

Parameter estimates have been elaborated for numerous freshwater fish species (Hughes, 1972, 1984; Murphy and Murphy, 1971; Niimi and Morgan, 1980; Oikawa and Itazawa, 1985; Price, 1931; Saunders, 1962). Typical values for b are about 0.8, which means that bigger fish of a given species have a smaller relative gill area. The value of a represents the gill area of a fish of a given size. Fast-swimming, oceanic fishes tend to have high values for this parameter. Some representative relationships for different fish species (slow and fast swimming ones, smaller and larger species) are given in Figure 6. Note that fishes which can also breath air tend to have low values for a (Wotton, 1990).

When estimating the surface area available for diffusive contaminant exchange, it must be considered that only about 60% of a fish's secondary lamellae are routinely perfused; the effective gill area of fish is generally around 36–80% of their anatomical gill area (Booth, 1978; Gehrke, 1987). Even this estimate may be high, because lamellar channels are probably not equally ventilated (Paling, 1968). For estimating and modeling the effective gill area, a value of 50% of the total area seems to be reasonable (Barber et al., 1991). From an evolutionary point of view, the gill's surface has probably not increased beyond physiological need for oxygen supply, since any further increase would rise the energetic costs for osmoregulation in both freshwater and marine fishes.

The skin probably plays an important role for uptake in very young and generally in very small fishes (up to the size of, e.g., guppies), where the entire skin area is relatively large compared to the gill surface area (Saarikosi et al., 1986). However, the skin will never have the optimal combinations of the crucial physiological determinants for effective uptake rates (high permeability, ventilation volume, counter-current flow of water and blood), which are realized in the gills (Hayton and Barron, 1990). The

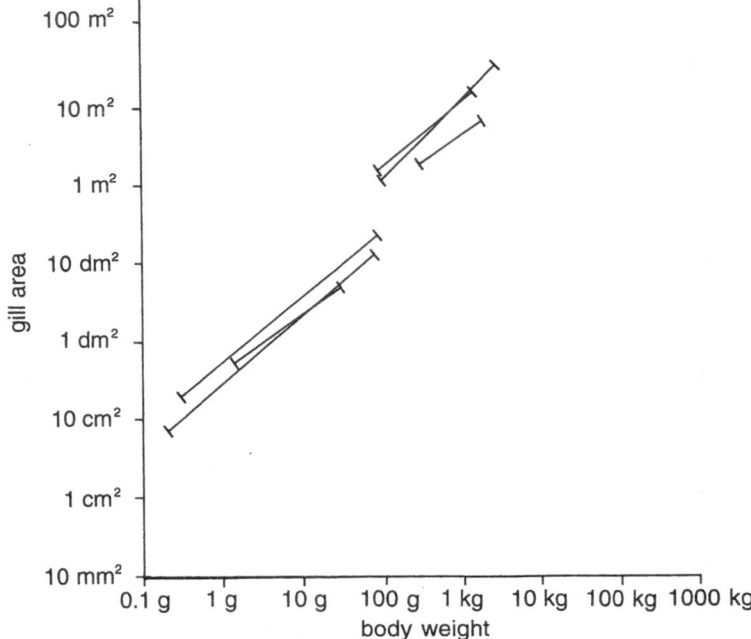

Figure 6. Relationship between gill area and body weight of various fish species (roach, toad-fish, *Micropterus*, tuna, *Coryphaena*). Adapted from Hughes (1984) and Wootton (1990).

relative contribution of the internal mouth skin for uptake and release does not seem to have been studied quantitatively yet.

Uptake by food ingestion

Fish occupy virtually every possible trophic role, although the majority are carnivores. These may be subdivided into benthivores (e.g., sticklebacks), zooplanktivores (e.g., whitefish); many are piscivores. Some feed on phytoplankton, some on higher plants, some are even detritivores and occasional scavengers. However, many fish species exhibit a great flexibility in their food preferences.

Swallowing is followed by digestion and absorption. Absorption efficiency of lipophilic contaminants increases with relative amounts of lipids within the food source: both natural lipids and lipophilic contaminants may be absorbed by passive absorption (see Fig. 4) as well as facilitated transport. The lipid-associated contaminants may be processed into lipoproteins, which are subsequently released into the blood circulation (Barron, 1995; Sire et al., 1981). The overall exchange of contaminants across the intestinal mucosa is usually assumed to be purely driven by diffusive gradients (Vetter et al., 1985).

The rate and the efficiency of contaminant absorption are basically determined by food disintegration, since desorption half-lives of contaminants from food may be longer than intestinal transit times (Schrap, 1991). Absorption efficiency of ingested and digested contaminants differs according to species and chemicals. For many lipophilic contaminants such as halogenated organic compounds, the overall efficiency was found to range between 0.5 and 0.8 (i.e., 50–80%). For other chemicals, the efficiency may differ considerably: Some chemicals are practically not absorbed (e.g., tetracycline); others are absorbed almost completely (e.g., quinoline; Dauble and Curtis, 1989; Plakas et al., 1988).

Blood circulation and contaminant storage

Circulation system
After absorption by gill or intestinal epithelia, contaminants will pass the external mucus layer and the outer epithelial membrane, then the epithelial cell, the internal epithelial membrane and the blood vessel cell, which is an endothelial cell. Contaminants will then be transferred with the blood stream to the various parts of the fish body. They are under constant sorption/desorption equilibria between various binding sites in the blood and in exchange with all vascularized tissues of the fish according to the respective local exchange equilibria.

The relative amount of blood in the circulation system of fish is lower than in mammals and accounts for roughly 3–4% of fish fresh weight in marine and freshwater Osteichthyes, with only few exceptions. In Chondrichthyes, blood volumes of 2.6–9% of fish fresh weight have been reported (Holmes and Donaldson, 1969). A lymphatic system comparable to that of terrestrial vertebrates does not exist. Overall circulation time of blood has been calculated to, e.g., about 60 seconds in rainbow trout (Davies, 1970).

Blood composition
Fish blood consists of suspended blood cells and a solution of micro- and macromolecules, which is important for the transport capacity of contaminants. Erythrocytes are nucleated and frequently measure $11-14\,\mu m$ in diameter. The amount of red cells in the blood varies greatly between species, and is indicated by the haematocrit, i.e., the volume of red cells, expressed as a percentage of a centrifuged blood sample. Some fish have almost no erythrocytes (e.g., icefish), most have haematocrit values between 10 and 25%. Fast-swimming predatory fish have around 50% (similar to humans with 47%), e.g., blue marlin, mackerel, and Atlantic salmon. The number of white cells exceeds that of man and is at several 100 000 per mm^3.

Beside cells, plasma proteins represent essential constituents of fish blood. Physiologically they are responsible for the colloid osmotic pressure of blood, functionally they can serve as transport vehicles for lipids and lipophilic chemicals, metals (e.g., iron, copper) and non-metals (e.g., iodine). Plasma proteins account for roughly 4 g/dl, i.e., less than in humans (7.2 g/dl). Elasmobranchs (sharks, rays) contain less plasma proteins than teleosts. The most important fraction of plasma proteins for chemical transport in fish blood are probably both albumin-like proteins and lipoproteins. Their major function (at least as found from mammalian studies) is to transport free fatty acids; an accessory function may lie in the transport of gases, e.g., in fish that lack erythrocytes. But they were also found to play an important role in the rapid and effective uptake of lipophilic contaminants from the water into the blood (Streit and Siré, 1993; Streit et al., 1991).

Other blood constituents are the globulins which serve fish as a defense to particular viral and bacterial pathogens (Satchell, 1991). They occur in elasmobranchs and teleosts. Glycopeptides are of a relatively low molecular weight ranging around 10000–30000. They occur especially in fish of cold environments and function as an antifreeze constituent.

Tissues of high and low vascularization and storage sites
Following uptake into the blood, hydrophobic contaminants are redistributed to high perfusion tissues such as the liver, then to low perfusion tissues including, e.g., skin and muscle, and finally to lipoidal tissues (Barron, 1990, 1995; Bickel, 1984). A diagram illustrating the various types of fish organs is given in Figure 7 (top) together with an outline of a detailed model of the gill subsystem and the protein binding of organic pollutants in fish (bottom). In the model presented the highly efficient transfer of xenobiotics from ambient water into fish could be shown to be actually based on the presence of albumin-like blood proteins (Streit and Siré, 1993). A similar model from a slightly different approach has been developed by Erickson and McKim (1990).

Elimination mechanisms

Elimination processes reduce bioaccumulation resulting from uptake and storage mechanisms. Elimination of organic contaminants includes

– active or facilitated excretion by specialized organs. These include the kidney and the biliary systems, where two mechanisms, filtration and active secretion are involved.
– biotransformation processes, which reduce the concentration of the original chemical, even if the product is not excreted. They can occur in any tissue, but are usually most effective in the liver.
– passive diffusional losses across epithelial membranes. They occur primarily across the gill epithelia, secondarily across other skin surfaces or the intestinal tract.

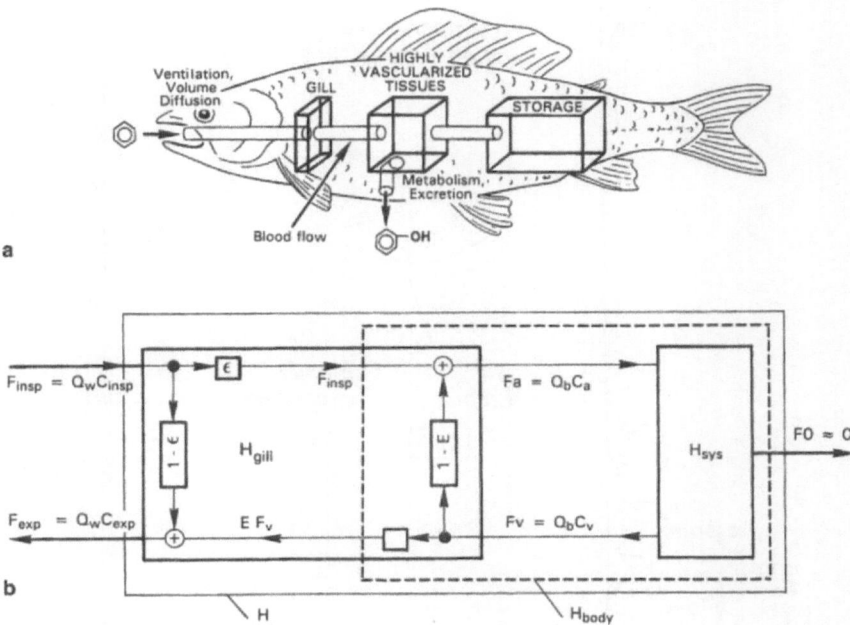

a

b

Figure 7. (a) Conceptual model of bioconcentration by fish from water, depicting the subsystems gills, highly vascularized tissues, and storage. Adapted from Barron (1990). (b) Schematic representation of the gills' mass transfer system across the epithelial diffusion barrier, including possible interactions of chemicals with albumin and blood cells in the secondary lamellar blood space. Calculations were performed based on two different types of models: a counter-current model on the one hand, assuming a space-dependent distribution of the compound along the lamellar longitudinal axis, and an arterial well-stirred model with well-mixed lamellar blood compartment on the other hand. Adapted from Streit and Siré (1993).

Generally, large nonpolar molecules and metabolites formed by hepatic biotransformation will be excreted *via* the hepatobiliary path of the liver. Factors to increase renal excretion are hydrophilicity and adequate molecular size. Filtration is most effective with small molecules, and active secretion is effective for anions and cations, secreted by the renal anion/cation transport systems (Pritchard and Bend, 1991). Factors to increase passive diffusion losses across membranes include proper lipid solubility (optimally K_{ow} between 1 and 3) and maximum gill clearance equal to cardiac output (Barron, 1995; Erickson and McKim, 1990; Hayton and Barron, 1990; Maren et al., 1968). Figure 8 may provide an idea of the relative importance of the various routes of elimination of an organic compound injected into a fish.

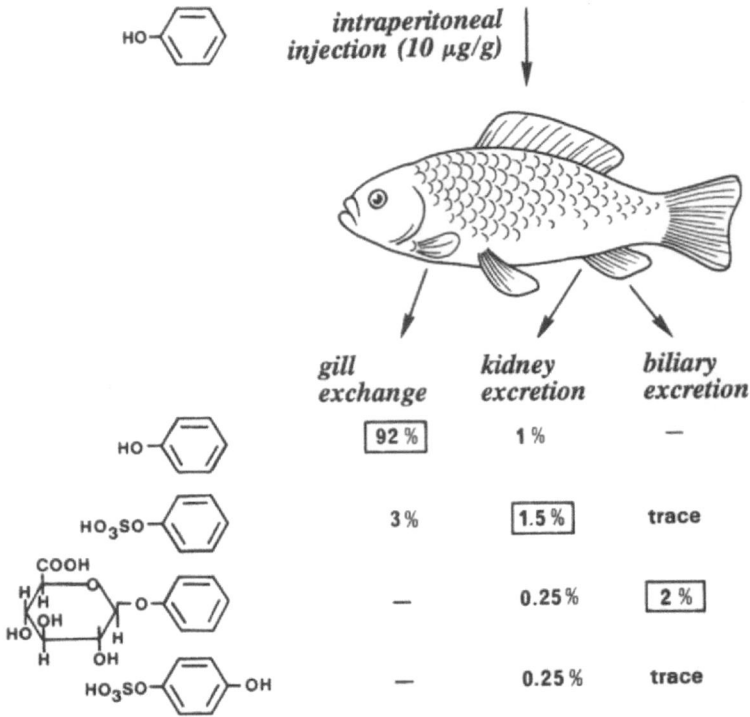

Figure 8. Metabolism and elimination balance of phenol in the goldfish, 4 h after intraperitoneal application (10 μg/g). From Streit (1992), based on data of Nagel (1981).

Bioaccumulation in fish: principles, kinetics, modeling

Metals

Concentrations of heavy metals in sediments frequently exceed those of the overlying water by a factor of 1000 to 10000 or more (Bryan and Langston, 1992; Streit, 1990). The concentrations in benthic invertebrates is often similar to the concentrations found in the sediment (Streit, 1990; Streit and Winter, 1993). In contrast to hydrophobic xenobiotics, the distribution of metals is not based on a partitioning between different phases (such as water and fat), but is the result of an equilibration between various adsorptive sites on the one hand and dissolution (either as free ions or as complexes) on the other. Negative surface charges of particles, algae and fish gills act in favor of metal adsorption.

Some metals (and isotopes, e.g., ^{90}Sr) accumulate predominantly within bony structures and underlie food chain effects to a lesser extent than others that are accumulated in soft tissues, which are more easily passed to higher trophic levels (e.g., caesium isotopes, zinc, mercury). The problem of biomagnification *vs.* bioconcentration for elements cannot be discussed in detail here. As far as lipophilic metal species are involved (alkyl metals, etc.), bioaccumulation principles similar to those outlined in the following section will apply. In cases where there is a strong homoeostatic regulation, as for copper, food-chain transfer is expected to be of less importance than in cases of less strictly regulated metals such as cadmium.

Organic compounds

Uptake and bioconcentration from water

Experimental uptake and clearance of persistent lipophilic contaminants usually reveal kinetics similar to the experimental values shown in Figure 9. Uptake and clearance of 2-*t*-butylphenol by zebrafish (*Danio rerio*) follow kinetics that nearly fit first-order kinetics. This means that the change of the concentration in the fish (C_{fish}) can be approximated according to the following simple relationship:

$$\frac{d\,C_{fish}}{dt} = k_{10} \times C_{water} - k_{01} \times C_{fish}$$

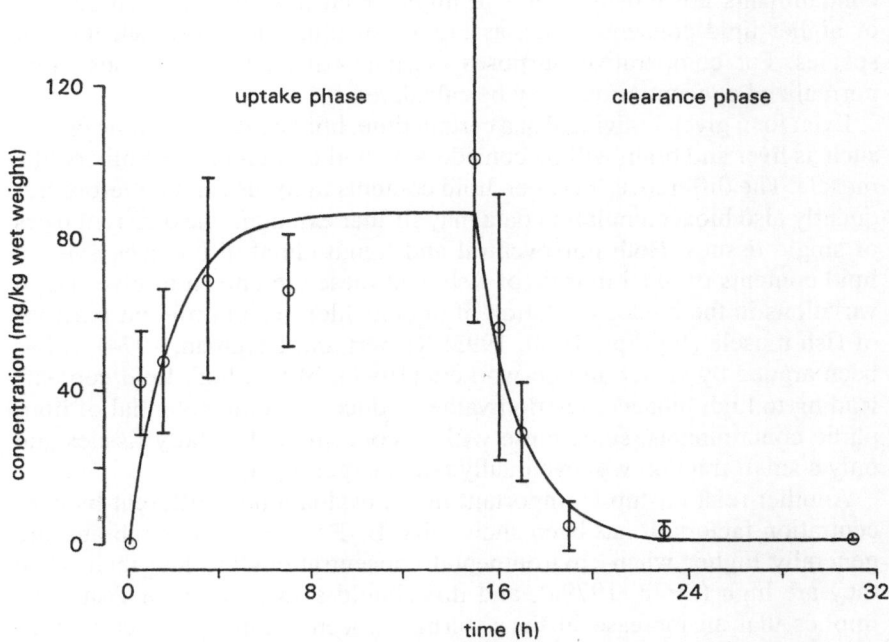

Figure 9. Rapid uptake of 2-*t*-butylphenol by zebrafish (*Danio rerio*): steady-state concentration is reached within about 5 h (wet weight basis). Adapted from Butte et al. (1988).

with C_{water} as concentration in the water (the water body is assumed to represent a very large reservoir, if compared to the fish body), and k_{10} and k_{01} as input and output flux rates, respectively. The notation follows the one used in Figure 3 (top). Integration of the equation results in:

$$C_{fish} = k_{10}/k_{01} \times C_{water} \times (1 - e^{-k_{01} \times t}).$$

One-compartment models are frequently used for the analyses of kinetic studies of uptake and elimination, although detailed studies have repeatedly revealed a two- or higher compartment system to fit data better (e.g., Nagel, 1988). For instance, a two-compartment model can be detected by a two-phase elimination kinetics. But any increase in system complexity makes calculations much more complicated, so that one-compartment approaches are frequently preferred. In short-term experiments of this type, uptake and exchange are dominated by fluxes across gill epithelia, so that food-uptake effects can be neglected.

Fish can differ considerably with respect to relative lipid contents, which affects both bioaccumulation and toxicity. Such differences in lipid contents occur within species as well as between species. In anadromous salmonid fish, appreciable differences have usually been found between migration to the spawning grounds (13–20% lipid content) and the time period after spawning (0.3–3.7% lipid; Geyer et al., 1994a, b). Lipophilic contaminants are mostly found in higher concentrations in fish species of higher lipid concentrations, as Figure 10 illustrates for a selection of species. For comparative purposes in standardized test situations, lipid normalized concentrations may be calculated.

Even for a given individual at a certain time, lipid contents in some organs such as liver and brain will be considerably higher than in some others like muscle. The differences between lipid contents in tissues and therefore frequently also bioaccumulation data may further extend to the different parts of single tissues. Both dorsoventral and longitudinal differences exist in lipid contents of axial muscle of fish, and these are known to give rise to variations in the bioaccumulation of organochlorines in different portions of fish muscle (Phillips, 1980, 1993; Reinert and Bergman, 1974). It has been argued by Geyer and co-workers (1994a, b) that high lipid contents leading to high bioaccumulation values reduce the toxic potential of lipophilic contaminants, since these will be concentrated in fatty tissues and only a small fraction will eventually reach target organs.

Another relationship is important in the evaluation of different bioconcentration factors. It has been shown that BAF values in invertebrates are generally higher when environmental concentrations are low than when they are high (Streit, 1979a), and this should also be true for fish. This implies that an increase in the external concentration by a factor of 10 would increase the equilibrium concentration in fish by a smaller factor, frequently about 8. The reason lies in the sorption isotherm characteristics

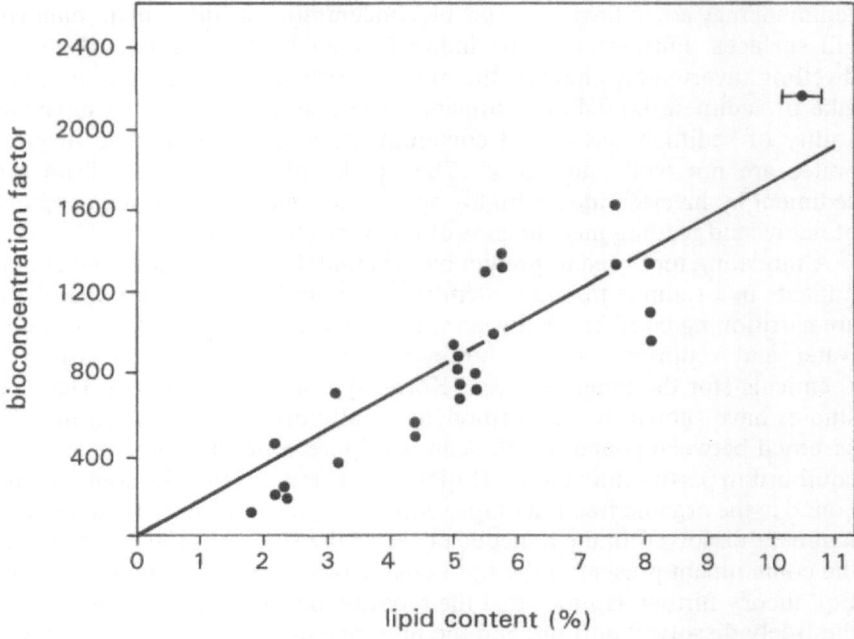

Figure 10. Correlation between lipid contents and the bioconcentration factor (on a wet weight basis) of trichlorobenzene in eight different fish species. The regression line follows the relationship: BCF = 166 × L with BCF = bioconcentration factor on a wet weight basis, and L = lipid contents in% (N = 26, r = 0.873). Adapted from Geyer et al. (1985).

(Freundlich or Langmuir adsorption/absorption isotherms), which reduce the maximum possible BAF at increasing external concentrations.

Uptake from sediments and equilibrium partitioning theory
Sediments can act as both a sink and a source of pollutants. Long-term contaminant input into aquatic ecosystems leads to sediment concentrations that can exceed the water concentration by several orders of magnitude because of partitioning of chemicals onto sediment binding sites. Sediments will thus act as a sink of pollutants, but they are also a source for pollutants, since particles are ingested by various benthic organisms. Further, sediments function as a source, whenever direct release into the water phase occurs after changes in the physico-chemical conditions at the sediment/water interface. Adsorption and absorption of contaminants to sediments depends on the relative contents of organic matter and clay per surface area. For hydrophilic compounds, cation exchange capacity and pH are important.

Bioaccumulation of sediment-associated contaminants is especially common for several invertebrates. For fish, the direct contact with sediments may be of some importance, as contaminant-rich water in or at the sediment may act in favor of direct bioconcentration of the contaminant *via* gill surfaces. Furthermore, the indirect effect by feeding upon bottom-dwelling invertebrates may be the most important source of (indirect) uptake of sediment-bound contaminants. The determinants of the bioavailability of sediment-associated contaminants to fish *via* aquatic invertebrates are not well understood. The uptake of contaminants from the sediment by invertebrates is highly species-dependent due to the diversity of habits and feeding mechanisms of the respective species.

A modeling tool used to predict bioaccumulation of hydrophobic contaminants in a sediment/biota system is based on the observation that there are partitioning equilibria between water and organism, as well as between water and sediment, which, however, vary considerably for different chemicals (for the latter see, e.g., Kornmayer and Streit, 1978). Detailed studies have shown that a thermodynamically driven partitioning may be assumed between contaminant, sediment, pore water and organisms. The equilibrium partitioning theory (EqP theory) assumes that the contaminant bound to the organic fraction of the sediment is in equilibrium with the contaminant dissolved in the aqueous phase of the sediment (pore water) and the contaminant present in the lipid phases of the exposed organisms. The EqP theory further assumes that the bioavailable concentration in water is the freely dissolved portion, and the presence of dissolved organic matter (DOM) is usually neglected. The model allows to accurately predict organism concentrations from contaminant concentrations freely dissolved in pore water or the organic carbon-normalized sediment concentrations. The equilibrium level accumulated by benthic organisms is assumed to be independent of the exposure route, i.e., by ingestion of sediment particles or exposure to pore water (Lake, 1990; Bierman, 1990; diToro et al., 1991), which is an assumption for this group. Finally, particle size effects are assumed to be minimal. The sediment-to-sediment variation in bioavailability could be reduced to a factor of 2 to 3 by application of the EqP theory (diToro et al., 1991).

Biomagnification and size effects

That part of bioaccumulation in fish that is observed in excess of the concentration resulting from aqueous exposure can be assumed to result from dietary uptake. The evidence for biomagnification is therefore mostly indirect by comparing concentrations in fish resulting from pure bioconcentration and from simultaneous bioconcentration and biomagnification processes. Direct intake of contaminated food particles in an elsewhere uncontaminated aquatic environment is impracticable and unrealistic: The contaminants at and in the particles would rapidly equilibrate with the aqueous phase, as would the contaminants taken up orally by fish as soon as they passed the gills. Thus, there will always result an overall equilibration between bioconcentration, bioaccumulation and the various elimination mechanisms. This is the main reason why biomagnification in fish has been estimated basically from model calculations (Barber et al., 1991; Thomann, 1989).

Evidence for a significant biomagnification in aquatic-based food webs has generally been restricted to contaminants that are resistant to biotransformation, halogenated, and very lipophilic, with log K_{OW} values of 5–8 (Suedel et al., 1994; Thomann, 1989). Food chain biomagnification will not really be significant for chemicals with a log K_{OW} less than 5 (Thomann et al., 1992). With regard to fish in natural environments, the biomagnification route of uptake may be important especially for benthic-based food webs and highly lipophilic contaminants.

Figure 11 represents the results of complex bioaccumulation modelings for organic contaminants (from Cowan et al., 1995). The figures are based on a compartment- and physiology-based model, the 'Food and Gill Exchange of Toxic Substances' model (FGETS; Barber et al., 1991), used by the U.S. Environmental Protection Agency. Figure 11 illustrates the contribution of food to total exposure for two fish types, a planktivorous fish

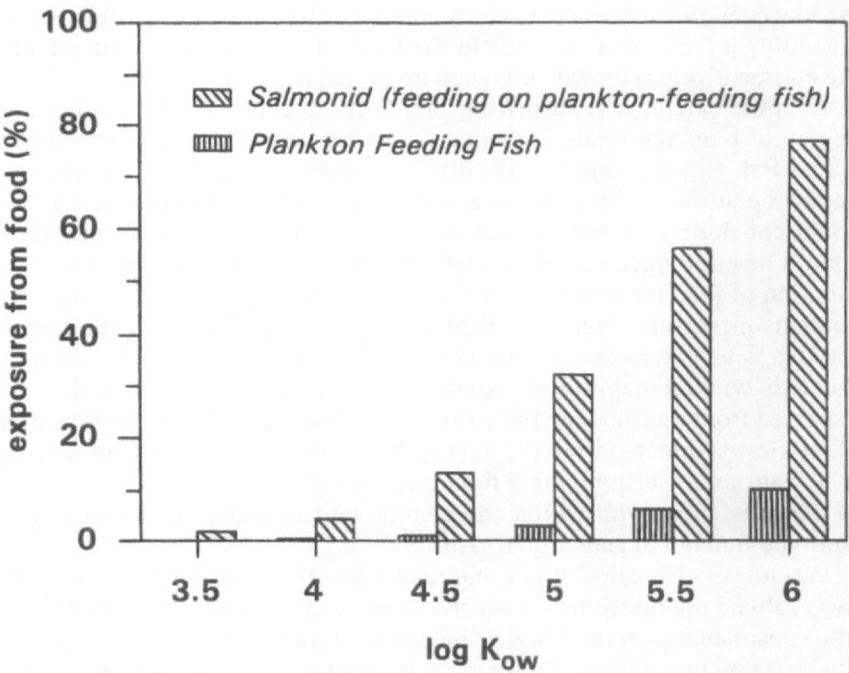

Figure 11. Contribution of food to total exposure for two freshwater fish – a plankton feeder and a salmonid feeding itself on plankton feeder for substances with log K_{OW} values from 3.5 to 6. Values reflect model simulation with alewife (*Alosa pseudoharengus*) feeding on plankton (basically *Mysis*) and salmonids (salmon and various trout species) feeding on alewife. Adapted from Cowan et al. (1995), based on the FGETS model set up by Barber et al. (1991) for the Lake Ontario situation.

(alewife, *Alosa pseudoharengus*) and salmonids feeding on alewife. The salmonids represent any of four species co-occurring in Lake Ontario: coho salmon as well as rainbow, brown and lake trout. They all behave similarly with respect to their feeding on alewife, so that they were treated as a single guild or compartment. Bioaccumulation in alewife was modeled assuming feeding of zooplankton species (*Mysis*, a planktonic crustacean). This latter was assumed to be thermodynamically equilibrated with the lake water.

Substances modeled for bioaccumulation exhibited log K_{OW} values from 3.5 to 6. It can be seen that for a low-molecular-weight substance with a log K_{OW} of 5.5 the diet would account for only 6% of the total exposure for a small plankton feeder and the final BAF would be 29 000. However, for a predator consuming the plankton feeder, diet would account for 56% of the total exposure with a final BAF of 48 000. These results indicate that the diet can be a significant source of exposure if the substance has low water solubility, high lipid solubility, and is slowly metabolized or eliminated by the prey organism (Cowan et al., 1995). The reason for this significant contribution of food-borne contaminants to the overall bioaccumulation lies in the low contaminant concentrations per unit volume of water, which is surpassed by a factor of thousands in the food from which the contaminants are consecutively released and taken up during digestion.

To some extent, it is surprising that FGETS does actually predict any degree of biomagnification, as one might argue that chemical exchanges across fish gills and intestine are driven by chemical gradients (thermodynamical equilibria). Since these are also incorporated into the model, one could conclude that any surplus of contaminant uptake in the intestines should be eliminated *via* blood and gills. The solution is that during assimilation of food the concentration of chemicals remaining in the intestinal contents increases. Thus a net diffusive flux into the fish is maintained. If the fish is unable to excrete this chemical across the gills rapidly enough, the fish will biomagnify its whole-body residues above those that are expected from thermodynamic partitioning (Barber et al., 1991). Since the blood flow is directed from the gut capillaries to the various organs, among which are strong accumulating tissues such as the liver, and only afterwards to the gills, biomagnification can also be understood as resulting in part from the functional anatomy of fish.

Various model calculations based on kinetic considerations for surface/volume relationships in fish and other aquatic animals of variable size have postulated generally higher bioconcentration values in larger individuals, if compared to smaller ones under various environmental scenarios. An increase in lipid content also favors bioconcentration and biomagnification. In aquatic communities, these two relationships are frequently interconnected with food-chain position much more than in terrestrial communities, because (1) predators are mostly larger than their prey items, and (2) large aquatic organisms frequently exhibit higher lipid concentrations than smaller ones (for example: seals with higher concentrations than their fish diet, fish frequently higher than their plankton diet, etc.). An increase in size-specific bioaccumulation figures can therefore also arise independently of biomagnification and simply result from allometric or lipid relationships (Streit, 1992).

Sophisticated models such as that presented by Barber et al. (1991) do not contradict this argumentation. In their model, larger fish also showed higher concentrations, as demonstrated with PCB congeners and calculations that included both the uptake route *via* gills and the one *via* intestines. In the approach by Barber and co-workers (1991), growth was simulated by using an allometric feeding model. Even for high K_{OW} chemicals, where uptake of food-borne contaminants was dominant in the model calculations and a negative net flux of contaminants across the gills was calculated, increasing fish sizes were parallel to bioaccumulation figures for any model calculations and any scenario presented by the authors. Thus, allometric effects of both water-borne and food-borne contaminants seem to be important determinants for bioaccumulation values, and the relative effects of growth and allometry as well as of the various uptake and release routes remain to be studied and modeled in more detail for a variety of aquatic communities and realistic scenarios.

References

Anderson, D.H. (1983) Compartmental modeling and tracer kinetics. *In: Lecture notes in bio-mathematics*, Vol. 50. Springer, Berlin, 302 pp.

Banerjee, S., G.L. and Baughman (1991) Bioconcentration factors and lipid solubility. *Environ. Sci. Technol.* 25:536–539.

Barber, M.C., Suárez, L.A. and Lassiter, R.R. (1988) Modeling bioconcentration of nonpolar organic pollutants in fish. *Environ. Toxicol. Chem.* 7:545–558.

Barber, M.C., Suárez, L.A. and Lassiter, R.R. (1991) Modeling bioaccumulation of organic pollutants in fish with an application to PCBs in Lake Ontario salmonids. *Can. J. Fish. Aquat. Sci.* 48:318–337.

Barron, M.G. (1990) Bioconcentration. *Environ. Sci. Technol.* 24:1612–ff.

Barron, M.G. (1995) Bioaccumulation and bioconcentration in aquatic organisms. In: Hoffman, D.J., Rattner, B.A., Burton, G.A. and Cairns, J. (eds.) *Handbook of ecotoxicology*. Lewis Publ., Boca Raton, pp. 652–666.

Baughman, G.L. and Paris, D.F. (1982) Microbial bioconcentration of organic pollutants from aquatic systems – a critical review. *CRC Crit. Rev. Microbiol.* 205–227.

Bierman, V.J. (1990) Equilibrium parititioning and biomagnification of organic chemicals in benthic animals. *Environ. Sci. Technol.* 24:1407–1412.

Boland, E.J. and Olson, K.R. (1979) Vascular organization of the catfish gill filament. *Cell Tissue Res.* 198: 487–500.

Booth, J.H. (1978) The distribution of blood flow in the gills of fish: application of a new technique to rainbow trout (*Salmo gairdneri*). *J. Exp. Biol.* 73:119–129.

Bruggeman, W.A., Martron, L.B.J.M., Kooiman, D. and Hutzinger, O. (1981) Accumulation and elimination kinetics of di-, tri- and tetra-chlorobiphenyls by goldfish after dietary and aqueous exposure. *Chemosphere* 10:811–832.

Bryan, G.W. and Langston, W.J. (1992) Bioavailability, accumulation and effects of heavy metals in sediments with special reference to United Kingdom estuaries: A review. *Environ. Pollut.* 76:89–131.

Butte, W., Paul, C., Willig, A. and Zauke, G.-P. (1988) *Beziehungen zwischen der Struktur von Phenolen und ihrer Akkumulation gemessen im Flow-through Fisch-Test (OECD No. 305E)*. Forschungsbericht im Auftrag des Umweltbundesamtes (FKZ 106 02 053), Oldenburg.

Carson, E.R., Cobelli, C. and Finkelstein, L. (1983) *The mathematical modeling of metabolic and endocrine systems*. John Wiley and Sons, New York, 394 pp.

Chiou, C.T. (1985) Partition coefficients of organic compounds in lipid-water systems and correlations with fish bioconcentration factors. *Environ. Sci. Technol.* 19:57–62.

Cobelli, C. and Goffolo, G. (1985) Compartmental and noncompartmental models as candidate classes for kinetic modeling, theory and computational aspects. *In:* Eisenfeld J. and DeLisi, C. (eds.) *Mathematics and computers in biomedical applications*. Elsevier Science Publishers B.V. (North-Holland).

Connell, D.W. (1998) Bioaccumulation of chemicals by aquatic organisms. *In:* Schüürmann, G. and Markert, B. (eds.) *Ecotoxicology*. John Wiley and Sons Inc., New York, and Spektrum Akademischer Verlag, Heidelberg, p. 439–450.

Covell, D.G., Berman, M. and Charles, D. (1984) Mean residence time – theoretical development, experimental determination, and practical use in tracer analysis. *Math. Biosci.* 72: 213–244.

Cowan, C.E., Versteeg, D.J., Larson, R.J. and Kloepper-Sams-P.J. (1995) Integrated approach for environmental assessment of new and existing substances. *Reg. Toxicol. Pharmacol.* 21: 3–31.

Dauble, D.D. and Curtis, L.R. (1989) Rapid branchial excretion of dietary quinoline by rainbow trout *(Salmo gairdneri)*. *Can. J. Fish. Aquat.* Sci. 46:705–713.

Davies, J.C. (1970) Estimation of circulation time in rainbow trout, *Salmo gairdneri*. *J. Fish. Res. Bd. Can.* 27:1860–1863.

diToro, D.M., Zarba, C.S., Hansen, D.J., Berry, W.J., Swartz, R.C., Cowan, C.E., Pavlou, S.P., Allen, H.E., Thomas, N.A. and Paquin, P.R. (1991) Technical basis for establishing sediment quality criteria for nonionic organic chemicals using equilibrium partitioning. *Environ. Toxicol. Chem.* 10:1541–ff.

Donkin, P. (1994) Quantitative structure-activity relationships. In: Calow, P. (ed.) *Handbook of ecotoxicology*. Vol. 2, Blackwell, Oxford, p. 321–347.

Erickson, R.J. and McKim, J.M. (1990) A model for exchange of organic chemicals at fish gills: flow and diffusion limitations. *Aquat. Toxicol.* 18:175–198.

Fent, K., Lovas, R. and Hunn, J. (1991) Bioaccumulation, elimination and metabolism of triphenyltin chloride by early life stages of minnows *Phoxinus phoxinus*. *Naturwissenschaften* 78:125–127.

Gehrke, P.C. (1987) Cardio-respiratory morphometrics of the spangled perch *Leiopotherapon unicolor* (Gunter, 1859) (Percoidei, Teraponidae). *J. Fish. Biol.* 31:617–623.

Geyer, H., Sheehan, D., Kotzias, D., Freitag, D. and Korte, F. (1982) Prediction of ecotoxicological behaviour of chemicals: relationship between physicochemical properties and bioaccumulation of organic chemicals in the mussel. *Chemosphere* 11:1121–1134.

Geyer, H., Scheunert, I. and Korte, F. (1985) Relationship between the lipid content of fish and their bioconcentation potential of 1,2,4-trichlorobenzene. *Chemosphere* 14:545–555.

Geyer, H.J., Muir, D.C.G., Scheunert, I., Steinberg, C.E.W. and Kettrup, A.A.W. (1992) Bioconcentration of octachlorodibenzo-*p*-dioxin (OCDD) in fish. *Chemosphere* 25:1257–1264.

Geyer, H., Muir, D.C.G., Scheunert, I., Steinberg, C.E.W. and Kettrup, A.W. (1994a) Bioconcentation of superlipophilic persistent chemicals – Octachlorodibenzo-*p*-dioxin (OCDD) in fish. *Environ. Sci. Pollut. Res.* 1:75–80.

Geyer, H.J., Scheunert, I., Brüggemann, R., Matthies, M., Steinberg, C.E.W., Zitko, V., Kettrup, A. and Garrison, W. (1994b) The relevance of aquatic organisms´ lipid content to the toxicity of lipophilic chemicals: Toxicity of lindane to different fish species. *Ecotox. Environ. Safety* 28:53–70.

Gobas, F.A.P.C. and Mackay, D. (1987) Dynamics of hydrophobic organic chemical bioconcentration in fish. *Environ. Toxicol. Chem.* 6:495–504.

Gunkel, G. and Streit, B., (1980) Mechanisms of bioaccumulation of a herbicide (atrazine, s-triazine) in a freshwater mollusc *(Ancylus fluviatilis* Müll.) and a fish *(Coregonus fera* Jurine). *Water Res.* 14:1574–1584.

Hawker, D.W. and Connell, D.W. (1986) Bioconcentration of lipophilic compounds by some aquatic organisms. *Ecotox. Environ. Safety* 11:184–197.

Hayton, W.L. and Barron, M.G. (1990) Rate limiting barriers to xenobiotic uptake by the gill. *Environ. Toxicol. Chem.* 9:151–ff.

Holmes, W.N. and Donaldson, E.M. (1969) The body compartments and the distribution of electrolytes. *In:* Hoar, W.S. and Randall D.J. (eds.) *Fish physiology.* Vol. I. Academic Press, London, p. 1–89.

Hughes, G.M. (1966) The dimensions of fish gills in relation to their function. *J. Exp. Biol.* 45: 177–195.

Hughes, G.M (1972) Morphometrics of fish gills. *Respir. Physiol.* 14:1–25.

Hughes, G.M. (1984) General anatomy of the gills. *In:* Hoar, W.S. and Randall D.J. (eds.) *Fish physiology.* Vol. X Part A. Academic Press, London, p. 1–72.

Karickhoff, S.W., Brown, D.S. and Scott, T.A. (1979) Sorption of hydrophobic pollutants on natural sediments and soil. *Water Res.* 13:241–248.

Kenaga, E.E. and Goring, C.A. (1980) Relationship between water solubility, soil sorption, octanol-water partitioning and bioconcentration of chemicals in biota. In: Eaton J.G. et al. (eds.) *Aquatic toxicology,* Vol. 7. Amer. Soc. Test. Mat. STM, Philadelphia.

Kornmayer, R. and Streit, B., (1978) Adsorption und Anreicherung von Atrazin und seinen Abbauprodukten an Flußwassersediment. *Arch. Hydrobiol. Suppl.* 55:186–210.

Lake, J.L. (1990) Equilibrium partitioning and bioaccumulation of sediment-associated contaminants by infaunal organisms. *Environ. Toxicol. Chem.* 9:1095–1106.

MacKay, D. (1982) Correlation of bioconcentration factors. Environ. Sci. Technol. 16:274–278.

MacKay, D. (1991) *Multimedia environmental models: the fugacity approach.* Lewis, Chelsea, 257 pp.

Mackay, D. (1998) Multimedia mass balance models of chemical distribution and fate. *In:* Schüürmann, G. and Markert, B. (eds.) *Ecotoxicology.* John Wiley and Sons Inc., New York, and Spektrum Akademischer Verlag, Heidelberg, p. 237–257.

Maren, T.H., Embry, R., Broder and L.E. (1968) The excretion of drugs across the gill of the dogfish, *Squalus acanthias. Comp. Biochem. Physiol.* 26:853–864.

Motais, R., Isaia, J., Rankin, J.C. and Maetz, J. (1969) Adaptive changes of the water permeability of the teleostean gill epithelium in relation to external salinity. *J. Exp. Biol.* 51:529–546.

Murphy, P.G. and Murphy, J.V. (1971) Correlations between respiration and direct uptake of DDT in the mosquito fish *Gambusia affinis. Bull. Environ. Contam. Toxicol.* 6:581–588.

Nagel, R. (1988) *Umweltchemikalien und Fische – Beiträge zu einer Bewertung.* Habilitationsschrift, Univ. Mainz (FRG), 256 pp.

Neely, W.B., Branson, D.R. and Blau, G.E. (1974) Partition coefficients to measure bioconcentration potential of organic chemicals in fish. *Environ. Sci. Technol.* 8:1113–1115.

Nendza, M. (1991) QSARs of bioconcentration: Validity assessment of log P_{ow}/log BCF correlations. *In:* Nagel, R. and Loskill, R. (eds.) *Bioaccumulation in aquatic systems. Contributions to the assessment.* VCH, Weinheim, p. 43–66.

Niimi, A.J. and Morgan, S.L. (1980) Morphometric examination of the gills of walleye, *Stizostedion vitreum vitreum* (Mitchell) and rainbow trout, *Salmo gairdneri* (Richardson). *J. Fish Biol.* 16:685–692.

Ogata, M., Fujisawa, K., Ogino, Y. and Mano, E. (1984) Partition coefficients as a measure of bioconcentration potential of crude oil compounds in fish and shellfish. *Bull. Environ. Contom. Toxicol.* 33:561–567.

Oikawa, S. and Itazawa, Y. (1985) Gill and body surface areas of the carp in relation to body mass, with special reference to the metabolic-size relationship. *J. Exp. Biol.* 117:1–14.

Oliver, B.G. and Niimi, A. (1983) Bioconcentration of chlorobenzenes from water to rainbow trout: correlation with partition coefficients and environmental residues. *Environ. Sci. Technol.* 17:287–291.

Opperhuizen, A. (1991) Bioconcentration and biomagnification: is a distinction necessary? *In:* Nagel, R. and Loskill, R. (eds.) *Bioaccumulation in aquatic systems. Contributions to the assessment.* VCH, Weinheim, p. 67–80.

Opperhuizen, A., Velde, E.W. van den, Gobas, F.A.P.C., Liem, D.A.K. and Steen, J.M.D. van den (1985) Relationships between bioconcentration in fish and steric factors of hydrophobic chemicals. *Chemosphere* 14:1871–1896.

Paling, J.E. (1968) A method of estimating the relative volumes of water flowing over the different gills of a freshwater fish. *J. Exp. Biol.* 48:533–544.

Phillips, D.J.H. (1980) *Quantitative aquatic biological indicators – their use to monitor trace metal and organochlorine pollution.* Applied Science Publishers, London, 488 pp.

Phillips, D.J.H. (1993) Bioaccumulation. *In:* Calow, P. (ed.) *Handbook of ecotoxicology.* Vol. 1, Blackwell, Oxford, p. 378–396.

Piiper, J. and Scheid, P. (1984) Model analysis of gas transfer in fish gills. *In:* Hoar, W.S. and Randall D.J. (eds.) *Fish physiology.* Vol. X, Part A. Academic Press, London, p. 229–262.

Piiper, J., Scheid, P., Perry, S.F. and Hughes, G.M. (1986) Effective and morphological oxygen-diffusing capacity of the gills of the elasmobranch *Scyliorhinus stellaris. J. Exp. Biol.* 123:27–41.

Plakas, S.M., McPhearson, R.M. and Guarino, A.M. (1988) Disposition and bioavailability of H^3-tetracycline in the channel catfish (*Ictalurus punctatus*). *Xenobiotica* 18:83–93.

Price, J.W. (1931) Growth and gill development in the small mouthed black bass, *Micropterus dolomieui* Lacepede. *Stud. Ohio State Univ.* 4:1–46.

Pritchard, J.B. and Bend, J.R. (1991) Relative roles of metabolism and renal excretory mechanisms in xenobiotic elimination by fish. *Environ. Health Perspect.* 90:85–92.

Reinert, R.E. and Bergman, H.L. (1974) Residues of DDT in lake trout (*Salvelinus namaycush*) and coho salmon (*Oncorhynchus kisutch*) from the Great Lakes. *J. Fish Res. Bd. Can.* 31:191–199.

Saarikosi, J., Lindstrom, R., Tyynela, M. and Viluksela, M. (1986) Factors affecting the absorption of phenolics and carboxylic acids in the guppy *(Poecilia reticulata). Ecotox. Environ. Safety* 11:158–173.

Satchell, G.H. (1991) *Physiology and form of fish circulation.* Cambridge University Press, Cambridge, 235 pp.

Saunders, R.L. (1962) The irrigation of the gills in fishes. II. Efficiency of oxgen uptake in relation to respiratory flow acitvity and concentrations of oxygen and carbon dioxide. *Can. J. Zool.* 40:817–862.

Schrap, S.M. (1991) Bioavailability of organic chemicals in the aquatic environment. *Comp. Biochem. Physiol.* 100C:13–16.

Sidell, B.D. and Hazel, J.R. (1987) Temperature affects the diffusion of small molecules through cytosol of fish muscle. *J. Exp. Biol.* 129:191–203.

Sire, M.F. Lutton, C. and Vernier, J.M. (1981) New views on intestinal absorption of lipids in teleostean fishes: An ultrastructural and biochemical study in the rainbow trout. *J. Lipid Res.* 22:81–94.

Södergren, A. and Svensson, B. (1973) Uptake and accumulation of DDT and PCB by *Ephemera danica* (Ephemeroptera) in continuous-flow systems. *Bull. Environ. Contam. Toxicol.* 9:345–350.

Spigarelli, S.A., Thommes, M.M., and Prepejchal, W. (1983) Thermal and metabolic factors affecting PCB uptake by adult brown trout. *Environ. Sci. Technol.* 17:88–94.

Streit, B. (1979a) Uptake, accumulation and release of organic pesticides by benthic invertebrates. 2. Reversible accumulation of lindane, paraquat and 2,4-D from aqueous solution by invertebrates and detritus. *Arch. Hydrobiol. Suppl.* 55:324–348.

Streit, B. (1979b) Uptake, accumulation and release of organic pesticides by benthic invertebrates. 3. Distribution of ^{14}C-atrazine and ^{14}C-lindane in an experimental 3-step food chain microcosm. *Arch. Hydrobiol. Suppl.* 55:374–400.

Streit, B. (1990) Chemikalien im Wasser: Experimente und Modelle zur Bioakkumulation bei Süßwassertieren. *In:* Kinzelbach, R. and Friedrich, G. (eds.) *Limnologie aktuell*, Band 1: Biologie des Rheins. G. Fischer, Stuttgart, p. 107–130.

Streit, B. (1992) Bioaccumulation processes in ecosystems. Review. *Experientia* 48:955–970.

Streit, B. (1994) *Lexikon Ökotoxikologie.* 2. Aufl., VCH, Weinheim, 899 pp.

Streit, B. (1998) Community ecology and population interactions in freshwater systems. *In:* Schüürmann, G. and Markert, B. (eds.) *Ecotoxicology.* John Wiley and Sons Inc., New York, and Spektrum Akademischer Verlag, Heidelberg, p. 133–161.

Streit, B. and Schwoerbel, J. (1976/77) Experimentelle Untersuchungen über die Akkumulation von Herbiziden bei benthischen Süsswassertieren. *Verh. Ges. Ökol.* 1976:371–383.

Streit, B. and Siré, E.-O. (1993) On the role of blood proteins for uptake, distribution, and clearance of waterborne lipophilic xenobiotics by fish: A linear system analysis. *Chemosphere* 26:1031–1039.

Streit, B. and Winter, S. (1993) Cadmium uptake and compartmental time characteristics in the freshwater mussel *Anodonta anatina. Chemosphere* 26:1479–1490.

Streit, B., Siré, E.-O., Kohlmaier, G.H., Badeck, F.W. and Winter, S. (1991) Modeling ventilation efficiency of teleost fish gills for pollutants with high affinity to plasma proteins. *Ecol. Model.* 57:237–262.

Suedel, B.C., Boraczek, J.A., Peddicord, R.K., Clifford, P.A. and Dillon, T.M. (1994) Trophic transfer and biomagnification potential of contaminants in aquatic ecosystems. *Rev. Environ. Contam. Toxicol.* 136:21–89.

Thomann, R.V. (1989) Bioaccumulation model of organic chemical distribution in aquatic food chains. *Environ. Sci. Technol.* 23:699–707.

Thomann, R.V. and Connolly, J.P. (1984) Model of PCB in the Lake Michigan lake trout food chain. *Environ. Sci. Technol.* 18:65–71.

Thomann, R.V., Connolly, J.P. and Parkerton, T.F. (1992) An equilibrium model of organic chemical accumulation in aquatic food webs with sediment interaction. *Environ. Toxicol. Chem.* 11:615–629.

Tjalve, H., Gottofrey, J. and Borg, K. (1988) Bioaccumulation, distribution and retention of Ni in the brown trout *(Salmo trutta)*. *Water Res.* 22:1129–1136.

Vetter, R.D., Carey, M.C., Patton and J.S. (1985) Coassimilation of dietary fat and benzo[a]pyrene in the small intestine: an absorption model using the killifish. *J. Lipid Res.* 26:428–434.

Veith, G.D. and Kosian, P. (1983) Estimating bioconcentration potential from octanol/water partition coefficients. *In:* MacKay, D. et al. (eds.) Physical behavior of PCBs in the Great Lakes. Ann Arbor Science Publishers, Ann Arbor.

Veith, G.D., DeFoe, D.L. and Bergstedt, B.V. (1979) Measuring and estimating the bioconcentration factor of chemicals in fish. *J. Fish. Res. Board Can.* 36:1040–1048.

Weininger, D. (1978) Accumulation of PCBs by lake trout in Lake Michigan. Ph.D. thesis, University of Wisconsin-Madison, Madison, WI. 1–232.

Woodwell, G.M. (1967) Toxic substances and ecological cycles. *Sci. Amer.* 216:24–31.

Wootton, R.J. (1990) *Ecology of teleost fishes*. Chapman and Hall, London.

Subject index

deSalle R., American Museum of Natural History, New York, USA / **Schierwater B.,**
University of Frankfurt, Germany (Ed.)

Molecular Approaches to Ecology and Evolution

1998. Approx. 270 pages. Hardcover
ISBN 3-7643-5725-8
Due in September 1998

The last ten years have seen an explosion of activity in the application of molecular biological
techniques to evolutionary and ecological studies. This volume attempts to summarize advances in the field and place into context the wide variety of methods available to ecologists and
evolutionary biologists using molecular techniques. Both the molecular techniques and the variety
of methods available for the analysis of such data are presented in the text.

The book has three major sections - populations, species and higher taxa. Each of these sections
contains chapters by leading scientists working at these levels, where clear and concise discussion
of technology and implication of results are presented.

The volume is intended for advanced students of ecology and evolution and would be a suitable
textbook for advanced undergraduate and graduate student seminar courses.

BioSciences with Birkhäuser

(Prices are subject to change without notice. 8/98)

For orders originating from all over the   For orders originating in the USA and
world except USA and Canada: Canada:

Birkhäuser Verlag AG   **Birkhäuser Boston, Inc.**
P.O. Box 133   **333 Meadowland Parkway**
CH-4010 Basel / Switzerland   **USA-Secaurus, NJ 07094-2491**
Fax: +41 / 61 / 205 07 92   **Fax: +1 / 201 348 4033**
e-mail: orders@birkhauser.ch   **e-mail: orders@birkhauser.com**

EXS 82

Streit B. / Städler T., University of Frankfurt, Germany / **Lively C.M.,**
Indiana University, Bloomington, IN, USA (Ed.)

Evolutionary Ecology of Freshwater Animals
Concepts and Case Studies

1997. 384 pages. Hardcover
ISBN 3-7643-5694-4

Evolutionary ecology includes aspects of community structure, trophic interactions, life-history
tactics, and reproductive modes, analyzed from an evolutionary perspective. Freshwater environments often impose spatial structure on populations, e.g. within large lakes or among habitat
patches, facilitating genetic and phenotypic divergence. Traditionally, freshwater systems have
featured prominently in ecological research and population biology.

This book brings together information on diverse freshwater taxa, with a mix of critical review,
synthesis, and case studies. Using examples from bryozoans, rotifers, cladocerans, molluscs, teleosts
and others, the authors cover current conceptual issues of evolutionary ecology in considerable
depth.
The book can serve as a source of critically evaluated ideas, detailed case studies, and open
problems in the field of evolutionary ecology. It is recommended for students and researchers
in ecology, limnology, population biology, and evolutionary biology.

BioSciences with Birkhäuser

(Prices are subject to change without notice. 8/98)

For orders originating from all over the
world except USA and Canada:

For orders originating in the USA and
Canada:

Birkhäuser Verlag AG
P.O. Box 133
CH-4010 Basel / Switzerland
Fax: +41 / 61 / 205 07 92
e-mail: orders@birkhauser.ch

Birkhäuser Boston, Inc.
333 Meadowland Parkway
USA-Secaurus, NJ 07094-2491
Fax: +1 / 201 348 4033
e-mail: orders@birkhauser.com

Birkhäuser